Physical Processes in Circumstellar Disks around Young Stars

Physical Processes in Circumstellar Disks around Young Stars

Edited by Paulo J. V. Garcia

THE UNIVERSITY OF CHICAGO PRESS · CHICAGO AND LONDON

Paulo J. V. Garcia is associate professor in the Department of Engineering Physics and researcher with the Laboratory for Systems, Instrumentation and Modeling in Science and Technology for Space and the Environment, both at the University of Porto in Portugal.

The University of Chicago Press, Chicago 60637
The University of Chicago Press, Ltd., London

Printed in the United States of America

20 19 18 17 16 15 14 13 12 11   1 2 3 4 5

ISBN-13: 978-0-226-28228-2 (cloth)
ISBN-13: 978-0-226-28229-9 (paper)
ISBN-10: 0-226-28228-7 (cloth)
ISBN-10: 0-226-28229-5 (paper)

Library of Congress Cataloging-in-Publication Data

Physical processes in circumstellar disks around young stars / edited by Paulo J. V. Garcia.
p. cm.

ISBN-13: 978-0-226-28228-2 (cloth : alk. paper)
ISBN-10: 0-226-28228-7 (cloth : alk. paper)
ISBN-13: 978-0-226-28229-9 (pbk. : alk. paper)
ISBN-10: 0-226-28229-5 (pbk. : alk. paper) 1. Disks (Astrophysics)
2. Stars—Formation. I. Garcia, Paulo J. V.
QB466.D58P498 2011
523.8—dc22   2010026629

♾ The paper used in this publication meets the minimum requirements of the American National Standard for Information Sciences—Permanence of Paper for Printed Library Materials, ANSI Z39.48-1992.

Contents

1

PAULO J. V. GARCIA, ANTONELLA NATTA, AND MALCOLM WALMSLEY

CIRCUMSTELLAR DISKS AROUND YOUNG STARS

Circumstellar disks around young stars have a long history. Kant, Laplace, and others realized in the 18th century that our solar system probably arose in something that one might call a "rotating disk." About 30 years ago, with the birth of infrared and millimeter astronomy, it became possible to observe direct emission from the gas and dust particles in disks surrounding young pre-main-sequence solar-mass stars. It was found that these might be the forerunners of "planetary systems" like our own. For fairly obvious reasons, studying the evolution of such disks became a subject of prime interest for many astronomers. Today, thousands of disks have been detected in nearby star-forming regions, and we have developed a variety of observational tools, over a wide range of wavelengths from X-rays to millimeter, to study their properties. Disk studies have become sufficiently important that they have driven the design of several new large facilities, such as the first worldwide telescope, the Atacama Large Millimeter Array (ALMA), whose operations will start in the 2010 decade.

Before we go into the details, it is worth pointing out that an important motivation for much of the work on disks surrounding young and forming stars is the wish to understand the origins of planetary systems. These, of course, include our own solar system, but also the hundreds of extrasolar planets that have been discovered in the last 15 years. The striking diversity of the planetary systems associated with nearby stars suggests that variations in the initial conditions of stellar birth give rise to a wide variety of resultant systems. Since planets form from the dust and gas in the disks, understanding disk structure and evolution is therefore the key to understanding planet formation.

This book is dedicated to the physical and chemical processes that take place in circumstellar disks around young stars after the early stage of star formation, when the central star is formed by the collapse of a molecular core, and before the disk dissipates entirely, leaving behind, perhaps, a planetary system and a tenuous disk of dust, formed by collisions among larger bodies.

Understanding disks requires a knowledge of dust and gas properties, chemical processes, hydrodynamics and magnetohydrodynamics (MHD), radiation transfer, and stellar evolution. This book provides a comprehensive and systematic discussion of the physical phenomena, feedback mechanisms, and observational probes that control disk physics and evolution. Advanced undergraduates, graduate students, and researchers in the field will find it useful. The book assumes a standard undergraduate knowledge of astronomy and star formation as presented in the excellent monographs by Stahler and Palla (2005) and Hartmann (2009).

What Does a Circumstellar Disk Look Like?

The most direct evidence for a disklike distribution of the matter around young stars came from direct imaging with the *Hubble Space Telescope* (*HST*), which showed that some young stars in the Orion Nebula Cluster were surrounded by dusty disks observed in silhouette against the background nebular emission.

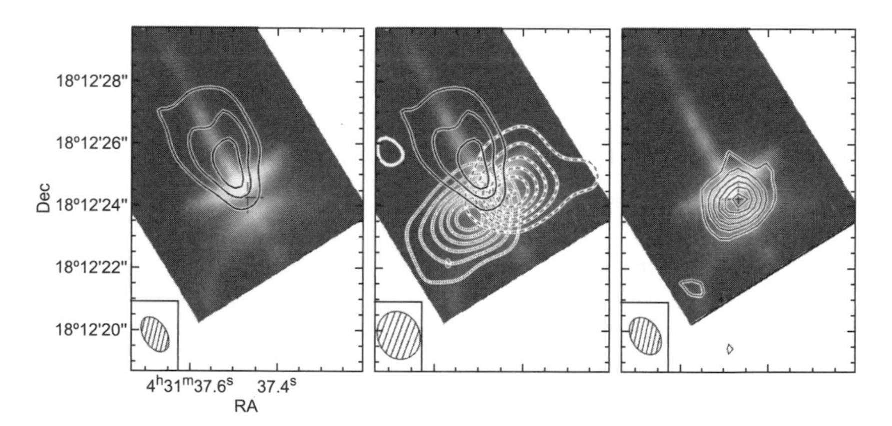

Fig. 1.1. HH 30 imaged by *HST* and the Plateau de Bure Interferometer (from Pety et al., 2006). In all panels, the gray scale is a red ($\sim 0.7\ \mu$m) image from *HST* ; the butterfly-shaped emission is reflected starlight from the disk surface, while the straight line of emission to the North East (NE) is the jet emission in [SII]. Left: Contours are the outflow emission traced by the ^{12}CO $J = 2-1$ line integrated only at extreme blue (≤ 4 km/s) and red (≥ 11 km/s) velocity to avoid contamination. Center: superimposition of the outflow emission as in the left panel and of the blueshifted (from 4.6 to 7.2 km/s; dotted line) and redshifted (from 7.2 to 9.8 km/s; dashed line) emission of the disk as traced by the ^{13}CO(2-1) line, which show the evidence of the disk Keplerian rotation. Right: the contours show the 1.30 mm dust continuum emission from the disk.

The disks imaged in Plate 1 have in a sense been observed by accident. They happened to be interposed between the Sun and the background emission from the Orion Nebula, and they are seen by virtue of the extinction that they cause to this background.

However, it has been possible in the last few years to image disk emission directly using a variety of techniques. One is to observe reflected starlight from the disk (see, e.g., Fig. 1.1). Another is to observe the emission from the dust and gas in the disk directly at infrared or radio wavelengths. Here the use of mm wavelength interferometers in particular has allowed imaging the outer disk (e.g., Fig. 1.1, right) in the dust-emitted continuum and in molecular lines, such as CO.

How Massive Are Disks?

Disk masses are the most important quantity that one would like to derive directly from observations. Unfortunately, this is a very difficult task because most of the disk mass is in cold molecular hydrogen, which cannot be directly detected. In general, the disk mass is indirectly determined from the dust mass, assuming a dust-to-gas mass ratio similar to that in the diffuse interstellar medium (ISM). In turn, dust masses are uncertain since they depend on the dust properties, mostly composition and sizes. With these caveats, one finds that the disk mass is generally small relative to the mass of the central star, and in fact current estimates are that the ratio of disk mass $M_{\rm disk}$ to stellar mass M_* is roughly a few percent in the early pre-main-sequence phases (Fig. 1.2). As discussed in the chapter by Bergin, this agrees with evidence from line profiles, both at millimeter wavelengths, where one sees the emission of the outer disk, and in the near infrared (IR), where one sees gas in the inner disk; in particular, both the profiles of the rotational transition of CO at 1–3 mm and of the vibrationally excited transitions at 4.6 μm are roughly consistent with Keplerian rotation around the central star.

Although the best-studied disks are associated with T Tauri stars, i.e., in the mass range $\sim$ 0.1–1 $M_\odot$, disks exist around intermediate-mass stars and around brown dwarfs. This is illustrated in Fig. 1.2, which shows a plot of observed disk masses for the Taurus region. One sees that objects below the hydrogen-burning mass limit of 0.1 solar masses (i.e., brown dwarfs) have disks with a disk-to-stellar-mass ratio similar to that of higher-mass stars. The number of brown dwarfs with disks is increasing quickly, as more sensitive surveys in the mid-IR are coming out, especially from the *Spitzer Space Telescope*. The disk emission of young brown dwarfs can be reproduced by

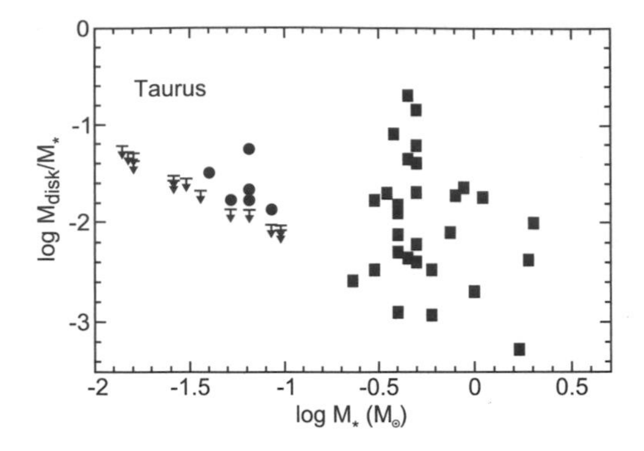

Fig. 1.2. Observed ratios of disk mass M_{disk} to stellar mass M_* for Taurus pre-main-sequence stars and brown dwarfs, plotted as function of stellar mass (from Natta and Testi, 2008). Data for stars are from Andrews and Williams (2005), and for brown dwarfs from Scholz et al. (2006); in all cases, a value of the opacity of $\kappa_{300\mu m} = 10$ cm^2/g has been used.

scaled-down T Tauri star disks, which reprocess the radiation emitted by the central objects. This similarity speaks in favor of a similar process for the formation of objects of all masses, although more and better data (for example, improved estimates of the brown-dwarf disk masses) are required.

What Are Disks Made Of?

Essentially, most of what we know about disks concerns the grains. This is because grains, being efficient radiators, are the most easily observable components of young pre-main-sequence disks. However, the gas is important both because it is the major component and also because it emits spectral lines and thus allows us to examine disk kinematics and excitation. Unfortunately, most of the gas is molecular hydrogen, which is difficult to observe except in special circumstances. As discussed by Bergin in this book, the next most abundant constituent of most pre-main-sequence disks is thought to be carbon monoxide (CO), which, as mentioned earlier, is an important tracer of both inner and outer disk properties. Obviously, however, it is important to fully characterize the disk chemistry, which is likely to play a role in the formation of planets and, possibly, survive in objects such as comets in the early solar system.

Are the Disks Actively Accreting?

It is very likely that the disk structure and its evolution in time are controlled primarily by accretion of the matter onto the central star. The strong magnetic

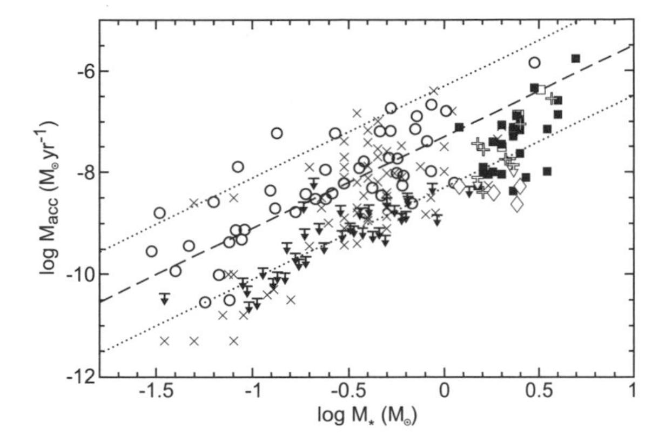

Fig. 1.3. Observed accretion rates of pre-main-sequence stars are plotted as a function of stellar mass. Crosses are brown dwarfs and T Tauri stars in Taurus; circles and upper limits area objects in Ophiuchus (Natta et al., 2006; and references therein); plus signs and squares are G-F–type stars and Herbig Ae stars (from Calvet et al., 2004; Garcia Lopez et al., 2006; and unpublished data). Accretion rates are derived from "veiling" and emission-line intensities and/or profiles.

fields measured in these stars point to a magnetic influence in the fate of the material reaching the inner end of the disk. As shown in the chapter by Calvet and D'Alessio, accretion rates can be measured because of the excess (over the photospheric radiation) radiation emitted from the accretion shock caused by disk material impinging on the stellar surface. This radiation essentially corresponds to the potential energy liberated (GM_*^2/R_*, where M_* is stellar mass and R_* is stellar radius) and hence is a measure of accretion rate. In Fig. 1.3, we show accretion rates as a function of stellar mass for a large number of young stars. The figure suggests rapidly increasing accretion rates for higher masses, leading one to wonder whether disk lifetimes are a function of stellar mass. Indeed, a trend of shorter disk lifetime for stars of increasing mass is observed in many star-forming regions, where, as in the much-studied Orion Nebula Cluster, the intermediate-mass stars lack any evidence of disks. In fact, most of the well-known young, active stars of intermediate mass (so-called Herbig Ae stars, the higher-mass brothers of T Tauri stars) are not in star-forming regions; one thus wonders why their disks have survived, in contrast to their counterparts in star clusters.

What physical mechanism allows accretion to occur? A classical phenomenological description, which has been extremely successful in understanding disk observations, is addressed in the chapter by Durisen. It is assumed that the angular momentum is dissipated locally by viscous friction between contiguous rings of gas, so that at time approaching infinity all the matter has accreted onto the central star and all the angular momentum is at infinite distance. The timescale is determined by the disk viscosity; however, the physical nature of viscosity is a much-debated question, and many of the contributions in this book attempt to provide elements for a response. The magnetorotational instability discussed by Balbus describes one possible solution involving small-scale magnetic fields. Another possibility discussed by Durisen in his contribution is that gravitational instabilities drive accretion and hence disk disappearance, as well as giving rise to planets and multiple systems. Finally, large-scale magnetic fields could play a role in transferring angular momentum by exerting torques on the disk, allowing for simultaneous accretion and wind ejection, as exposed by Königl and Salmeron. Pinning down the physical process (or "viscosity") responsible for accretion will be a major accomplishment in this field.

What about Disk Structure?

Matter in disks spans several orders of magnitude in density and a large range of temperatures. One can get a crude idea of this by looking at the predictions of the simple viscous models mentioned previously. In these so-called α-disks, the viscosity ν is given by

$$\nu = \alpha c_s H_p,$$

where c_s is the sound speed, H_p is the vertical pressure scale height, and α is a parameter (probably in the range 0.1–0.001). In these models, the gas motions are dominated by the Keplerian rotation, which is supersonic at all radii; there is no motion in the vertical direction; and the radial drift toward the star, with velocity of the order of ν/R, is subsonic. At 1 AU from a solar-mass star, the Keplerian rotation velocity is $v_k \sim 30$ km/s, the sound speed is $c_s \sim 1$ km/s, and the accretion velocity v_R is only 30 m/s. In steady state, the mass accretion rate $\dot{M}_{acc}$ is

$$\dot{M}_{acc} \sim 2\pi \Sigma R v_R,$$

where Σ is the surface density through the disk at R.

Viscous α-disks are geometrically thin ($H_p/R \sim c_s/v_k$), and most of the mass is concentrated on the disk midplane. If we assume hydrostatic equilibrium in the vertical direction, density decreases as $\exp(-z^2/2H_p^2)$. If one neglects the heating due to the stellar radiation, the surface radial temperature profile due to the viscous dissipation is $T \propto R^{-3/4}$; α-disks are hotter in the midplane than on the surface, so that lines can be seen only in absorption, as in a plane-parallel stellar atmosphere. The radiation emitted by these disks is powered by accretion, and the disk luminosity is directly proportional to the mass-accretion rate.

Is Irradiation from the Central Star Important?

In general, viscosity is not the only source of heating, because the disk surface and, as an indirect consequence, the whole disk will be heated by radiation from the star. This is often the dominant effect in observed disks, and thus disk emission can be considered to be reprocessed stellar radiation generated by the star contraction. In this situation, disks are heated from the outside; the temperature decreases from the surface to the disk midplane, and gas and dust features are seen in emission, rather than in absorption, as would be the case if the disk heating were dominated by accretion energy dissipated in the midplane.

Even the simplest disk models include the heating from the central star in the calculation of the sound speed c_S and of the disk structure. Fig. 1.4 shows an example of the range of densities and temperatures involved in a

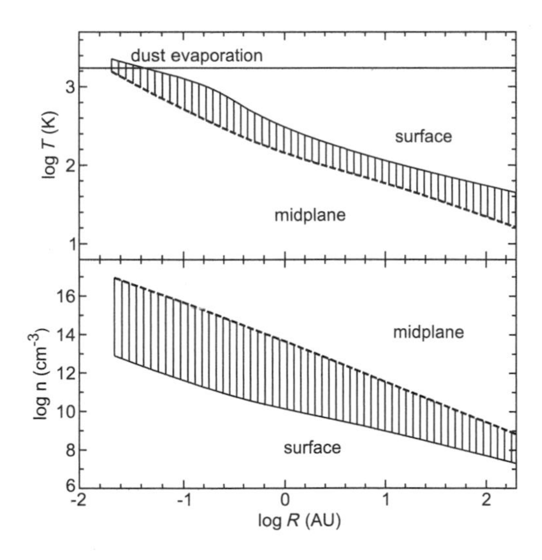

Fig. 1.4. Typical density (lower panel) and temperature (upper panel) in a disk around a T Tauri star. The range shown by the hatching is that between the temperature or density on the disk surface and at the midplane. The horizontal line on the top panel shows the dust evaporation temperature (about 1,500 K), which provides a natural inner cutoff to the dust (but not the gas) disk.

typical α-disk around a T Tauri star when the heating due to the central star is included. There is a large range in both parameters, and one can infer, for example, as a result of the temperature variation, that grains emit at different wavelengths depending on the distance from the star. Moreover, one sees that typical midplane densities vary by many orders of magnitude, from the inner disk 0.1 AU from the star to the outer disk at 100 AU. There are also large gradients from midplane to surface. Finally, one should note that very close to the star, at around 0.03 AU in the example of Fig. 1.4, the dust is expected to evaporate, and within this radius, the disk will be purely gaseous.

In irradiated disks, the amount of radiation reprocessed depends on the disk geometry, i.e., on the geometrical shape of the disk. If the disk is flared, the fraction of solid angle covered by the disk increases with radius (see the top panel of Fig. 1.5 and note that "flat" in this context implies a constant ratio of scale height to radius). This gives origin to the shape seen in the *HST* images, and also to the relatively weak dependence of the disk emission an wavelength observed in most T Tauri star disks. An example showing the importance of disk geometry is given in Fig. 1.5, where one sees (center panel) the fraction of stellar light intercepted by the disk as a function of radius for both a flat-disk model and a flared model. The bottom panel shows the corresponding spectral energy distributions (SEDs). The flared disk, and in particular its outer regions, intercepts a greater fraction of the stellar luminosity and is consequently a much stronger source of radiation. One notes also in these model spectral energy distributions the "bumps" at 10 and 20 μm due to emission by silicate grains on the disk surface, as discussed by Henning and Meeus. Most of the observed SEDs (Fig. 1.6) can be accounted for by models with varying degrees of flaring, from flared to flat disks.

Disk models will be discussed in detail in this book, from detailed viscous models (Calvet and D'Alessio) to more physical models, which include a physical description of accretion based on hydrodynamic processes (Durisen) and local magnetohydrodynamic processes (Balbus), as well as large-scale magnetic fields (Königl and Salmeron). In all cases, disks present a wide variation of conditions, whose characterization requires many different complementary observational techniques. As will be shown in the chapters by Calvet and D'Alessio and by Clarke, the millimeter continuum observations sample the cold outer disk and allow estimates of the disk mass, whereas the near-infrared continuum is sensitive to the hottest grains and thus tells us about the structure of the inner disk (radii of less than a few AUs). Continuum multiwavelength observations probe the spatial distributions of the grains and

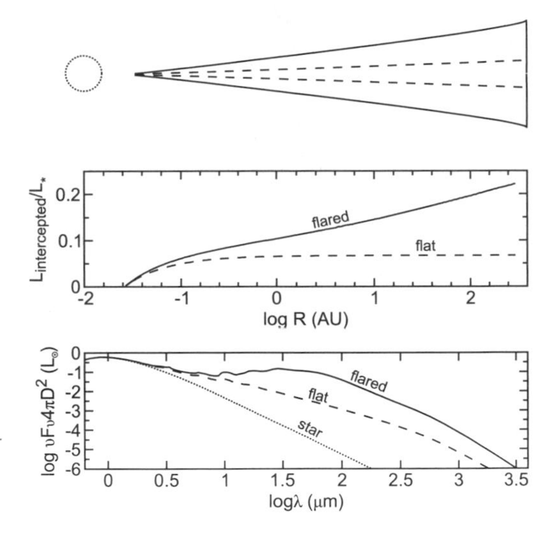

Fig. 1.5. The upper panel is a sketch showing the difference between a flared (disk opening angle increases with radius, dashed line) and a flat (full-line) disk. The center panel shows the fraction of stellar luminosity intercepted by a flat and a flared disk as a function of radius. The bottom panel shows the corresponding spectral energy distributions (SEDs); the dotted curve shows the contribution from the stellar photosphere. The flared disk intercepts a larger fraction of stellar light, especially at larger radii; correspondingly, it has a much stronger emission than the flat disk at far-infrared and mm wavelengths.

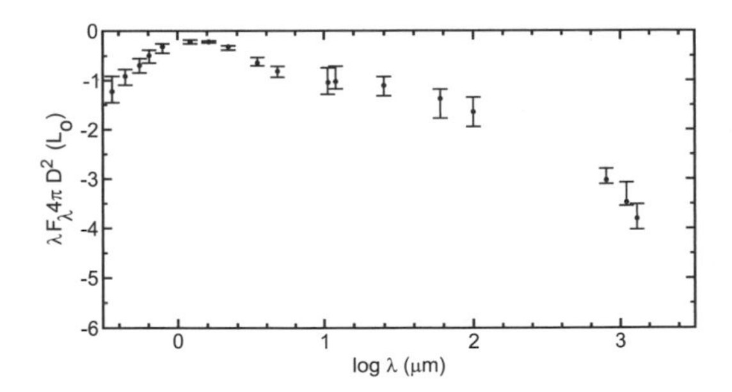

Fig. 1.6. Median SED (points) and quartiles (error bars) of Taurus-Auriga pre-main-sequence sources normalized at $\lambda = 1.6\ \mu m$ (adapted from D'Alessio et al., 1999).

(implicitly) gas and thus disk geometry. As explained in the chapters by Calvet and D'Alessio and by Bergin, lines from gaseous species of different excitation, covering the whole range of wavelengths from IR to millimeter, can be used to obtain information on disk kinematics (tests of Keplerian rotation, for example) and atomic and molecular gas content and properties.

How Do Disks Evolve?

Disk properties change with time as the star forms and evolves toward the main sequence. We believe that stars form from "dense cores" of cold molecular gas with sizes of roughly 0.1 parsec in clouds where initially, at least, centrifugal and magnetic forces are not capable of preventing gravitational collapse. On timescales of roughly 10^5 years (essentially the "free-fall time" corresponding to a core density of typically 10^4–10^5 H nuclei cm^{-3}), protostars form. Angular momentum breaks the spherical symmetry of the system, and because of its conservation such protostars are surrounded by a disk of gas and dust, as well as by an infalling "envelope." Powerful jets of matter are generally present, and a large-scale magnetic field is the prime candidate for driving them. At this stage, the disk is continually fed by matter from the surrounding envelope, and its properties depend on the properties of the core (density distribution, angular momentum, and magnetic field) from which the protostar forms. Because these disks are embedded in the envelope, which is optically thick (at visible and infrared wavelengths), such objects can be observed only at mm and cm wavelengths using interferometry, to detect emission from structures with dimensions of hundreds of astronomical units (AUs). The disk is difficult to distinguish at this point from infalling gas close to the star, and as a result, we have rather little detailed information. Nevertheless, the development of mm interferometers such as the Institut de Radioastronomie Millimétrique (IRAM) Plateau de Bure instrument in France and the Submillimeter Array (SMA) in Hawaii, as well as the ALMA interferometer in Chile, suggest that this situation will change in the not-too-distant future. This is of importance because the young protostellar disks and envelopes represent the initial conditions for the more evolved objects that can be studied at optical and infrared wavelengths. Sections of the chapters by Durisen and by Königl and Salmeron address this earlier stage of core collapse into a disklike structure from a hydrodynamic and magnetohydrodynamic perspective.

On timescales of a few 10^5 years, the envelope accretes onto the disk and young protostar, and the pre-main-sequence star with its accompanying disk is revealed. At this point, the disk lifetime depends mainly on the disk mass and on the accretion rate $\dot{M}_{acc}$ onto the central star. Roughly speaking, one expects disk age to be $M_{disk}/\dot{M}_{acc} \sim 10^6$ years. The observational evolution of the disk fraction is presented in Fig. 1.7. As will be detailed in the chapters by Calvet and D'Alessio and by Clarke, the evolution of disks with time seems also to depend on the properties of the central object. In addition, disk evolution can

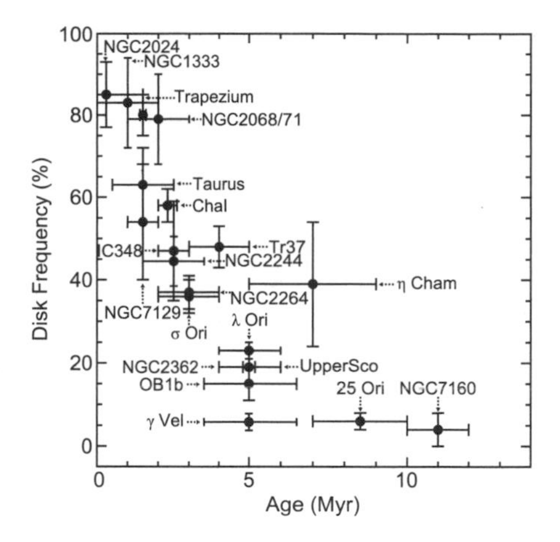

Fig. 1.7. Disk frequency as a function of the age of the population. Each population is named in the figure. The disk frequency is measured by the ratio of stars with excess in the near infrared (either JHK or in the *Spitzer*/Infrared Array Camera [IRAC] bands) to the total number of stars. From Hernández et al., 2007.

be affected by the disk's environment. If, for example, the star is not isolated but is part of a dense stellar cluster, at least two effects are worth considering: one is the effect of the radiation field of the higher-mass objects in the cluster, which cause evaporation of disks along the lines discussed by Clarke; the other is that the high stellar density makes interactions between different star-disk systems more frequent. Overall, these effects argue for faster disk evolution in clusters.

A specific aspect of disk evolution concerns how the solid disk component changes with time, from the small, submicron size of grains in the ISM to planetesimals. This process can be followed theoretically and, to some degree, observationally. As explained in the chapter by Henning and Meeus, the silicate features are a tracer of the existence of small micron–sized grains in the disks because larger grains will not show a distinct feature. Grain growth is expected to occur and to be followed by settling of larger grains toward the midplane. Observed spectral energy distributions show great variety, and this can be understood if the dust population undergoes growth and settling on timescales comparable to the T Tauri starses lifetimes (see the chapters by Calvet and D'Alessio and by Henning and Meeus). The statistical studies that are now available show that T Tauri stars with disks live a few million years, although there is clearly much dispersion. It is quite plausible that this is the timescale on which planets form and thus that the process of grain growth that gives rise to spectral energy distribution variations is the first signature of planet formation.

Final Comments

In this brief introduction, we have summarized some basic properties of pre-main-sequence stellar disks, setting the stage for the more detailed presentations of the chapters that follow. Most of the theoretical chapters are aimed at understanding how disks evolve. The contributions of Clarke, Balbus, and Durisen attack this problem from different angles, and probably all the processes that they discuss contribute to producing the observed characteristics of "real disks." An essential input to any understanding of real disks comes from the evolution of the dust grains, as discussed by Henning and Meeus. Then, coming closer to the observations, the contributions of Calvet and D'Alessio and of Bergin allow one to transform the theoretical description into expected observational consequences. Last but not least, the contribution of Königl and Salmeron addresses the effects of large-scale magnetic fields in pre-main-sequence disks whose main signatures are winds and jets. All in all, we are at a fascinating moment in the research of the properties of these objects, with many unanswered questions, and it is our hope that this book provides a starting point for discovering some answers.

References

Andrews, S. M., and Williams, J. P. (2005). Circumstellar Dust Disks in Taurus-Auriga: The Submillimeter Perspective. *ApJ*, 631:1134–1160.

Calvet, N., Muzerolle, J., Briceño, C., Hernández, J., Hartmann, L., Saucedo, J. L., and Gordon, K. D. (2004). The Mass Accretion Rates of Intermediate-Mass T Tauri Stars. *AJ*, 128:1294–1318.

D'Alessio, P., Calvet, N., Hartmann, L., Lizano, S., and Cantó, J. (1999). Accretion Disks around Young Objects. II. Tests of Well-Mixed Models with ISM Dust. *ApJ*, 527:893–909.

Garcia Lopez, R., Natta, A., Testi, L., and Habart, E. (2006). Accretion Rates in Herbig Ae Stars. *A&A*, 459:837–842.

Hartmann, L. (2009). *Accretion Processes in Star Formation*. Cambridge, Cambridge University Press.

Hernández, J., Calvet, N., Briceño, C., Hartmann, L., Vivas, A. K., Muzerolle, J., Downes, J., Allen, L., and Gutermuth, R. (2007). *Spitzer* Observations of the Orion OB1 Association: Disk Census in the Low-Mass Stars. *ApJ*, 671:1784–1799.

McCaughrean, M. J., and O'Dell, C. R. (1996). Direct Imaging of Circumstellar Disks in the Orion Nebula. *AJ*, 111:1977–1986.

Natta, A., and Testi, L. (2008). The Study of Young Substellar Objects with ALMA. *Ap&SS*, 313:113–117.

Natta, A., Testi, L., and Randich, S. (2006). Accretion in the ρ-Ophiuchi Pre-Main Sequence stars. *A&A*, 452:245–252.

Pety, J., Gueth, F., Guilloteau, S., and Dutrey, A. (2006). Plateau de Bure Interferometer Observations of the Disk and Outflow of HH 30. *A&A*, 458:841–854.

Scholz, A., Jayawardhana, R., and Wood, K. (2006). Exploring Brown Dwarf Disks: A 1.3 mm Survey in Taurus. *ApJ*, 645:1498–1508.

Stahler, S. W., and Palla, F. (2005). *The Formation of Stars.* Weimheim, Germany, Wiley-VCH.

2

NURIA CALVET AND PAOLA D'ALESSIO

PROTOPLANETARY DISK STRUCTURE AND EVOLUTION

1. Introduction

The standard picture of star formation, based on several decades of observations and models, can be summarized as follows. Stars are born in dense cores inside molecular clouds, which are the densest regions of the interstellar medium (ISM). These dense cores collapse under their self-gravity, and because they have some angular momentum, they end up forming disklike structures and/or multiple stellar systems. When a single star's disk is formed, the lowest-angular-momentum core material falls toward its center, where the star builds up. The rest of the material falls onto the circumstellar disk, which surrounds the young star during its first few million years of life. These young disks are called *accretion disks* because their central star acquires mass from them.

Disks evolve during their lifetime. Viscosity and gravitational torques transfer angular momentum to a fraction of the disk material that ends up moving toward the outer regions, increasing the disk radius. On the other hand, the fraction of material that has lost angular momentum falls toward the star, increasing the stellar mass. Disk mass is replenished by the molecular dense core until the latter is dissipated by a disk wind or some other mechanisms. Gravitational torques must be important in transferring angular momentum when a disk has a large mass compared with its central star (see the chapter by Durisen, this volume). Simultaneously, dust grains grow inside the disk, from micrometer sizes like grains found in the diffuse interstellar medium to millimeter, meter, and even kilometer sizes, eventually to form planetesimals. The disk is a natural place to grow a planetary system if it happens to last long enough. However, how long a disk should live to actually form planets depends on the details of the planet-formation process, which are a matter of debate. Recently, timescales have been observationally constrained, helping define the finer details of how a planetary system is built up. What seems to be clear, on the basis of the number of extrasolar planets found so far, is that the formation of planetary systems around stars is not a rare process. This is why accretion disks around young stars are frequently called protoplanetary disks, reflecting their potential to form planets. The final destiny of the disk mass is to be part of planets, to be accreted by the star, or to be lost in a photodissociated wind or by a stellar encounter.

In this chapter we will describe what has been learned about disk structure and evolution from the combination of observations and modeling. In § 2 we give a summary of the observations on which models rely. In § 3 we list evidence for accretion; review magnetospheric accretion, the current paradigm for accretion; an describe determinations of accretion luminosity and mass-accretion rate, concentrating in low-mass Young Stellar Objects YSOs In § 4 we describe the physics of irradiated accretion disks and how it relates to the properties of solids in the disk and to accretion. Finally, in § 5, we describe the effects of dust evolution in disks and evidence that the expected evolution is taking place.

2. Observational Overview

Observations give us quantitative information about the reality we are able to measure. This makes them the basis of any modeling effort, not only by motivating the generation of new sets of models but also by constraining models, verifying their value as an appropriated description of reality. Frequently, one is able to explain a particular observation with many models, but only a few give a consistent picture of the whole set of observations available and allow predictions that guide new observations for further tests. In this section we summarize observations of young stars, related to the disk-model interpretation. This is not a review of all the observations involving young stars, but only an attempt to introduce some basic observational information and to illustrate its importance for constructing the models described in the following sections.

When a protoplanetary disk model is calculated, the importance of the central star is sometimes underestimated. The star is the ultimate source of the disk energy, heating it by irradiation and producing the gravitational field where the disk mass falls, releasing energy during the accretion process. Thus the properties of the central stars are an essential input for modeling disks. Fig. 2.1 shows the location in the Hertzsprung-Russell (HR) diagram of the most commonly studied visible YSOs, together with evolutionary tracks

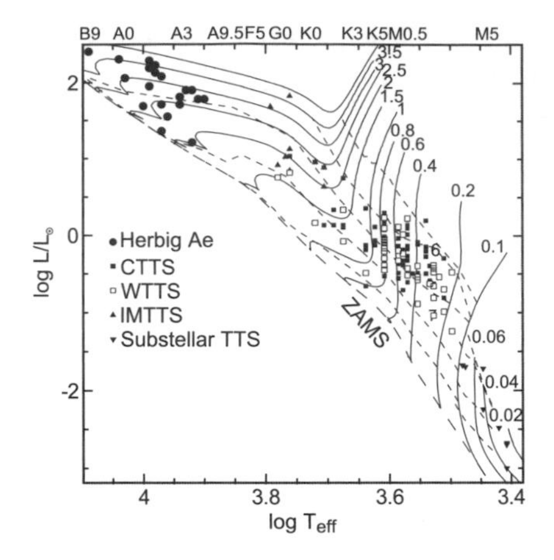

Fig. 2.1. Location of young stellar objects in HR diagram. Shown are substellar T Tauri stars (YBDs), classical T Tauri stars (CTTSs), weak-line T Tauri stars (WTTSs), Herbig Ae stars and their predecessors, and intermediate-mass T Tauri stars (IMTTSs). Sources for the data: substellar objects [15], Tauru's CTTSs and WTTSs [114], and Herbig Ae [95]. Zero-age main sequence (ZAMS, long-dashed line), evolutionary tracks (solid lines, labeled by the corresponding stellar mass in solar masses), and isochrones (dashed lines, corresponding to 0.3, 1, 3, 10, and 30 Myr from top to bottom) from [161] and [13].

and isochrones from [161] and [13]. These objects are given different names depending on their mass:

- **Herbig Ae/Be stars (HAeBe)**, A and B stars with emission lines [90], with masses 1 $M_\odot < M_* < 8\ M_\odot$.
- **T Tauri stars (TTSs):**, characterized by late-type spectra superimposed by strong emission lines [90], with masses $0.08 < M_* \leq 1\ M_\odot$;
- **Young brown dwarfs (YBDs)**, substellar objects with masses $M_* < 0.08\ M_\odot$ that will never reach a central temperature high enough to burn hydrogen.

As shown in Fig. 2.1, TTSs and YBDs tend to be on the Hayashi track, while HAeBe are much closer to the main sequence, and all have ages $\sim$ 1–10 Myr. However, note that some TTSs have masses comparable to the HAeBe (the intermediate-mass TTSs, IMTTSs) and will end up as HAeBe as they evolve along the radiative tracks.

Young stars are subject to different classification schemes, based on observational criteria. These categories are usually related to evolutionary stages, but we will see that the picture is not as clear as it seems at first sight. For instance, according to the slope $dlog(\lambda F_\lambda)/dlog\lambda$ of their spectral energy distribution (SED) in the 2.2 to 25 μm range [117, 116], YSOs have been classified as

- **Class I**, with positive slope. These are thought to be still surrounded by infalling material from which they form.
- **Class II**, with negative slope, but still flatter than the Rayleigh-Jeans slope expected for stellar photospheres, $\lambda F_\lambda \propto \lambda^{-3}$. Their excess is explained as produced by the disk.
- **Class III**, with photospheric slopes [117]. They appear to have dissipated their disks.

This classification scheme, which is based on the slope of the SED, is applicable to YSOs of all masses. Another classification scheme, applied only to T Tauri stars, is based on the strength of their emission lines. According to this scheme, TTSs can be classified as follows:

- **Classical TTSs (CTTSs)**, with an equivalent width of Hα larger than 10 Å [92]. These stars also show strong excesses in line and continua above the intrinsically photospheric fluxes and are thought to be accreting mass from the disk (see § 3).
- **Weak-line TTSs (WTTSs)**, with Hα equivalent widths lower than this limit [92]. More recent studies indicate that the Hα equivalent width separating both classes depends on spectral type [179]. These stars are thought not to be accreting mass from the disk.

The identification of CTTSs with Class II objects and WTTSs with Class III objects is usually made under the assumption that the disk that produces the near-IR emission characterizing the Class II objects is an accretion disk. In this case, the potential energy released by the accretion process is responsible for the excess in line and continuum emission seen in CTTSs in the optical and shorter wavelengths. However, observations of the *Spitzer Space Telescope* in the last 5 years have altered this simple picture. Although most CTTSs are Class II, some WTTSs may show emission from remnant disks; also, some stars with clear signs of being accreting may show no excess in the 1–10μm range [35] and thus could not be classified as Class II. Thus an easy one-to-one correspondence cannot be made in all cases, and all indicators have to be examined to determine the physical properties of a given object.

Another inference usually drawn from these classifications is that because WTTSs have already lost their disks, they should be older than CTTSs; however, examination of Fig. 2.1 shows that CTTSs and WTTSs coexist in the HR diagram; that is, they have similar ages, indicating that age is not the only factor in determining disk dissipation.

The accretion-based classification of CTTSs and WTTSs does not extend to the HAeBe. These stars are selected because they have emission lines in their spectra, while most stars in the corresponding spectral range show absorption lines only. Moreover, they have infrared excesses consistent with the presence of disks [97]. This means that all are Class II objects and, if anything, would be intermediate-mass analogs of the CTTS. Thus a classification scheme often used in the literature for these objects is based on the shape of their infrared SED [123], and the differences are thought to be due to different properties in the disk:

- **Group I**, objects with a continuum that can be reconstructed by a power law and a blackbody. Members of the subgroup Ia have solid-state bands present in their SEDs, and those of Ib have no solid-state bands.
- **Group II**, objects with a continuum that can be reconstructed by only a power law. Again, members of subgroup IIa have solid-state bands, and those of IIb have no solid-state bands in their spectra.

The large midinfrared continuum excess observed in SEDs of HAeBe in group I is usually associated with the emission of flared disks that are heated by stellar irradiation. The smaller excess shown by stars in Group II is associated with a flat outer disk in the shadow of a puffed-up inner region [123, 52] or with the settling of dust grains [55].

CTTSs and WTTSs show emission lines in their spectra, but the lines are much broader and generally stronger in CTTSs. In addition, CTTS show *veiling* of their photospheric absorption lines [79]; i.e., these lines are less deep than those of main-sequence stars of the same spectral type. This is interpreted in terms of an excess in continuum that adds to the intrinsic photospheric flux. Veiling is measured in terms of the veiling parameter $r_\lambda = F_\nu / F_{ph}$, where F_ν is the excess flux and F_{ph} is the photospheric flux at the same wavelength. The veiling parameter increases as wavelength decreases [75, 181]; the excess flux dominates in the ultraviolet (UV) and shorter wavelengths because the intrinsic photospheric flux drops [76]. The excess luminosity in CTTSs is typically $\sim$ 10% of the stellar luminosity, but it can be comparable with or higher than the stellar luminosity for a few CTTSs (see [85]). WTTSs also emit at short wavelengths, from X-rays to UV, but their excess luminosity at these wavelengths is comparable with or slightly higher than that of active main-sequence stars [12]. Similar to these, WTTS excesses are thought to be powered by magnetic activity at the stellar surface; however, magnetic activity cannot produce a luminosity comparable to that of the star, as seen in some CTTSs. A source of energy external to the star is required; this is the main justification for expecting CTTSs to be accreting matter from their disks and in the process releasing gravitational potential energy that powers the excess.

Broad emission lines [142, 136] and veiling of absorption lines and bands in the blue and ultraviolet regions of the spectra [98] have been detected in YBDs, and as in CTTSs, the source of the excess has been attributed to accretion energy. It is more difficult to measure veiling in HAeBe; however, measurements of an excess of flux in the Balmer discontinuity have also been interpreted as evidence of an excess of energy produced probably by accretion [135]. Similarly, association with jets and UV excess have been interpreted as due to accretion in these objects [74].

In the near IR, CTTSs show characteristic excesses as well. In particular, in the J-H versus H-K diagram, most CTTSs fall in a well-defined region, "the CTTS locus" (left panel of Fig. 2.2; [122]), while WTTSs have colors consistent with dwarf stars. The HAeBe also show characteristic colors in this diagram [97], which helps distinguish them from classical Be stars. Around 2.5–3 μm, HAeBe stars show a characteristic "bump" in their SEDs [99], and it has been found that CTTSs also show a similar kind of excess, although it is less apparent because of the relatively larger photospheric emission at those bands [134]. This emission has been interpreted as produced by the inner wall of the disk made by gas and dust, located at the dust-sublimation radius [139, 171]. With near-infrared photometry, the inner radii have been measured for CTTSs and HAeBe stars [125] and have been found to be consistent with the radius expected for dust sublimation (§ 4.3).

Observations from the *Infrared Space Observatory* (*ISO*) telescope in space provided a wealth of information on the IR spectra of YSOs. Observations from the instruments on board *Spitzer* have now characterized the excesses

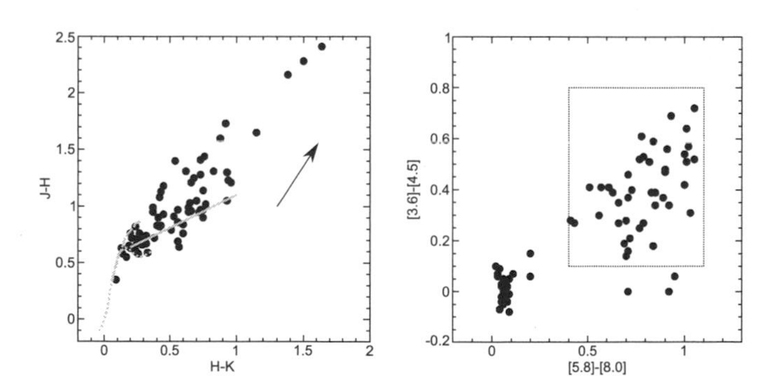

Fig. 2.2. Left panel: Taurus stars in the J-H versus H-K diagram. The dwarf sequence and the CTTS locus (solid gray line) are shown. Observations have not been corrected for reddening. Right panel: Taurus stars in the IRAC [3.6]–[4.5] versus [5.8]–[8] diagram. The dotted square indicates the region covered by colors of irradiated accretion-disk models. Data from [82].

of TTSs, HAeBe and YBDs in these wavelength regions [66, 67, 162, 112, 120, 119, 128]. For instance, in color-color diagrams constructed with combinations of the four bands of the Infrared Array Camera (IRAC) instrument, at 3.6, 4.5, 5.8, and 8 μm, and the Multiband Imaging Photometer (MIPS) 24μm band, most CTTSs fall in a region well separated from the WTTSs. The right panel of Fig. 2.2 shows one of these diagrams. The WTTSs have colors ~ 0, while most CTTSs populate a distinct region of the [3.6]–[4.5] versus [5.8]–[8.0] diagram, which corresponds to the emission expected from optically thick disks [5, 82] (§ 4). In addition, spectra of CTTSs between 5 and 30 μm obtained with the Infrared Spectrograph (IRS) on *Spitzer* have shown the large diversity of these spectra, including fluxes and profiles of the silicate features and the slope of the SED in this wavelength region [66, 120, 119, 128]. As discussed in § 5, these observations give direct information on the spacial distribution and evolutionary state of the solid component in the disks.

Another important observational development of the last 10 years has been interferometric observations in the near and mid-IR of YSOs; with resolutions of a few milliarcseconds, these observations are probing the structure of the innermost disk regions and providing invaluable information about their structure [125]. Optical and near-IR scattered light disk images, combined with detailed modeling of the transfer of stellar radiation through the disk dust, have been an important tool to understand the disk geometry and to constrain the properties of its atmospheric dust [187]. In addition, observations with single dish and interferometers in the submillimeter and millimeter range are probing the midplane regions of the disks from ~ 10 AU and beyond [9, 170]. With the *Herschel Observatory*, which is observing the region between ~ 50 and 600 μm, we will have complete wavelength coverage of the SEDs of YSOs against which we should confront our models.

As a summary of the emission properties of YSOs, Fig. 2.3 shows the SED of the CTTS BP Tau. The excess over the photospheric fluxes, shown with dashed lines, is clearly apparent from the UV to the mm. Shown are the observed fluxes and fluxes corrected for reddening.

With the wealth of information of the last 10 years, we have learned much about disk structure, for example, what the main disk-heating mechanisms are, the role of the dust, and how matter is accreted by the central star. There are still debates on the details of the accretion mechanisms and how to quantify the viscosity. We have also learned about disk evolution. We now know that mass-accretion rate and disk emission decrease with age, as expected in general terms from evolutionary models of the gas [86] and the solids [177, 55] in the disk. However, many points in this evolution from the primordial gas-rich

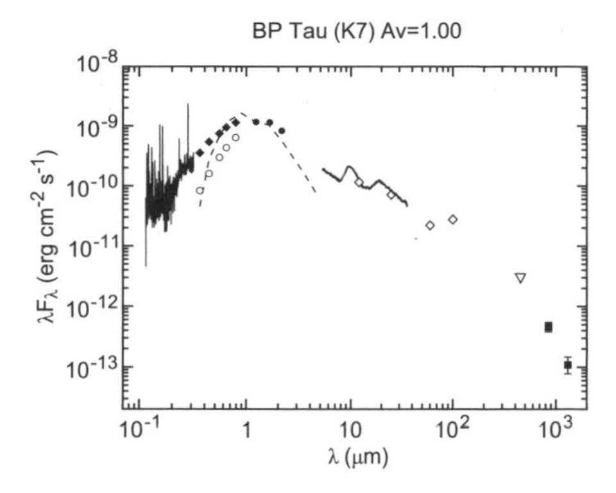

Fig. 2.3. Spectral energy distribution of the CTTS BP Tau as an illustration of the typical SED of an accreting late-type star. The photosphere is shown with a dashed line. In the near IR, the solid circles correspond to the observations and the open circles to the data corrected for reddening. This correction is not important beyond the near IR for most nonembedded objects. The *HST*/ Space Telescope Imaging Spectrograph (STIS) UV data are from [19]; the optical photometry, Infrared Astronomy Satellite (IRAS) fluxes, and spectral type are from KH95; the IRS spectrum is from [66]; the millimeter data are from [7].

disks to the debris disks with planets are yet to be understood. In this contribution, we will outline some of the basic physical principles characterizing disk structure and emission, hoping to give the reader insight to understand and perhaps help clarify some of the many remaining problems. Although we will concentrate on CTTSs, the principles outlined are general enough to be applicable to stars in other mass ranges.

3. Magnetospheric Accretion and Mass-Accretion Rate

The collapse of slowly rotating molecular cores forms stars surrounded by disks by conservation of angular momentum [166]. Most of the mass of the cores resides at large distances, which have the highest angular momentum, so it falls onto the disk and not onto the star. This matter has to be transported through the disk to the center and onto the star to make up the bulk of the stellar mass. One can define the disk mass-accretion rate $\dot{M}$ onto the star as the mass per unit time going through a cylindrical control surface centered in the star. The buildup of the stellar mass occurs mostly in thc Class I phase, while the protostar is still embedded in its envelope and receiving matter from it; it is thought to occur in episodes of high mass accretion due to disk instabilities [85], which have been identified with the the outbursts in *FU Ori objects* [88, 191].

The YSOs we have discussed in § 2 are at a later evolutionary phase, when most of the mass of the star has already been built. Still, as we have seen, the total luminosity of some of these objects exceeds that expected from release of magnetic energy on the stellar surface or even from the expected quasi-static contraction of a star in the pre-main-sequence phase, implying the existence of an external source of energy. For stars surrounded by disks, which is the case for the objects with large excesses, a readily available energy source is gravitational potential energy if the disks are accreting matter onto the star. But can we confirm that this is the case, and if so, how is potential energy actually released?

In standard *steady* accretion disks, matter is transfered from the disk to the star through a narrow boundary layer in which $\sim 1/2$ of the accretion luminosity $L_{acc} = GM_* \dot{M}/R_*$ is released as material slows down from the Keplerian velocity at R_* to the much lower stellar rotational velocity [85]. Although this model could explain in general terms the continuum excess [16, 115], other crucial observations could not be understood with it. As noted, CTTSs have strong emission lines; the peaks of these lines are close to the line center, and their wings extend to hundreds of km s^{-1} [80, 131, 130, 133]. These emission lines exhibit blueshifted absorption components, usually attributed to the matter ejected from the star, so early models sought to explain the emission lines as formed in winds [81, 141]. However, the profile of emission lines in CTTSs are different from those produced in a spherically symmetric wind for expected wind-velocity profiles. In particular, they are nearly centrally peaked, and, moreover, a redshifted absorption component with characteristic velocities of ~ 100 km s^{-1} is seen in some cases [63], in addition to the blueshifted component. These redshifted absorptions, indicative of high-velocity *infalling* material, which coexists with outflowing material, cannot be understood with the standard accretion-disk model.

The present-day paradigm for transferring matter from the disk to the star is *magnetospheric accretion*. In this model, the stellar magnetic field truncates the inner disk, and matter falls onto the star along magnetic-field lines, merging with the photosphere through an accretion shock at the stellar surface. Simultaneously, matter is ejected through open field lines, most likely in the innermost regions of the disk. The virtue of this model is that it can consistently explain a number of observations. To start with, disks are expected to be truncated at a few stellar radii, given the typical mass-accretion rate and the strength of the stellar magnetic field. In spherical infall onto a magnetized body, if the gas is sufficiently ionized, matter cannot move freely inside a given distance r, where the infall velocity and density are v and ρ, respectively, such that $B^2/8\pi > 1/2\ \rho v^2$; rather, matter couples to the magnetic field B, and accretion may even be stopped [65]. The radius where the magnetic pressure equals the ram pressure is

$$\frac{r_i}{R_*} = 7\left(\frac{B}{1\ KG}\right)^{4/7}\left(\frac{\dot{M}}{10^{-8}\ \mathrm{M_\odot\ yr^{-1}}}\right)^{-2/7}\left(\frac{M_*}{0.5\ \mathrm{M_\odot}}\right)^{-1/7}\left(\frac{R_*}{2\ \mathrm{R_\odot}}\right)^{5/7}. \quad (1)$$

In accretion disks, the truncation radius is a fraction of this value, $\sim 1/3 - 2/3$ [65]. Surface magnetic-field strengths in TTSs are of the order of kG [109], and the average mass-accretion rate is of the order of $10^{-8}\ \mathrm{M_\odot\ yr^{-1}}$ [89, 180], so disks should be truncated at a few stellar radii.

Observed emission-line profiles are naturally explained if these lines form in the magnetic infall region, as shown schematically in Fig. 2.4. The bulk of the line forms in the region where matter is just lifted from the disk; this matter has very low velocities but large emitting volumes. In contrast, the wings of the line form near the star where matter is approaching the surface at free-fall velocities of a few $\times$ 100 km s^{-1}. Moreover, this high-velocity matter may absorb the background accretion-shock emission for appropriate line-of-sight inclinations, which naturally explains the redshifted absorption component. Detailed models confirm these expectations [131, 133]. Fig. 2.5 shows observed profiles of Hα and Na I 5876 for three CTTSs in increasing order of mass-accretion rate from bottom to top (determined from their UV excess and veiling); observations are compared with the magnetospheric-model predictions. The comparison indicates that emission lines form in the magnetospheric infall flows for all but the CTTSs with the highest accretion rates. For the high accretors, high opacity lines like Hα have a typical wind profile, but lower-opacity lines like Na I D are magnetospheric in nature [133].

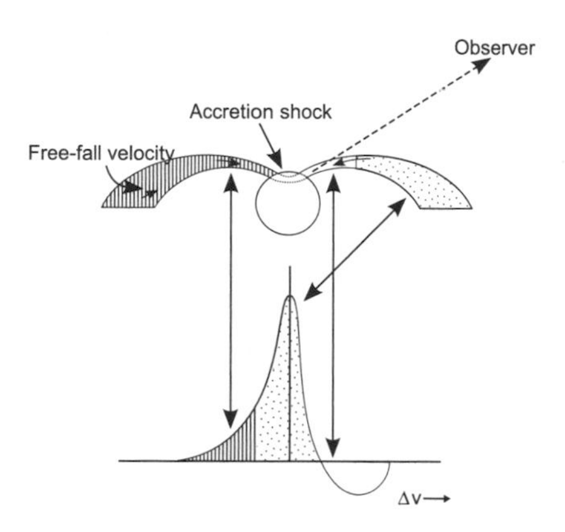

Fig. 2.4. Schematic of line formation in magnetospheric flow. Adapted from [23].

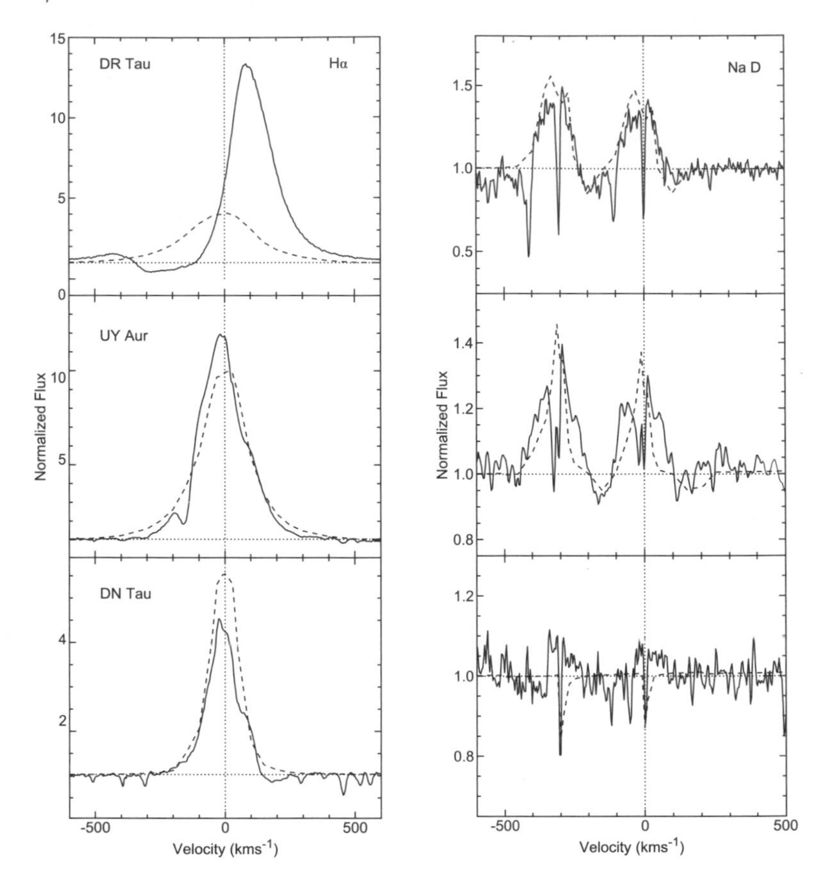

Fig. 2.5. Observed line profiles (solid lines) compared with predictions of the magnetospheric accretion model (dashed lines). Stars have increasing mass-accretion rates in their disks from bottom to top, determined from veiling measurements. The magnetospheric accretion model explains fairly well the observed profiles except in high-opacity lines like Hα in high accretors like DR Tau (upper left panel), for which the line is formed mainly in the wind [4]. Even in those cases, the profiles of lower-opacity lines like Na D (upper right panel) indicate that those lines form in the magnetosphere. Adapted from [133].

Finally, the magnetospheric accretion model can explain the flux excess observed in the optical and UV wavelengths in CTTSs as formed in the accretion shock at the stellar surface. Matter approaches the photosphere at free-fall velocities and forms an accretion shock very near the surface, where matter heats to temperatures

$$T_s = 8.6 \times 10^5 \left(\frac{M_*}{0.5\ \mathrm{M}_\odot}\right) \left(\frac{R_*}{2\ \mathrm{R}_\odot}\right)^{-1} K. \tag{2}$$

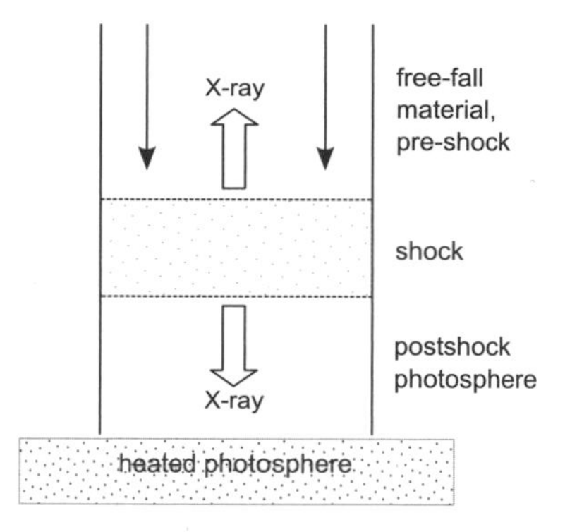

Fig. 2.6. Schematic of accretion column. Matter falling at free-fall velocities reaches the stellar surface, with which it merges through an accretion shock. Kinetic energy thermalizes, and for the expected temperatures, the shock emits soft X-rays, which heat the preshock region and the photosphere below the shock. These regions reprocess the shock emission and emit mostly in the UV and optical wavelengths [26].

Soft X-ray radiation from the shock heats the preshock region and the photosphere just below the shock, as indicated schematically in Fig. 2.6; the reprocessed emission from these regions adds to the photospheric emission as an excess continuum [26]. Plate 2 shows that the predicted emission from the accretion-shock model fits fairly well the *HST* and ground-based observations of a typical CTTS. Recent modeling of the X-ray spectra of one CTTS, TW Hya, also shows that the accretion-shock model can explain the features in at least this star [62], although this may not be the case for all CTTSs [169].

The models discussed above are very simple in the sense that they are axially symmetric, assuming homogeneous flows. Reality, of course, is much more complicated, as the variability observed in excess continuum and emission lines readily shows. Magnetic field and rotation axes are most likely not aligned, and flows are not uniform sheets onto the star. But the main physical principles are present in these models. In particular, the energy budget is modified from that of the standard model, in which the excess continuum could have at most half of the accretion luminosity, dissipated in the boundary layer, while the other half was emitted by the disk. In contrast, in the magnetospheric model the intrinsic disk emission is drastically reduced because matter is being lifted from it at a few stellar radii, depriving the disk of the contributions of regions deep in the stellar potential well. This potential energy, rather, comes out in the magnetospheric infalling material. This is fortunate because the luminosity in the excess continuum is then of the order of the accretion

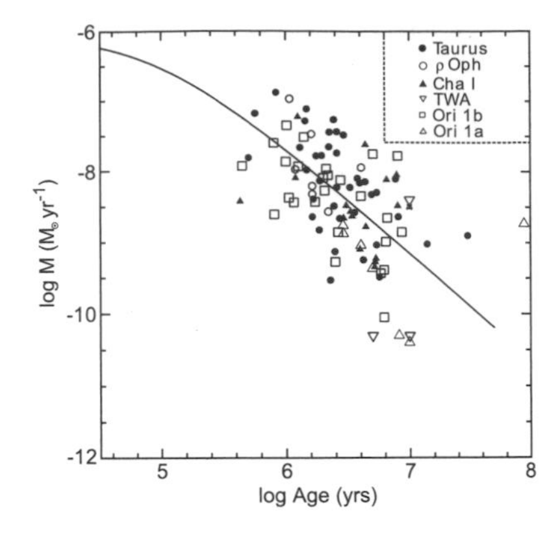

Fig. 2.7. Mass-accretion rates for stars of a number of populations indicated in the insert with ages between ~ 1 and 10 Myr. These include stellar clusters and distributed populations. From [30].

luminosity L_{acc}; for known M_* and R_*, given by the position of the star in the HR diagram, measurements of the excess luminosity allow us to estimate the mass-accretion rate onto the star, an important quantity for understanding disk structure.

On the basis of these principles, measurements of the excess luminosity in the 3,200–5,300 Å range and conversion to L_{acc} have yielded mass-accretion rates for a sample of 18 stars [75]. These measurements have been used to establish secondary calibrations of the accretion luminosity in terms of observables easier to obtain, such as the excess luminosity in the U band [75] and the luminosity in Ca II 8542, Paschen β, and Brγ lines [130, 129, 22, 28]. These calibrations have been used to determine accretion luminosities and mass-accretion rates in numerous samples of stars [86, 180]. The average mass-accretion rate in CTTSs belonging to ~1 Myr old populations is ~ 10^{-8} $M_\odot$ yr^{-1}. Moreover, determinations of mass-accretion rates for populations in the 1–10 Myr age range using these methods indicate that the accretion rate decreases with age roughly as age$^{-1.5}$, as shown in Fig. 2.7, in agreement with predictions of viscous evolution [86].

One of the main caveats of this procedure is the so-called bolometric correction, that is, the conversion from the excess luminosity measured in the observed wavelength range to the total excess luminosity, which relies on the model used. Ref. [75] and others adopted a ~ 10^4 K slab model with variable optical depth. Using an accretion-shock model, [26] found corrections that were consistent with those of [75]; this agreement is due to the fact that the emission from a hot slab with optical depth ≤ 1 is not very different from the emission of the shock, which essentially consists of the sum of the blackbody-like emission from the photosphere below the shock, which reaches similarly hot temperatures, and a smaller contribution from the optically thin pre shock region. In any case, recent determinations [181, 64] of the wavelength dependence of the veiling have shown that the excess flux in the 5,000–10,000 Å range is higher than predicted by the single accretion column of [26]. A more realistic model, including a diversity of accretion columns carrying different energy fluxes, can better explain the veiling observations; the excess at longer wavelengths could be due to accretion columns carrying low-energy fluxes and thus heating the photosphere to lower temperatures than the high-energy columns that contribute to the UV [103]. The resultant mass-accretion rates are higher by a factor of $\sim$2 to 3, which is within the uncertainties of the determinations [103].

The determination of the mass-accretion rate from the veiling excess or by direct measurements of the UV excess continuum is limited by the intrinsic chromospheric emission of the star. WTTSs show levels of magnetic activity comparable to those of the most active stars. This is seen in emission lines as Ca II triplet [12] and in their X-ray luminosities [60]. CTTSs are in the same evolutionary stage, and thus their chromospheres and transition regions must be similar. Therefore, emission from accretion flows with mass-accretion rates much lower than the average are not detectable in the continuum. However, accretion can be detected in high-opacity emission lines as Hα by the presence of high-velocity wings sometimes superimposed on the narrow chromospheric emission component. Thus [179] proposed that the 10% half width of Hα is a much better discriminant between CTTSs and WTTSs than the usually used Hα equivalent width (see § 2). Similarly, modeling of the emission lines produced in the low-density flows of extremely low accretors, which are not affected by saturation effects, is the best way to estimate the mass-accretion rates. Such models have been done for the Hα and Br γ lines of very low-mass stars and YBDs [142, 136] and have yielded mass-accretion rates on the order of $10^{-12} - 10^{-11}$ $M_\odot$ yr^{-1}.

The situation is also more complicated for HAeBe stars. In this case, the intrinsic photospheric emission is much higher than the excess; in addition, the photospheric emission peaks at wavelengths similar to those of the expected shock emission because both photosphere and shock have similar temperatures. These factors make the intrinsic shock emission much more difficult to detect, except by a small filling in of the Balmer jump [135]. Emission-line profiles of at least the HAeBe (least likely to be confused with classical Be stars) seem to be consistent with magnetospheric infall [135], but the contribution from the wind seems to be much more important than in the TTS case. Secondary calibrations of L_{acc} versus L(Br γ), determined from measurements of mass accretion in intermediate mass CTTSs, predecessors of the HAeBe and still cool enough that the excess can be measured in the UV [28], are commonly used [69], but justification for their application from first principles is still to be given.

In any event, measurements of the accretion luminosity now exist from the substellar limit to the intermediate-mass range. These measurements indicate a dependence $\dot{M} \propto M_*^2$ [136]. Several models have been proposed to explain this dependence [83, 53, 37], but none has yet been widely accepted.

4. Disks in YSOs

The first images of disks around CTTSs came from the *Very Lange Array* (*VLA*) at 7 mm [154, 185] and from *HST* observations [21, 145, 146]. Observations by millimeter interferometers have now imaged a substantial number of disks [9, 170], from which we are getting more and more detailed information on their velocity field, which is usually Keplerian, and on their surface density distribution.

Several groups now have codes for sophisticated calculations of disk structure and emission (see [52]). Here we focus on the general physical principles governing the structure and emission of disks around young stars and use our models to illustrate the results.

4.1. Irradiated Accretion Disks

Even though disks around YSOs are accreting (see § 3), their SEDs do not agree with the predictions of standard accretion disks; specifically, they are flatter than $\lambda F_\lambda \propto \lambda^{-3/4}$ [115, 85]. However, for an average mass-accretion rate of 10^{-8} $M_\odot$ yr^{-1} (§ 3) and typical stellar parameters, L_{acc} ~0.1 $L_\odot \approx 0.1$ L_*, the stellar luminosity is higher than the accretion luminosity in most cases, indicating that stellar irradiation must be an important heating agent.

For a flat disk, the irradiation flux can be estimated as

$$F_{irr} \sim I_* \Omega_* \cos\theta_0 \sim \frac{I_* \pi R_*^2}{R^2} \frac{h_*}{R} \sim \frac{2I_*}{3} \left(\frac{R_*}{R}\right)^3, \quad (3)$$

where θ_0 is the angle between the line connecting from a representative point on the stellar surface (at a height $h_* \sim 2R_*/3\pi$ above the disk midplane) and a point on the disk surface at distance R and the vector normal to this surface. Here we have approximated the stellar radiation as impinging on the disk surface in a single beam. If we take $F_{irr} \sim \sigma T^4$, the temperature is $T \propto R^{-3/4}$, which is similar to the temperature distribution of an accretion disk, and thus the flat irradiated disk results in SEDs that cannot explain the observations either.

Kenyon & Hartmann (1987) [115] proposed that disks of CTTSs are not flat but flared, which means that they can capture more stellar radiation than flat disks. The equation of hydrostatic equilibrium in the vertical direction for a geometrically thin disk in the central gravitational potential well of the star can be written as

$$\frac{1}{\rho}\frac{dp}{dz} = -\frac{GM_*z}{R^3}, \tag{4}$$

where $\rho(z, R)$ and $p(z, R)$ are the mass density and pressure at height z and radius R. If the disk is isothermal in the vertical direction, then $p = \rho c_s^2$, where $c_s = (kT/m)^{1/2}$ is the sound speed, which does not depend on z, and m is the mean mass of the gas particles. With this approximation

$$\rho(z, R) = \rho_m e^{-\frac{z^2}{2H^2}}, \tag{5}$$

where $\rho_m(R)$ is the density at the midplane and H is the gas scale height, given by

$$H = \frac{c_s}{(GM_*/R^3)^{1/2}} = \frac{c_s}{\Omega_K} \propto T^{1/2}R^{3/2}. \tag{6}$$

If temperature decreases with distance more slowly than R^{-3}, then the disk gas scale height increases with distance. Then, if the height of the disk surface that intercepts the stellar radiation is proportional to the gas scale height, it is curved, capturing more stellar flux than a flat disk. With these approximations, $T \sim R^{-3/7}$, and the corresponding SEDs are much flatter than a viscous disk or a flat irradiated disk, i.e., they have a larger excess at longer wavelengths.

Disks are not vertically isothermal. Stellar radiation enters the disk at an angle θ_0 to the local normal to the stellar surface (again approximating the stellar radiation as coming in a single beam), so the energy flux captured by the disk is $\sim(\sigma T_*^4)(R_*/R)^2\mu_0$, with $\mu_0 = \cos\theta_0$ [24, 25, 121, 33, 43, 44, 49]. For simplicity, the radiation field can be separated into two frequency ranges, the *stellar* range, given by the stellar energy distribution, which is related to the stellar effective temperature, and the *disk* range, given by the local emissivity and related to the disk local temperature. As stellar flux enters the disk, a fraction $d\tau_*/\mu_0$ is absorbed at each z, where τ_* is the mean optical depth at the stellar range. This energy reemerges at the wavelength characterizing the local temperature, the disk range, so the direct impinging stellar flux decreases with height. The main opacity source in CTTS disks is dust grains, because temperatures are low enough that dust is not sublimated in most of the disk, except in regions very close to the central star. Dust opacity increases as λ decreases, and stellar radiation is emitted at a shorter wavelength than the radiation emitted by the disk, so the hotter the star, the larger the opacity at the stellar range, and the higher the energy capture and thus the heating.

The vertical temperature profile can be obtained from the equation of conservation of energy,

$$\int \rho\kappa_\nu B_\nu(T)d\nu - \int \rho\kappa_\nu J_d d\nu = \frac{1}{4\pi}\frac{dF_d}{dz}, \tag{7}$$

where κ_ν is the monochromatic absorption coefficient per unit of (gas plus dust) mass, J_d is the mean intensity of the local radiation at the disk wavelength range, and F_d is the local radiative flux. This equation indicates that the emitted energy is due to the absorption of photons from the radiation field plus the change in local flux F_d.

If we neglect viscous heating, the local flux changes only by deposition of stellar energy, so

$$\frac{dF_d}{dz} \sim 4\pi\kappa^*\rho J_* = 4\pi\kappa^*\rho J_{*,0}e^{-\tau_*/\mu_0}, \tag{8}$$

where κ^* is the mean opacity at the stellar wavelength range, and $J_{*,0}$ is the mean stellar intensity at the disk surface, given by $J_{*,0} = 1/4\pi \int Id\Omega \sim I_*\Omega_*/4\pi$. If we perform the integrals, equation (7) can be written as

$$\kappa_P(T)\frac{\sigma T^4(z)}{\pi} = \kappa_P(T)J_d(z) + \kappa_P(T_*)\frac{\sigma T_*^4}{4\pi}\left(\frac{R_*}{R}\right)^2 e^{-\tau_*/\mu_0}, \tag{9}$$

where $\kappa^* \sim \kappa_P(T_*)$, and κ_P is the Planck mean opacity ([49, 44]).

In the surface, where the local field is much smaller than the stellar field, $J_d << J_{*,0}$; in addition, the medium is optically thin, so $\tau_*/\mu_0 << 1$, and we can write

$$\kappa_P(T_0)T_0^4(z) \sim \kappa_P(T_*)T_*^4\left(\frac{R_*}{2\,R}\right)^2. \tag{10}$$

This is an implicit equation for the surface temperature T_0, corresponding to the *optically thin limit*. Note that this is the "hot-layer" temperature in the 2-layer approximation [33].

Stellar heating and thus the temperature decrease as radiation penetrates the disk, because the optical depth increases (cf. eq. (9)). The actual T profile depends on μ_0, and through this, on the actual shape of the surface of the disk, defined as the surface where $\tau_*/\mu_0 \sim 1$, that is, where most of the stellar energy is deposited. The mass surface density of the upper optically thin region above the surface is given by $\Delta\Sigma \sim \mu_0/\kappa^*$; the more flared the disk surface, the larger the μ_0 and the higher the mass of the optically thin region [140].

The stellar flux intersected by the disk surface $z_s(R)$ can be written as [115]

$$F_{irr}(R, z_s) \sim \frac{\sigma T_*^4}{\pi} \left[\frac{2}{3} \left(\frac{R_*}{R} \right)^3 + \pi \left(\frac{R_*}{R} \right)^2 \frac{Rd(z_s/R)}{dR} \right] \quad (11)$$

for $R >> R_*$.

A flat disk has a negligible $d(z_s/R)/dR$; thus

$$F_{irr}^{flat} = \frac{2\sigma T_*}{3\pi} \left(\frac{R_*}{R} \right)^3, \quad (12)$$

and $T^{flat} \sim (F_{irr}^{flat}/\sigma)^{1/4} \sim R^{-3/4}$, as we already have shown.

For a flared disk, if we assume that the height of the surface is a fixed number of scale heights, z_s becomes a function of T, and

$$\frac{Rd(z_s/R)}{dR} = \frac{1}{2} \frac{z_s}{T} \frac{dT}{dR} - \frac{5}{2} \frac{z_s}{R}. \quad (13)$$

With $F_{irr} = \sigma T^4$, eqs. (11) and (13) form a system for $T(R)$ in the isothermal approximation, resulting in $T(R) \propto R^{3/7}$. If T increases, the gas scale height increases, then the cross section to capture stellar photons increases, and the disk heating increases. This $T(R) \propto R^{3/7}$ solution has been found to be stable under perturbations in temperature or scale height as long as the thermal timescale of the disk is longer than the vertical dynamic timescale [43]. However, if the opposite is true, as could be the case for the outer disk, this simple solution turns out to be unstable [57]. The simple flared-disk solution is based on several simplifying assumptions. Effects like viscous dissipation, radial energy transfer, and scattering of the incident radiation tend to stabilize the disk. Moreover, disks are not isothermal, and the height of the surface is not a fixed number of scale heights.

Results of the detailed solution of the set of equations of vertical structure, including viscous dissipation for a disk with typical CTTS parameters are shown in Figs. 2.8 and 2.10 [44]. Figs. 2.8 and 2.9 shows the height of the surface z_s, the scale height H, and the photospheric height z_{phot}, where the Rosseland mean optical depth $\tau_{Ross} \sim 1$. Note that the height z_{phot} is defined only in the region where the disk is optically thick to its own radiation, determined by the Rosseland mean opacity. In the example shown in Figs. 2.8 and 2.9, the disk becomes optically thin to its own radiation ($\tau_{Ross} < 1$) for $R \gtrsim 20$ AU. In contrast, since the opacity at the wavelength where the stellar radiation is absorbed ($\lambda_* \sim 1\mu$m for $T_* \sim 4000$K) is large, the disk remains optically thick to the stellar radiation, and the surface is flared out to a few hundred AU.

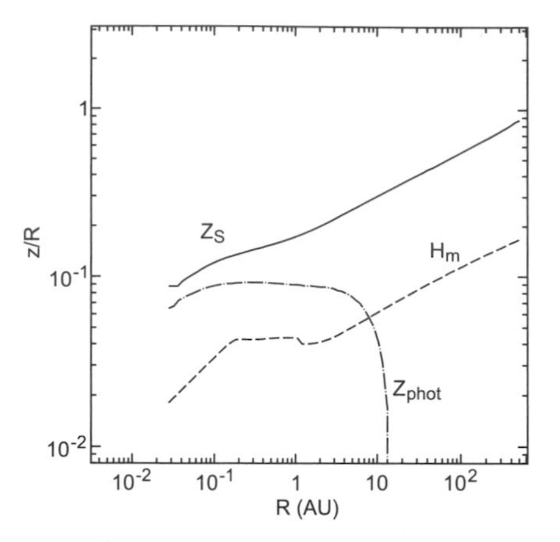

Fig. 2.8. Characteristic heights in the disk. Height z_s where stellar radiation is absorbed (solid line); scale height H_m (dashed line); photospheric height z_{phot} (dot-dashed line), defined where the disk is optically thin to its own radiation. Model parameters are $M_* = 0.5$ $M_\odot$, $R_* = 2$ $R_\odot$, $T_* = 4{,}000$ K and $\dot{M} = 10^{-8}$ $M_\odot$ yr^{-1}. Dust is uniformly mixed with gas and has an ISM composition and size distribution. Adapted from [44, 48].

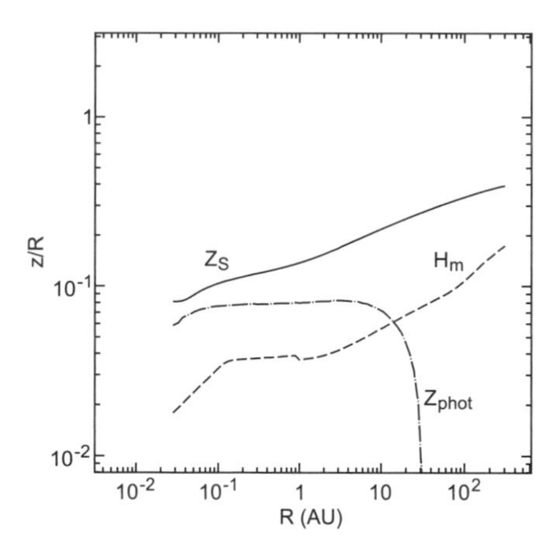

Fig. 2.9. Same as Fig. 2.8, but for grains with a size distribution characterized by $a_{max} = 1$ mm.

Figs. 2.10 and 2.11 shows characteristic temperatures in the disk. It can be seen that the midplane temperature T_m is higher than the photospheric temperature $T_{phot} = T(z_{phot})$ in the inner disk, where it is optically thick to its own radiation. This can be understood by using the diffusion approximation

$$\frac{dJ_d}{dz} = \frac{\sigma}{\pi}\frac{dT^4}{dz} = -\frac{3}{4\pi}\chi_R \rho F_d, \tag{14}$$

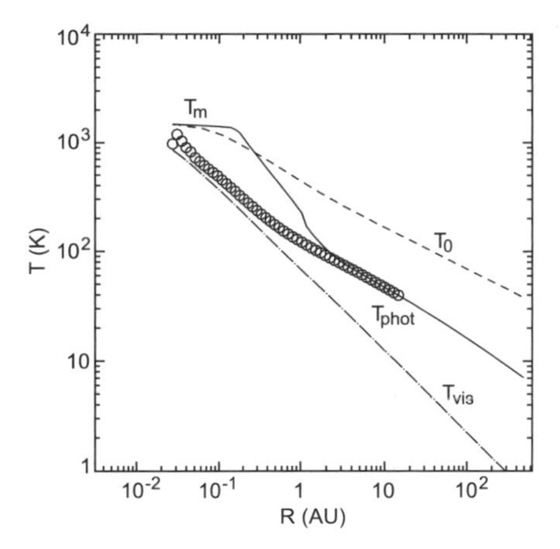

Fig. 2.10. Characteristic temperatures of the disk. Midplane temperature T_m (solid line); upper-layers temperature T_0 (dashed line); photospheric temperature T_{phot} (circles), defined where the disk is optically thin to its own radiation; viscous temperature T_{vis} (dot-dashed line). Model parameters are $M_* = 0.5\ \mathrm{M}_\odot$, $R_* = 2\ \mathrm{R}_\odot$, $T_* = 4,000\ K$, and $\dot{\mathrm{M}} = 10^{-8}\ \mathrm{M}_\odot\,\mathrm{yr}^{-1}$. Dust is uniformly mixed with gas and has an ISM composition and size distribution. Adapted from [44, 48].

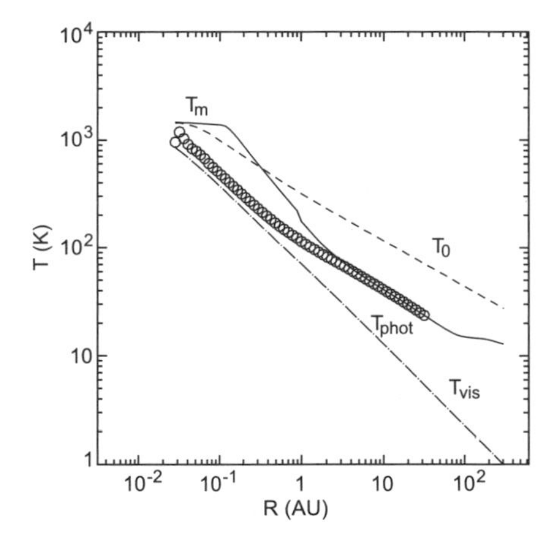

Fig. 2.11. Same as Fig. 2.10, but for grains with a size distribution characterized by $a_{max} = 1$ mm.

from which wc can write

$$\Delta(\sigma T^4) \sim \frac{3}{4}\tau_R F_d. \tag{15}$$

In the inner annuli, $\tau_R >> 1$, and $T_m > T_{phot}$; the temperature gradient allows the flux (viscous plus local radiation) to emerge from the disk. In the outer regions, when the disk becomes optically thin to its own radiation, $\tau_R << 1$, the disk becomes nearly isothermal in the regions near the midplane. As shown in Figs. 2.10 and 2.11, the midplane temperature remains at the dust-destruction temperature ($\sim$1,500 K) in the innermost regions; if the temperature increases, the dust gets destroyed and the opacity drops; as a result, the disk becomes optically thin and cools below the dust-destruction temperature, at which point the opacity increases again. This thermostat effect makes the temperature at the midplane stay at the dust-destruction temperature [189].

The surface temperature T_0 is higher than T_{phot} in the inner regions and than T_m in the outer optically thin regions, as predicted by eq. (9). By comparison of T_{vis} and T_{phot} in Figs. 2.10 and 2.11, it can be seen that viscous heating is important only in regions inside 1 AU, given the low $\dot{M}$ characteristic of the typical CTTS. Finally, it can be seen that temperatures behave as $1/R^{1/2}$ for $R >> R_*$. As a summary, the upper panel of Fig. 2.12 shows isocontours of temperature for the disk model in Figs. 2.8 and 2.10. It also shows isocontours of number density for the same cases, and the surface z_s where stellar radiation is absorbed.

The particular shape of the temperature profile has important observational implications. For one thing, features formed in the optically thin upper regions will appear in emission, even if the disk is optically thick, because the local temperature is so much higher than that of deeper regions where the continuum forms. As a result of this "chromospheric effect," features like those from silicates that form in the atmosphere of the optically thick inner disk regions appear in emission [140, 66]. Molecular features formed in the upper layers appear in emission as the CO near infrared [24, 20] and water lines [31]. The higher temperatures of the upper layers also imply that molecules can exist in the gas phase in the disk even when the midplane temperatures are so low that molecules are settled onto grain surfaces [188, 190, 3, 2]. For example, if the disk shown in Fig. 2.10 were isothermal at T_m, the CO molecules would be on grain surfaces for $R > 60$ AU, for which $T < 20$ K (see also Fig. 2.12). However, the hot upper layers are kept at a temperature high enough for molecules to be in the gas phase out to $\sim$400 AU, in agreement with millimeter molecular observations [59, 61, 149, 102].

An additional important effect further affects molecular equilibrium and emission. The results discussed so far assume that the temperatures of the gas and dust are the same. However, in the low-density uppermost layers of the disk, the collisional coupling between gas and dust becomes much less effective, and other factors become more important in heating and cooling the gas. Heating factors include absorption of X-rays and UV radiation, grain

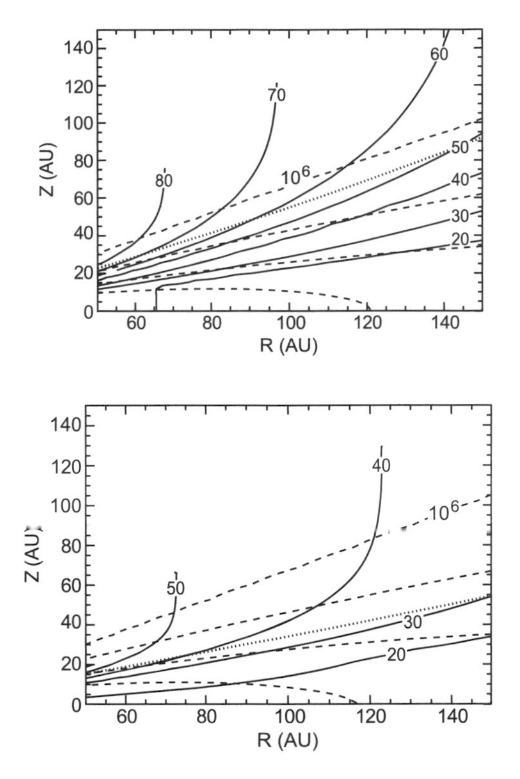

Fig. 2.12. Isocontours of temperature and number density in the disk. Isocontours are shown for temperatures (solid lines) 80 K, 70 K, 60 K, 50 K, 40 K, 30 K, and 20 K, and densities (dashed lines) $10^6, 10^7, 10^8$ and 10^9 cm^{-3}. The height where most of the energy in the stellar radiation is absorbed is indicated by a dotted line. The disk model parameters are $M_* = 0.5$ M$_\odot$, $R_* = 2$ R$_\odot$, $T_* = 4{,}000$ K, and $\dot{M} = 10^{-8}$ M$_\odot$ yr^{-1}. Upper panel: ISM dust. Lower panel: $a_{max} = 1$ mm. Adapted from [44, 48].

photoelectric heating, exothermic chemical reactions, and collisions with warm grains, while cooling factors include line emission and collisions with cooler dust grains. Several groups are working on the calculation of gas temperatures [70, 71, 72, 110, 73, 144], and not all groups include all these factors. The gas temperature in the low-density, uppermost layers of the disk becomes much higher than the dust temperature, reaching $\sim$ 5,000 K at 1 AU and $\sim$300 K at 100 AU. These elevated temperatures imply that molecules can be in the gas phase even in the uppermost layers of the outer disk, in agreement with observations.

The surface density of the disk can be self-consistently calculated from the equations of an irradiated accretion disk. A geometrically thin accretion disk with a steady mass-accretion rate $\dot{M}$ has a surface-density distribution given by the conservation of the angular momentum flux,

$$\Sigma = \frac{\dot{M}}{3\pi\nu}\left[1 - \left(\frac{R_*}{R}\right)^{1/2}\right], \tag{16}$$

where ν is the viscosity. In the parametric α prescription [157], the viscosity can be written as $\nu = \alpha c_s H = \alpha c_s^2/\Omega_K$ by using eq. (6). With $c_s \propto T^{1/2} \propto R^{-1/4}$ and $\Omega_K \propto M_*^{1/2} R^{-3/2}$, we obtain at large radii

$$\Sigma \sim 4 \left(\frac{\dot{M}}{10^{-8}\,\frac{\mathrm{M}_\odot}{\mathrm{yr}}}\right)\left(\frac{\alpha}{0.01}\right)^{-1} \times \left(\frac{T_{100\,\mathrm{AU}}}{10\,K}\right)^{-1}\left(\frac{R}{100\,\mathrm{AU}}\right)^{-1}\left(\frac{M_*}{1\,\mathrm{M}_\odot}\right)^{1/2} \mathrm{gr\,cm}^{-2}, \tag{17}$$

using values of α found in modeling CTTS disks and expected from theories (see the Balbus chapter, this volume), and typical temperatures at 100 AU. The surface-density dependence on radius of irradiated accretion disks, $\Sigma \propto R^{-1}$, is much flatter than the usually assumed dependence $\Sigma \propto R^{-1.5}$, and it has been confirmed by observations [184, 8].

If we assume that this dependence holds at all radii, the disk mass would be given by

$$\frac{M_d}{M_\odot} = 0.03 \left(\frac{\dot{M}}{10^{-8}\,\frac{\mathrm{M}_\odot}{\mathrm{yr}}}\right)\left(\frac{R_d}{100\,\mathrm{AU}}\right)\left(\frac{\alpha}{0.01}\right)^{-1} \left(\frac{T_{100\,\mathrm{AU}}}{10\,K}\right)^{-1}\left(\frac{M_*}{1\,\mathrm{M}_\odot}\right)^{1/2}, \tag{18}$$

in agreement with values determined from dust millimeter emission. We can see that the mass-accretion rates, determined from the inner disk properties, are consistent with large-scale properties like the disk mass. Note that if $\dot{M}$ is known for a given star, then M_d and α are complementary parameters, since the temperature at the outer disk radii, T_{Rd}, is fixed by stellar irradiation, and sizes can be estimated from observations.

4.2. *Effects of Dust Properties*

The temperature of a volume element inside the disk, assumed to be in thermal equilibrium (i.e., such that its temperature does not change with time), is mostly controlled by the balance between the heating produced through absorption of stellar radiation, accretion-shock radiation, and the fraction of disk radiation that reaches the volume element (§ 4.1), and the cooling due to the radiative losses of the element. There are different mechanisms that transport energy between disk regions, contributing to local heating and cooling, but radiation is the most relevant of these mechanisms [49]. Another important aspect to keep in mind is that the disk temperature is not an isolated quantity but is connected to the disk viscosity, gas scale height, density, and other factors, so the whole disk structure is an interdependent phenomenon.

The crucial ingredient in the absorption and emission of radiation is the opacity of the disk material, and given the low temperatures in almost the whole disk (around a young low- or intermediate-mass star), dust happens to be its main opacity source, despite the fact that it represents only around 1% of the disk mass. In this subsection we summarize important dust properties and the general effects they have on the structure and emission of circumstellar disks.

The dust opacity depends on the shape, size distribution, abundance, and constitution of the dust grains. Schematically, a spherical grain with radius a has a cross section for absorption of radiation at wavelength λ of the order of the geometric cross section πa^2, if $\lambda << 2\pi a$, and it decreases as $\sim\lambda^{-2}$ for $\lambda >> 2\pi a$. It is common to use the size parameter $x = \lambda/2\pi a$ and an efficiency factor Q_a, defined as the cross section for absorption divided by the geometric cross section. Thus $Q_a \sim 1$ for $x < 1$, and $Q_a \propto x^{-2}$ for $x > 1$. The dust opacity, $\kappa_\nu(a)$, given in cm^2 per gram of dust, is given by the cross section over the mass of the grain, $m_g = \rho_g 4/3\pi a^3$, where ρ_g is the bulk density of the grain. Thus $\kappa_\nu \propto 1/a$, which means that the bigger the grain, the lower the opacity at short wavelengths and the higher at long wavelengths, close to where $x \sim 1$. Note that the transition to the λ^{-2} regime occurs at a wavelength that increases with a. This is shown schematically in Fig. 2.13.

In general, dust grains are not of a single size, and typically there are different numbers of grains in different size intervals, between a minimum radius a_{min} and a maximum radius a_{max}. This is described by a *size-distribution function*, for example, a power law $n(a)da \propto a^{-p}da$, where a is the grain size and the exponent p is usually taken as 3.5, describing the properties of the ISM dust [51], or 2.5 if there has been some degree of coagulation [126]. The coefficient in the dust size distribution is proportional to the dust-to-gas mass ratio, ζ, which specifies the mass in dust in a given disk mass element. If a_{min} and ζ are fixed, the larger the a_{max}, the less the number of smaller particles, because the larger particles take more mass, and therefore, the lower the opacity at short wavelengths and the higher at long wavelengths. For a mixture of sizes, the transition from the flat to the $\propto \lambda^{-2}$ regime occurs over a large range of wavelengths (see Fig. 2.13), so that the local slope of the function $\kappa(\lambda)$ versus λ changes slowly from ~ 0 to ~ -2. Therefore, the form usually assumed to represent the dust opacity $\kappa \propto \lambda^{-\beta}$ is not actually valid, since β depends on

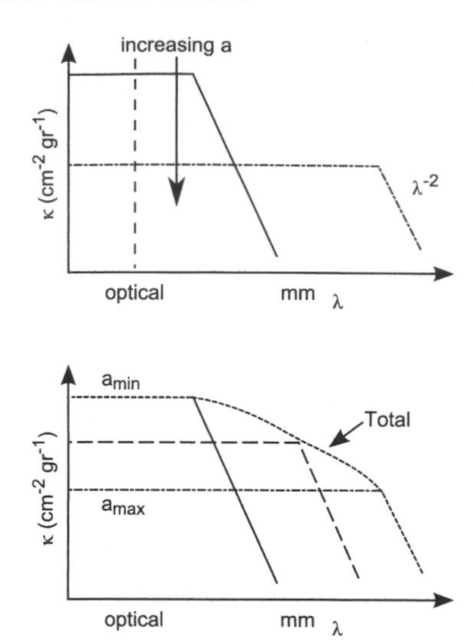

Fig. 2.13. Dependence of opacity on grain size. Upper panel: opacity for single grains. As the grain size a increases, κ decreases at short λ and increases at long λ. Lower panel: opacity for a grain size distribution. The slope β of κ versus λ depends on λ. From [23].

λ [126, 48], although it may be applicable over a sufficiently small λ interval. For a detailed study of dust properties, see [126].

But how does all this affect the disk? Most of the energy of the stellar radiation intercepted by the disk is deposited at the "irradiation surface," where the mean radial optical depth of the stellar radiation is unity. The actual fraction of stellar flux absorbed and reprocessed by the disk is proportional to the solid angle subtended by the irradiation surface as seen from the star (see [85]). Thus a higher and more curved surface intercepts more stellar radiative flux than a lower and flatter surface, and this is why a flared disk is hotter and has an SED with a larger infrared excess than a flat disk (see § 4.1).

The location of this surface depends on the dust opacity at the wavelength range where most of the stellar radiation is emitted. For instance, a T Tauri star, with an effective temperature of $T_* = 4{,}000$ K, has an SED that peaks around $\lambda \sim 1 \mu$m. If the number of grains in the disk with sizes much smaller than $\sim$0.1 μm decreases, then the height of the irradiation surface decreases, and the fraction of stellar flux reprocessed by the disk decreases too. As we have mentioned above, if a_{max} increases, the number of small grains decreases. Thus a disk with a dust mixture characterized by $a_{max} = 0.25 \mu$m—a size distribution typical of the ISM—has a higher irradiation surface (compare Figs. 2.8 and 2.9) and is hotter (compare Figs. 2.10 and 2.11) than a disk characterized by a_{max} ~1 mm and the same dust-to-gas mass fraction. Thus the resulting SED of the disk with small grains has a larger infrared excess than the disk with big grains. On the other hand, the millimeter emission of the disk with $a_{max} = 0.25\ \mu$m will be lower than that of the disk with $a_{max} =$ 1 mm because disks with mm grains are more efficient emitters at those wavelengths. Therefore, the SED gives direct information about the dust properties. In particular, the median SED of the CTTSs in Taurus can be much better explained by a disk with a uniform grain mixture characterized by $a_{max} \sim$ 1 mm dust than by ISM dust ($a_{max} = 0.25\mu$m) [48].

Another way to decrease the irradiation surface and make a disk model with an infrared SED more consistent with observations than the SED predicted using a disk model with ISM-like grains is to decrease the number of small grains in the disk atmosphere by decreasing the dust-to-gas mass ratio. This happens naturally when grains grow and settle toward the disk midplane, since a fraction of the dust mass disappears from the disk atmosphere, decreasing locally the dust-to-gas mass ratio and, therefore, the opacity. At the same time, the missing dust mass increases the dust-to-gas mass ratio at the disk midplane, where grains can grow even more and the millimetric emissivity of the disk increases. In § 5.1 the evolution of dust and its effects on the disk are discussed in more detail.

4.3. Inner Disk

The picture of the inner disk around young stars has changed considerably over the years. As discussed in § 3, the disk was thought to extend all the way to the star and to be separated from it by a thin boundary layer. This picture evolved into the present paradigm of magnetospheric accretion, where the disk is truncated at a few stellar radii by the stellar magnetosphere. Since the temperature obtained from eq. (9) at the disk photosphere was below the dust-sublimation temperature, it was assumed that dust was present in the disk all the way to the truncation radius. It then became clear that something was wrong with this picture. References [139, 56] showed that the peculiar 3 μm "bump" in the SEDs of HAeBe, first discussed in [99] and which had defeated explanation until then, could be explained as emission from a "wall" in the inner disk, located at the dust-destruction radius and illuminated by the star with a normal impinging angle. The gas disk inside this radius was assumed to have a low optical depth, an assumption confirmed at first approximation by [135]. At the same time, first interferometric results showed that indeed there was a sharp cutoff in the inner disks of HAeBe stars also associated with the dust-sublimation radius [124, 171].

The existence of this wall can be easily understood in the context of irradiated accretion disks. For radii just outside the magnetospheric truncation radius, the temperature in the uppermost layers above the disk photosphere is higher than that in the photosphere (see eq. [10]), and dust could sublimate there. But if those upper regions were clean of dust, stellar radiation would reach the dust down at the photosphere and sublimate it too. Stellar radiation could then erode the dusty disk from the magnetospheric radius to a radius where the temperature of dust grains illuminated directly by the star (and the accretion shocks at the stellar surface) are equal to the sublimation temperature; for larger radii, dust grains are cold and could not be destroyed. Thus in the present picture of the inner disk there is a region free of dust between the magnetospheric truncation radius and the dust-destruction radius. Plate 3 shows an artistic representation of the inner disk.

Disk models usually start the calculation of the structure at the dust-sublimation radius under the assumption that the inner gaseous disk does not contribute to the emission. The emission from the wall adds to the disk emission to make the resultant SED; wall emission is comparable to or even dominates the rest of the disk emission in the near IR, specially in the *Spitzer*/IRAC bands (3.6, 4.5, 5.8, and 8 μm), with important observational implications [45, 66].

The contribution of the wall to the SED depends on its emitting area, $\sim 4\pi R_{sub} Z_{wall}$. The sublimation radius can be written as

$$R_{sub} = \left[\left(\frac{L_* + L_{acc}}{16\pi\sigma_R} \right) \left(2 + \frac{\kappa_*}{\kappa_d} \right) \right]^{1/2} \frac{1}{T_{sub}^2} \qquad (19)$$

[47], neglecting scattering. In this expression, L_* and L_{acc} are the stellar and accretion-shock luminosities, κ_* and κ_d are mean opacities weighted by the Planck function at the stellar and local disk temperatures, respectively, and T_{sub} is the dust-sublimation temperature. The accretion luminosity is important only for the high accretors among the CTTSs [134]. The kind of dust at the inner disk affects R_{sub} through the sublimation temperature. In general, dust at the wall is assumed to be mostly silicates, the dust with the highest sublimation temperature among the expected dust composition in disks, as indicated by solar system studies [148]. Sublimation temperatures can be determined observationally from interferometric sizes, and they range from 1,300 to 1,900 K [127], consistent first approximation with silicates. However, sublimation temperatures increase with the local density. Thus dust at the dense midplane can survive at a smaller radius than the dust at the less dense upper regions, and the wall curves outward [106]. In addition, the size distribution of the dust

is important because the smaller the dust, the larger the ratio κ_*/κ_d (§ 4.2), and thus the larger the R_{sub}. Moreover, if there is settling at the wall, large grains at the midplane will survive at smaller radii than the small grains at the upper layers, again curving the surface [168].

The height of the wall Z_{wall} can be estimated as the height where the optical depth to the stellar radiation becomes unity [56]. If the wall is assumed to be vertical, then the height can be calculated from the vertical density profile and the assumed dust properties, and it is usually 3 to 4 scale heights [56]. The scale height is calculated with the local temperature, the dust-destruction temperature. In the case of "passive disks," i.e., those in which accretion heating is neglected, the wall is hotter than the disk behind it, so it is "puffed up" and shadows the disk behind. However, accretion heating keeps the midplane of CTTS disks behind the wall at the dust-sublimation temperature through the thermostat effect (§ 4.1). In this case it seems unlikely that the inner disks are shadowed.

5. Dust Evolution

5.1. Dust Settling

With the launch of IR missions to space in the last decade, such as *ISO* and *Spitzer*, the near- to mid-IR region has became accessible, showing disk spectra of YSOs characterized by strong silicate features in emission [140, 186, 158, 1, 66]. Fig. 2.14 shows a sample of spectra of CTTSs in Taurus obtained with the Infrared Spectrograph (IRS) on board *Spitzer*. At first, these observations seem to contradict the prediction from fits to the overall SED that grains must have grown in the disk to millimeter sizes [48], because the opacity due to grains larger than $\sim 1\mu$m is featureless [126]. However, the silicate features form in the upper layers of the inner disk, while the long-wavelength SED comes mostly from regions closer to the midplane; therefore, this apparent contradiction is actually directly indicating that large grains must have settled to the midplane, leaving behind a population of small grains, as predicted by dust-evolution theories [177, 55].

Grain growth and settling toward the midplane are natural consequences of the environmental conditions in disks [177, 55, 54]. Particles tend to collide and stick together, growing in size; gravity pulls them toward the midplane. The timescales for these processes to occur scale as the orbital period, and they are thus faster for smaller disk radii. As a consequence of these processes, the upper disk layers get depleted with time, while the dust-to-gas mass ratio builds up at the midplane (for a good description of the physics of these processes, see the introductory lectures in [10]).

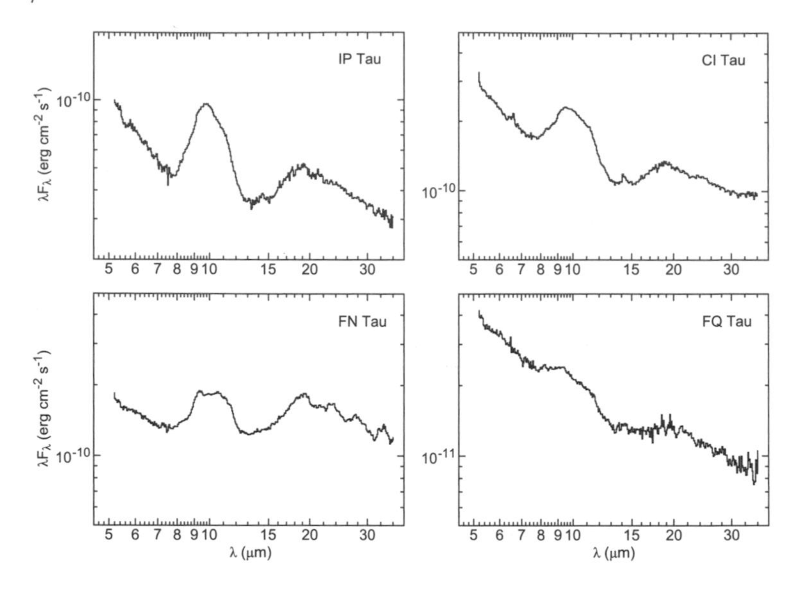

Fig. 2.14. IRS spectra of Taurus stars. Note the diversity of profiles in the silicate emission features. Adapted from [66].

The depletion of the upper layers expected from dust evolution has direct observational consequences. As a result of depletion, the absorption of stellar radiation by the upper layers decreases, the surface where stellar radiation is absorbed becomes flatter, so less radiation is absorbed, and the disks emit less [55, 54, 45]. Fig. 2.15 shows the effects on the temperature and disk structure of dust settling. The models shown have two populations of grains: ISM dust, which resides mostly in the upper layers, and a mixture with $a_{max} = 1$ mm, at the midplane. In the left panel, the dust-to-gas mass ratio of the small grains is equal to the standard. In the lower panel, it has been reduced to 1% of the standard (depletion is parameterized by ϵ, the dust-to-gas mass ratio of the small grains relative to the standard). The disk with a strong depletion has a much flatter surface ($z_s(R)$, see § 4) than the disk without depletion. Note the warm and dense regions above $z_s(R)$ in the depleted disk, important for molecular formation ([144]; chapter 3 this volume).

Fig. 2.16 shows the noticeable effects of upper-layer depletion due to dust settling in the SED. Dust settling decreases the opacity in the upper layers as increasing dust size did, but there is an important difference. According to eq. (10), the surface temperature depends only on the ratio $\kappa_P(T_*)/\kappa_P(T_0)$, which is independent of the dust-to-gas mass ratio. This ratio is large for the small grains remaining in the upper layer, and thus these layers are hot,

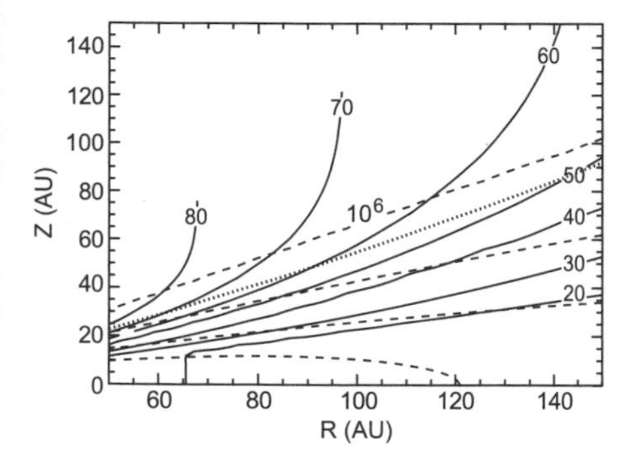

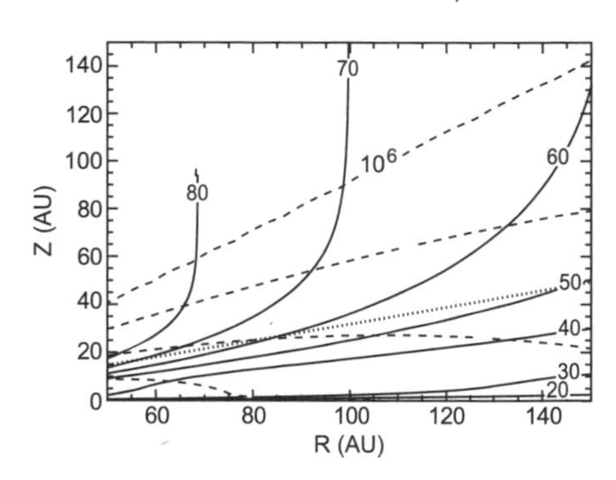

Fig. 2.15. Effects of dust settling on the disk structure: isocontours of temperature (solid lines) and number density (dashed lines) for disks with $\dot{M} = 10^{-8}\ M_{\odot}\ yr^{-1}$, $\alpha = 0.01$, typical stellar parameters, and depletion parameters $\epsilon = 1$ (left) and $\epsilon = 0.01$ (right). The dotted line shows the disk surface, $z_s(R)$. Adapted from [45].

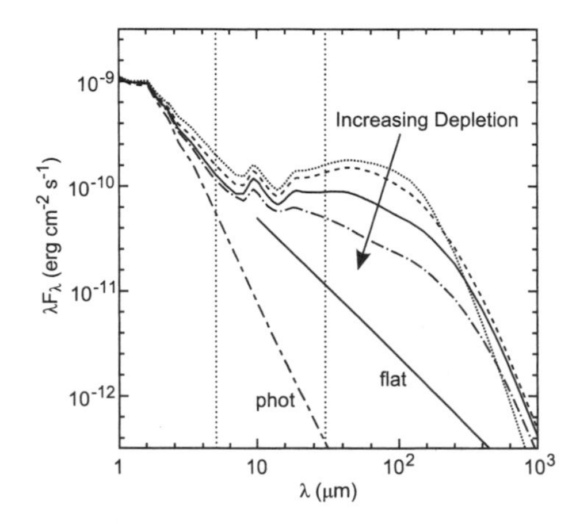

Fig. 2.16. Effects of settling on the SEDs. Disk models are calculated for the same stellar parameters and accretion rate and vary by the degree of depletion of small grains in the upper disk layers, from the standard dust-to-gas ratio (uppermost dotted curve) to 0.001 times the standard ratio (dash-dotted line). Even with this degree of depletion, the slope of the SED is not as steep as that of the flat disk. The photosphere is shown for comparison. From [45].

regardless of depletion. As a result, the silicate features appear in emission, even though the opacity of the upper layers decreases and so does the IR flux [45].

Observations obtained with the IRS instrument on board *Spitzer* allow us to make inferences about the degree of depletion in the upper layers in typical disks. As can be seen in Fig. 2.16, the slope of the SED in the range

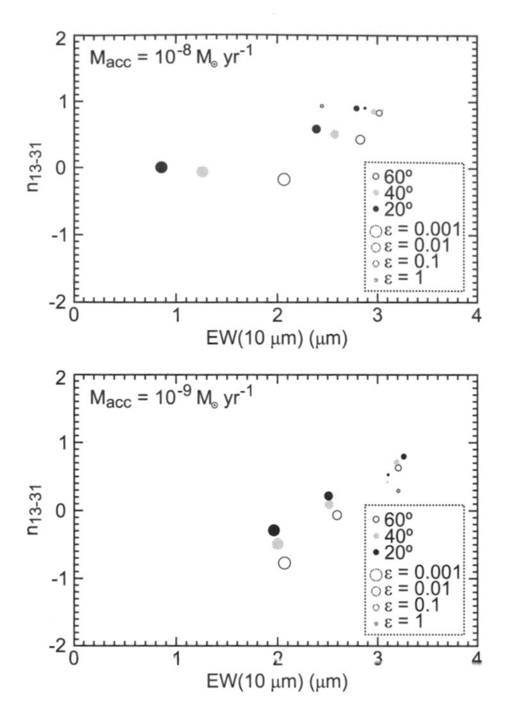

Fig. 2.17. Predicted midinfrared slope of the SED between 13 and 31 μm versus the equivalent width of the silicate feature at 10 μm for two mass-accretion rates, shown in the upper left corner, several values of the depletion parameter ϵ, which measures the dust-to-gas mass ratio of small grains in the upper layers relative to the standard dust-to-gas mass ratio (lower values of ϵ correspond to larger symbol sizes), and several inclinations, as indicated in the inset. From [68].

13μm–30 μm, covered by IRS spectra, is a good indicator of settling. Fig. 2.17 shows this slope versus the equivalent width of the 10μm silicate feature calculated with irradiated accretion-disk models from [45] for two values of the mass-accretion rate, $\dot{M} = 10^{-9}$ and 10^{-8} $M_{\odot}$ yr^{-1}, and typical values for the stellar mass and radius of CTTSs. The models are calculated for different amounts of dust settling, measured by the depletion parameter ϵ. Values of these quantities determined from IRS spectra of CTTSs in the Taurus clouds are shown in Fig. 2.18. Most of the observations fall in the region explained by the models, indicated in the figure, and among these, comparison with models indicates large factors of depletion in the upper layers; namely, observations seem to indicate that the dust abundance in the upper layers of the Taurus disks is 1% or even less of the standard ISM abundance [66, 68]. A number of disks fall outside the region explained by the models. These will be discussed in § 5.3.

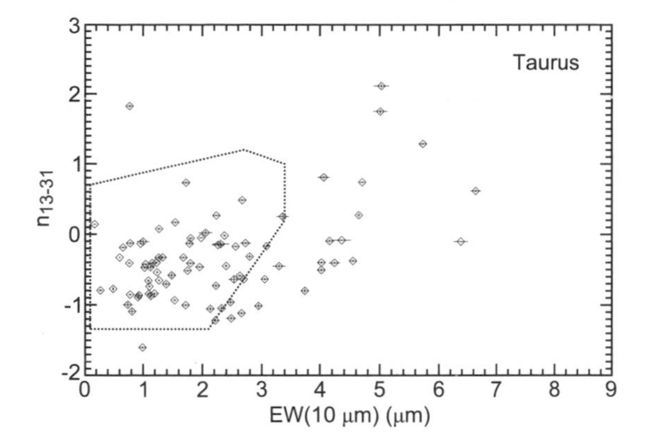

Fig. 2.18. Observed midinfrared slope of the SED between 13 and 31 μm versus the equivalent width of the silicate feature at 10 μm in the CTTSs in Taurus. The curved line encloses the region covered by the models in Fig. 2.17. From [68].

5.2. Evolution of Dust Properties

Studies of dust growth and settling [177, 55, 54] describe in general terms the expected evolution of the solids in disks; however, the microphysics of dust growth and the effects of turbulence in disks are complicated, weakening the predictive power of those studies. Observations of disks over a large range of ages and environments can provide necessary input to guide theoretical studies. Such observations have been vigorously conducted in the last few years, specially with *Spitzer*. They focus on populations of a given age, rather than individual stars, because one important aspect that is emerging from these studies is that age is not the only factor controlling evolution; there is a large spread of properties within a population of a given age, and median properties are better indicators of the processes that may be happening.

Several important points have been established. First, the disk frequency decreases with the age of the population. Fig. 2.19 shows the disk frequency in late-type mass stars (late G or later) as a function of age for a number of populations, including clusters and associations, some without high mass formation. The disk frequency is measured by the presence of excesses over the photosphere in the near-IR bands, and therefore it refers to the inner disk. It is apparent that $\sim$ 70–80% of the inner disks have dissipated by 5 Myr. The disk frequency is also mass dependent, with disks around intermediate-mass stars (types F and early) evolving much faster than for later types [97, 94].

In addition, the excess itself decreases with age. Fig. 2.20 shows the median slope of the SED as a function of age. The slope is measured between the K band, representative of the photospheric levels, and three bands: 24 μm,

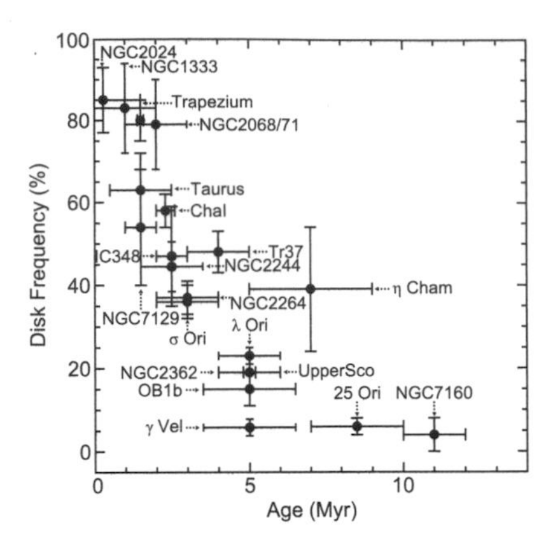

Fig. 2.19. Disk frequency as a function of the age of the population. Each population is named in the figure. The disk frequency is measured by the ratio of stars with excess in the near infrared (either JHK or in the *Spitzer*/IRAC bands) to the total number of stars. From [94].

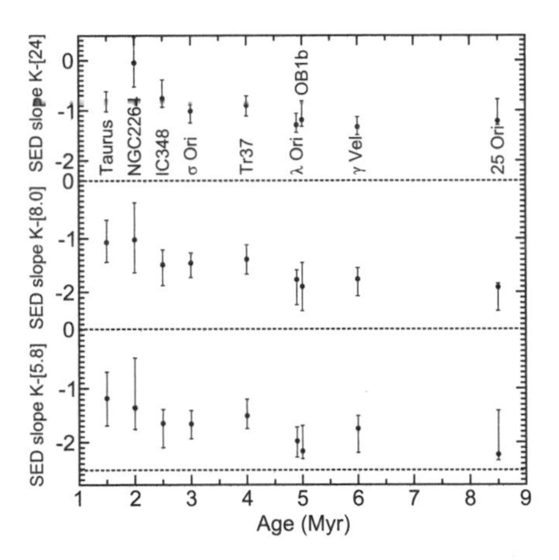

Fig. 2.20. Slope of the near-IR SED for populations with ages from ~ 1 to 10 Myr. Top panel: slope between K and 24 μm; middle panel: slope between K and 8 μm; bottom panel: slope between K and 5.8 μm. It is apparent that the innermost disk, emitting at the shorter wavelengths, clears up first. The dotted line indicates the photospheric level. Adapted from [93].

8 μm, and 5.8 μm (the last two come from IRAC measurements, the first from MIPS measurements). The error bars represent the first quartiles; that is, 50% of the disks have slopes within those bars. The photospheric level is shown for comparison. Although there is a large spread at a given age, it is apparent that the excess over the photosphere decreases as the age of the population increases. Moreover, the slope becomes steeper and closer to the

photospheric slope sooner at the shorter wavelengths, indicating that the inner disk becomes optically thin first.

Present evidence thus indicates that the older the population, the fewer disks remain, and those remaining disks show less excess above the photosphere. In the near IR, the contributors to the flux are the wall at the dust-destruction radius and the innermost disk. The first evolutionary effect that must be taken into account is that the mass-accretion rate decreases with age (§3); this implies that the surface density decreases (eq. [16]), and thus the optical depth of the disk decreases. However, [94] shows that the decrease of mass-accretion rate alone cannot account for the decrease of the excess. Additional factors, such as dust growth and settling in the inner disk and wall, are required to explain the decrease. The observed inside-out clearing (Fig. 2.20) is consistent qualitatively with predictions of dust evolution [177, 55]. Extreme cases of inner disk clearing are the transitional disks.

5.3. Transitional and Pretransitional Disks

Transitional disks have been characterized observationally as having little excess above the photosphere in the near IR below $\sim 10\mu$m and excesses consistent with optically thick disks at longer wavelengths. This characteristic SED can be understood in terms of an optically thick outer disk truncated at some radius, with or without a small amount of dust remaining in the inner disk region inside the truncation radius [27, 46, 29, 41, 67]; the edge of the truncated disk, frontally illuminated by the star, is responsible in this model for the flux above 10 μm, while the small excess above the photosphere at shorter wavelengths, and often the silicate feature, are due to the remaining dust in the inner region. A sketch of the model and a typical SED are shown in Figs. 2.21 and 2.22. Although the inner truncation of the disk was originally inferred from the SED, inner edges to disks at radii in agreement with SED predictions have been confirmed by millimeter interferometric data [101, 147], which probe the large grains at the midplane of the outer disk. The inner disk regions are not actually empty in most known cases, because the star is accreting mass from the disk.

The observational signatures of transitional disks are consistent with the SEDs of binary objects surrounded by a circumbinary, optically thick disk with the inner edge illuminated by the stars, and several cases have been identified so far [172, 83, 67, 104]. However, in other cases no stellar companion has been found, or the companion cannot explain the location of the truncation radius. Alternative theories have been truncation by a planetary-mass object [153, 150] (which would have been missed in surveys for binary

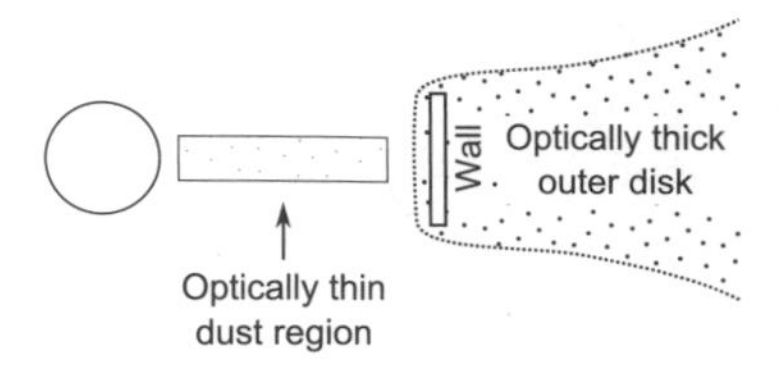

Fig. 2.21. Schematic representation of a transitional disk. The outer optically thick disk is truncated at a given radius, inside which some optically thin dust may or may not be detectable. The wall of the truncated outer disk is illuminated directly by the star.

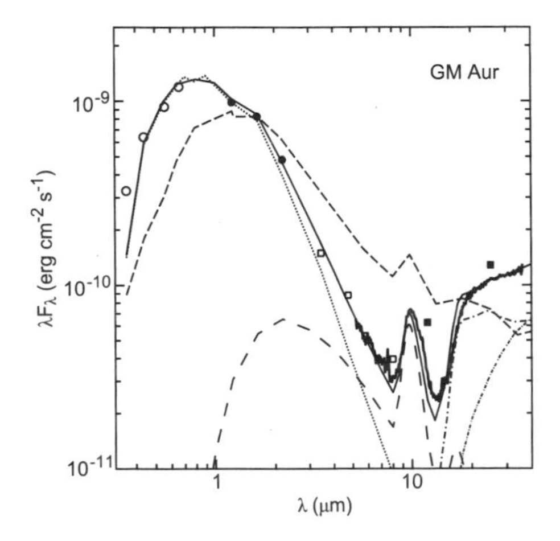

Fig. 2.22. SED of the transitional disk GM Aur. The median SED of the CTTSs in Taurus is shown by short dashed lines. The near-IR flux deficit and the fluxes beyond $\sim$10 μm comparable to or higher than the median SED are apparent. The disk model includes the contribution from the outer disk (short dash-dot-dotted lines), the wall at the edge of the truncated disk, at $\sim$20 AU (long dash-dotted lines), and an optically thin region inside the hole (wide dashed lines), in which small dust particles are responsible for most of the observed silicate feature. Adapted from [29].

companions), photoevaporation [36], or erosion of the disk wall by induced by magnetorotational instability mass loss [34].

High-energy radiation fields, and in particular external ultraviolet radiation, from the central objects are expected to heat the uppermost layers of the disk, raising the thermal velocities above the escape velocity [100]; this effect is important outside a given radius, the gravitational radius, which scales with the mass of the star, and results in mass loss from the disk surface. Ref. [36] proposed that as the mass-accretion rate decreases with age, at some time it becomes equal to the rate of mass loss. From this point on, matter is lost through the wind and does not reach the inner disk, which quickly dumps its mass onto the star, clearing up the inner region. Further photoevaporation of the exposed edge of the disk makes the inner hole grow with time ([6]; chapter 8 in this volume). This model predicts low mass-accretion rates and disk masses in transitional disks, which are not observed [6]. Thus photoevaporation does not seem to be the explanation for the clearing in transitional disks, although it may be responsible for getting rid of the remaining disk gas in the latest phases of evolution.

Planets opening gaps in disks emerge as the most likely cause of transitional disks without stellar binary companions. A further step in this direction has been taken with the identification of pretransitional disks. In these disks, optically thick material remains inside the cleared region, as shown by near-IR excesses comparable to optically thick disks [40, 18, 17]. A compelling proof comes from spectroscopy in the 2–4 μm band; the spectrum of the excess is that of a blackbody at the dust-destruction temperature [39], similar to that of full disks [134]. These disks may be in the first stages of gap opening by forming planets.

We have given positive steps toward understanding how planets form, and how disks actually evolve and dissipate. Still, much more work remains to be done to achieve these goals, which is a good thing.

References

1. Acke, B., & van den Ancker, M. E. 2004, AA, 426, 151.
2. Aikawa, Y., & Nomura, H. 2006, ApJ, 642, 1152.
3. Aikawa, Y., van Zadelhoff, G. J., van Dishoeck, E. F., & Herbst, E. 2002, AA, 386, 622.
4. Alencar, S. H. P., Basri, G., Hartmann, L., & Calvet, N. 2005, AA, 440, 595.
5. Allen, L. E., et al. 2004, ApJS, 154, 363.
6. Alexander, R. D, & Armitage, P. J. 2007, MNRAS, 375, 500.
7. Andrews, S. M., & Williams, J. P. 2005, ApJ, 631, 1134.
8. Andrews, S. M., & Williams, J. P. 2007, ApJ, 659, 705.
9. Andrews, S. M., & Williams, J. P. 2008, Astrophysics and Space Science, 313, 119.
10. Armitage, P. J. 2007, ArXiv Astrophysics e-prints, arXiv:astro-ph/0701485.
11. Artymowicz, P., & Lubow, S. H. 1994, ApJ, 421, 651.
12. Batalha, C. C., & Basri, G. 1993, ApJ, 412, 363.
13. Baraffe, I., Chabrier, G., Allard, F., & Hauschildt, P. H. 1998, AA, 337, 403.
14. Baraffe, I., et al. 2002, A&A, 382, 563.
15. Briceño, C., Luhman, K. L., Hartmann, L., Stauffer, J. R., & Kirkpatrick, J. D. 2002, ApJ, 580, 317.
16. Bertout, C., Basri, G., & Bouvier, J. 1988, ApJ, 330, 350.
17. Brown, J. M., Blake, G. A., Qi, C., Dullemond, C. P., & Wilner, D. J. 2008, ApJL, 675, L109.
18. Brown, J., et al. 2007, ApJL, 664, L107.

19. Bergin, E., et al. 2004, ApJ, 614, L133.
20. Brittain, S. D., Simon, T., Najita, J. R., & Rettig, T. W. 2007, ApJ, 659, 685.
21. Burrows, C. J., et al. 1996, ApJ, 473, 437.
22. Calvet, N., Hartmann, L., & Strom, S. E. 2000, Protostars and Planets IV, 377.
23. Calvet, N. 2004, Accretion Discs, Jets and High Energy Phenomena in Astrophysics, ed.: Vassily Beskin, Gilles Henri, Franois Menard, et al., Les Houches Summer School, 78, 521.
24. Calvet, N., et al. 1991, ApJ, 380, 617.
25. Calvet, N., Magris, G. C., Patiño, A., & D'Alessio, P. 1992, Rev. Mex. Astron. Astrofis. 24, 27.
26. Calvet, N., & Gullbring, E. 1998, ApJ, 509, 802.
27. Calvet, N., D'Alessio, P., Hartmann, L., Wilner, D., Walsh, A., & Sitko, M. 2002, ApJ, 568, 1008.
28. Calvet, N., et al. 2004, AJ, 128, 1294.
29. Calvet, N., et al. 2005, ApJL, 630, L185, C05.
30. Calvet, N., Briceño, C., Hernández, J., Hoyer, S., Hartmann, L., Sicilia-Aguilar, A., Megeath, S. T., & D'Alessio, P. 2005, AJ, 129, 935.
31. Carr, J. S., & Najita, J. R. 2008, Science, 319, 1504.
32. Chiang, E. I., & Goldreich, P. 1999, ApJ, 519, 279.
33. Chiang, E. I., & Goldreich, P. 1997, ApJ, 490, 368.
34. Chiang, E. I., & Murray-Clay, R. A. 2007, Nature Physics, 3,604.
35. Cieza, L., et al. 2007, ApJ, 667, 308.
36. Clarke, C. J., Gendrin, A., & Sotomayor, M. 2001, MNRAS, 328, 485.
37. Clarke, C. J., & Pringle, J. E. 2006, MNRAS, 370, L10.
38. Cushing, M. C., Vacca, W. D., & Rayner, J. T. 2004, PASP, 116, 362.
39. Espaillat, C., Calvet, N., Luhman, K. L., Muzerolle, J., & D'Alessio, P. 2008, ArXiv e-prints, 807, arXiv:0807.2291.
40. Espaillat, C., Calvet, N., D'Alessio, P., Hernández, J., Qi, C., Hartmann, L., Furlan, E., & Watson, D. M. 2007, ApJL, 670, L135.
41. Espaillat, C., et al. 2007, ApJL, 664, L111.
42. D'Alessio, P., Calvet, N., & Hartmann, L. 1999, ApJ, 527, 893.
43. D'Alessio, P. 1999, PhD Thesis, Universidad Nacional Autónomade México.
44. D'Alessio, P., Cantó, J., Hartmann, L., Calvet, N., & Lizano, S. 1999, ApJ, 511, 896.
45. D'Alessio, P., Calvet, N., Hartmann, L., Franco-Hernández, R., & Servín, H. 2006, ApJ, 638, 314.
46. D'Alessio, P., et al. 2005, ApJ, 621, 461.
47. D'Alessio, P., Calvet, N., Hartmann, L., Muzerolle, J., & Sitko, M. 2004, Star Formation at High Angular Resolution, 221, 403.
48. D'Alessio, P., Calvet, N., & Hartmann, L. 2001, ApJ, 553, 321.
49. D'Alessio, P., Cantó, J., Calvet, N., & Lizano, S. 1998, ApJ, 500, 411.
50. D'Alessio, P., Calvet, N., & Hartmann, L. 1997, ApJ, 474, 397.
51. Draine, B. T., & Lee, H. M. 1984, ApJ, 285, 89.
52. Dullemond, C. P., Hollenbach, D., Kamp, I., & D'Alessio, P. 2007, Protostars and Planets V, 555.
53. Dullemond, C. P., Natta, A., & Testi, L. 2006, ApJL, 645, L69.
54. Dullemond, C. P., & Dominik, C. 2005, AA, 434, 971.
55. Dullemond, C. P., & Dominik, C. 2004, AA, 421, 1075.

56. Dullemond, C. P., Dominik, C., & Natta, A. 2001, ApJ, 560, 957.
57. Dullemond, C. P. 2000, AA, 361, L17.
58. Dutrey, A., Guilloteau, S., Duvert, G., Prato, L., Simon, M., Schuster, K., & Menard, F. 1996, AA, 309, 493.
59. Dutrey, A., Guilloteau, S., Prato, L., Simon, M., Duvert, G., Schuster, K., & Menard, F. 1998, AA, 338, L63.
60. Feigelson, E., Townsley, L., Güdel, M., & Stassun, K. 2007, Protostars and Planets V, 313.
61. Guilloteau, S., Dutrey, A. 1998, AA, 339, 467.
62. Günther, H. M., Schmitt, J. H. M. M., Robrade, J., & Liefke, C. 2007, AA, 466, 1111.
63. Edwards, S., Hartigan, P., Ghandour, L., & Andrulis, C. 1994, AJ, 108, 1056.
64. Edwards, S., Fischer, W., Hillenbrand, L., & Kwan, J. 2006, ApJ, 646, 319.
65. Frank, J., King, A., & Raine, D. Accretion Power in Astrophysics 2002, Cambridge, UK: Cambridge University Press.
66. Furlan, E., et al. 2006, ApJS, 165, 568.
67. Furlan, E., et al. 2007, ApJ, 664, 1176.
68. Furlan, E., et al. 2009, ApJ, 703, 1964.
69. Garcia Lopez, R., Natta, A., Testi, L., & Habart, E. 2006, AA, 459, 837.
70. Glassgold, A. E., Najita, J., & Igea, J. 1997, ApJ, 480, 344.
71. Glassgold, A. E., Najita, J., & Igea, J. 2004, ApJ, 615, 972.
72. Glassgold, A. E., Najita, J., & Igea, J. 2007, ApJ, 656, 515.
73. Gorti, U., & Hollenbach, D. 2004, ApJ, 613, 424.
74. Grady, C. A., et al. 2004, ApJ, 608, 809.
75. Gullbring, E., Hartmann, L., Briceño, C., & Calvet, N. 1998, ApJ, 492, 323.
76. Gullbring, E., Calvet, N., Muzerolle, J., & Hartmann, L. 2000, ApJ, 544, 927.
77. Haisch, K. E., Lada, E. A., & Lada, C. J. 2001, ApJL, 553, L153.
78. Harker, D. E., Woodward, C. E., Wooden, D. H., & Temi, P. 2005, ApJ, 622, 430.
79. Hartigan, P., Kenyon, S. J., Hartmann, L., Strom, S. E., Edwards, S., Welty, A. D., & Stauffer, J. 1991, ApJ, 382, 617.
80. Hartigan, P., Edwards, S., & Ghandour, L. 1995, ApJ, 452, 736.
81. Hartmann, L., Avrett, E., & Edwards, S. 1982, ApJ, 261, 279.
82. Hartmann, L., Megeath, S. T., Allen, L., Luhman, K., Calvet, N., D'Alessio, P., Franco-Hernández, R., & Fazio, G. 2005, ApJ, 629, 881.
83. Hartmann, L., et al. 2005b, ApJL, 628, L147.
84. Hartmann, L., D'Alessio, P., Calvet, N., & Muzerolle, J. 2006, ApJ, 648, 484.
85. Hartmann, L. 2009, Accretion Processes in star formation, Cambridge Astrophysics Series, 32. Cambridge, UK: Cambridge University Press.
86. Hartmann, L., Calvet, N., Gullbring, E., & D'Alessio, P. 1998, ApJ, 495, 385.
87. Hartmann, L., & Kenyon, S. J. 1985, ApJ, 299, 462.
88. Hartmann, L., & Kenyon, S. J. 1996, Annual Review of Astronomy and Astrophysics, 34, 207.
89. Hartmann, L., Calvet, N., Gullbring, E., & D'Alessio, P. 1998, ApJ, 495, 385.
90. Herbig, G. H. 1962, Advances in Astronomy and Astrophysics, 1, 47.
91. Herbig, G. H. 1960, ApJS, 4, 337.
92. Herbig, G. H., & Bell, K. R. 1988, Lick Observatory Bulletin, Santa Cruz: Lick Observatory.

93. Hernández, J., Calvet, N., Hartmann, L., Muzerolle, J.,Gutermuth, R. & Stauffer, J. 2009, APJ, 707, 705.
94. Hernández, J., et al. 2007, ApJ, 671, 1784.
95. Hernández, J., Calvet, N., Briceño, C., Hartmann, L., & Berlind, P. 2004, AJ, 127, 1682.
96. Hernandez, J. et al. 2006, ApJ, 652, 472.
97. Hernández, J., Calvet, N., Hartmann, L., Briceño, C., Sicilia-Aguilar, A., & Berlind, P. 2005, AJ, 129, 856.
98. Herczeg, G. J., & Hillenbrand, L. A. 2008, ApJ, 681, 594.
99. Hillenbrand, L. A., Strom, S. E., Vrba, F. J., & Keene, J. 1992, ApJ, 397, 613.
100. Hollenbach, D. J., Yorke, H. W., & Johnstone, D. 2000, Protostars and Planets IV, 401.
101. Hughes, A. M., et al. 2007, ApJ, 664, 536.
102. Hughes, A. M., Wilner, D. J., Qi, C., & Hogerheijde, M. R. 2008, ApJ, 678, 1119.
103. Ingleby, L., & Calvet, N. 2010 (in preparation).
104. Ireland, M. J., & Kraus, A. L. 2008, ApJ, 678, 59.
105. Isella, A., Testi, L., Natta, A., Neri, R., Wilner, D., & Qi, C. 2007, AA, 469, 213.
106. Isella, A., & Natta, A. 2005, AA, 438, 899.
107. Johns-Krull, C. M., Valenti, J. A., & Koresko, C. 1999, ApJ, 516, 900.
108. Johns-Krull, C. M., Valenti, J. A., Hatzes, A. P., & Kanaan, A. 1999, ApJ, 510, L41.
109. Johns-Krull, C. M. 2007, ApJ, 664, 975.
110. Kamp, I., & Dullemond, C. P. 2004, ApJ, 615, 991.
111. Keller, Ch., & Gail, H. P., 2004, A&A, 415, 1177.
112. Keller, L. D., et al., 2008, ApJ, 684, 411.
113. Kenyon, S. J., Dobrzycka, D., & Hartmann, L. 1994, AJ, 108, 1872.
114. Kenyon, S. J., & Hartmann, L. 1995, ApJS, 101, 117.
115. Kenyon, S. J., & Hartmann, L. 1987, ApJ, 323, 714.
116. Wilking, B. A., & Lada, C. J. 1983, ApJ, 274, 698.
117. Lada, C. J. 1987, Star Forming. Regions, 115, 1
118. Leinert, C., Zinnecker, H., Weitzel, N., Christou, J., Ridgway, S. T., Jameson, R., Haas, M., & Lenzen, R. 1993, AA, 278, 129.
119. Luhman, K. L., et al. 2007, ApJ, 666, 1219.
120. Luhman, K. L., Adame, L., D'Alessio, P., Calvet, N., Hartmann, L., Megeath, S. T., & Fazio, G. G. 2005, ApJL, 635, L93.
121. Malbet, F., & Bertout, C. 1991, ApJ, 383, 814.
122. Meyer, M. R., Calvet, N., & Hillenbrand, L. A. 1997, AJ, 114, 288.
123. Meeus, G., Waters, L. B. F. M., Bouwman, J., van den Ancker, M. E., Waelkens, C., & Malfait, K. 2001, AA, 365, 476.
124. Millan-Gabet, R., Schloerb, F. P., & Traub, W. A. 2001, ApJ, 546, 358.
125. Millan-Gabet, R., Malbet, F., Akeson, R., Leinert, C., Monnier, J., & Waters, R. 2007, Protostars and Planets V, 539.
126. Miyake, K., & Nakagawa, Y. 1993, Icarus, 106, 20.
127. Monnier, J. D., & Millan-Gabet, R. 2002, ApJ, 579, 694.
128. Morrow, A. L., et al. 2008, ApJL, 676, L143.
129. Muzerolle, J., Hartmann, L., & Calvet, N. 1998, AJ, 116, 2965.
130. Muzerolle, J., Hartmann, L., & Calvet, N. 1998, AJ, 116, 455.
131. Muzerolle, J., Calvet, N., & Hartmann, L. 1998, ApJ, 492, 743.
132. Muzerolle, J., et al., 2000, ApJ, 535, L47.
133. Muzerolle, J., Calvet, N., & Hartmann, L. 2001, ApJ, 550, 944.

134. Muzerolle, J., Calvet, N., Hartmann, L., & D'Alessio, P. 2003, ApJL, 597, L149.
135. Muzerolle, J., D'Alessio, P., Calvet, N., & Hartmann, L. 2004, ApJ, 617, 406.
136. Muzerolle, J., Luhman, K. L., Briceño, C., Hartmann, L., & Calvet, N. 2005, ApJ, 625, 906.
137. Najita, J., Carr, J. S., & Mathieu, R. D. 2003, ApJ, 589, 931.
138. Najita, J. R., Carr, J. S., Glassgold, A. E., & Valenti, J. A. 2007, Protostars and Planets V, eds. B. Reipurth, D. Jewitt and K. Keil. Tucson: University of Arizona Press, p. 507.
139. Natta, A., Prusti, T., Neri, R., Wooden, D., Grinin, V. P., & Mannings, V. 2001, AA, 371, 186.
140. Natta, A., Meyer, M. R., & Beckwith, S. V. W. 2000, ApJ, 534, 838.
141. Natta, A., Giovanardi, C., Palla, F., & Evans, N. J., II. 1988, ApJ, 327, 817.
142. Natta, A., Testi, L., Muzerolle, J., Randich, S., Comerón, F., & Persi, P. 2004, AA, 424, 603.
143. Neuhauser, R., et al. 1995, A&A, 297, 391.
144. Nomura, H., Aikawa, Y., Tsujimoto, M., Nakagawa, Y., & Millar, T. J. 2007, ApJ, 661, 334.
145. O'Dell, C. R. 1998, AJ, 115, 263.
146. Stapelfeldt, K. R., Krist, J. E., Menard, F., Bouvier, J., Padgett, D. L., & Burrows, C. J. 1998, ApJ, 502, L65.
147. Piètu, V., Dutrey, A., Guilloteau, S., Chapillon, E., & Pety, J. 2006, A&A, 460, L43.
148. Pollack, J. B., Hollenbach, D., Beckwith, S., Simonelli, D. P., Roush, T., & Fong, W. 1994, ApJ, 421, 615.
149. Qi, C., et al. 2004, ApJL, 616, L11.
150. Quillen, A.C., et al. 2004, ApJ, 612, L137.
151. Ratzka, T., Leinert, C., Henning, T., Bouwman, J., Dullemond, C. P., & Jaffe, W. 2007, AA, 471, 173.
152. Rayner, J. T., Toomey, D. W., Onaka, P. M., Denault, A. J., Stahlberger, W. E., Watanabe, D. Y., & Wang, S. I. 1998, SPIE 3354, 468.
153. Rice et al. 2003, MNRAS, 342, 79.
154. Rodriguez, L. F., Canzó, J., Torrelles, J. M., Gomez, J. F., Anglada, G., & Ho, P. T. P. 1994, ApJL, 427, L103.
155. Sargent, B., et al. 2006, ApJ, 645, 395.
156. Setiawan, J., Henning, T., Launhardt, R., Mueller, A., Weise, P., & Kuerster, M. 2008, Nature, 451, 38.
157. Shakura, N. I., & Sunyaev, R. A. 1973, AA, 24, 337.
158. Siebenmorgen, R., Prusti, T., Natta, A., & Muller, T. G. 2000, AA, 361, 258.
159. Simon, M., Dutrey, A., & Guilloteau, S. 2001, ApJ, 545, 1034.
160. Simpson, J. P., Witteborn, F. C., Price, S. D., & Cohen, M. 1998, ApJ, 508, 268.
161. Siess, L., Dufour, E., & Forestini, M. 2000, AA, 358, 593.
162. Sloan, G. C., et al. 2005, ApJ, 632, 956.
163. Strom, K. M., Strom, S. E., Edwards, S., Cabrit, S., & Skrutskie, M. F. 1989, AJ, 97, 1451.
164. Su, K. Y. L., et al. 2006, ApJ, 653, 675.
165. Takami, M., Dailey, J., & Chrysostomou, A. 2003, A&A, 397, 675.
166. Terebey, S., Shu, F. H., & Cassen, P. 1984, ApJ, 286, 529.
167. Tannirkulam, A., et al. 2008, ApJL, 677, L51.
168. Tannirkulam, A., Harries, T. J., & Monnier, J. D. 2007, ApJ, 661, 374.

169. Telleschi, A., Güdel, M., Briggs, K. R., Skinner, S. L., Audard, M., & Franciosini, E. 2007, AA, 468, 541.
170. Testi, L., & Leurini, S. 2008, New Astronomy Review, 52, 105.
171. Tuthill, P. G., Monnier, J. D., & Danchi, W. C. 2001, Nature, 409, 1012.
172. Uchida, K. I., et al. 2004, ApJS, 154, 439.
173. Varnière, P., Blackman, E. G., Frank, A., & Quillen, A. C. 2006, ApJ, 640, 1110.
174. Watson, D. M. 2009, ApJs, 180, 84.
175. Weaver, W. B., & Jones, G. 1992, ApJS, 78, 239.
176. Webb, R. A. 1999, ApJ, 512, 63.
177. Weidenschilling, S. J. 1997, Icarus, 127, 439.
178. Werner, M. W., et al. 2004, ApJS, 154, 1.
179. White, R., & Basri, G. 2003, ApJ, 582, 1109.
180. White, R. J., & Ghez, A. M. 2001, ApJ, 556, 265.
181. White, R. J., & Hillenbrand, L. A. 2004, ApJ, 616, 998.
182. Whittet, D. C. B., et al. 1997, A&A, 327, 1194.
183. Wilner, D. J., D'Alessio, P., Calvet, N., Claussen, M. J., & Hartmann, L. 2005, ApJL, 626, L109.
184. Wilner, D. J., Ho, P. T. P., Kastner, J. H., & Rodríguez, L. F. 2000, ApJ, 534, L101.
185. Wilner, D. J., Ho, P. T. P., & Rodriguez, L. F. 1996, ApJL, 470, L117.
186. van den Ancker, M. E., Bouwman, J., Wesselius, P. R., Waters, L. B. F. M., Dougherty, S. M., & van Dishoeck, E. F. 2000, AA, 357, 325.
187. Watson, A. M., Stapelfeldt, K. R., Wood, K., & Ménard, F. 2007, Protostars and Planets V, 523.
188. Willacy, K., & Langer, W. D. 2000, ApJ, 544, 903.
189. Woolum, D. S., & Cassen, P. 1999, Meteoritics and Planetary Science, 34, 897.
190. van Zadelhoff, G.-J., van Dishoeck, E. F., Thi, W.-F., & Blake, G. A. 2001, AA, 377, 566.
191. Zhu, Z., Hartmann, L., Calvet, N., Hernández, J., Muzerolle, J., & Tannirkulam, A.-K. 2007, ApJ, 669, 483.

3

EDWIN A. BERGIN

THE CHEMICAL EVOLUTION OF PROTOPLANETARY DISKS

1. Introduction

The origins of planets, and perhaps of life itself, are intrinsically linked to the chemistry of planet formation. In astronomy these systems are labeled protoplanetary disks—disks on the incipient edge of planet formation. For our Sun, the rotating ball of gas and dust that collapsed to a disk has been called the Solar Nebula. In this chapter I will attempt to explore the chemistry of planet-forming disks from the perspective of knowledge gained from decades of solar system study, combined with our rapidly growing knowledge of extrasolar protoplanetary disks. This chapter is not written in the form of a review. Rather, I survey our basic knowledge of chemical/physical processes and the various techniques that are applied to study solar/extrasolar nebular chemistry. Therefore, my reference list is limited to works that provide a direct discussion of the topic at hand.

A few aspects of general terminology and background are useful. I will use $n_{\rm H}$ to refer to the space density (particles per cubic centimeter) of atomic hydrogen, $n_{\rm H_2}$ for the molecular hydrogen density, and n for the total density ($n_{\rm H} + n_{\rm H2}$). Similar terminology will be used for the column density (particles per square centimeter), which is denoted by N. In addition, T_{gas} will refer to the gas temperature and T_{dust} to the dust/grain temperature, which are not always coupled and have separate effects on the chemistry. At high densities, where they are coupled, I will simply refer to the temperature, T.

Stars are born in molecular clouds with typical densities of a few thousand H_2 molecules per cubic centimeter, gas temperatures of $\sim$10–20 K, and 10^3–10^5 $M_\odot$. These clouds exist over a scale of tens of parsecs (1 pc = 3×10^{18} cm) but exhibit definite substructure, with stars being born in denser ($n > 10^5$ cm^{-3}) cores with typical sizes of 0.1 pc. The chemistry of these regions is dominated by reactions between ions and neutrals, but also by the freeze-out of molecules onto the surfaces of cold dust grains. A summary of the properties of the early stages can be found in the reviews of Bergin & Tafalla (2007) and di Francesco et al. (2007). Molecular cloud cores are rotating, and upon gravitational collapse, the infalling envelope flattens to a disk. This stage is labeled Class 0. As the forming star begins to eject material, the envelope is disrupted along the poles, and the star/disk system begins the process of destroying its natal envelope. In this Class I stage the star accretes material from the disk, and the disk accretes from the envelope (see Adams et al., 1987; André et al., 1993; for a discussion of astronomical classification). As the envelope dissipates, the system enters the Class II stage, wherein material accretes onto the star from the disk. The disk surface is directly exposed to energetic ultraviolet (UV) radiation and X-rays from the star, but also from the interstellar radiation field. Observational systems in the Class II stage are often called T Tauri stars. These disks have strong radial and vertical gradients in physical properties, with most of the mass residing in the middle of the disk, labeled the midplane. It is the midplane that is the site of planet formation. The density of the midplane significantly exceeds that of the dense natal core, by many orders of magnitude. The midplane is, in general, colder than the disk upper layers, which can also be called the disk atmosphere. Finally, the outer regions of the disk ($r > 10$ AU) have reduced pressure and temperature but contain most of the mass. Bergin et al. (2007) and Ciesla & Charnley (2006) provide recent summaries of disk chemical evolution. This chapter will provide greater detail on the methodology and techniques used to explore the chemistry of protoplanetary disks, and the reader is referred to the previous reviews for additional information regarding both observations and theory.

I will start this study by discussing some key observational results, beginning with bodies in our solar system and extending to circumstellar disk systems at a typical distance of 60–140 pc. I expand the exploration to the basic facets of theoretical studies of disk chemistry: thermodynamic equilibrium and chemical kinetics. This is followed by an outline of our knowledge of key physical processes. In the final sections, I synthesize these results to summarize our theoretical understanding of the chemistry of protoplanetary disks.

2. Observational Constraints

There exists a wealth of data on the chemical evolution of our own solar system regarding the composition of planetary bodies, moons, asteroids, comets, and meteorites. In addition, there is a rapidly growing observational sample of molecular and atomic transitions in extrasolar protoplanetary disks. The current astronomical data set is limited to only a few objects that have been subjected to deep searches. However, the upcoming Atacama Large Millimeter Array (ALMA) will dramatically expand our current capabilities and offers great promise for gains in our understanding of planet formation. This will directly complement knowledge gained from the remnants and products of planet formation within the solar system. For more detailed views of disk chemistry, the reader is referred to the recent compilations found in *Meteorites and the Early Solar System* II (Lauretta & McSween, 2006), the *Treatise on Geochemistry* (vol. 1, A. M. Davis et al., 2005), and *Protostars and Planets* V (Reipurth et al., 2007).

2.1. Planets, Comets, and Meteorites: Tracing the Midplane

This chapter cannot do justice to the gains from the decades of study of bodies within our solar system. Instead, I will highlight a few basic facets of our understanding and a few examples where the chemical composition is a fossil remnant that tracks conditions that existed billions of years ago.

PLANETS. The standard model for the formation of the Earth and other rocky planets is that the temperature of the nebula was initially hot enough that all primordial grains sublimated. As the nebula cooled, various species condensed out of the hot gas, depending on their condensation temperature (defined in § 3.1). The nebula also had strong radial temperature gradients such that different species could first condense at larger radii. This model and various issues are summarized in Palme et al. (1988) and A. M. Davis (2006). Fig. 3.1 shows the condensation temperatures of various minerals and ices, along with the radii where these species potentially condensed in the nebula, as estimated by Lewis (1990). The fraction of rock/ices in planets is also provided on this plot and clearly follows this trend. It is a basic astronomical fact that the rocky planets reside in the inner nebula, with gas giants in the outer nebula; however, this is supplemented by Uranus and Neptune having incorporated a greater percentage of ices. Thus basic molecular properties (i.e., temperature of condensation) work in tandem with the nebular thermal structure to determine the composition of bodies in the nebula, supplemented by dynamic evolution (e.g., Morbidelli et al., 2000). Other facets of chemical composition of bodies in the solar system in relation to nebular thermal structure are discussed by Lewis (1974), who provides some of the basis for the condensation temperatures and radii placements in Fig. 3.1.

METEORITES. Meteorites are the most directly accessible primitive material in the solar system. Meteorites come in a number of classes, based, in part,

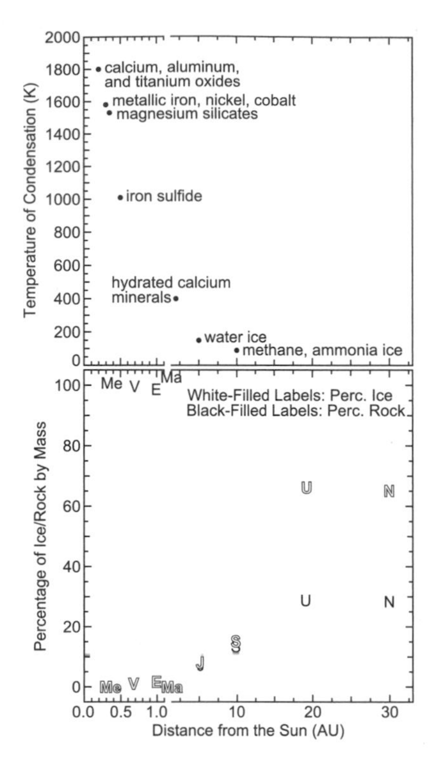

Fig. 3.1. *Top:* Plot of temperature of condensation for materials that represent some of the major contributors to planetary composition as a function of distance from the Sun. This part of the figure is adapted from NASA's Genesis mission educational materials. Condensation temperature and estimated position from Lewis (1990). *Bottom:* Plot of percentage of planetary mass constituted by rocks (black-filled labels) and ices (white-filled labels). Labels represent each of the 8 major planets in the solar system. Jupiter, Saturn, Uranus, and Neptune estimates are from Guillot (1999a, 1999b). Values for the giant planets have larger uncertainties than those for the terrestrial planets.

on whether the material is undifferentiated and therefore more primitive or is part of the mantle or core of some larger planetesimal (see Krot et al., 2005). The most primitive meteorites are the carbonaceous chondrites (labeled C). Among the various types, the CI carbonaceous chondrite contains a composition that is quite similar to that of the Sun (Palme & Jones, 2003). In fact, in many instances abundances from CI chondrites provide a better value for the solar composition than spectroscopy of the solar photosphere (Lodders, 2003; Palme & Jones, 2003). This is illustrated in Fig. 2.1, where I plot abundances measured in CI chondrites against those estimated in the Sun.

The exceptions to the general agreement in Fig. 2.1 are telling. Li is underabundant in the Sun because of convective mixing into layers where Li is

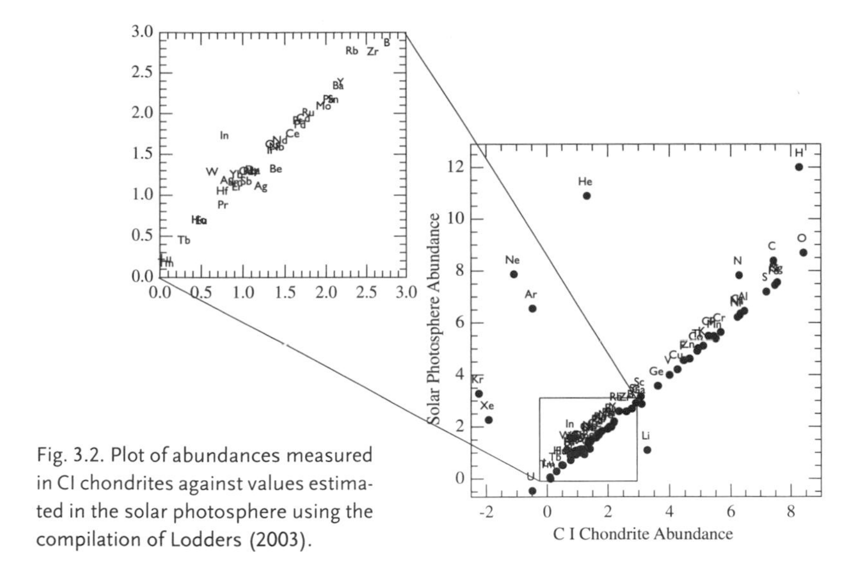

Fig. 3.2. Plot of abundances measured in CI chondrites against values estimated in the solar photosphere using the compilation of Lodders (2003).

destroyed. Li is inherited from the interstellar medium (ISM), and stars begin their evolution with abundant Lithium. Primordial Li is gradually destroyed (Bodenheimer, 1965; Skumanich, 1972); thus the presence of Li in a stellar photosphere is generally taken as a sign of youth. The most volatile elements have low abundances in chondrites; these include the noble gases, along with H, N, C, and to some extent O. Unreactive noble gases reside for the most part in the nebular gas along with H (which was in molecular form). Noble gases can be trapped inside the lattices of rocky and icy planetesimals, and their abundances provide key clues to the evolution of terrestrial planet atmospheres (Pepin, 1991; Owen & Bar-Nun, 1995; Dauphas, 2003). In the cases of C and N, thermodynamic equilibrium would place these molecules in the form of CO/CH_4 or N_2/NH_3 (depending on the pressure/temperature). Thus the temperature in the nebula where these rocks formed must have been above the sublimation temperature for these molecules. In the case of O I, the oxygen not consumed by refractory material and CO likely resides in the form of gas-phase H_2O.

An additional clue to the chemistry of rock formation in the inner nebula lies in the relative abundances of elements within the various classes of meteorites. Fig. 3.3 shows a sample of these results, plotting the abundances of CM, CO, and CV carbonaceous chondrites normalized to CI as a function of condensation temperature. For a more expanded plot, the reader is referred to A. M. Davis (2006). This plot demonstrates that elements with higher condensation temperatures are more likely to be incorporated into planetesimals.

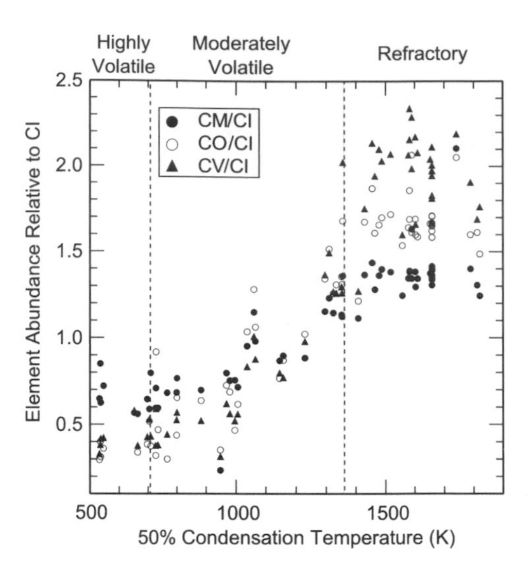

Fig. 3.3. Plot of elemental abundances measured in the CM/CO/CV chondrites normalized to CI, shown as a function of the 50% condensation temperature for that element. Figure adapted from Alexander et al. (2001). Elemental abundances within chondrites are taken from Wasson & Kallemeyn (1988), with condensation temperatures from Lodders (2003).

Hence rock formation in the Solar Nebula was in some sense a volatility-controlled process. This can involve aspects of both condensation in a cooling nebula and, potentially, evaporation during heating events (Palme et al., 1988; A. M. Davis, 2006).

COMETS. Because comets harbor volatile ices, they are likely the most pristine remnants of the formation of the solar system, tracing conditions that existed in the outer tens of AUs in the Solar Nebula. Long-period comets, with periods > 200 yrs, originate in the Oort cloud, while short-period comets (< 200 yrs) are believed to originate in the Kuiper belt. Oort cloud comets are believed to have formed predominantly in the region between Uranus and Neptune and to have scattered to distant orbits via interactions with the giant planets (Fernandez, 1997; Dones et al., 2004). The chemical composition of volatiles in cometary comae can be studied via remote sensing techniques with the basic physics summarized in § 2.2. In general, there are some gross similarities in the abundances of volatiles in cometary comae to those of interstellar ices, which are suggestive of a direct link (e.g., Bockelée-Morvan et al., 2000; Ehrenfreund et al., 2004). In particular, water, CO, and CO_2 are the most abundant ices in comets and in the ISM, with smaller contributions from CH_4, NH_3, and others. However, there is also a wide degree of chemical inhomogeneity in the cometary inventory. In particular, a study of 85 short-period comets by A'Hearn et al. (1995) found that a fraction appeared to be depleted of the carbon-chain precursors that produce C_2. Such diversity is not limited to short-period comets, because Oort cloud comets also exhibit variations in CO, C_2H_6, and CH_4 relative to H_2O, along with other compositional differences (see Bockelée-Morvan et al., 2004; Disanti & Mumma, 2008; and references therein). Methanol also stands out as another species where variations are found within the Oort cloud population of comets and also short-period comets (Disanti & Mumma, 2008). These differences likely point to systematic changes in the chemistry throughout the cometary formation zone, perhaps due to a diversity of formation radii for cometary nuclei.

One potential clue to the formation of cometary volatiles lies in the ortho/para ratio of water ice. The spin pairing of the hydrogen atoms for H_2O leads to two independent forms, para-H_2O (spins antiparallel, with total nuclear spin $I_p = 0$), and ortho-H_2O (spins parallel, with total nuclear spin $I_p = 1$). The lowest energy level of para-H_2O is $\sim$34 K below that of ortho-H_2O, and if water forms at temperatures below this difference, the water will preferentially be in the form of para-H_2O. The ortho/para ratio depends on the distribution of population within the energy levels that are characterized by three quantum numbers (J, K_+, K_-) used for asymmetric tops (§ 2.2). In thermodynamic equilibrium the ratio is given by (Mumma et al., 1987)

$$\text{o/p} = \frac{2I_o + 1}{2I_p + 1} \frac{\Sigma(2J+1)e^{-E_o(J,K_+,K_-)/kT}}{\Sigma(2J+1)e^{-E_p(J,K_+,K_-)/kT}} \tag{1}$$

and at high temperature is set by the ratio of the spin statistical weights ($2I + 1$), o/p = 3:1. When water forms in the gas via exothermic reactions, the energy is much greater than the ground-state energy difference, and the ortho/para ratio reflects the 3:1 ratio of statistical weights between these species. When water forms on the surfaces of cold dust grains, it is possible that the excess energy in the reaction is shared with the grain, and the water molecules equilibrate to an ortho/para ratio at the grain temperature (e.g., Limbach et al., 2006). If the grain temperature is below $\sim$ 50 K, then the ortho/para ratio will lie below 3:1. In Comet P/Halley, Mumma et al. (1987) measured an o/p ratio of 2.73 ± 0.17. This is consistent with an equilibrium spin temperature of $\sim$30 K, which is similar to that measured for water and other species (ammonia, methane) in other cometary comae (Kawakita et al., 2006). Bonev et al. (2007) provide a summary of water ortho/para ratios measured in cometary comae that span a range of spin temperatures from $\sim$20 to 40 K. One potential interpretation of this ratio is that cometary ice formed at temperatures near 30 K, possibly in the interstellar medium, although other explanations exist (see Bockelée-Morvan et al., 2004; for a discussion).

ISOTOPIC CHEMISTRY. A clear chemical constraint on planet formation is the deuterium enrichment observed in the solar system. In Fig. 3.4 I provide a compilation of D/H ratios seen in a variety of solar system bodies and molecules in the ISM. This figure essentially summarizes salient issues as determined by cosmochemical and geological studies over the past decades. (1) Gas in the early Solar Nebula (proto-Sun, Jupiter) is believed to have had a ratio comparable to that seen in the main mass reservoir in the natal cloud (i.e., ISM H_2). (2) As we move toward bodies that formed at greater distances, there is a clear trend toward deuterium enrichment in trace gas components (Uranus, Neptune). (3) The material that formed directly from ices and rocks (comets, interplanetary dust particles, meteorites) exhibits large enrichments. (4) The Earth's oceans are enriched in deuterium.

From astronomical studies of the dense ($n > 10^5$ cm^{-3}) cores of molecular clouds—the sites of star and planet formation—we know that these enrichments are readily initiated by a sequence of reactions that are favored for $T_{gas} \lesssim 30$ K (§ 3.2; Geiss & Reeves, 1981; Millar et al., 1989). Indeed, large D/H ratios are observed inside molecular cloud cores, both in the gas phase and in evaporated ices (Fig. 3.4; Ceccarelli et al., 2007). The same cold ($T_{gas} \lesssim 30$ K) nonequilibrium gas chemistry that gives rise to the deuterium fractionation in molecular cloud cores is also observed to be active in the outer reaches of extrasolar protoplanetary disks (Bergin et al., 2007; and references therein).

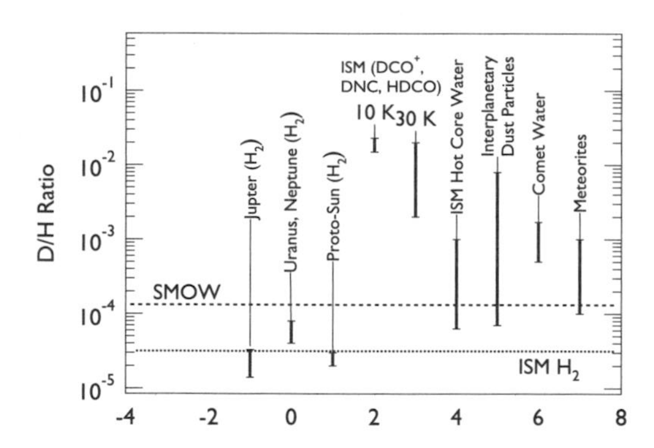

Fig. 3.4. D/H ratios measured in key species in solar system bodies and molecules in the ISM. Plot adapted from Messenger et al. (2003) and using additional ratios taken from the tabulation of Robert (2006) with references therein. The Jupiter value is an in situ measurement from the *Galileo* probe (Mahaffy et al., 1998) and Infrared Space Observatory observations (Lellouch et al., 1996), while the protosolar ratio is estimated from measurements of ^{3}He in the solar wind (Geiss & Gloeckler, 1998). SMOW—standard mean ocean water—is the reference standard for the Earth's ocean. The ISM value for H_2 is based on measurements of atomic D/H in clouds within 1 kpc of the Sun (Wood et al., 2004).

These inferences hint that the Earth received a contribution to its water content from some cold reservoir (e.g., comets, icy asteroids).

Beyond hydrogen, meteorites also exhibit a number of other isotopic anomalies (see A. M. Davis et al., 2005; Lauretta & McSween, 2006; for a complete discussion). Of particular interest is the isotopic enhancement seen in carbon, oxygen, and nitrogen, which is potentially related to kinetic chemical effects early in nebular evolution or perhaps in the cloud that collapsed to form the Sun (Clayton, 1993; Floss et al., 2004; Yurimoto et al., 2007; Rodgers & Charnley, 2008).

2.2. Astrophysical Techniques

Astronomy offers several methods to explore the predominantly molecular chemistry of protoplanetary disks. In general, molecules are fantastic probes of their environment. In the following I will outline a few basic facets of molecular astrophysics and discuss the current status of observations. A listing of molecules and transitions that have been detected in protoplanetary disks is provided in Table 3.1.

MOLECULAR ASTROPHYSICS. For atomic species the electronic states are discrete, or quantized, and their emission spectra have been successfully characterized by using the principles of quantum mechanics (e.g., Cowley et al., 2000; or an astronomical perspective). In comparison to atoms, molecules have more complicated structures where rotational and vibrational motions are coupled to electronic states. Similar to atoms, the various motions/states are quantized. A more complete description of the molecular physics and spectroscopy is provided by Townes & Schawlow (1955) and Herzberg (1950 and subsequent volumes). Evans (1999) provides a more descriptive outline regarding the use of molecules as astrophysical probes.

The Born-Oppenheimer approximation (Born & Oppenheimer, 1927) allows the separation of the nuclear and electronic motions in the wave function. Thus molecular transitions are characterized by three different energies: rotational, vibrational, and electronic. Electronic transitions have energies typically $\sim$4 eV (40,000 K), with lines found at visible and ultraviolet (UV) wavelengths. Vibrational levels probe energies of $\sim$0.1 eV (1,000 K) and are, for the most part, found in the near to mid-infrared ($\sim$2–20μm). Rotational transitions of the ground vibrational and electronic states have energies below 0.01 eV (100 K) and emit at millimeter and submillimeter wavelengths. The wide range of temperatures and densities present in protoplanetary disks has led to detections of electronic, vibrational, and low-/high-energy rotational lines of a variety of molecular species (Table 3.1).

Table 3.1: A Sample of Current Astrophysical Probes

Species	$\lambda(\mu m)$	Transition	E_u (K)	Radius probed	Notes[a]
H_2	0.10–0.15	Lyman-Werner bands	10^5	r < 1 AU	(1)
H_2	2.12	$v = 1-0$ S(0)	6,471	r ~ 10–40 AU	(2)
CO	2.23	$v = 2-0$	6,300	r ~ 0.05–0.3 AU	(4)
H_2O	~2.9	$\nu_3 = 1-0$	5,000–10,000	r ~ 1 AU	(7)
OH	~3	$v = 1-0$ P branch	>5,000	r ~ 1 AU	(7)
CO	4.6	$v = 1-0$	3,000	r ~ <0.1–2 AU	(5)
H_2	8.0–17.0	$v = 0-0$ S(1), S(2), S(4)	1,015–3,474	r ~ 10–40 AU	(3)
H_2O	10–30	$J > 4$	>500	r ~ 1–2 AU	(6)
C_2H_2	~13.7	$\nu_5 = 1-0$ Q branch	1,000	$r \sim 1$ AU	(8)
HCN	~14	$\nu_2 = 1-0$ Q branch	1,000	$r \sim 1$ AU	(8)
CO_2	14.98	$\nu_2 = 1-0$ Q branch	1,000	$r \sim 1$ AU	(8)
Ne II	12.81	${}^2P_{3/2} - {}^2P_{1/2}$	1,100	r ~ 0.1 AU[b]	(9)
CO	460–2,600	$6-5, 3-2, 2-1, 1-0$	5–116	$r > 20$ AU[c]	(10)
HCO^+	1,000–3,300	$3-2, 1-0$	5–25	$r > 20$ AU[c]	(11)
CS	1,000–3,000	$2-1, 3-2, 5-4$	5–30	$r > 20$ AU[c]	(12)
N_2H^+	3,220	$1-0$	5	$r > 20$ AU[c]	(13)

H_2CO	1,400–2,000	$3_{13} - 2_{12}, 2_{12} - 1_{11}, 3_{12} - 2_{11}$	20–32	$r > 20$ AU[c]	(14)
CN	1,000–2,500	$3 - 2, 2 - 1$	5–30	$r > 20$ AU[c]	(15)
HCN	850–3,300	$4 - 3, 2 - 1, 1 - 0$	5–40	$r > 20$ AU[c]	(16)
HNC	3,300	$1 - 0$	5	$r > 20$ AU[c]	(17)
H_2D^+	805	$1_{10} - 1_{11}$	104	$r > 20$ AU[c]	(18)
DCO^+	830–1,400	$5 - 4, 3 - 2$	20–50	$r > 20$ AU[c]	(19)
DCN	1,381	$3 - 2$	20	$r > 20$ AU[c]	(20)

[a] References: (1) Herczeg et al. (2002); (2) Bary et al. (2003, 2008); (3) Bitner et al. (2007); (4) Carr et al. (1993); (5) Najita et al. (2003); Brittain et al. (2007); (6) Carr & Najita (2008); Salyk et al. (2008); (7) Salyk et al. (2008); (8) Lahuis et al. (2006); Carr & Najita (2008); (9) Espaillat et al. (2007) Lahuis et al. (2007); Herczeg et al. (2007); (10) Dutrey et al. (1996); Kastner et al. (1997); Qi (2001); van Zadelhoff et al. (2001); Qi et al. (2006); (11) Dutrey et al. (1997); Kastner et al. (1997); van Zadelhoff et al. (2001); Qi et al. (2008); (12) Dutrey et al. (1997); (13) Dutrey et al. (1997, 2007b); (14) Dutrey et al. (1997); Thi et al. (2004); (15) Dutrey et al. (1997); Kastner et al. (1997); van Zadelhoff et al. (2001); (16) Dutrey et al. (1997); Kastner et al. (1997); van Zadelhoff et al. (2001); Qi et al. (2008); (17) Dutrey et al. (1997); Kastner et al. (1997); (18) Ceccarelli et al. (2004); (19) van Dishoeck et al. (2003); Qi et al. (2008); (20) Qi et al. (2008).

[b] If the [Ne II] emission arises from a photoevaporative wind, then the emission can arise from greater distances (Herczeg et al., 2007).

[c] It is important to note that many of these species will have rotational emission inside 20 AU, particularly in the high-*J* transitions. However, the observations are currently limited by the spatial resolution, which will be overcome to a large extent by ALMA.

For the bulk of the disk mass, which has temperatures below 10s of Kelvin, the disk is best traced by molecular rotational lines. Solutions of the wave equation for a "rigid" rotating linear molecule yield rotational energy levels

$$E_J = \frac{\hbar^2}{2I} J(J+1), \tag{2}$$

where $I = \mu a_0^2$ is the moment of inertia (μ is the reduced mass, and a_0 is the bond length, typically ~ 1 Å) and J is the angular momentum quantum number. The quantity $\hbar^2/2I$ is redefined as the rotation constant, B_0, which is an intrinsic property of a given molecule. For linear molecules the excitation is simplified because there is symmetry about all rotational axes and the selection rules for electric dipole transitions are $\Delta J \pm 1$. For these molecules the frequency of a rotational transition from level J to level $J-1$ is

$$\nu = E_J(J) - E(J-1) = 2B_0 J. \tag{3}$$

Heavy linear molecules with a large moment of inertia (e.g., HCN, CS, CO) have small rotational constants and, in consequence, have more transitions at longer millimeter/centimeter wavelengths. Lighter molecules, such as H_2 or HD, have larger energy spacings and emit primarily in the near to far infrared (2–115μm). Molecules have three degrees of rotational freedom, and the simplest case is the linear rigid rotor. Some nonlinear molecules have a degree of rotational symmetry. For a symmetric-top molecule, two of the principal moments of inertia are equal. In this case the rotational energy spacing is charactered by two quantum numbers (J, K). Because of their energy-level spacing and excitation properties, symmetric tops have been used as probes of the gas temperature. Prominent examples are NH_3 or CH_3C_2H (Ho & Townes, 1983; Bergin et al., 1994). Nonlinear molecules with no symmetry are called asymmetric tops. The energy levels for these species are described by three quantum numbers (J, K_+, K_-). Water is the most notable species in this category, which also includes most complex organic molecules.

The spontaneous emission arises from a rotating dipole, which from an upper to a lower J state of a linear molecule is given by the Einstein A-coefficient:

$$A_{ul} = \frac{S}{2J_u+1} \frac{64\pi^4}{3hc^3} \nu^3 \mu_d^2. \tag{4}$$

Here μ_d is the permanent electric dipole moment of the molecule. A common unit for the dipole moment is the Debye (1 D $= 10^{-18}$ esu cm). S is the line strength of the transition. For linear rotors, $S = J_u$.

From eqs. 3 and 4, the methodology of using molecular emission to probe the physical environment can be demonstrated. In a disk with density n_{H_2}, molecules are excited through collisions with molecular hydrogen at a typical rate $n_{H_2}\gamma_{ul}$, where the collision rate coefficient, $\gamma_{ul} \sim 10^{-11}$ cm^3 s^{-1} (see Flower, 2003; for a thorough discussion of molecular collisions). This leads to a critical density required for significant excitation of a given transition, which is $n_{cr} \sim A_{ul}/\gamma_{ul}$. If we use eqs. 3 and 4, the critical density scales as $n_{cr} \propto B_0^3\mu_d^2 J^3$. Of these, the rotation constant and the dipole moment are molecular properties, while J depends on the observed transition. Heavy molecules with large dipole moments, due to charge disparities, have higher critical densities and are excited in denser gas. Thus HCN ($\mu_d = 2.98$ D) and H_2O ($\mu_d = 1.85$ D) are tracers of dense ($n_{H_2} > 10^5$ cm^{-3}) gas, and CO ($\mu_d =$ 0.11 D) traces lower-density material. Higher J transitions probe successively denser (and warmer layers), presumably at the disk surface, or radially closer to the star.

Molecules with identical nuclei have no permanent electric dipole but in some instances can emit via weaker quadrupole (H_2) or magnetic dipole (O_2) transitions. In the case of H_2, which is the dominant gas constituent, the first rotational level (J = 2-0 for a quadrupole) is 510 K above the ground state. The population in this state is $\propto \exp(-E_u/kT_{gas})$, which hinders H_2 emission from the cold ($T_{gas} \sim$ 10–30 K) midplane. Hence, similar to the ISM, other molecules with dipole moments are used as proxies to trace the molecular gas. These species are generally heavier molecules consisting of H, C, O, and N (typically CO).

Vibrational modes are also commonly detected in warm and dense disk systems. Molecules with N atoms have $3N$ degrees of freedom. In a linear molecule 2 degrees are rotational and 3 translational. Thus there are $3N - 5$ degrees of freedom ($3N - 6$ for nonlinear). In the case of H_2O there are 3 modes, 2 stretching and 1 bending. Vibrational modes are quantized. However, the stretching frequency can be roughly approximated in a model where the two atoms are attached via a spring. Thus, if we use Hooke's law, where the vibration of the spring depends on the reduced mass (μ_r) and the force constant k (the bond),

$$\nu_0 = \frac{1}{2\pi c}\sqrt{\frac{k}{\mu_r}}. \quad (5)$$

The force constant for bonds is $\sim 5 \times 10^5$ Dynes cm^{-1} for single bonds, 10×10^5 Dynes cm^{-1} for double bonds, and 15×10^5 Dynes cm^{-1} for triple bonds. The energy of vibration depends on the vibrational quantum number, n:

$$E_{vib} = \left(n + \frac{1}{2}\right) h\nu_0. \tag{6}$$

Thus heavier molecules have vibrational modes at shorter wavelengths when compared with lighter molecules. Each vibrational state has a spectrum of rotational states called rovibrational transitions. The frequency of these transitions from an upper state $'$ to a lower state $''$ is given by (ignoring some higher-order terms):

$$\nu = \nu_0 + B_{\nu'} J'(J'+1) - B''_{\nu''} J(J''+1). \tag{7}$$

B_ν contains the rotational constant along with a correction due to the interaction between vibration and rotation (i.e., the intermolecular distance can change). The selection rule is $\Delta J = \pm 1$, so there are 2 cases: $J' = J'' + 1$, the R branch, and $J' = J'' - 1$, the P branch. These are labeled as R(J) or P(J), where J is the lower state. In some instances $\Delta J = 0$ is allowed, which is called the Q branch. Quadrupole transitions $\Delta J \pm 2$ are labeled S(J) for $\Delta J = 2$ and O(J) for $\Delta J = -2$.

BRIEF SUMMARY OF OBSERVATIONAL RESULTS. Some features of the observational picture become clear in Table 3.1. Molecular transitions that probe higher-temperature gas have been found to predominantly probe the innermost regions of the disk. This is inferred via spatially unresolved but spectrally resolved line profiles, assuming that the line broadening is due to Keplerian rotation. Spatially resolved observations of the lower-energy rotational transitions trace the outer disk and confirm that protoplanetary disks are in Keplerian rotation (Koerner & Sargent, 1995; Simon et al., 2000) and that the temperature structure has sharp radial gradients (as expected). In addition, analysis of high-J molecular transitions by van Zadelhoff et al. (2001) and Aikawa et al. (2002) confirms that the surface of the irradiated disk is warmer than the midplane, as suggested by analysis of the dust spectral energy distribution (Calvet et al., 1991; Chiang & Goldreich, 1997). It is important to state that current submillimeter and millimeter wave observational facilities are limited by resolution and sensitivity and, therefore, only probe size scales of tens of AUs at the nearest star-forming regions (e.g., TW Hya at 65 pc and Taurus at 140 pc). This limitation will be significantly reduced with the advent of ALMA (and the *James Webb Space Telescope* [*JWST*]) in the coming decade, and we can expect the listing in Table 3.1 to increase drastically.

The chemistry of protoplanetary disks is slowly being cataloged. Molecular emission depends on physical parameters (density, temperature, velocity field) and the molecular abundance. The radial distribution of density and temperature is estimated via thermal dust emission (e.g., Beckwith et al.,

1990). Power-law fits provide $\Sigma(r) \propto r^{-p}$ and $T(r) \propto r^{-q}$ with $p = 0$–1 and $q = 0.5$–0.75. Temperatures are estimated to be ~ 100–200 K at the midplane at 1 AU with surface densities of ~ 500–1500 g cm^{-2}. Modern models now take into account the disk vertical structure (§ 4.1; Calvet, this volume; Calvet et al., 1991; Chiang & Goldreich, 1997; D'Alessio et al., 1998). Thus the determination of chemical composition requires untangling the physical and chemical structures, both of which can vary in the radial and vertical directions. Abundances were first calculated using emission assuming local thermodynamic equilibrium in a disk with radial variations in density and temperature, but researchers now adopt more sophisticated models capable of exploring two-dimensional-parameter space (Monte Carlo, accelerated lambda iteration: Bernes, 1979; Rybicki & Hummer, 1991).

With these techniques, a simple understanding of the chemistry is emerging. One underlying result is that the molecular abundances are generally reduced when compared with those measured in the interstellar medium (Dutrey et al., 1997; Bergin et al., 2007; Dutrey et al., 2007a). This is due to the freeze-out of molecules in the dense cold midplane, which will be discussed later in this chapter. Closer to the star there exist species-specific "snow lines" where molecular ices evaporate. In this regard, the presence of hot abundant water in the inner (<2 AU) disk has been confirmed in some systems (Carr & Najita, 2008; Salyk et al., 2008). A sample of these results is shown in Fig. 3.5, where a *Spitzer* spectrum of a T Tauri star is shown. This spectrum reveals not only the presence of water vapor inside the snow line but also abundant prebiotic organics (see also Lahuis et al., 2006). It is also becoming possible to resolve chemical structure within protoplanetary systems (e.g., Dutrey et al.,

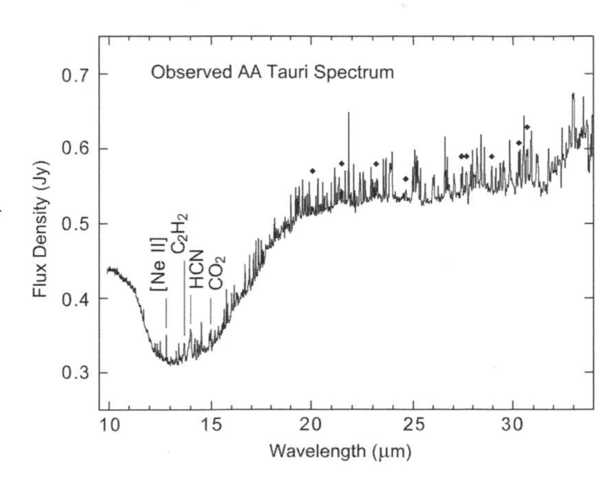

Fig. 3.5. Infrared spectrum of the classical T Tauri star AA Tau taken by the *Spitzer Space Telescope*. Vibrational modes of C_2H_2, HCN, and CO_2, along with a transition of [Ne II] are labeled. Diamonds mark detections of OH rotational transitions. Numerous rotational transitions of water vapor (not marked) are spread throughout the spectrum. Figure and discussion published by Carr & Najita (2008).

2007a). A sample of what is currently possible is shown in Plate 4, where I show the integrated emission distributions of HCO^+, DCO^+, HCN, and DCN. As can be seen, the emission of deuterium-bearing DCO^+ is off-center compared with HCO^+, which is indicative of an active deuterium chemistry, at least for that species. Finally, in the case in which most species containing heavy elements (with dipole moments) are frozen onto grains in the dense midplane, two species, H_2D^+ and D_2H^+, will remain viable probes (Ceccarelli et al., 2004). Some complex organics have been detected in the millimeter wave, and a summary of disk organic chemistry is given by Henning & Semenov (2008).

One key factor for modeling disk chemistry is the dissipation timescale of the gas-rich disk. The molecular detections listed in Table 3.1 are predominantly from systems with ages of $\sim 1-3$ Myr, demonstrating that the gas-rich disk clearly exists for at least a few Myr. In addition, some gas-rich systems, such as TW Hya, are observed at 10 Myr. There is growing evidence of increased dust evolution (i.e., grain growth and possible planet formation) in the inner disk ($r < 10$–30 AU) of some systems (Calvet & D'Alessio, this volume; Najita et al., 2007). Systematic surveys of disk dust emission in stellar clusters have provided some of the strongest limits on the evolutionary timescales for solids. From these studies it is estimated that the timescale for grain growth, settling, and potentially planetesimal formation is $\sim$ 1–10 Myr (e.g., Haisch et al., 2001; Cieza, 2008). The disk gas-accretion rate decays over similar timescales, although there is a large scatter in the data (Muzerolle et al., 2000). Beyond disk accretion, direct observations of gas dissipation are limited by our observational capabilities. However, a *Spitzer* search for gas emission lines in systems spanning a wide range of ages has suggested that the gas disk dissipates on timescales similar to those seen for dust evolution (i.e., $\sim$ 1–10 Myr; Pascucci et al., 2006). It is worth noting that the available data, while showing a general decline with age, exhibit a large degree of scatter. Thus there are systems that have lost their disk in $\sim$ 1 Myr, while others exhibit rich gaseous disks at 10 Myr (Cieza et al., 2007).

3. Equilibrium and Kinetic Chemistry

In the following I will outline the basic methods and chemical processes that are generally used to examine chemistry in protoplanetary disks. In the past, the primary method to explore disk chemistry involved assuming a gas-solid mixture at fixed pressure and temperature with solar composition and determining the resulting composition in thermodynamic equilibrium. This method has been quite successful in explaining various trends observed in meteorites and planetary bodies. For a summary of these efforts, see Prinn (1993); A. M. Davis & Richter (2004; and references therein). More recently, in part because of observations of an active chemistry on the disk surface, it has been recognized that for much of the disk mass, the temperatures and pressures are too low for the gas to reach equilibrium, and the gas kinetic chemistry must be considered. In the following I will describe the basic techniques for both of these cases, with a brief discussion regarding the regimes where each is relevant.

3.1. Thermodynamic Equilibrium

Thermodynamic equilibrium is defined as the state of maximum entropy or the state of minimum Gibbs's free energy of the system. It is the state toward which all chemical systems are being driven if one waited "forever and a day." A key aspect of these calculations is that the final composition depends on the pressure, temperature, and elemental composition, but not the initial chemical state. For the following the author is grateful to C. Cowley for numerous discussions and for providing an outline of the methodology (see Cowley, 1995).

The simplest method involves using the equilibrium constants. For every reaction an equilibrium can be defined. For example, in the following reaction,

$$2\mathrm{H} + \mathrm{O} \rightleftharpoons \mathrm{H_2O}, \tag{8}$$

the ratio of partial pressures ($p_X = n_X kT$) can be determined from the equilibrium constant $K_p(\mathrm{H_2O})$, which is a tabulated quantity (cf. JANAF tables from the National Institute of Standards and Technology NIST). The equilibrium constant can be expressed as

$$K_p(\mathrm{H_2O}) = \frac{\mathrm{p(H_2O)}}{\mathrm{p^2(H)p(O)}} = exp[-\Delta G^\circ(\mathrm{H_2O})/RT], \tag{9}$$

where R is the gas constant and ΔG is the Gibbs's free energy, which for this reaction is

$$\Delta G^\circ(\mathrm{H_2O}) = \Delta G_f^\circ(\mathrm{H_2O}) - 2\Delta G_f^\circ(\mathrm{H}) - \Delta G_f^\circ(\mathrm{O}). \tag{10}$$

The $^\circ$ superscript denotes that the substance is in its standard state given the temperature (i.e., liquid, gas, solid), and the subscript f refers to "of formation." In a system with multiple species at fixed pressure and temperature, we can use the equilibrium constants to determine the composition. For an example, let us take the following system:

$$\mathrm{CO} \rightleftharpoons \mathrm{C} + \mathrm{O}, \tag{11}$$

$$\mathrm{CN} \rightleftharpoons \mathrm{C} + \mathrm{N}. \tag{12}$$

We know the following:

$$K_{\mathrm{CO}}(T) = \frac{p_{\mathrm{C}}p_{\mathrm{O}}}{p_{\mathrm{CO}}}, \tag{13}$$

$$K_{\mathrm{CN}}(T) = \frac{p_{\mathrm{C}}p_{\mathrm{N}}}{p_{\mathrm{CN}}}. \tag{14}$$

We can now define "fictitious pressures," P_{C}, P_{O}, and P_{N}, which can be expressed as

$$P_{\mathrm{O}} = p_{\mathrm{O}} + p_{\mathrm{CO}} = p_{\mathrm{O}} + \frac{p_{\mathrm{C}}p_{\mathrm{O}}}{K_{\mathrm{CO}}}, \tag{15}$$

$$P_{\mathrm{N}} = p_{\mathrm{N}} + p_{\mathrm{CN}} = p_{\mathrm{N}} + \frac{p_{\mathrm{N}}p_{\mathrm{C}}}{K_{\mathrm{CN}}}, \tag{16}$$

$$P_{\mathrm{C}} = p_{\mathrm{C}} + p_{\mathrm{CN}} + p_{\mathrm{CO}} = p_{\mathrm{C}} + \frac{p_{\mathrm{C}}p_{\mathrm{N}}}{K_{\mathrm{CN}}} + \frac{p_{\mathrm{C}}p_{\mathrm{O}}}{K_{\mathrm{CO}}}, \tag{17}$$

and the pressure (mass) balance equation,

$$P = p_{\mathrm{C}} + p_{\mathrm{N}} + p_{\mathrm{O}} + p_{\mathrm{CO}} + p_{\mathrm{CN}}. \tag{18}$$

From these we can solve for the partial pressures and composition.

There are some other more commonly used methods in the literature, which I will outline below. The Gibbs's free energy (G) is defined by

$$G = H - TS = E + PV - TS, \tag{19}$$

where H is the enthalpy, E is the energy, and S is the entropy. P, T, and V are, respectively, the pressure, temperature, and volume. From this the change in energy is

$$dG = dE + PdV + VdP - TdS - SdT, \tag{20}$$

which, by use of the first law of thermodynamics, reduces to

$$dG = VdP - SdT. \tag{21}$$

If this is an ideal gas, then $PV = nRT$, and

$$(dG)_T = VdP = \frac{nRT}{P}dP, \tag{22}$$

$$G_B - G_A = \int_{p_A}^{p_B} \frac{nRT}{P}dP = nRTln\left(\frac{p_B}{p_A}\right). \tag{23}$$

In the case of departures from an ideal gas,

$$G_B - G_A = \int_{p_A}^{p_B} \left(V - \frac{nRT}{P}\right)dP + nRTln\left(\frac{p_B}{p_A}\right) = nRTln\left(\frac{f_B}{f_A}\right), \tag{24}$$

where f is the fugacity, which is essentially a pressure term for a nonideal (real) gas. It is equivalent to the partial pressure if the gas is ideal. It is now common to rewrite this in terms of some known or measured state (e.g., 1 atm where $f = f_0$):

$$G_B = G^0 + nRTln\left(\frac{f_B}{f_0}\right) = G^0 + nRTln(a), \quad (25)$$

where a is the activity, which is defined as f/f_0. If we have an ensemble of atoms and molecules, we can write out the total Gibbs's free energy:

$$G = \Sigma n_{i,p}(\Delta G_{i,p} + RTln\, a_{i,p}), \quad (26)$$

where n is the number of moles of species i in phase p (either pure condensed, condensed, or gaseous). The final equation is for mass balance normalized to the assumed initial composition. Given these sets of equations the minimized value of the Gibbs's free energy determines the equilibrium composition. Thermodynamic data are available via the JANAF tables (Chase, 1998), although the tables are not complete. Some useful references that outline this procedure and more directly discuss issues regarding thermodynamic data are Burrows & Sharp (1999) and Lodders & Fegley (2002).

CONDENSATION. The condensation temperature of an element, as defined by Richardson & McSween (1989), is "the temperature at which the most refractory solid phase containing that element becomes stable relative to a gas of solar composition." A full discussion of this topic is beyond the scope of this chapter, and the reader is referred to Grossman (1972), Richardson & McSween (1989), and Lodders (2003) (and references therein). However, as discussed in § 2, these calculations lie close to the heart of understanding the composition of some meteorites and therefore warrant a brief discussion. The essential idea is that as the nebula gas cools (at constant pressure), refractory minerals condense from the gas, and this sequence depends on the condensation temperature of the various potential refractory minerals. When a given element condenses, one must pay attention to mass balance because some (or all) of its abundance is now unavailable for the gas in equilibrium. As an example, corundum (Al_2O_3) is one of the first minerals to condense. This is consistent with the detection of this mineral in Ca-Al–rich inclusions in chondrites and the fact that Ca-Al inclusions provide the oldest age measured in the solar system (Amelin et al., 2002). An excellent further discussion of this topic is found in the references above and also in A. M. Davis (2006).

FISCHER-TROPSCH CATALYSIS. Fischer-Tropsch catalysis is the name of a subset of high-temperature reactions wherein CO (and CO_2) with H_2 is converted to methane and other hydrocarbons on the surfaces of transition metals, such as Fe and Ni. If Fe/Ni grains are present, then Fischer-Tropsch catalysis is an effective method to create hydrocarbons from gas-phase CO at temperatures where the reaction is inhibited because of high activation energies ($T < 2{,}000$ K). The conversion of CO to organics via this process has been explored by numerous authors, and the reader is referred to Fegley & Prinn (1989), Kress & Tielens (2001), and Sekine et al. (2005), and references therein. The efficiency of Fischer-Tropsch catalysis is highly dependent on the gas temperature and pressure. The efficiency, measured at pressures relevant to a giant-planet subnebula[1] (0.09–0.53 bar), peaks at ~550 K, with reduced efficiency at either lower on higher temperatures (Sekine et al., 2005), but see also Llorca & Casanova (2000) for measurements at lower pressures. At higher temperatures the surface of the catalyst is poisoned by the conversion of surface carbide to unreactive graphite. At lower temperatures the efficiency will ultimately be limited by the incorporation of Fe into silicates, the formation of Fe_3O_4 coatings (Prinn & Fegley, 1989), or coatings by organic and water ice (Sekine et al., 2005). It is also worth noting that ammonia could also be formed from N_2 via a similar process called Haber-Bosch catalysis.

CLATHRATE HYDRATES. Lunine & Stevenson (1985) provided a detailed analysis of the thermodynamics and kinetics of clathrate hydrates at relevant conditions for the Solar Nebula. A clathrate hydrate is a water ice crystal in which the lattice is composed of water molecules with open cavities that are stabilized by the inclusion of small molecules of other chemical species. The inclusion of additional molecules in the lattice is actually key to the stability of the clathrate (Buffett, 2000; Devlin, 2001). Because of this support, the clathrate latices are more open than either the regularly structured lattice of crystalline ice or the irregularly structured lattice of amorphous ice (see, e.g., Tanaka & Koga, 2006).

The relevance of these compounds for nebula chemistry is that they provide a means of trapping volatile species (noble gases, CO, N_2) into ices at temperatures well above their nominal sublimation temperature. Thus volatiles can be provided to forming planets to be observed later in their atmospheres.

1. A giant-planet subnebula refers to the collapsing pocket of the surrounding disk gas out of which the giant planet forms. Both the gas/ice giant and its satellites form within the subnebula.

Lunine & Stevenson (1985) suggest that clathrate hydrates can be important, but only if fresh ice is exposed to the surrounding gas via frequent collisions or perhaps if ices evaporate and condense upon an accretion shock (e.g., Lunine et al., 1991).

KINETIC INHIBITION. It has been recognized that the conditions in the Solar Nebula may not always be in thermodynamic equilibrium. In particular, the timescale of nebular evolution (e.g., the gas lifetime) could be shorter than reaction timescales, and therefore equilibrium would not be attained. Alternately, the dynamic mixing of gases from colder regions within the nebula with the warmer central regions could preclude the formation of certain species that would otherwise be favored, given local conditions. Thus knowledge of the reaction kinetics is required. Cases where kinetics limit the outcome of thermodynamic equilibrium are labeled kinetic inhibition by Lewis & Prinn (1980). It is also referred to as quenching in the literature. As an example, in equilibrium CH_4 should dominate over CO in the inner Solar Nebula and also within giant-planet subnebulae. However, the chemical timescale for CO conversion is $\sim 10^9$ s at $T_{gas} \sim 1{,}000$ K and longer at lower temperatures (Lewis & Prinn, 1980). Estimates suggest that dynamic timescales are shorter, and the zone where CH_4 dominates therefore shrinks considerably (e.g., Fegley & Prinn, 1989; Mousis & Alibert, 2006). In general, these models derive the equilibrium results and then explore whether kinetics would change the predicted results (see Smith, 1998; for a discussion).

3.2. Chemical Kinetics

For much of the disk mass, the density and temperature are low enough that the gas does not have sufficient time to reach equilibrium, and one needs to explore the time dependence of the chemistry. The exact temperature and density (and various combinations) where equilibrium calculations are not valid are presently unclear, and I include a brief discussion of this in § 3.3. However, it is certain that within regions of the disk where the gas temperature is below 100 K (which is most of the disk mass), the chemical kinetics should be considered.

Fortunately, observations (see § 2) appear to suggest that the observed chemistry on tens of AU scales is similar to that seen in the dense ($n > 10^6$ cm^{-3}) regions of the interstellar medium exposed to energetic radiation fields. Models developed for application in this regime (Herbst, 1995) have been readily applied to protoplanetary disks (see, e.g., Aikawa et al., 1997; Bauer et al., 1997). A key feature of the interstellar models is that the cold

($T_{gas} < 100$ K) temperatures require exothermic reactions, while the low ($n \sim 10^3$–10^6 cm^{-3}) density limits reactions to two bodies. At these energies, typical reactions between two neutrals are endothermic or have an activation barrier. Therefore, exothermic reactions between ions and neutrals are believed to dominate (Herbst & Klemperer, 1973; Watson, 1976). However, given the strong radial gradients in density and temperature, it is certain that in some instances reactions with barriers are important, and three-body reactions can also be activated (Willacy et al., 1998; Tscharnuter & Gail, 2007).

In the disk context, the temperature is often low enough that water and other more volatile species will freeze (or even be created in situ) on grain surfaces. Thus the inclusion gas-grain interaction is a key piece of any successful disk chemistry model. Below I will first describe the basic gas-phase chemistry and how networks are created and solved. Then I extend the discussion to include grain chemistry.

GAS-PHASE CHEMISTRY. In Table 3.2 I provide a listing of common chemical reaction types and typical rates. To be active, the gas-phase chemistry requires a source of ionization, which in the disk can be cosmic rays, X-rays, or active radionuclides (§ 4.2). The ionization of H_2 ultimately produces H_3^+, which powers the ion-molecule chemistry. A general feature of this chemistry is the subsequent creation of complex molecular ions that undergo dissociative recombination to create both simple and complex molecules. To illustrate the

Table 3.2: Typical Reaction Types and Coefficients

Reaction type	Example	Typical rate coefficient
Ion-neutral	$H_3^+ + CO \rightarrow HCO^+ + H_2$	$\sim 10^{-9}$ cm^3 s^{-1}
Radiative association[a]	$C^+ + H_2 \rightarrow CH_2^+ + h\nu$	$\sim 10^{-17}$ cm^3 s^{-1}
Neutral exchange	$O + OH \rightarrow O_2 + H$	$\sim 10^{-11} - 10^{-10}$ cm^3 s^{-1}
Charge transfer	$C^+ + S \rightarrow S^+ + C$	$\sim 10^{-9}$ cm^3 s^{-1}
Radiative recombination[a]	$C^+ + e^- \rightarrow C + h\nu$	$\sim 10^{-12}$ cm^3 s^{-1}
Dissociative recombination[a]	$H_3O^+ + e^- \rightarrow H_2O + H$	$\sim 10^{-6}$ cm^3 s^{-1}
Photoionization[b]	$C + h\nu \rightarrow C^+ + e^-$	$\sim 10^{-9} e^{-\gamma A_V}$ s^{-1}
Photodissociation[b]	$H_2O + h\nu \rightarrow O + OH$	$\sim 10^{-9} e^{-\gamma A_V}$ s^{-1}

[a] Sample rate coefficients derived at a gas temperature of 40 K using the UMIST database (Woodall et al., 2007a).

[b] The depth dependence of photoionization rates and photodissociation rates is often parameterized by $e^{-\gamma A_V}$. In this form γ includes factors such as the shape and strength of the radiation field, the absorption and scattering due to dust, and the wavelength dependence of the photoionization or photodissociation cross sections. For more discussion, see van Dishoeck (1988), Roberge et al. (1991), and van Dishoeck et al. (2006).

various features of the chemistry, I will examine the chemistry of water, which has significance in the disk context.

The oxygen chemistry begins with the ionization of H_2 and the formation of the trihydrogen ion (H_3^+). H_3^+ will react with O, and a sequence of rapid reactions produces H_3O^+. The next step involves the dissociative recombination of H_3O^+ with electrons:

$$\begin{aligned} H_3O^+ + e^- &\rightarrow H + H_2O \quad (f_1) \\ &\rightarrow OH + H_2 \quad (f_2) \\ &\rightarrow OH + 2H \quad (f_3) \\ &\rightarrow O + H + H_2 \quad (f_4), \end{aligned} \tag{27}$$

where f_{1-4} are the branching ratios.[2] Via this sequence of reactions water vapor is then created. The primary *gas-phase* destruction pathway in shielded gas is via a reaction with He^+ (another ionization product); at the disk surface photodissociation also plays a key role. It should be stated that the example above provides only the primary formation and destruction pathways under typical conditions (i.e., where ionizing agents are available). In addition, water participates in numerous other reactions both as reactant and product, and some water can be created via other pathways, albeit at reduced levels (e.g., Bergin et al., 2000).

At temperatures below a few hundred degrees Kelvin, ion-neutral chemistry dominates. Above this temperature, two neutral-neutral reactions rapidly transform all elemental oxygen in the molecular gas into water vapor (Elitzur & de Jong, 1978; Wagner & Graff, 1987):

$$O + H_2 \rightarrow OH + H \ (E_a = 3,160 \text{ K}), \tag{28}$$

$$OH + H_2 \rightarrow H_2O + H \ (E_a = 1,660 \text{ K}). \tag{29}$$

This mechanism (along with evaporation of water ice) will keep water vapor the dominant oxygen component in the gas within a few AU of the forming star.

To model these reactions, one creates a series of ordinary differential equations, one for each species included, accounting for both formation rates and destruction rates: $dn/dt = formation - destruction$. For water (excluding the numerous reactions not mentioned above) the equation is as follows:

2. Two measurements using a storage ring give roughly consistent results (Vejby-Christensen et al., 1997; Jensen et al., 2000). I provide here the latest measurement: $f_1 = 0.25$, $f_2 = 0.14$, $f_3 = 0.60$, and $f_4 = 0.01$. Using a different technique (flowing afterglow), Williams et al. (1996) find $f_1 = 0.05$, $f_2 = 0.36$, $f_3 = 0.29$, and $f_4 = 0.30$.

$$\frac{dn(\mathrm{H_2O})}{dt} = n(\mathrm{H_3O^+})n(\mathrm{e^-})\alpha_r + n(\mathrm{OH})n(\mathrm{H_2})k_1 - n(\mathrm{H_2O})[n(\mathrm{He^+})k_2 + k_{\gamma,A_v=0}e^{-1.7A_v}]. \quad (30)$$

In eq. (30) α_r is the recombination rate of H_3O^+ (Jensen et al., 2000), and $k_{\gamma,A_v=0}$ is the unshielded photodissociation rate. Values for this rate and others can be found in either of the two available astrochemical databases (one originally maintained by E. Herbst at Ohio State and the other at UMIST: Woodall et al., 2007a). These rate networks have thousands of reactions occurring over a wide range of timescales, and thus there is a stiff system of ordinary differential equations (ODEs). A number of ODE solvers have been developed to solve these systems with the output of composition as a function of time. A summary of some of the available codes for astrochemical applications is given by Nejad (2005).

GAS-GRAIN CHEMISTRY. In the dense ($n > 10^5$ cm^{-3}), cold ($T < 20$ K) centers of condensed molecular cores, which are the stellar birth sites, there is ample evidence for the freeze-out of many neutral species onto grain surfaces (Bergin & Tafalla, 2007). These even include the volatile CO molecule, which freezes onto grains only when temperatures are below 20 K (Collings et al., 2003), as opposed to water, which freezes at 110 K for interstellar pressures (Fraser et al., 2001). The outer disk midplane ($r > 20$–40 AU) is believed to be similarly cold, with densities that are several orders of magnitude higher than in prestellar cores where volatile CO freeze-out is readily detected. Thus to model disk kinetic chemistry, it is necessary to model the deposition of species onto grain surfaces in tandem with the mechanisms that desorb frozen ices. A number of such desorption mechanisms have been identified in the literature, including thermal evaporation, X-ray desorption, and UV photodesorption. Each of these will be described below and is generally included in the rate equations (eq. [30]) as a source or loss term with rate coefficients given below.

Gas-Phase Freeze-Out. Atoms and molecules freeze onto grain surfaces or adsorb with a rate of $k_{fo} = n_g \sigma_g v S$ s^{-1}. Here σ_g is the grain cross section, v is the mean velocity of the Maxwellian distribution of gaseous particles, n_g is the space density of grains, and S is the sticking coefficient (i.e., how often a species will remain on the grain upon impact). Sticking coefficients for molecular species are generally thought to be unity at low T, a value that is supported by theoretical (Burke & Hollenbach, 1983) and laboratory work

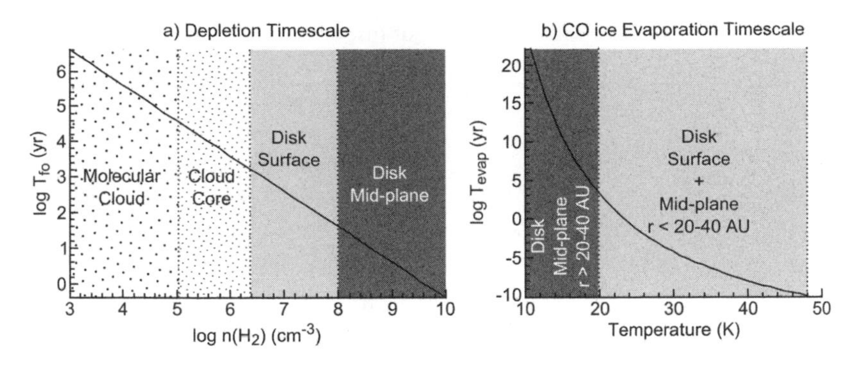

Fig. 3.6. (a) CO freeze-out timescale as a function of density of molecular hydrogen. (b) Thermal evaporation timescale for CO molecules as a function of dust temperature.

(Bisschop et al., 2006). For an interstellar grain size distribution[3] with a lower cutoff of 20 Å, $n_g \sigma_g \simeq 2 \times 10^{-21} n$ cm^{-1} (see discussion in Hollenbach et al., 2008). In disk systems the process of grain coagulation and settling will locally reduce the grain surface area, and this value is only a starting point for potential situations. If we assume this surface area, the freeze-out timescale, τ_{fo}, is

$$\tau_{fo} = 5 \times 10^3 \text{yrs} \left(\frac{10^5 cm^{-3}}{n} \right) \left(\frac{20K}{T_{gas}} \right)^{\frac{1}{2}} \sqrt{A}, \tag{31}$$

where A is the molecule mass in atomic mass units. Fig. 3.6 demonstrates that within the disk the freeze-out time for CO is on the order of tens of years or below. Thus, without effective means to remove molecules from the grain surface, most heavy molecules freeze onto grain surfaces. In the astronomical literature this timescale is often referred to as the depletion timescale. Because molecular abundances can "deplete" by freeze-out or gas-phase destruction, I adopt here the more descriptive terminology.

Grain Binding Energies. The removal of ices from the refractory grain core requires overcoming the strength of the bond to the surface. Because low temperatures preclude the breaking of chemical bonds, molecules are not expected to be chemically bound to grain surfaces (although chemical bonds may be important at higher temperatures; Cazaux & Tielens, 2004). Instead, the approaching molecule has an induced dipole interaction with the grain surface or mantle and is bound through weaker Van

3. In the interstellar medium grains are inferred to exist with a size distribution that follows a power law of $dn(a)/da \propto a^{-3.5}$, with a as the grain radius (Mathis et al., 1977).

Table 3.3: Binding Energies of Important Ices[a]

Species	E_b/k (k)	Reference
H_2O	5,800	Fraser et al. (2001)
CO	850	Collings et al. (2003)
N_2	800	Bisschop et al. (2006)
NH_3	2,800	Brown & Bolina (2007)
CH_3OH	5,000	Brown & Bolina (2007)

[a] Values are from laboratory work only using pure ices.

der Waals–London interactions called physical adsorption (Kittel, 1976). Van der Waals interactions are proportional to the product of the polarizabilities (α) of the molecule and the nearby surface species ($E_b \propto \alpha_{mol}\alpha_{surface}$). Because of this property, binding energies on a silicate surface have been calculated by using the extensive literature on the physical adsorption of species onto graphite. With accounting for differences in polarizability, these can be scaled to silicates (Allen & Robinson, 1977). Hasegawa et al. (1992) reexamined this issue and provide a sample list of estimated values that have widely used. These estimates need to be revised for cases where actual experimental information exists. Measurements have now been performed for a number of key species; a listing is given in Table 3.3.

Thermal Evaporation. The rate of evaporation from a grain with a temperature of T_{dust} is given by the Polanyi-Wigner relation (Tielens, 2005):

32
$$k_{evap,i} = \nu_{0,i} exp(-E_{b,i}/kT_{dust}) \ (s^{-1}),$$

where ν_0 is the vibrational frequency of molecule i in the potential well,

33
$$\nu_i = 1.6 \times 10^{11} \sqrt{(E_{b,i}/k)(m_H/m_i)} \ \mathrm{s}^{-1},$$

with $E_{b,i}$ the binding energy of species i. From Hollenbach et al. (2008) the sublimation or freeze-out temperature can be derived by setting the flux of desorbing molecules ($F_{td,i}$) equal to the flux of species adsorbing from the gas. Thus

34
$$F_{td,i} \equiv N_{s,i} k_{evap,i} = 0.25 n_i v_i,$$

where $N_{s,i}$ is the number of adsorption sites per cm^2 ($N_{s,i} \sim 10^{15}$ cm^{-2}), and n_i and v_i are the gas-phase number density and thermal velocity of species i. Solving for the dust temperature, we can derive the sublimation temperature for species i (Hollenbach et al., 2008):

35
$$T_{sub,i} \simeq \left(\frac{E_{b,i}}{k}\right)\left[57 + ln\left[\left(\frac{N_{s,i}}{10^{15}\mathrm{cm}^{-2}}\right)\left(\frac{\nu_i}{10^{13}\ \mathrm{s}^{-1}}\right)\left(\frac{1\ \mathrm{cm}^{-3}}{n_i}\right)\left(\frac{10^4\ \mathrm{cm\ s}^{-1}}{v_i}\right)\right]\right]^{-1}.$$

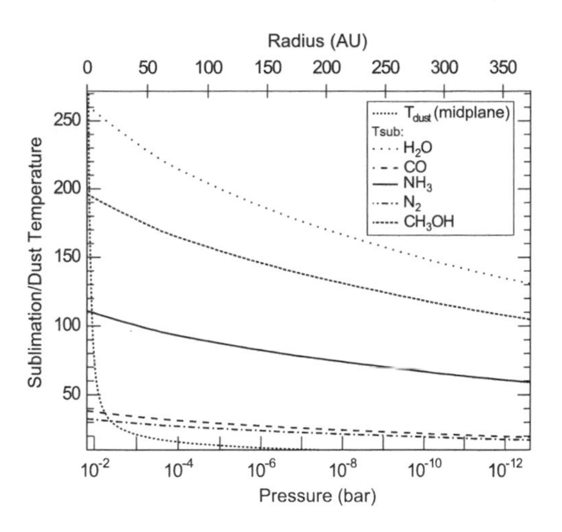

Fig. 3.7. Plot of the sublimation temperature for various molecular ices, and the midplane dust temperature, as a function of disk pressure and radius for a disk model taken from D'Alessio et al. (2001).

Eq. (35) assumes that there is at least one monolayer of the particular species present on the grain surface. In the case where only a partial monolayer exists, then the thermal desorption rate per unit grain surface area is $f_{s,i}N_{s,i}k_{evap,i}$, where $f_{s,i}$ is the fraction of the $N_s\pi a^2$ surface sites occupied by species i. In Fig. 3.6(b) I show a plot of the CO evaporation timescale as a function of grain temperature. Comparison with the freeze-out timescale in Fig. 3.6(a) suggests that CO will be frozen in the form of ice throughout much of the outer ($r > 20$–40 AU) disk midplane. However, on the warmer disk surface CO will remain in the gas phase. This is consistent with observations (van Zadelhoff et al., 2001).

Another interesting aspect of eq. (35) is the dependence on gas density. In Fig. 3.7 I provide a plot of the sublimation temperature (and dust temperature) as a function of disk pressure and radius for a model solar analog disk. This calculation uses the binding energies given in Table 3.3. Thus most volatile ices are frozen beyond tens of AUs, with various species returning to the gas in a fashion that depends on their relative binding strength to the grain surface. It should also be noted that experiments suggest that the return of species to the gas is not entirely as simple as described above. The evaporation also depends on the composition of the mantle and the relative disposition of its components (Viti et al., 2004).

Photodesorption. The direct absorption of an ultraviolet (UV) photon by a molecule adsorbed on the grain surface puts an electron in an excited state. If the interaction between the excited state and the binding surface is repulsive, then the excited molecule may be ejected from the surface (Watson, 1976).

The rate of photodesorption per adsorbed molecule is given by (Boland & de Jong, 1982; Hollenbach et al., 2008)

$$R_{pd} = n_{gr} \pi a_d^2 \epsilon Y G_0 F_0 exp(-1.8A_v). \text{ cm}^{-3}\text{s}^{-1}. \quad (36)$$

In this equation Y is the photodesorption yield (number of molecules desorbed per incident photon), ϵ is the fraction of species that are found in the top few monolayers, a_d is the grain radius, and G_0 is the UV field enhancement factor in units of the mean interstellar UV field of $F_0 = 10^8$ photons cm^{-2} s^{-1} (Habing, 1968). The stellar UV field and its relation to the interstellar radiation field are discussed in § 4.2. For most molecules, the yield is unknown and is assumed to be $\sim 10^{-5}$–10^{-3} (Bergin et al., 1995). More recently there have been measurements for CO ($Y = 3 \times 10^{-3}$ mol/photon; Öberg et al., 2007) and H_2O ($Y = 10^{-3}(1.3 + 0.032 \times T)$ mol/photon; Öberg et al., 2009).

X-ray Desorption. Young stars are known to be prolific emitters of X-ray photons with emission orders of magnitude above typical levels for main-sequence stars (Feigelson & Montmerle, 1999). This emission is capable of penetrating much deeper into the disk atmosphere than UV photons and hence can potentially provide energy to release molecular ices from grains on the disk surface. In this chapter I will focus on the effects of these X-rays on the chemistry and refer the reader to Feigelson & Montmerle (1999) for a discussion of the origin and characteristics of X-ray emissions. X-ray photons are absorbed by the atoms that compose the grain refractory core and mantle with a cross section that depends on species and wavelength (Morrison & McCammon, 1983; Henke et al., 1993). The absorption of an X-ray produces a hot primary electron that generates a number of secondary electrons (s) via a process known as the Auger effect. Depending on the energy of these electrons and the grain size, they then deposit all or a fraction of their total energy on the grain (Dwek & Smith, 1996). This heat diffuses through the grain lattice on the basis of the thermal physics of the material. Fortunately, experiments have demonstrated that the thermal conductivity and specific heats of amorphous solids tend have near-universal values to within an order of magnitude over 3 decades in temperature (Pohl, 1998). This is encouraging for interstellar studies because we can more reliably adopt laboratory estimates for studies of grain heating (see the appendix in Najita et al., 2001).

Léger et al. (1985) explored grain heating by X-rays and cosmic rays by adopting the specific heat measurements of amorphous silica from Zeller & Pohl (1971) in the following form:

$$\rho C_V(T) = 1.4 \times 10^{-4} T^2 \ J \ cm^{-3} \ K^{-1} \quad (10 < T < 50 \ K), \tag{37}$$

$$\rho C_V(T) = 2.2 \times 10^{-3} T^{1.3} \ J \ cm^{-3} \ K^{-1} \quad (50 < T < 150 \ K). \tag{38}$$

If we use these expressions, the energy increase due to an energy impulse raises the temperature by (using the specific heat appropriate for $T < 50$ K)

$$\Delta E(eV) = 3.4 \times 10^4 \left(\frac{a}{0.1 \mu m} \right)^3 \left(\frac{T_f}{30 \ K} \right)^3 ,$$

where T_f is the final grain temperature and a is the grain size (Léger et al., 1985). Najita et al. (2001) explored the question of X-ray desorption in protoplanetary disks by using the results of Dwek & Smith (1996) and Léger et al. (1985). As one example, they found that a 1 keV photon will deposit $\sim$ 700eV in a 0.1 μm grain. The warm grain cools via evaporation and thermal radiation. This energy impulse is insufficient to raise a cold ($T_{dust} \sim 10$ K) grain to above the threshold for desorption of the volatile CO molecule. However, smaller grains ($a < 0.03 \mu$m) have reduced volume for heat diffusion and equilibrate to temperatures that evaporate CO molecules. Thus X-ray desorption from the whole grain is not feasible, because grains with sizes > 0.1 μm are present in the ISM and in disks. A more efficient method for desorption is found when one considers the fact that grains are likely porous fractal aggregates composed of a number of small grains with a range of sizes. In this case the absorption of an X-ray by a portion of the grain is capable of local evaporation before the heat diffuses throughout the aggregate (Najita et al., 2001). This is efficient for desorbing the most volatile ices (e.g., CO, N_2) on the disk surface, but not water ice. There is some evidence that this process is active (Greaves, 2005), although the situation is complex because one additional effect of X-ray ionization is the excitation of H_2, which generates a local UV field that could release molecules via photodesorption.

Cosmic-Ray Desorption. Like X-rays, cosmic rays absorbed by dust grains can provide non thermal energy to desorb molecules frozen on grain surfaces. Cosmic rays also have greater penetrating power than X-rays and UV and hence can have a greater impact on the chemistry in well-shielded regions. The penetration of cosmic rays to the inner tens of AUs of the disk is highly uncertain; however, it is likely that cosmic rays can affect the chemistry of the outer disk. This issue will be discussed in § 4.2; here I will focus on the effects of nonthermal desorption via cosmic-ray impact. Cosmic rays have energies between 0.02 and 4 GeV per nucleon, but the energy deposition goes as Z^2 such that the less abundant heavier cosmic rays have greater impact on the grain thermal cycling and ice evaporation (Léger et al., 1985). Bringa & Johnson (2004) noted that the impact of the cosmic ray on a grain is not exactly a question of thermal diffusion (as normally treated), but rather one of a pressure pulse and a melt. In general, the effect of cosmic rays is to efficiently remove all ices (including water) from small (0.01 μm) grains because of large thermal impulses. Cosmic-ray impacts can also release ices from larger grains via local spot heating. This is thought to be capable of removing the most volatile ices (CO, N_2), but not water ice (Léger et al., 1985; Herbst & Cuppen, 2006). A number of groups have looked at this process, and there is disparity in the calculations. A summary of the various approximations is provided by Roberts et al. (2007).

Explosive Desorption. Experiments with ices irradiated in the lab suggest that when an ice mantle formed at $\sim$ 10 K is exposed to radiation and then warmed to $\sim$ 25 K, the rapid movement and reactivity of radicals lead to an explosion of chemical energy and desorption (Schutte & Greenberg, 1991). There are two issues with this mechanism that are not insurmountable but make it more difficult to readily place it into models: (1) The generation of radicals on the surface needs the ice-coated grain to be exposed to the UV radiation field. A criticism of this is that the UV flux in the lab was significantly greater than typical exposure in the well-shielded ISM where ice mantles form. However, the strength of the UV (Ly α) field at 10 AU in TW Hya, the closest young star with a circumstellar disk, is $\sim 10^{14}$ photons cm^{-2} s^{-1} (Herczeg et al., 2002), comparable to that used in the laboratory. The UV radiation may also come from external high-mass stars if a young solar-type star is born in a cluster. Thus it is likely that on the disk surface, radicals are being created in the frozen mantle. (2) Once the radicals have been generated, a cold ($\sim$ 10 K) grain needs to be warmed to above $\sim$ 25 K. This could be possible if the grain absorbs an X-ray on a cosmic ray, undergoes a grain-grain collision, or advects to warmer layers.

GRAIN SURFACE CHEMISTRY. A full treatment of grain surface chemistry is beyond the scope of this chapter, and the reader is referred to the review by Herbst et al. (2005). There are a few issues specific to disk chemistry that should be discussed. First, at low dust temperatures ($<$ 20 K) H atom addition will dominate because H (and D) atoms can more rapidly scan the surface than heavier atoms or molecules. The scanning timescale is (Tielens, 2005)

$$\tau_m = \nu_m^{-1} e^{(-E_m/kT_{dust})}, \tag{39}$$

where ν_m is the vibration frequency of the migrating molecules, which is equivalent to ν_0 (eq. [33]). It is typical to assume that $E_m \sim 0.3 - 0.5 E_b$; thus at $T_{dust} \sim 10$ K, atoms with low binding energies (H, D) migrate rapidly, and heaver molecules tend to remain in place.

The H atom abundance in molecular gas is set by the balance between H_2 ionization (which ultimately produces a few H atoms per ionization) and reformation of H_2 on grains. Since the ionization rate in the disk midplane is likely small (see § 4.2) and cold grains are present, there may be few H atoms (or other atoms) present in the gas to allow for an active surface chemistry. To be more explicit, the sequence of H formation and destruction reactions in a well-shielded molecular medium is as follows:

$$\mathrm{H} + \mathrm{gr} \rightarrow \mathrm{H_{gr}}, \tag{40}$$

$$\zeta_{\mathrm{CR,XR,RN}} + \mathrm{H_2} \rightarrow \mathrm{H_2^+} + \mathrm{e}, \tag{41}$$

$$\mathrm{H_2^+} + \mathrm{H_2} \rightarrow \mathrm{H_3^+} + \mathrm{H}. \tag{42}$$

The limiting step in H atom formation is the ionization from cosmic rays (CRs), X-rays (XRs), or radionuclides (RNs). In essence, every ionization of H_2 ultimately produces $f_H \sim 2$–3 H atoms per ionization. In equilibrium the H atom concentration is

$$n_H = \frac{f_H \zeta n_{H_2}}{n_g \sigma_g v_H S_H \eta} = \frac{f_H \zeta}{10^{-21} v_H S_H \eta}. \tag{43}$$

In this equation v_H is the thermal velocity of hydrogen atoms, and η is an efficiency factor that can be set to unity provided that there is more than one H atom on each grain. S_H is the sticking coefficient for H atoms. Burke & Hollenbach (1983) and Buch & Zhang (1991) provide analytical expressions for the sticking coefficient that can be used with values of $S_H \sim 0.7$ at 20 K.

I have also used the expression $n_g \sigma_g = 10^{-21} n_{H_2}$. Each of the relations in eq. (43) is constant (given a temperature); thus the space density of H atoms is a constant in the molecular gas. If we assume that cosmic rays penetrate to the midplane with an ionization rate of $\zeta = 1 \times 10^{-17}$ s^{-1} (the most extreme high-ionization case for the midplane) and the gas temperature is 30 K, then $n_H \sim 5$ cm^{-3}. The H atom abundance is therefore inversely proportional to the density. The average grain in the ISM with a size of 0.1 μm has an abundance of $\sim 10^{-12}$. This sets the initial conditions for the abundance of grains. If $\eta = 1$, then for much of the disk midplane, where $n > 10^{12}$ cm^{-3}, there will be less than 1 H atom per grain, significantly reducing H atom surface chemistry. However, it may be the case that there is less than one H atom on each grain and $\eta \ll 1$, which could significantly reduce the grain surface formation rate. This would require more gas-phase H atoms and a potential to power chemical processes with longer timescales.

Second, at higher temperatures heavier atoms and molecular radicals can begin to diffuse on the grain surface and create complex species. Given the abundance of radiation on the disk surface, it is highly likely that bonds are broken in the ices, creating radicals (from CH_3OH, H_2O, CO_2, CH_4, NH_3, and so on). Upon warm-up this can produce more complex species (see the models of Garrod et al., 2008; as an example), unless it is limited by explosive desorption. Thus grain chemistry may be important on the disk surface, and, depending on the strength of vertical mixing, its products may reach the disk midplane.

DEUTERIUM FRACTIONATION. Enhancements of deuterium-bearing molecules relative to the hydrogen counterparts have been known for some time in the solar system and in the interstellar medium. In cold ($T_{dust} \sim$ 10–20 K) cores of clouds where stars are born, enrichments of 2–3 orders of magnitude are observed above the atomic hydrogen value of $D/H \geq (2.3 \pm 0.2) \times 10^{-5}$ estimated within 1 kpc of the Sun (Linsky et al., 2006). Because of the lower zero-point energy for deuterium than for hydrogen, at low ($T < 30$ K) temperatures, conditions become favorable for transfer the D bond, as opposed to the H bond. Ion-molecule reactions in the dense ISM are thought to be the mechanism responsible for these enrichments (Millar et al., 1989). Deuterium fractionation can also be powered via reactions on the surfaces of dust grains (Tielens, 1983; Nagaoka et al., 2005), which are also discussed below.

In the gas, deuterium chemistry is driven by the following reaction:

$$H_3^+ + HD \leftrightarrow H_2D^+ + H_2 + 230K. \quad (44)$$

The forward reaction is slightly exothermic, favoring the production of H_2D^+ at 10 K and enriching the D/H ratio in the species that lie at the heart of ion-molecule chemistry (Millar et al., 1989).[4] These enrichments are then passed forward along reaction chains to DCO^+, DCN, HDO, HDCO, and others.

At densities typical of the dense interstellar medium ($n \sim 10^5$ cm^{-3}), pure gas-phase models without freeze-out cannot produce significant quantities of doubly (NHD_2; Roueff et al., 2000) and triply deuterated ammonia (ND_3; Lis et al., 2002; van der Tak et al., 2002). This motivated a reexamination of the basic deuterium chemistry. The primary advance in our understanding is twofold: (1) Deuteration reactions do not stop with H_2D^+; rather, they continue toward

4. Reaction (44) has been measured in the lab at low temperatures, where it has been found that there is an additional dependence on the ortho/para ratio of H_2 (Gerlich et al., 2002).

the formation of both D_2H^+ and D_3^+ via a similar reaction sequence (Phillips & Vastel, 2003; Roberts et al., 2003; Walmsley et al., 2004):

$$H_2D^+ + HD \leftrightarrow D_2H^+ + H_2 + 180K, \quad (45)$$

$$H_2D^+ + HD \leftrightarrow D_3^+ + H_2 + 230K. \quad (46)$$

(2) The freeze-out of heavy species in the disk midplane, in particular CO, a primary destroyer of both H_3^+ and H_2D^+, increases the rate of the gas-phase fractionation reactions. Thus in vertical gas layers, before potential complete freeze-out of heavy molecules, there can exist active deuterium fractionation, provided the gas temperature is below 30 K (see Aikawa & Herbst, 2001).

At high densities in the inner nebula, thermodynamic equilibrium can apply. At temperatures $T > 500$ K [HDO]/[H_2O] closely follows [HD]/[H_2], but at $T < 500$ K fractionation can proceed via the following reaction (Richet et al., 1977):

$$HD + H_2O \leftrightarrow HDO + H_2. \quad (47)$$

This reaction can provide a maximum enrichment of a factor of 3 (Lecluse & Robert, 1992).

Grain surface chemistry can also produce deuterium fractionation because the gas-phase chemistry (at low temperatures) leads to an imbalance and enhancement in the D/H ratio of atomic H. The efficiency of this process depends on the number of atoms that are generated by ionization. At low temperatures more deuterium is placed in H_2D^+ relative to H_3^+. These species create H and D atoms by reacting with neutrals. Thus $f_D > f_H$ (see eq. [43]), and there is the potential for these atoms to freeze onto grains and react, thereby fractionating molecular ices. However, as noted earlier, it is possible that in the dense midplane there will be less than one D atom generated per grain, and the efficiency of D fractionation on grain surfaces could be reduced.

To summarize, deuterium chemistry can be active in the disk, both in the gas and on grains, but it requires low temperatures and sufficient levels of ionization.

3.3. Kinetics and Thermodynamic Equilibrium

At present it is not clear exactly where in the disk thermodynamic equilibrium is valid. It is clear that some aspects of meteoritic composition are consistent with equilibrium condensation (§ 2.1). However, there are also aspects that are inconsistent, such as the detection of unaltered presolar grains (Huss & Lewis, 1995) and deuterium enrichments in cometary ices, which suggest that not all material cycled to high temperatures (above evaporation). Fegley (2000) provides a schematic of the relative importance of equilibrium to kinetics. He proposes that thermodynamic equilibrium is of greater significance in the inner nebula and within giant-planet subnebulae. To some extent this must be the case, but divining the exact transition is not entirely clear because, for example, the inner nebula and surfaces of giant-planet subnebulae will be exposed to radiation that powers nonequilibrium photochemistry. Given evidence from trends in meteoritic composition (e.g., A. M. Davis & Richter, 2004), it is clear that in some instances thermochemical equilibrium is valid. However, one must pay careful attention, where possible, to the relevant kinetic timescales.

4. Physical Picture and Key Processes

4.1. Density and Temperature Structure

Estimates of the physical structure of circumstellar disks can come from the solar system and also from extrasolar systems. Within the Solar Nebula, estimates are based on the so-called minimum-mass Solar Nebula: the minimum amount of mass required to provide the current distribution of planetary mass as a function of distance from the Sun, with a correction for a gas-to-dust ratio. Hayashi (1981) estimates that $\Sigma(r) \sim 1,700r^{-3.2}$, which is commonly used, but see also Weidenschilling (1977) and S. S. Davis (2005). It is important to note that this value is a lower limit and does not account for any potential solids that have been lost from the system during the early evolutionary phases. Limits on the gas temperature have been estimated by Lewis (1974), based on the differences in planetary composition. For instance, the temperature at 1 AU is suggested to be $\sim$ 500–700 K on the basis of the estimation that the Earth formed below the condensation temperature of FeS, but above the end point for oxidation of metallic iron in silicates (Lewis, 1974). If we use these estimates, the nebular thermal structure would decay as R^{-1}, which is much steeper than in pure radiative equilibrium, $R^{-0.5}$ (see, e.g., Hayashi, 1981). However, as discussed below, radiation does dominate the thermal structure of protoplanetary disks.

More detailed estimates as a function of position (both radial and vertical) are available from extrasolar protoplanetary systems. Models take into account the energy generated by accretion and by stellar irradiation of a flared disk structure. Disk flaring is due to hydrostatic equilibrium (Kenyon & Hartmann, 1987) and is characterized by a pressure scale height $H_p = \sqrt{R^3 c_s^2 / GM_*}$, with the sound speed $c_s = \sqrt{kT/\mu m_H}$ (μ is the mean molecular weight, $\mu \sim 2.3$). The vertical density then follows $\rho = \rho_0 \exp(-z^2/2H_p^2)$, with ρ_0 a function of radius. With radial density profile as a variable and some assumptions regarding dust composition and size distribution (to set the optical properties), the spectral energy distribution of dust emission from UV/visible ($\sim$ 4,000 Å) to millimeter (1,000 μm) can be predicted and compared with observations (Calvet et al., 1991; Chiang & Goldreich, 1997; D'Alessio et al., 1998). These models have found that disk surfaces, which are directly heated by stellar irradiation, are warmer than the midplane. The midplane is heated by reprocessed stellar radiation and, for the inner few AU, by viscous dissipation of accretion energy. Plate 5 shows an example of the thermal structure from one such model (D'Alessio et al., 2001).

An important issue is the effect of the initial steps of planetesimal formation, grain growth and settling, on the disk physical structure. Smaller grains ($\langle a \rangle \sim 0.1$ μm) preferentially absorb the energetic short wavelength, and as these grains grow to larger sizes, their optical depth decreases (Dullemond & Dominik, 2004; D'Alessio et al., 2006). The presence of small grains in the disk atmosphere leads to higher temperatures and greater flaring than in disks with larger grains (or more settling to the midplane). Thus the disk physical structure evolves with the evolution of solids. Moreover, there is an observed decay in accretion rate over similar timescales (Muzerolle et al., 2000). Thus the energy input from accretion should similarly decline. In sum, both the midplane temperature and pressure should decline with evolution (D'Alessio et al., 2005). As I will show in § 5, this structure and evolution are important components of time-dependent disk chemical models.

4.2. Disk Ionization

The gas-phase chemistry is powered by ionizing photons. The characteristics of the chemical kinetics therefore depend on the flux and wavelength dependence of ionizing photons. The molecular ions generated by photo- or cosmic-ray absorption are also the key to linking the mostly (by many orders of magnitude) neutral gas to the magnetic field, which is thought to have a direct relation to the physics of disk accretion. In the following I will explore each of the various components of disk ionization. In Table 3.4 I provide a listing of the major contributors to the ion fraction.

COSMIC RAYS. The interstellar cosmic-ray ionization rate has been directly constrained by the Voyager and Pioneer spacecraft at $\sim$ 60 AU from the Sun (Webber, 1998), and estimates in the ISM have been summarized by Dalgarno (2006). On the basis of these studies, the ionization rate is believed to be $\zeta_{H_2} \sim 5 \times 10^{17} s^{-1}$ in the dense ISM, with growing evidence for higher rates

Table 3.4: Disk Ionization Processes and Vertical Structure[a]

Layer/carrier	Ionization mechanism	$\Sigma_{\tau=1}$ (g cm^{-2})[b]	α_r (cm^{-3}s^{-1})[c]	x_e[d]
Upper surface	UV photoionization of H[e]	6.9×10^{-4}	$\alpha_{H^+} = 2.5 \times 10^{-10} T^{-0.75}$	$> 10^{-4}$
H^+	$k_{H^+} \sim 10^{-8}$ s^{-1}			
Lower surface	UV photoionization of C[f]	1.3×10^{-3}	$\alpha_{C^+} = 1.3 \times 10^{-10} T^{-0.61}$	$\sim 10^{-4}$
C^+	$k_{C^+} \sim 4 \times 10^{-8}$ s^{-1}			
Warm mol.	Cosmic-[g] and X-ray[h] ionization	96 (CR)	$\alpha_{H_3^+} = -1.3 \times 10^{-8} +$	$10^{-11 \to -6}$
H_3^+, HCO^+	$\zeta_{cr} = \frac{\zeta_{cr,0}}{2} [\exp(-\frac{\Sigma_1}{\Sigma_{cr}}) + \exp(-\frac{\Sigma_2}{\Sigma_{cr}})]$		$1.27 \times 10^{-6} T^{-0.48}$	
	$\zeta_X = \zeta_{X,0} \frac{\sigma(kT_X)}{\sigma(1keV)} L_{29} J(r/\mathrm{AU})^{-2}$	0.008 (1 keV)	$\alpha_{HCO^+} = 3.3 \times 10^{-5} T^{-1}$	
		1.6 (10 keV)		
Midplane	Cosmic ray[g] and radionuclide[i]			
	$\zeta_R = 6.1 \times 10^{-18}$ s^{-1}			
Metal$^+$/gr	($r < 3$ AU)		$\alpha_{Na^+} = 1.4 \times 10^{-10} T^{-0.69}$	$< 10^{-12}$
HCO^+/gr	($3 < r < 60$ AU)		α_{gr} (see text)	$10^{-13,-12}$
$H_3^+ - D_3^+$	($r > 60$ AU)		$\alpha_{D_3^+} = 2.7 \times 10^{-8} T^{-0.5}$	$> 10^{-11}$

[a] Table originally from Bergin et al. (2007).

[b] Effective penetration depth of radiation (e.g., $\tau = 1$ surface).

[c] Recombination rates from UMIST database (Le Teuff et al., 2000), except for H_3^+, which is from McCall et al. (2003), and D_3^+, from Larsson et al. (1997).

[d] Ion fractions estimated from Semenov et al. (2004) and Sano et al. (2000). Unless noted, values are relevant for all radii.

[e] Estimated at 100 AU, assuming 10^{41} s^{-1} ionizing photons (Hollenbach et al., 2000) and $\sigma = 6.3 \times 10^{-18}$ cm^2 (H photoionization cross section at threshold). This is an overestimate because I assume that all ionizing photons are at the Lyman limit.

[f] Rate at the disk surface at 100 AU using the radiation field from Bergin et al. (2003).

[g] Taken from Semenov et al. (2004); $\zeta_{cr,0} = 5 \times 10^{-17}$ s^{-1}, and $\Sigma_1(r, z)$ is the surface density above the point with height z and radius r, with $\Sigma_2(r, z)$ the surface density below the same point; $\Sigma_{cr} = 96$ g cm^{-2}, as given above in the text (Umebayashi & Nakano, 1981).

[h] X-ray ionization formalism from Glassgold et al. (2000); $\zeta_{X,0} = 1.4 \times 10^{-10}$ s^{-1}, while $L_{29} = L_X/10^{29}$ erg s^{-1} is the X-ray luminosity, and J is an attenuation factor, $J = A\tau^{-a} e^{-B\tau^b}$, where $A = 0.800$, $a = 0.570$, $B = 1.821$, and $b = 0.287$ (for energies around 1 keV and solar abundances).

[i] ^{26}Al decay from Umebayashi & Nakano (1981). If ^{26}Al is not present, ^{40}K dominates, with $\zeta_R = 6.9 \times 10^{-23}$ s^{-1}.

in the diffuse gas (McCall et al., 2003; Indriolo et al., 2007). It is this rate that impinges on the surface of the disk. Umebayashi & Nakano (1981) explored penetration depth of ionizing protons as a function of gas column and found that cosmic rays penetrate to a depth of $\Sigma \sim 96\,\mathrm{g\,cm^{-2}}$. Table 3.4 compares this exponential-folding depth with that of other potential contributors to the ion fraction. Excluding radionuclides, cosmic rays represent the best mechanism to ionize and to power chemistry in the disk midplane. However, if present, even cosmic rays will have difficulty penetrating to the midplane in the inner several AUs of disks around solar-type stars. Another important question is whether ionizing cosmic rays are present in the inner tens of AUs at all. Within our own planetary system, the solar wind limits ionizing cosmic rays to beyond the planet-formation zone. Estimates of mass loss rates from young star winds significantly exceed the solar mass-loss rate (Dupree et al., 2005) and may similarly exclude high-energy nuclei. Cosmic rays will contribute to the chemistry for the outer disk, and the detection of ions potentially provides some confirmation (Ceccarelli & Dominik, 2005).

ULTRAVIOLET RADIATION. T Tauri stars are known to have excess UV fluxes much higher than their effective temperature, $T_{eff} \sim 3,000$ K (Herbig & Goodrich, 1986), which are generated, at least in part, by accretion (Calvet & Gullbring, 1998). Observations have suggested that the UV radiation field and X-ray ionization play a key role in the observed molecular emission (Willacy & Langer, 2000; Aikawa et al., 2002). UV radiation is important because the primary molecular photodissociation bands lie below 2,000 Å, and this radiation both dissociates molecules and can power a rich chemistry on the disk surface (see, for example, Tielens & Hollenbach, 1985).

In this context there are two photon fields to consider, the stellar radiation field and the external radiation field, which can be enhanced when compared with the standard interstellar radiation field (ISRF). It is common to place both the stellar and the interstellar UV radiation fields in the context of the ISRF, which is defined by the measurements of Habing (1968) and Draine (1978). If we use the measurements of Habing, the energy density of the ISRF is $1.6 \times 10^{-3}\,\mathrm{erg\,cm^{-2}\,s^{-1}}$, which is defined as $G_0 = 1$ (in this context the Draine field is $G_0 = 1.7$). As an example, the FUV flux below 2,000 Å for T Tau is $\sim 3 \times 10^{-13}\,\mathrm{erg\,cm^{-2}\,s^{-1}\,Å^{-1}}$. Sealing to 100 AU at 140 pc and integrating over 1,000 Å (the Lyman limit to 2,000 Å, where most molecular photoabsorption cross sections lie) give $23.5\,\mathrm{erg\,cm^{-2}\,s^{-1}}$, or $G_0 = 1.5 \times 10^4$ at 100 AU (scaling as $1/r^2$). T Tau has a particularly high mass-accretion rate of $dM/dt \sim 10^{-7}$ $M_\odot$/yr that is about an order of magnitude higher than typically observed. Most

T Tauri stars have lower accretion rates and weaker UV fields with values in the range of $G_0(100\ \mathrm{AU}) = 300\text{–}1,000$, according to observations from the *Fare Ultraviolet Spectroscopic Explorer* (*FUSE*) and the *Hubble Space Telescopes* (*HST*) (Bergin et al., 2004). Bergin et al. (2003) pointed out the importance of Ly α emission, which is present in the stellar radiation field but absent in the ISRF. In one star that is unobscured by interstellar absorption, TW Hya, Ly α contains as much as 85% of the stellar UV flux (Herczeg et al., 2002). Even in systems where Ly α is absent because of absorption and scattering by interstellar H I, Ly α–pumped lines of H_2 are detected, testifying to the presence of Ly α photons in the molecular gas (Herczeg et al., 2002; Bergin et al., 2004; Herczeg et al., 2006). A sample of the UV field of typical T Tauri stars is shown in Fig. 3.8.

Stellar photons irradiate the flared disk with a shallow angle of incidence, while external photons impinge on the disk at a more normal angle. Interstellar photons therefore have greater penetrating power (Willacy & Langer, 2000). van Zadelhoff et al. (2003) employed a two-dimensinal model of UV continuum photon transfer and demonstrated that scattering of stellar photons is key to understanding how UV radiation influences disk chemistry. Fig. 3.9 illustrates the geometry of various contributors to the radiation field on the disk surface. An analytical approximation of the transfer of the stellar field is provided by Bergin et al. (2003). In Taurus the average extinction toward T Tauri stars is $A_V \sim 1^m$ (Bergin et al., 2003); under these circumstances the stellar field will dominate over the ISRF ($G_0 = 1$) for much of the disk.

Using the parameterization of the stellar field with respect to the ISRF allows the use of photodissociation rates calculated on the basis of the strength

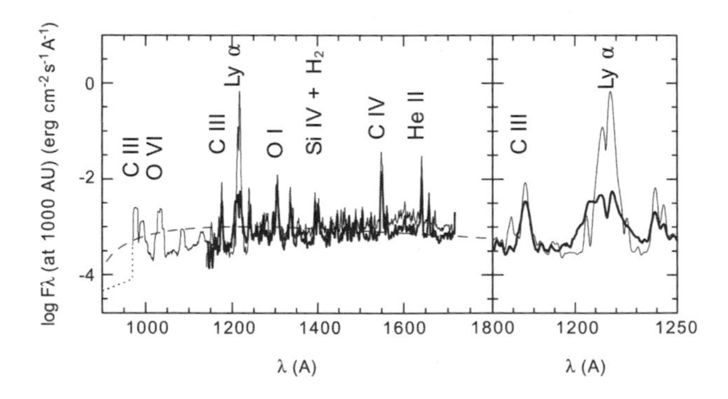

Fig. 3.8. UV spectra of T Tauri stars. Heavy solid lines and light solid lines represent the spectra of BP Tau and TW Hya, respectively. The spectrum of TW Hya is scaled by 3.5 to match the BP Tau continuum level. The long dashed line represents the interstellar radiation field of Draine (1978) scaled by a factor of 540. The region around the Ly α line is enlarged in the right panel. Taken from Bergin et al. (2003).

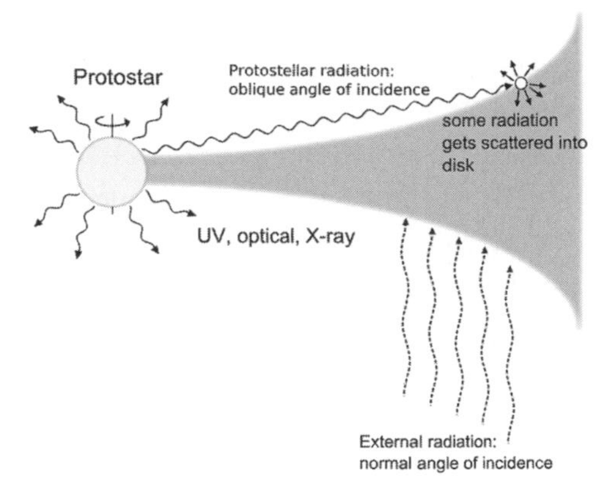

Fig. 3.9. Schematic showing the difference in angle of incidence between stellar and external radiation impinging on a flared disk. Radiation with a shallow angle of incidence will have less penetration into the disk. Scattering of stellar radiation increases its importance, particularly if the external radiation field suffers from any local extinction from a surrounding molecular cloud.

and shape of the IRSF, assuming albedo and scattering properties estimated for interstellar grains (van Dishoeck, 1988; Roberge et al., 1991). This correctly treats the direct attenuation of the field, but not the contribution from scattering, which dominates the deep surface (van Zadelhoff et al., 2003). The shape of the stellar field, which is primarily line emission, is also different from that of the ISRF (see Fig. 3.8). This is particularly acute for Ly α because some species are sensitive to dissociation by Ly α emission (e.g., H_2O, HCN), while others are not, such as CO and CN (Bergin et al., 2003; van Dishoeck et al., 2006). In this case, it is better to calculate the photodissociation and ionization rates directly from the flux and the cross sections. This can be done by using the observed UV field, or a proxy blackbody, in the following fashion:

$$k_{photo} = \int \frac{4\pi\lambda}{hc}\sigma(\lambda)J_{\lambda,0}(r)e^{-\tau_\lambda}d\lambda \ (\mathrm{s}^{-1}), \quad (48)$$

where $\sigma(\lambda)$ is the photodissociation cross section and $J_{\lambda,0}$ is the mean intensity of the radiation field at the surface ($F_\lambda = 4\pi J_\lambda = \int I_\lambda d\Omega$, where Ω is the solid angle of the source). The main source of opacity at these wavelengths is dust grains, and some assumptions must be made regarding the composition, optical properties, and size distribution (see § 4.3 for a more complete discussion of dust-grain absorption of radiation). van Dishoeck et al. (2006) provide photodissociation-rates for various blackbodies and also photodissociation cross sections coincident with Ly α. In addition, a Web site for cross sections is available at http://www.strw.leidenuniv.nl/ewine/photo.

It is important to note that the photodissociation of two key species, H_2 and CO, requires additional treatment. These species are dissociated via a line-mediated process, as opposed to a dissociation continuum typical for most other molecules. In this regard the optical depth of the lines depends on the total column. When the lines become opaque, molecules closer to the radiation source can shield molecules downstream in a process that is called self-shielding. In addition, H_2 molecules can partially shield some of the CO predissociation lines. Self-shielding rates are time and depth dependent. In general, approximations based on the interstellar UV field are adopted from Draine & Bertoldi (1996) for H_2, while for CO the shielding functions provided by van Dishoeck & Black (1988) and Lee et al. (1996) can be used. One caveat of these approximations is that the shielding functions are calculated by assuming that molecules predominantly reside in the ground state with a line profile appropriate for cold interstellar conditions. In regions with high temperatures, these approximations can potentially break down.

X-RAY IONIZATION. X-rays are a key source of ionization and heating of protoplanetary disks. The typical X-ray luminosity (based on protostars in Orion) is $L_x = 10^{28.5}$–10^{31} ergs s^{-1}, with a characteristic temperature of the X-ray-emitting plasma of $T_x \sim 1$–2 keV (e.g., Feigelson & Montmerle, 1999). X-ray photons interact directly with the atoms in the disk; i.e., the inclusion of the atom in a molecule does not affect the photoabsorption cross section. In particular, the opacity is dominated by inner-shell ionization of heavier elements. For atoms heaver than Li, the inner-shell ionization is followed by the Auger effect, which generates a number of primary electrons. Each primary electron (p) produces secondary electrons (s) by impact ionization of the gas, with $N_s \gg N_p$. Therefore, for the X-ray ionization rate, ζ_x,

$$\zeta_x = \zeta_p + \zeta_s = \zeta_s. \tag{49}$$

This is a well-known result and is similar to that seen for cosmic rays. In general, there are $N_s \sim 30$ atoms and molecules ionized per keV (Aikawa & Herbst, 2001).

The X-ray ionization rate can be derived by an integration of the X-ray flux, $F(E)$, and the cross section $\sigma_i(E)$ of element i (in this case hydrogen) over energy:

$$\zeta_{x,i} = N_s \int_{1\ \mathrm{keV}}^{30\ \mathrm{keV}} \sigma_i(E) F(E) dE. \tag{50}$$

Table 3.5: Coefficients for Fit to X-Ray Cross Section[a,b]

Energy Range (keV)	c_0	c_1	c_2
0.030–0.100	17.3	608.1	−2,150.0
0.100–0.284	34.6	267.9	−476.1
0.284–0.400	78.1	18.8	4.3
0.400–0.532	71.4	66.8	−51.4
0.532–0.707	95.5	145.8	−61.1
0.707–0.867	308.9	−380.6	294.0
0.867–1.303	120.6	169.3	−47.7
1.303–1.840	141.3	146.8	−31.5
1.840–2.471	202.7	104.7	−17.0
2.471–3.210	342.7	18.7	0.0
3.210–4.038	352.2	18.7	0.0
4.038–7.111	433.9	−2.4	0.75
7.111–8.331	629.0	30.9	0.0
8.331–10.000	701.2	25.2	0.0

[a] Fit and tabulation reprinted from Morrison & McCammon (1983).
[b] Cross section per hydrogen atom, $\sigma_T(E)$, is $(c_0 + c_1 E + c_2 E^2)E^{-3} \times 10^{-24}$ cm^2 (E in keV).

The X-ray flux can be estimated from the X-ray luminosity by assuming a characteristic temperature for the plasma (see Glassgold et al., 1997; for an example). The X-ray flux decays with depth into the disk, $F(E) = F_{X,0}e^{-\tau(E)}$, where $\tau(E) = N\sigma_T(E)$. X-ray photons are attenuated by photoabsorption at the atomic scale. In this regard it is customary to use the total X-ray cross section, $\sigma_T(E)$, computed using solar abundances (Morrison & McCammon, 1983). With this assumption the total hydrogen column is the only information required. Table 3.5 lists the fits to the X-ray cross section computed by Morrison & McCammon (1983). These assume that the heavy elements, which control the opacity, have solar abundances and are uniformly distributed in the disk.

It is well known that dust grains suspended in the disk atmosphere will sink to the midplane, reducing the dust-to-gas mass ratio in the upper disk atmosphere (Weidenschilling & Cuzzi, 1993). This will reduce the optical depth of the upper layers to both X-rays and UV radiation. Compared with UV transfer, dust coagulation will not reduce the X-ray opacity as dramatically unless grains grow to large sizes. Instead, the mass of heavy elements must be redistributed, as in the case of dust settling. More recent calculations account for these differences and provide tabulated values similar to that provided in Table 3.5 (Bethell & Bergin, 2010). In the extreme limit only H and He are present in the upper atmosphere, and the cross section at 1 keV is reduced by a factor of ~4.5 (i.e., the H and He absorption limit; Morrison & McCammon, 1983). In the case of nonsolar abundances or mass redistribution the cross section for each individual element can be used (Henke et al., 1993). X-ray ionization is crucial for the inner disk, where cosmic rays likely do not penetrate. X-ray photons also have a significantly greater penetration depth than UV photons and thus can power chemistry (and perhaps accretion; Gammie, 1996) on the disk surface. The characteristics of the X-ray-driven chemistry can extend beyond the simple ionization of H_2 because X-rays can produce short-lived, but reactive, doubly ionized molecules. For greater detail, the reader is referred to Langer (1978), Krolik & Kallman (1983), and Stäuber et al. (2005).

ACTIVE RADIONUCLIDES. In case the midplane of the disk is completely shielded from cosmic-ray radiation, radionuclides provide a baseline level of ionization. A primary issue in this regard is the overall abundance of various radionuclides and whether these elements are uniformly distributed throughout the system. Key elements in this regard are ^{26}Al and ^{60}Fe, which have short lifetimes [$\tau_{1/2}(^{26}\text{Al}) = 0.72$ Myr; $\tau_{1/2}(^{60}\text{Fe}) = 2.6$ Myr] and have been inferred to be present in the solar system (e.g., Wasserburg et al., 2006; Wadhwa et al., 2007). If present, they are a strong source of ionization for several million years. Table 3.2 provides an estimate of this ionization rate, which is dominated by ^{26}Al. This is sufficient to allow for an active ion-driven gas chemistry in regions where neutral molecules (excluding H_2 and HD) are not completely frozen on grains. This ionization rate will decay with time according to respective half-lives. If ^{60}Fe or ^{26}Al is not present, then this rate gets substantially reduced. Finocchi & Gail (1997) provide a short summary of the relevant ionization processes, including the most important radionuclides.

DISSOCIATIVE RECOMBINATION. A key to the question of the overall ionization fraction is the balance between the ionization rate and the recombination rate of atoms and molecules. In equilibrium, the ion fraction, $x_e = n_e/n$, can be expressed by $x_e = \sqrt{\zeta/(\alpha_r n)}$, where α_r is the electron recombination rate. The ion fraction therefore depends not only on the flux of ionizing agents, but also on the recombination rate of the most abundant ions. There are some key differences in this regard that are summarized in Table 3.4. For the midplane, a key question is whether molecular ions, metal ions, or grains are the dominant charge carriers, and this serves as a useful foil to discuss some differences in recombination.

In general, atomic species have longer recombination timescales than molecular ions. Thus the presence of abundant slowly recombining metals can have key consequences for magnetic-field coupling (Ilgner & Nelson, 2006). However, metal ions are not present in the dense star-forming cores that set the initial conditions for disk formation (Maret & Bergin, 2007), and they are likely not present in the disk gas. If grains are the dominant charge carrier, the recombination rate, α_{gr}, is the grain collisional timescale with a correction for long-distance Coulomb focusing: $\alpha_{gr} = \pi a_d^2 n_{gr} v(1 + e^2/ka_d T_{dust})$. At $T_{dust} = 20$ K, Draine & Sutin (1987) show that for molecular ions, grain recombination will dominate when $n_e/n < 10^{-7}(a_{min}/3\text{Å})^{-3/2}$. Grains can be positive or negative and can carry multiple charges: Sano et al. (2000) find that the total grain charge is typically negative, while the amount of charge is 1–2e^-, varying with radial and vertical distance. Florescu-Mitchell & Mitchell (2006) provide a summary of molecular ion recombination rates determined in the laboratory.

4.3. Grain Growth and Settling

The onset of grain evolution within a protoplanetary disk consists of collisional growth of submicron-sized particles into larger grains; the process continues until the larger grains decouple from the gas and settle to an increasingly dust-rich midplane (Nakagawa et al., 1981; Weidenschilling & Cuzzi, 1993).

Grain coagulation can alter the chemistry through the reduction in the total geometric cross section, lowering the adsorption rate and the Coulomb force for ion-electron grain recombination. Micron-sized grains couple to the smallest scales of turbulence (Weidenschilling & Cuzzi, 1993) and have a thermal, Brownian velocity distribution. Thus the timescale of grain-grain collisions is $\tau_{gr-gr} \propto a_d^{5/2}/(T_{dust}^{1/2}\xi n)$, where ξ is the gas-to-dust mass ratio and T_{dust} is the dust temperature (Aikawa et al., 1999). In this fashion grain coagulation proceeds faster at small radii where the temperatures and densities are higher. Aikawa et al. (1999) note that the longer timescale for adsorption on larger grains leaves more time for gas-phase reactions to drive toward a steady-state solution; this involves more carbon trapped in CO, as opposed to other more complex species.

Overall, the evolution of grains, by both coagulation and sedimentation, can be a controlling factor for the chemistry. As grains grow and settle to the midplane, the UV opacity, which is dominated by small grains, decreases, allowing greater penetration of ionizing/dissociating photons. As an example, in the coagulation models of Dullemond & Dominik (2004) the integrated vertical UV optical depth at 1 AU decreases over several orders of magnitude toward being optically thin over the entire column (see also Weidenschilling, 1997).

This can be understood by exploring the question of dust extinction of starlight from the perspective of the interstellar medium. In this case the amount of dust absorption is treated in terms of magnitudes of extinction, labeled A_λ.[5] At any given point in the disk (or in the ISM), the intensity can be characterized by the intensity (ergs s^{-1} cm^{-2} $Å^{-1}$ sr^{-1}) that impinges on the surface, $I_{\nu,0}$, modified by the dust absorption, $I_\nu = I_{\nu,0}\exp(-\tau_\lambda)$. The opacity, τ_λ, is given by $\tau_\lambda = N_{dust} Q_\lambda \sigma$, where N_{dust} is the total dust column along the line of sight, Q_λ is the extinction efficiency, and σ is the geometric cross section of a single grain. The wavelength-dependent extinction, placed in terms of magnitudes, is defined as follows:

$$A_\lambda = -2.5\log(I_\nu / I_{\nu,0}) \tag{51}$$

$$= 2.5\log(e)\tau_\lambda \tag{52}$$

$$= 1.086 N_{dust} Q_\lambda \sigma \tag{53}$$

$$= 1.086 \frac{n_{dust}}{n} N Q_\lambda \sigma. \tag{54}$$

The total dust column can be related to the gas column (N) via the gas-to-dust mass ratio, ξ, which is measured to be 1% for interstellar grains (i.e., the primordial condition):

$$\frac{\rho_{dust}}{\rho_{gas}} = \xi = \frac{n_d m_{gr}}{n \mu m_H}, \tag{55}$$

$$\frac{n_d}{n} = \frac{3}{4\pi} \frac{\xi \mu m_H}{\rho_{gr} a_{gr}^3}. \tag{56}$$

If we place the dust abundance (n_{dust}/n) into eq. (54) and explore the extinction at visible wavelengths where $Q_V \sim 1$, assuming a typical grain size of 0.1 μm, and $\rho_{gr} = 2$ g cm^{-3}, then

$$A_V = 1.086 \frac{3}{4} \frac{\xi \mu m_H}{\rho_{gr} a_{gr}} N Q_V \sim \left(\frac{\xi}{0.01}\right)\left(\frac{Q_V}{1}\right) 10^{-21} N. \tag{57}$$

Thus a total gas column of 10^{21} cm^{-2} provides 1 mag of extinction at visible wavelengths for a standard gas-to-dust mass ratio. As grains settle to the midplane, the gas remains suspended in the upper layers, and the gas-to-dust ratio decreases. This requires larger columns for the dust to absorb stellar radiation. In my example I have explored extinction in the visual range, while most molecules are dissociated by ultraviolet radiation. Dust extinction is larger in the UV (requiring a smaller gas column); however, the dependence on the gas-to-dust ratio and grain size remains.

5. Magnitudes in astronomy are defined such that a difference of 5 mag between two objects corresponds to a factor of 100 in the ratio of the fluxes. Thus an extinction of 1 mag corresponds to a flux decrease of $(100)^{1/5} \sim 2.512$.

As grains evolve, there will be a gradual shifting of the warm molecular layer deeper into the disk and eventually into the midplane (Jonkheid et al., 2004). Because the grain emissivity, density, and temperature will also change, the chemical and emission characteristics of this layer may be altered (Aikawa & Nomura, 2006). These effects are magnified in the inner disk, where there is evidence for significant grain evolution in a few systems (Furlan et al., 2005) and deeper penetration of energetic radiation (Bergin et al., 2004). A key question in this regard is the number of small grains (e.g., polycyolic aromatic hydrocarbons) present in the atmosphere of the disk during times when significant coagulation and settling have occurred.

4.4. Turbulence and Mixing

It is becoming increasingly clear that mixing played an important role in the chemistry of at least the solids in the nebula. This awariness is due to the detection of crystalline silicates in comets (e.g., Brownlee et al., 2006) and the presence of chondritic refractory inclusions in meteorites (MacPherson et al., 1988). Since the smallest solids (micron-sized grains) are tied to the gas, it is likely that the gas is also affected to some extent (perhaps in a dominant fashion) by turbulent mixing. The question of mixing and its relevance has a long history in the discussions of solar nebula chemistry (e.g., Prinn, 1993; and references therein).

With regard to the dynamic movement of gas within a protoplanetary disk and its chemical effects, a key question is whether the chemical timescale, τ_{chem}, is less than the relevant dynamic timescale, τ_{dyn}, in which case the chemistry will be in equilibrium and unaffected by the motion. If $\tau_{dyn} < \tau_{chem}$, then mixing will alter the anticipated composition. These two constraints are the equilibrium and disequilibrium regions, respectively outlined in Prinn (1993). What is somewhat different in our current perspective is the recognition of an active gas-phase chemistry on a photon-dominated surface (§ 5). This provides another potential mixing reservoir in the vertical direction, as opposed to the radial direction, which was the previous focus.

It is common to parameterize the transfer of angular momentum in terms of the turbulent viscosity, $\nu_t = \alpha c_s H_p$, where ν_t is the viscosity; c_s is the sound speed, H_p is the disk scale height, and α is a dimensionless parameter (Shakura & Sunyaev, 1973; Pringle, 1981). Hartmann et al. (1998) empirically constrained the α-parameter to be $\lesssim 10^{-2}$ for a sample of T Tauri disks.

A number of authors have begun to explore the question of including dynamics in the chemistry. A brief and incomplete list is provided by Bergin et al. (2007). Although the details differ, a common approach is to use mixing-length theory, where the transport is treated as a diffusive process (Allen et al., 1981; Xie et al., 1995), but see also Ilgner et al. (2004); Tscharnuter & Gail (2007). In mixing-length models the fluctuations of abundance of species i can be determined by the product of the abundance fluctuations in a given direction and the mixing length (l), $dx_i \sim -l dx_i/dz$, where z denotes the direction where a gradient in abundance exists. The net transport is then (Willacy et al., 2006)

$$\phi_i(\mathrm{cm}^{-2}\ \mathrm{s}^{-1}) = n_{\mathrm{H}_2}\langle v_t dx_i\rangle = -Dn_{\mathrm{H}_2}\frac{dx_i}{dz} = -Dn_i\left[\frac{1}{n_i}\frac{dn_i}{dz} - \frac{1}{n_{\mathrm{H}_2}}\frac{dn_{\mathrm{H}_2}}{dz}\right], \quad (58)$$

where D is the diffusion coefficient and v_t is the turbulent velocity. The diffusivity is related to the viscosity ν_t by $D = \langle v_t l\rangle = \nu_t = \alpha c_s H_p$. If we use this description, the chemical continuity equation can be written as

$$\frac{\partial n_i}{\partial t} + \frac{\partial \phi_i}{dz} = P_i - L_i, \quad (59)$$

where P_i and L_i are the chemical production and loss terms for species i.

The radial disk viscous timescale is $\tau_\nu = r^2/\nu$ or

$$\tau_\nu \sim 10^4 \mathrm{yr}\left(\frac{\alpha}{10^{-2}}\right)^{-1}\left(\frac{T_{gas}}{100\ \mathrm{K}}\right)^{-1}\left(\frac{r}{1\ \mathrm{AU}}\right)^{\frac{1}{2}}\left(\frac{M_*}{M_\odot}\right)^{\frac{1}{2}}.$$

The diffusivity, D or K, is not necessarily the same as the viscosity, ν_t (e.g., Stevenson, 1990), as given above. Moreover, it is not entirely clear that radial mixing will be the same as vertical mixing. In the case of disks, several groups have explored this question, with potential solutions ranging from ν_t/D below 1 to near 20 (Carballido et al., 2005; Johansen et al., 2006; Turner et al., 2006; Pavlyuchenkov & Dullemond, 2007). For completeness, when $\nu_t/D > 1$, turbulent mixing is much less efficient than angular momentum transport.

5. Current Understanding

In this section I will synthesize the various physical and chemical processes in a discussion of our evolving understanding of disk chemistry. One important issue is that because of its low binding energy to grain surfaces (Hollenbach & Salpeter, 1971), molecular hydrogen will remain in the gas throughout the nebula. Provided that sufficient levels of ionizing agents reach the midplane or the deep interior of the disk surface, the ion-molecule chemistry outlined earlier can proceed. Surface chemistry based on diffusing hydrogen atoms will be more difficult to initiate, but in warmer regions heavier radicals can potentially migrate and react. Thus both gas and grain surface chemistry are important.

For other molecules we expect large compositional gradients in the vertical and radial directions. The dominant effect is the sublimation/freeze-out as molecules transition from warm to colder regions in a medium with high densities ($n \sim 10^7$–10^{15} cm^{-3}) and therefore short collisional timescales between gas and solid grains. In large part the overall chemical structure follows the thermal structure and is schematically shown in Fig. 3.10. This plot illustrates that beyond ~40 AU, the disk can be divided into three vertical layers. CO, one of the most volatile and abundant species, controls the gas-phase chemistry in the outer disk and sets the radial and vertical boundary of these layers. The top of the disk is dominated by stellar UV and X-ray radiation, which leads to molecular photodissociation in a photon-dominated layer. This layer will have several transitions. H/H_2 is the first transition because molecular hydrogen is strongest at self-shielding. This is followed by C II/C I/CO and subsequently the oxygen/nitrogen pools, which require

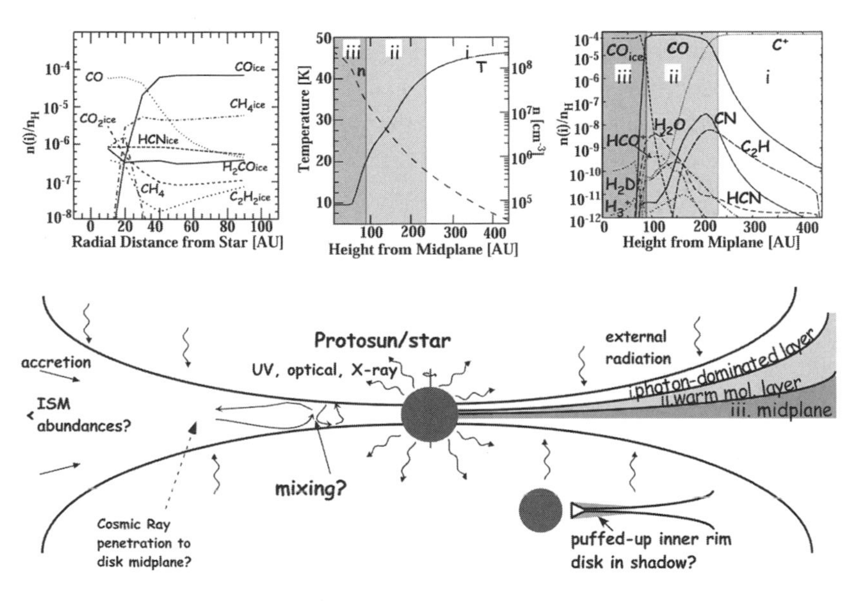

Fig. 3.10. Chemical structure of protoplanetary disks (taken from Bergin et al., 2007). Vertically the disk is schematically divided into three zones: a photon-dominated layer, a warm molecular layer, and a midplane freeze-out layer. The CO freeze-out layer disappears at $r \lesssim$ 30–60 AU as the midplane temperature increases inward. Various nonthermal inputs, cosmic rays, UV rays, and X-rays drive chemical reactions. Viscous accretion and turbulence will transport the disk material both vertically and radially. The upper panels show the radial and vertical distribution of molecular abundances from a typical disk model at the midplane (Aikawa et al., 1999) and $r \sim$ 300 AU (van Zadelhoff et al., 2003). A sample of the hydrogen density and dust temperature at the same distance (D'Alessio et al., 1999) is also provided.

dust shielding (e.g., O I/H_2O, N I/N_2 or NH_3; see Hollenbach & Tielens, 1999).

Inside this photon-dominated layer, the grain temperatures are warm enough for CO to exist in the gaseous state ($T_{dust} > 20$ K), but water will remain as an ice. Thus the C/O ratio ~ 1, leading to an active carbon-based chemistry in a vertical zone labeled the warm molecular layer. The ion-molecule chemistry of this layer is powered predominantly by X-rays and cosmic rays (if present). Reactions between CO and molecular ions (predominantly H_3^+) transfer a small fraction of this carbon into other simple and complex species (Aikawa et al., 1997). If CH_4 is present on grain surfaces and evaporates into the gas, it will also be a key precursor to the creation of larger hydrocarbons and carbon chains. If $T_{dust} < T_{sub}$ for any product of this gas-phase chemistry, then that molecule freezes onto grains. In a large sense the gas phase is acting as an engine for building complexity within molecular ices (Tscharnuter & Gail, 2007). As material advects inward, the region of the disk with $T_{dust} > 20$ K increases in depth, and, in addition, the more tightly bound complex species created earlier (e.g., HCN, C_2H_2, C_3H_4) will eventually evaporate. An additional issue that is likely important is the creation of frozen radicals via photodissociation on the grain surface. If the grain is warm enough, these radicals can migrate and react on grain surfaces. This can lead to a large increase in molecular complexity (Garrod et al., 2008) and perhaps contribute to organics detected in meteorites (see, e.g., Bernstein et al., 2002; and references therein).

Below the warm molecular layer lies the dense ($n \gg 10^8$ cm^{-3}), cold ($T \leq 20$ K) midplane where molecules are frozen on grain surfaces (Willacy & Langer, 2000; Aikawa & Herbst, 2001). This is the case for much of the outer disk (R $\gtrsim$ 20–40 AU) beyond the radial snow lines. If ionizing agents are still present and $T_{gas} < 30$ K, then the transition to the midplane and the midplane itself are the main layers for deuterium fractionation in the disk (Aikawa & Herbst, 2001). Gas-phase deuterium fractionation requires that heavy molecules be present in the gas. In the limit of total heavy-element (C, O, N) freeze-out, the chemistry will be reduced to the following sequence: $H_3^+ \rightarrow H_2D^+ \rightarrow D_2H^+ \rightarrow D_3^+$ (Ceccarelli & Dominik, 2005).

Fig. 3.11 shows the CO abundance structure in a fiducial disk model to illustrate some of these effects more directly. The three panels refer to different models of dust settling, with lower values of the parameter ϵ (defined in the figure caption) referring to a greater degree of dust settling. The top panel is for a disk with no settling. As can be seen, the chemical structure divides into three layers, as shown in Fig. 3.10. In addition, the interior evaporation zone

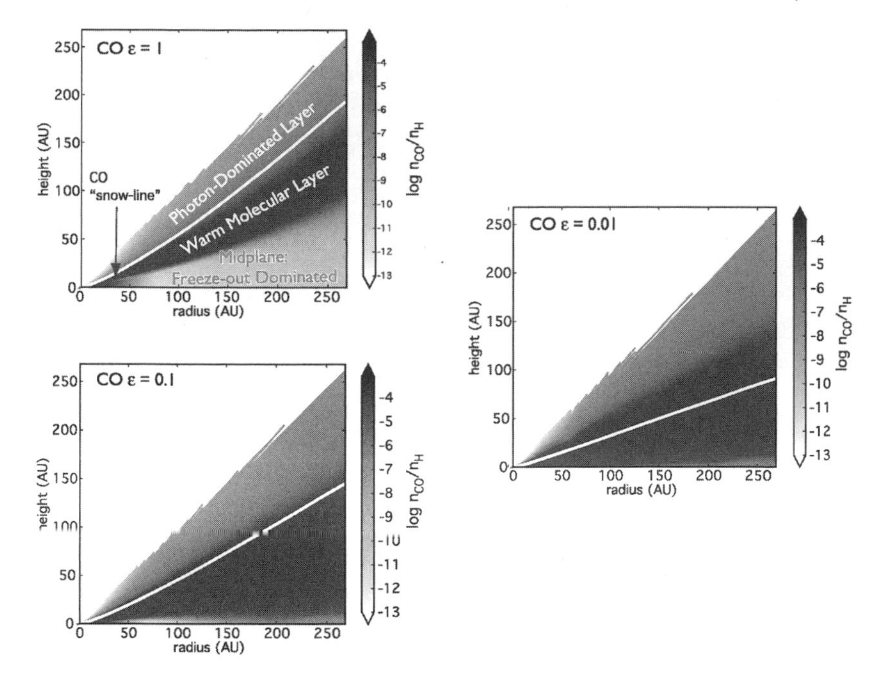

Fig. 3.11. Plot of the abundance of carbon monoxide as a function of radial and vertical height in three disk models with variable gas-to-dust ratios. The gas-to-dust ratio is parameterized by the parameter ϵ, which is the gas-to-dust ratio in the upper layers relative to the gas-to-dust ratio in the midplane (see Calvet and D'Alessio, this volume). Lower values of ϵ are representative of the effects of the settling of dust grains. Also shown is the $\tau = 1$ surface for stellar photons, which appears deeper in the disk as grains settle to the midplane. The top panel shows some of the basic disk chemical structure. Figure from Fogel et al. 2010.

(inside the snow line for CO) can be resolved. As grains settle into the disk, the radiation penetrates more deeply, as seen by the line delineating the $\tau = 1$ surface and also in the chemical structure, which moves the warm layer to lower depths (Jonkheid et al., 2004).

The case of water deserves special mention. Beyond the snow line at ~ 1–5 AU, water will exist mostly in the form of water ice (see Ciesla & Cuzzi, 2006; S. S. Davis, 2007). Dust grains suspended in the atmosphere will typically not be heated above the sublimation temperature (Fig. 3.7), and water will, for the most part, remain as ice. However, UV photons can photodesorb water ice from grain surfaces, producing a small layer where water exists with moderate abundance ($x_{H_2O,max} \sim 10^{-7}$, relative to H_2; Dominik et al., 2005). The water vapor abundance peaks at the layer where the rate of photodesorption is balanced by photodissociation. The depth where the abundance reaches this maximum value depends on the strength of the UV field, the local density, and grain opacity (Hollenbach et al., 2008). At lower depths photodissociation will destroy any desorbed H_2O molecules and erode the mantle. At greater depths the lack of UV photons leaves the ice mantle intact. Inside the snow line the water ice can evaporate from the grains and can also be produced in the gas via the high-temperature chemistry discussed in § 3.2. There is also a strong possibility of rapid movement of icy solids from the outer nebula that can seed additional water and other ices inside the snow line (Ciesla & Cuzzi, 2006).

Radially there exist large gradients in the gas/solid ice ratio of molecular species in the midplane. The snow line is species specific in the sense that CO should be present in the gas phase at greater radii than water (see Fig. 3.7). This may not be a continuous transition, with various species sublimating according to their molecular properties. Rather, some molecules may be frozen in the water lattice (possibly clathrates in denser regions) and may evaporate along with water, as seen in laboratory experiments (Sandford & Allamandola, 1993; Collings et al., 2004; Viti et al., 2004). In the dense interior a wide range of high-temperature kinetic reactions are likely active. These reactions can produce observed species, such as HCN and C_2H_2 (Gail, 2002; Agúndez et al., 2008), and perhaps even greater chemical complexity. However, the eventual products are tempered by the destructive influence of photons. As noted in § 3.3, thermodynamic equilibrium can potentially be reached in the midplane of the inner few AUs and in giant-planet subnebulae, with the removal of some elements from the gas as solids condense. However, the surface layers in the inner disk are dominated by the short-timescale effects of stellar radiation.

All these processes must be viewed in light of the likelihood of dynamic mixing both radially and vertically. Models including mixing generally show that the vertical structure illustrated in Fig. 3.10 is preserved with some widening of the warm molecular layer (Willacy et al., 2006). Beyond the inward advection of material, the outward diffusion of hot gas from the inner nebula can bring gas to cold layers where molecules freeze onto grain surfaces (Tscharnuter & Gail, 2007). Moreover, the grains are evolving via coagulation and settling, and the gas gradually becomes transparent to destructive radiation. Grain evolution ultimately results in the formation of planetesimals and planets (likely at different timescales for giant and terrestrial worlds), and the gas chemistry will continue until the gas disk dissipates.

6. Conclusion

We are approaching an age where studies of extrasolar systems can be informed by, and inform, studies of the chemical composition of bodies in our solar system. Some major gaps remain in our understanding that will benefit from this closer cooperation. These include the following: (1) What is the relative importance of thermodynamic equilibrium and kinetics in the inner disk, and how does this inform the meteoritic inventory? (2) Inclusive of presolar grains, does any material remain pristine and chemically unaltered from its origin in the parent molecular cloud? It is likely that the deuterium enrichments ultimately originate in the cold prestellar stages, but are these components dissociated and reassembled into different forms? (3) How deep inside the disk (radially) do cosmic rays penetrate, and how might this influence dynamic evolution? (4) How extensive is the complex chemistry in the inner disk, and how are these organics incorporated into planetesimals? (5) Diffusive mixing can be important, but how much and for how long is it active? These are just a sample of the many questions that remain. In this chapter I have outlined some of the basic physical and chemical processes that have laid a foundation for much of our current understanding. I hope that this chapter and this book inform the next generation of researchers in order to untangle long-standing questions regarding the initial conditions, chemistry, and dynamics of planet formation, the origin of cometary ices, and, ultimately, a greater understanding of the organic content of gas/solid reservoirs that produced life at least once in the Galaxy.

References

Adams, F. C., Lada, C. J., & Shu, F. H. 1987, *Astrophys. J.*, 312, 788.
Agúndez, M., Cernicharo, J., & Goicoechea, J. R. 2008, *Astron. & Astrophys.*, 483, 831.
A'Hearn, M. F., Millis, R. L., Schleicher, D. G., Osip, D. J., & Birch, P. V. 1995, *Icarus*, 118, 223.
Aikawa, Y., & Herbst, E. 2001, *Astron. & Astrophys.*, 371, 1107.
Aikawa, Y., & Nomura, H. 2006, *Astrophys. J.*, 642, 1152.
Aikawa, Y., Umebayashi, T., Nakano, T., & Miyama, S. M. 1997, *Astrophys. J. Letters*, 486, L51+.
———. 1999, *Astrophys. J.*, 519, 705.
Aikawa, Y., van Zadelhoff, G. J., van Dishoeck, E. F., & Herbst, E. 2002, *Astron. & Astrophys.*, 386, 622.
Alexander, C. M. O., Boss, A. P., & Carlson, R. W. 2001, *Science*, 293, 64.
Allen, M., & Robinson, G. W. 1977, *Astrophys. J.*, 212, 396.

Allen, M., Yung, Y. L., & Waters, J. W. 1981, *J. Geophys. Res*, 86, 3617.

Amelin, Y., Krot, A. N., Hutcheon, I. D., & Ulyanov, A. A. 2002, *Science*, 297, 1678.

André, P., Ward-Thompson, D., & Barsony, M. 1993, *Astrophys. J.*, 406, 122.

Bary, J. S., Weintraub, D. A., & Kastner, J. H. 2003, *Astrophys. J.*, 586, 1136.

Bary, J. S., Weintraub, D. A., Shukla, S. J., Leisenring, J. M., & Kastner, J. H. 2008, *Astrophys. J.*, 678, 1088.

Bauer, I., Finocchi, F., Duschl, W. J., Gail, H.-P., & Schloeder, J. P. 1997, *Astron. & Astrophys.*, 317, 273.

Beckwith, S. V. W., Sargent, A. I., Chini, R. S., & Guesten, R. 1990, *Astron. J.*, 99, 924.

Bergin, E., Calvet, N., D'Alessio, P., & Herczeg, G. J. 2003, *Astrophys. J. Letters*, 591, L159.

Bergin, E., Calvet, N., Sitko, M. L., Abgrall, H., D'Alessio, P., Herczeg, G. J., Roueff, E., Qi, C., Lynch, D. K., Russell, R. W., Brafford, S. M., & Perry, R. B. 2004, *Astrophys. J. Letters*, 614, L133.

Bergin, E. A., Aikawa, Y., Blake, G. A., & van Dishoeck, E. F. 2007, in *Protostars and Planets* V, ed. B. Reipurth, D. Jewitt, & K. Keil, (Tucson: University of Arizona Press), 751.

Bergin, E. A., Goldsmith, P. F., Snell, R. L., & Ungerechts, H. 1994, *Astrophys. J.*, 431, 674.

Bergin, E. A., Langer, W. D., & Goldsmith, P. F. 1995, *Astrophys. J.*, 441, 222.

Bergin, E. A., & Tafalla, M. 2007, *Ann. Rev. Astron. Astrophys.*, 45, 339.

Bergin, E. A., et al. 2000, *Astrophys. J. Letters*, 539, L129.

Bernes, C. 1979, *Astron. & Astrophys.*, 73, 67.

Bernstein, M. P., Elsila, J. E., Dworkin, J. P., Sandford, S. A., Allamandola, L. J., & Zare, R. N. 2002, *Astrophys. J.*, 576, 1115.

Bethell, T. J., & Bergin E. A. 2010, *Astrophys. J.*, submitted.

Bisschop, S. E., Fraser, H. J., Öberg, K. I., van Dishoeck, E. F., & Schlemmer, S. 2006, *Astron. & Astrophys.*, 449, 1297.

Bitner, M. A., Richter, M. J., Lacy, J. H., Greathouse, T. K., Jaffe, D. T., & Blake, G. A. 2007, *Astrophys. J. Letters*, 661, L69.

Bockelée-Morvan, D., Crovisier, J., Mumma, M. J., & Weaver, H. A. 2004, The Composition of Cometary Volatiles. In *Comets II*, ed. H. G. Keller, H. A. Weaver, & M. C. Festou (Tucson: University of Arizona Press), 391–423.

Bockelée-Morvan, D., et al. 2000, *Astron. & Astrophys.*, 353, 1101.

Bodenheimer, P. 1965, *Astrophys. J.*, 142, 451.

Boland, W., & de Jong, T. 1982, *Astrophys. J.*, 261, 110.

Bonev, B. P., Mumma, M. J., Villanueva, G. L., Disanti, M. A., Ellis, R. S., Magee-Sauer, K., & Dello Russo, N. 2007, *Astrophys. J. Letters*, 661, L97.

Born, M., & Oppenheimer, R. 1927, *Annalen der Physik*, 389, 457.

Bringa, E. M., & Johnson, R. E. 2004, *Astrophys. J.*, 603, 159.

Brittain, S. D., Simon, T., Najita, J. R., & Rettig, T. W. 2007, *Astrophys. J.*, 659, 685.

Brown, W. A., & Bolina, A. S. 2007, *MNRAS*, 374, 1006.

Brownlee, D., et al. 2006, *Science*, 314, 1711.

Buch, V., & Zhang, Q. 1991, *Astrophys. J.*, 379, 647.

Buffett, B. A. 2000, *Annual Review of Earth and Planetary Sciences*, 28, 477.

Burke, J. R., & Hollenbach, D. J. 1983, *Astrophys. J.*, 265, 223.

Burrows, A., & Sharp, C. M. 1999, *Astrophys. J.*, 512, 843.

Calvet, N., & Gullbring, E. 1998, *Astrophys. J.*, 509, 802.

Calvet, N., Patino, A., Magris, G. C., & D'Alessio, P. 1991, *Astrophys. J.*, 380, 617.

Carballido, A., Stone, J. M., & Pringle, J. E. 2005, *MNRAS*, 358, 1055.

Carr, J. S., & Najita, J. R. 2008, *Science*, 319, 1504.

Carr, J. S., Tokunaga, A. T., Najita, J., Shu, F. H., & Glassgold, A. E. 1993, *Astrophys. J. Letters*, 411, L37.

Cazaux, S., & Tielens, A. G. G. M. 2004, *Astrophys. J.*, 604, 222.

Ceccarelli, C., Caselli, P., Herbst, E., Tielens, A. G. G. M., & Caux, E. 2007, in *Protostars and Planets* V, ed. B. Reipurth, D. Jewitt, & K. Keil, 47–62.

Ceccarelli, C., & Dominik, C. 2005, *Astron. & Astrophys.*, 440, 583.

Ceccarelli, C., Dominik, C., Lefloch, B., Caselli, P., & Caux, E. 2004, *Astrophys. J. Letters*, 607, L51.

Chase, M. W., ed. 1998, *JANAF Thermochemical Tables*, 4th ed., Journal of Physical and Chemical Reference Data, Monograph, No. 9 (Woodbury: AIP).

Chiang, E. I., & Goldreich, P. 1997, *Astrophys. J.*, 490, 368.

Ciesla, F. J., & Charnley, S. B. 2006, The Physics and Chemistry of Nebular Evolution (*Meteorites and the Early Solar System* II), 209–230.

Ciesla, F. J., & Cuzzi, J. N. 2006, *Icarus*, 181, 178.

Cieza, L., Padgett, D. L., Stapelfeldt, K. R., Augereau, J.-C., Harvey, P., Evans, N. J., II, Merín, B., Koerner, D., Sargent, A., van Dishoeck, E. F., Allen, L., Blake, G., Brooke, T., Chapman, N., Huard, T., Lai, S.-P., Mundy, L., Myers, P. C., Spiesman, W., & Wahhaj, Z. 2007, *Astrophys. J.*, 667, 308.

Cieza, L. A. 2008, in Astronomical Society of the Pacific Conference Series, Vol. 393, "New Horizons in Astronomy. Frank N. Bash Symposium 2007," ed. A. Frebel, J. R. Maund, J. Shen, & M. H. Siegel, 35.

Clayton, R. N. 1993, *Annual Review of Earth and Planetary Sciences*, 21, 115.

Collings, M. P., Anderson, M. A., Chen, R., Dever, J. W., Viti, S., Williams, D. A., & McCoustra, M. R. S. 2004, *MNRAS*, 354, 1133.

Collings, M. P., Dever, J. W., Fraser, H. J., & McCoustra, M. R. S. 2003, *Astrophys. Space Sci.*, 285, 633.

Cowley, C., Wiese, W. L., Fuhr, J., & Kuznetsova, L. A. 2000, Spectra (*Allen's Astrophysical Quantities*), 53.

Cowley, C. R. 1995, *An Introduction to Cosmochemistry* (Cambridge; New York: Cambridge University Press).

D'Alessio, P., Calvet, N., & Hartmann, L. 2001, *Astrophys. J.*, 553, 321.

D'Alessio, P., Calvet, N., Hartmann, L., Franco-Hernández, R., & Servín, H. 2006, *Astrophys. J.*, 638, 314.

D'Alessio, P., Calvet, N., Hartmann, L., Lizano, S., & Cantó, J. 1999, *Astrophys. J.*, 527, 893.

D'Alessio, P., Calvet, N., & Woolum, D. S. 2005, in Astronomical Society of the Pacific Conference Series, Vol. 341, *Chondrites and the Protoplanetary Disk*, ed. A. N. Krot, E. R. D. Scott, & B. Reipurth, 353.

D'Alessio, P., Cantó, J., Calvet, N., & Lizano, S. 1998, *Astrophys. J.*, 500, 411.

Dalgarno, A. 2006, *Proceedings of the National Academy of Science*, 103, 12269.

Dauphas, N. 2003, *Icarus*, 165, 326.

Davis, A. M. 2006, Volatile Evolution and Loss (*Meteorites and the Early Solar System* II), 295–307.

Davis, A. M., Holland, H. D., & Turekian, K. K. 2005, *Treatise on Geochemistry, Meteorites, Comets, and Planets*. Vol. 1 (Amsterdari: Elsevier B. V.)

Davis, A. M., & Richter, F. M. 2004, *Condensation and Evaporation of Solar System Materials* (Oxford: Elsevier), 407–427.

Davis, S. S. 2005, *Astrophys. J. Letters*, 627, L153.
———. 2007, *Astrophys. J.*, 660, 1580.
Devlin, J. P. 2001, *J. Geophys. Res*, 106, 33333.
di Francesco, J., Evans, N. J., II, Caselli, P., Myers, P. C., Shirley, Y., Aikawa, Y., & Tafalla, M. 2007, in *Protostars and Planets* V, ed. B. Reipurth, D. Jewitt, & K. Keil (Tucson: University of Arizona Press), 17–32.
Disanti, M. A., & Mumma, M. J. 2008, *Space Science Reviews*, 138, 127.
Dominik, C., Ceccarelli, C., Hollenbach, D., & Kaufman, M. 2005, *Astrophys. J. Letters*, 635, L85.
Dones, L., Weissman, P. R., Levison, H. F., & Duncan, M. J. 2004, in Astronomical Society of the Pacific Conference Series, Vol. 323, *Star Formation in the Interstellar Medium: In Honor of David Hollenbach*, ed. D. Johnstone, F. C. Adams, D. N. C. Lin, D. A. Neufeld, & E. C. Ostriker, 371.
Draine, B. T. 1978, *Astrophys. J. Suppl.*, 36, 595.
Draine, B. T., & Bertoldi, F. 1996, *Astrophys. J.*, 468, 269.
Draine, B. T., & Sutin, B. 1987, *Astrophys. J.*, 320, 803.
Dullemond, C. P., & Dominik, C. 2004, *Astron. & Astrophys.*, 421, 1075.
Dupree, A. K., Brickhouse, N. S., Smith, G. H., & Strader, J. 2005, *Astrophys. J. Letters*, 625, L131.
Dutrey, A., Guilloteau, S., Duvert, G., Prato, L., Simon, M., Schuster, K., & Menard, F. 1996, *Astron. & Astrophys.*, 309, 493.
Dutrey, A., Guilloteau, S., & Guelin, M. 1997, *Astron. & Astrophys.*, 317, L55.
Dutrey, A., Guilloteau, S., & Ho, P. 2007a, in Protostars and Planets V, ed. B. Reipurth, D. Jewitt, & K. Keil, 495–506.
Dutrey, A., Henning, T., Guilloteau, S., Semenov, D., Piétu, V., Schreyer, K., Bacmann, A., Launhardt, R., Pety, J., & Gueth, F. 2007b, *Astron. & Astrophys.*, 464, 615.
Dwek, E., & Smith, R. K. 1996, *Astrophys. J.*, 459, 686.
Ehrenfreund, P., Charnley, S. B., & Wooden, D. 2004, From Interstellar Material to Comet Particles and Molecules. In *Comets II*, ed. H. G. Keller, H. A. Weaver, & M. C. Festou (Tucson: University of Arizona Press), 115–133.
Elitzur, M., & de Jong, T. 1978, *Astron. & Astrophys.*, 67, 323.
Espaillat, C., Calvet, N., D'Alessio, P., Bergin, E., Hartmann, L., Watson, D., Furlan, E., Najita, J., Forrest, W., McClure, M., Sargent, B., Bohac, C., & Harrold, S. T. 2007, *Astrophys. J. Letters*, 664, L111.
Evans, N. J. 1999, *Ann. Rev. Astron. Astrophys.*, 37, 311.
Fegley, B. J. 2000, Space Science Reviews, 92, 177.
Fegley, B. J., & Prinn, R. G. 1989, in *The Formation and Evolution of Planetary Systems*, ed. H. A. Weaver & L. Danly, 171–205.
Feigelson, E. D., & Montmerle, T. 1999, *Ann. Rev. Astron. Astrophys.*, 37, 363.
Fernandez, J. A. 1997, *Icarus*, 129, 106.
Finocchi, F., & Gail, H.-P. 1997, *Astron. & Astrophys.*, 327, 825.
Florescu-Mitchell, A. I., & Mitchell, J. B. A. 2006, *Phys. Rep.*, 430, 277.
Floss, C., Stadermann, F. J., Bradley, J., Dai, Z. R., Bajt, S., & Graham, G. 2004, *Science*, 303, 1355.
Flower, D. 2003, *Molecular Collisions in the Interstellar Medium* (Cambridge, UK: Cambridge University Press).

Fogel, J. K. J., Bethell, T. J., Bergin, E. A., Calvet, N. & Semenov, D. 2010, *Astrophys. J.*, submitted.

Fraser, H. J., Collings, M. P., McCoustra, M. R. S., & Williams, D. A. 2001, *MNRAS*, 327, 1165.

Furlan, E., et al. 2005, *Astrophys. J. Letters*, 628, L65.

Gail, H.-P. 2002, *Astron. & Astrophys.*, 390, 253.

Gammie, C. F. 1996, *Astrophys. J.*, 457, 355.

Garrod, R. T., Widicus Weaver, S. L., & Herbst, E. 2008, *ArXiv e-prints*, 803.

Geiss, J., & Gloeckler, G. 1998, *Space Science Reviews*, 84, 239.

Geiss, J., & Reeves, H. 1981, *Astron. & Astrophys.*, 93, 189.

Gerlich, D., Herbst, E., & Roueff, E. 2002, *Plan. Space Sci.*, 50, 1275.

Glassgold, A. E., Feigelson, E. D., & Montmerle, T. 2000, *Protostars and Planets* IV, 429.

Glassgold, A. E., Najita, J., & Igea, J. 1997, *Astrophys. J.*, 480, 344.

Greaves, J. S. 2005, *MNRAS*, 364, L47.

Grossman, L. 1972, *Geochim. Cosmochem. Acta*, 36, 597.

Guillot, T. 1999a, *Plan. Space Sci.*, 47, 1183.

———. 1999b, *Science*, 286, 72.

Habing, H. J. 1968, *Bull. Astron. Inst. Netherlands*, 19, 421.

Haisch, K. E., Lada, E. A., & Lada, C. J. 2001, *Astrophys. J. Letters*, 553, L153.

Hartmann, L., Calvet, N., Gullbring, E., & D'Alessio, P. 1998, *Astrophys. J.*, 495, 385.

Hasegawa, T. I., Herbst, E., & Leung, C. M. 1992, *Astrophys. J. Suppl.*, 82, 167.

Hayashi, C. 1981, Progress of Theoretical Physics Supplement, 70, 35.

Henke, B. L., Gullikson, E. M., & Davis, J. C. 1993, *Atomic Data and Nuclear Data Tables*, 54, 181.

Henning, T., & Semenov, D. 2008, *ArXiv e-prints*, 805.

Herbig, G. H., & Goodrich, R. W. 1986, *Astrophys. J.*, 309, 294.

Herbst, E. 1995, *Ann. Rev. Phys. Chem.*, 46, 27.

Herbst, E., Chang, Q., & Cuppen, H. M. 2005, *Journal of Physics Conference Series*, 6, 18.

Herbst, E. & Cuppen, H. M. 2006, *Proceedings of the National Academy of Science*, 103, 12257.

Herbst, E., & Klemperer, W. 1973, *Astrophys. J.*, 185, 505.

Herczeg, G. J., Linsky, J. L., Valenti, J. A., Johns-Krull, C. M., & Wood, B. E. 2002, *Astrophys. J.*, 572, 310.

Herczeg, G. J., Linsky, J. L., Walter, F. M., Gahm, G. F., & Johns-Krull, C. M. 2006, *Astrophys. J. Suppl.*, 165, 256.

Herczeg, G. J., Najita, J. R., Hillenbrand, L. A., & Pascucci, I. 2007, *Astrophys. J.*, 670, 509.

Herzberg, G. 1950, *Molecular Spectra and Molecular Structure*, Vol. 1: *Spectra of Diatomic Molecules* (New York: Van Nostrand Reinhold, 1950, 2nd ed.)

Ho, P. T. P., & Townes, C. H. 1983, *Ann. Rev. Astron. Astrophys.*, 21, 239.

Hollenbach, D., & Salpeter, E. E. 1971, *Astrophys. J.*, 163, 155.

Hollenbach, D. J., Kaufman, M. J., Bergin, E. A., & Melnick, G. J. 2008, *Astrophys. J.*, 690, 1497–1521.

Hollenbach, D. J., & Tielens, A. G. G. M. 1999, *Reviews of Modern Physics*, 71, 173.

Hollenbach, D. J., Yorke, H. W., & Johnstone, D. 2000, in *Protostars and Planets* IV, 401.

Huss, G. R., & Lewis, R. S. 1995, *Geochim. Cosmochem. Acta*, 59, 115.

Ilgner, M., Henning, T., Markwick, A. J., & Millar, T. J. 2004, *Astron. & Astrophys.*, 415, 643.

Ilgner, M., & Nelson, R. P. 2006, *Astron. & Astrophys.*, 445, 223.

Indriolo, N., Geballe, T. R., Oka, T., & McCall, B. J. 2007, *Astrophys. J.*, 671, 1736.

Jensen, M. J., Bilodeau, R. C., Safvan, C. P., Seiersen, K., Andersen, L. H., Pedersen, H. B., & Heber, O. 2000, *Astrophys. J.*, 543, 764.

Johansen, A., Klahr, H., & Mee, A. J. 2006, *MNRAS*, 370, L71.

Jonkheid, B., Faas, F. G. A., van Zadelhoff, G.-J., & van Dishoeck, E. F. 2004, *Astron. & Astrophys.*, 428, 511.

Kastner, J. H., Zuckerman, B., Weintraub, D. A., & Forveille, T. 1997, *Science*, 277, 67.

Kawakita, H., Dello Russo, N., Furusho, R., Fuse, T., Watanabe, J.-I., Boice, D. C., Sadakane, K., Arimoto, N., Ohkubo, M., & Ohnishi, T. 2006, *Astrophys. J.*, 643, 1337.

Kenyon, S. J., & Hartmann, L. 1987, *Astrophys. J.*, 323, 714.

Kittel, C. 1976, *Introduction to Solid State Physics* (New York: Wiley, 1976, 5th ed.).

Koerner, D. W., & Sargent, A. I. 1995, *Astron. J.*, 109, 2138.

Kress, M. E., & Tielens, A. G. G. M. 2001, *Meteoritics and Planetary Science*, 36, 75.

Krolik, J. H., & Kallman, T. R. 1983, *Astrophys. J.*, 267, 610.

Krot, A. N., Keil, K., Goodrich, C. A., Scott, E. R. D., & Weisberg, M. K. 2005, Classification of Meteorites (*Meteorites, Comets, and Planets: Treatise on Geochemistry*, Vol. 1), 83.

Lahuis, F., van Dishoeck, E. F., Blake, G. A., Evans, N. J., II, Kessler-Silacci, J. E., & Pontoppidan, K. M. 2007, *Astrophys. J.*, 665, 492.

Lahuis, F., van Dishoeck, E. F., Boogert, A. C. A., Pontoppidan, K. M., Blake, G. A., Dullemond, C. P., Evans, N. J., II, Hogerheijde, M. R., Jørgensen, J. K., Kessler-Silacci, J. E., & Knez, C. 2006, *Astrophys. J. Letters*, 636, L145.

Langer, W. D. 1978, *Astrophys. J.*, 225, 860.

Larsson, M., Danared, H., Larson, Å., Le Padellec, A., Peterson, J. R., Rosén, S., Semaniak, J., & Strömholm, C. 1997, *Physical Review Letters*, 79, 395.

Lauretta, D. S., & McSween, Jr., H. Y. 2006, *Meteorites and the Early Solar System* II (Tucson: University of Arizona Press)

Le Teuff, Y. H., Millar, T. J., & Markwick, A. J. 2000, *Astron. & Astrophys.* Suppl., 146, 157.

Lecluse, C., & Robert, F. 1992, *Meteoritics*, 27, 248.

Lee, H.-H., Herbst, E., Pineau des Forets, G., Roueff, E., & Le Bourlot, J. 1996, *Astron. & Astrophys.*, 311, 690.

Léger, A., Jura, M., & Omont, A. 1985, *Astron. & Astrophys.*, 144, 147.

Lellouch, E. et al. 1996, *Bulletin of the American Astronomical Society*, 28, 1148.

Lewis, J. S. 1974, *Science*, 186, 440.

———. 1990, Putting It All Together (*The New Solar System*), 281.

Lewis, J. S., & Prinn, R. G. 1980, *Astrophys. J.*, 238, 357.

Limbach, H. H., Buntkowsky, G., Matthes, J., Grundemann, S., Pery, T., Walaszek, B., & Chaudret, B. 2006, *ChemPhysChem*, 7, 551.

Linsky, J. L., Draine, B. T., Moos, H. W., Jenkins, E. B., Wood, B. E., Oliveira, C., Blair, W. P., Friedman, S. D., Gry, C., Knauth, D., Kruk, J. W., Lacour, S., Lehner, N., Redfield, S., Shull, J. M., Sonneborn, G., & Williger, G. M. 2006, *Astrophys. J.*, 647, 1106.

Lis, D. C., Roueff, E., Gerin, M., Phillips, T. G., Coudert, L. H., van der Tak, F. F. S., & Schilke, P. 2002, *Astrophys. J. Letters*, 571, L55.

Llorca, J., & Casanova, I. 2000, *Meteoritics and Planetary Science*, 35, 841.

Lodders, K. 2003, *Astrophys. J.*, 591, 1220.

Lodders, K. & Fegley, B. 2002, *Icarus*, 155, 393.

Lunine, J. I., Engel, S., Rizk, B., & Horanyi, M. 1991, *Icarus*, 94, 333.

Lunine, J. I., & Stevenson, D. J. 1985, *Astrophys. J. Suppl.*, 58, 493.

MacPherson, G. J., Wark, D. A., & Armstrong, J. T. 1988, Primitive Material Surviving in Chondrites—Refractory Inclusions. (In *Meteorites and the Early Solar System*), eds. J. F. Kerridge & M. S. Matthew (Tucson: University of Arizona Press), 746–807.

Mahaffy, P. R., Donahue, T. M., Atreya, S. K., Owen, T. C., & Niemann, H. B. 1998, *Space Science Reviews*, 84, 251.

Maret, S., & Bergin, E. A. 2007, *Astrophys. J.*, 664, 956.

Mathis, J. S., Rumpl, W., & Nordsieck, K. H. 1977, *Astrophys. J.*, 217, 425.

McCall, B. J., Huneycutt, A. J., Saykally, R. J., Geballe, T. R., Djuric, N., Dunn, G. H., Semaniak, J., Novotny, O., Al-Khalili, A., Ehlerding, A., Hellberg, F., Kalhori, S., Neau, A., Thomas, R., Österdahl, F., & Larsson, M. 2003, *Nature*, 422, 500.

Messenger, S., Stadermann, F. J., Floss, C., Nittler, L. R., & Mukhopadhyay, S. 2003, *Space Science Reviews*, 106, 155.

Millar, T. J., Bennett, A., & Herbst, E. 1989, *Astrophys. J.*, 340, 906.

Morbidelli, A., Chambers, J., Lunine, J. I., Petit, J. M., Robert, F., Valsecchi, G. B., & Cyr, K. E. 2000, *Meteoritics and Planetary Science*, 35, 1309.

Morrison, R., & McCammon, D. 1983, *Astrophys. J.*, 270, 119.

Mousis, O., & Alibert, Y. 2006, *Astron. & Astrophys.*, 448, 771.

Mumma, M. J., Weaver, H. A., & Larson, H. P. 1987, *Astron. & Astrophys.*, 187, 419.

Muzerolle, J., Calvet, N., Briceño, C., Hartmann, L., & Hillenbrand, L. 2000, *Astrophys. J. Letters*, 535, L47.

Nagaoka, A., Watanabe, N., & Kouchi, A. 2005, *Astrophys. J. Letters*, 624, L29.

Najita, J., Bergin, E. A., & Ullom, J. N. 2001, *Astrophys. J.*, 561, 880.

Najita, J., Carr, J. S., & Mathieu, R. D. 2003, *Astrophys. J.*, 589, 931.

Najita, J. R., Strom, S. E., & Muzerolle, J. 2007, *MNRAS*, 378, 369.

Nakagawa, Y., Nakazawa, K., & Hayashi, C. 1981, *Icarus*, 45, 517.

Nejad, L. A. M. 2005, *Astrophys. Space Sci.*, 299, 1.

Öberg, K. I., Fuchs, G. W., Awad, Z., Fraser, H. J., Schlemmer, S., van Dishoeck, E. F., & Linnartz, H. 2007, *Astrophys. J. Letters*, 662, L23.

Öberg, K. I., Linnartz, H., Visser, R., & van Dishoeck, E. F. 2009, *Astrophys. J.*, 693, 1209.

Owen, T., & Bar-Nun, A. 1995, *Icarus*, 116, 215.

Palme, H. & Jones, A. 2003, Treatise on Geochemistry. In *Meteorites and the Early Solar System*, eds. J. F. Kerridge & M. S. Matthews (Tucson: University of Arizona Press), 1, 41.

Palme, H., Larimer, J. W., & Lipschutz, M. E. 1988, Moderately Volatile Elements. In *Meteorites and the Early Solar System*, eds. J. F. Kerridge & M. S. Matthews (Tucson: University of Arizona Press), 436.

Pascucci, I., et al. 2006, *Astrophys. J.*, 651, 1177.

Pavlyuchenkov, Y., & Dullemond, C. P. 2007, *Astron. & Astrophys.*, 471, 833.

Pepin, R. O. 1991, *Icarus*, 92, 2.

Phillips, T. G., & Vastel, C. 2003, in *SFChem 2002: Chemistry as a Diagnostic of Star Formation*, ed. C. L. Curry & M. Fich, 3–+.

Pohl, R. O. 1998, Vol. 23, Vibrational States in Disordered Solids. In *Encyclopedia of Applied Physics* (Weinheim: Wiley-VCH), 223.

Pringle, J. E. 1981, *Ann. Rev. Astron. Astrophys.*, 19, 137.

Prinn, R. G. 1993, in *Protostars and Planets* III, 1005–1028.

Prinn, R. G. P., & Fegley, B., Jr. 1989, Solar Nebula Chemistry: Origins of Planetary, Satellite and Cometary volatiles. In *Origin and Evolution of Planetary and Satellite Atmospheres*, eds. S. K. Atreya, J. B. Pollack, & M. S. Matthews (Tucson: University of Arizona Press), 78.

Qi, C. 2001, Ph.D. thesis, Caltech.

Qi, C., Wilner, D. J., Aikawa, Y., Blake, G. A., & Hogerheijde, M. R. 2008, *ArXiv e-prints*, 803.

Qi, C., Wilner, D. J., Calvet, N., Bourke, T. L., Blake, G. A., Hogerheijde, M. R., Ho, P. T. P., & Bergin, E. 2006, *Astrophys. J. Letters*, 636, L157.

Reipurth, B., Jewitt, D., & Keil, K., eds. 2007, *Protostars and Planets* V.

Richardson, S. M., & McSween, Jr., H. Y. 1989, *Geochemistry: Pathways and Processes* (Englewood Cliffs, NJ: Prentice Hall).

Richet, P., Bottinga, Y., & Janoy, M. 1977, *Annual Review of Earth and Planetary Sciences*, 5, 65.

Roberge, W. G., Jones, D., Lepp, S., & Dalgarno, A. 1991, *Astrophys. J. Suppl.*, 77, 287.

Robert, F. 2006, Solar System Deuterium/Hydrogen Ratio (*Meteorites and the Early Solar System* II), 341–351.

Roberts, H., Herbst, E., & Millar, T. J. 2003, *Astrophys. J. Letters*, 591, L41.

Roberts, J. F., Rawlings, J. M. C., Viti, S., & Williams, D. A. 2007, *MNRAS*, 382, 733.

Rodgers, S. D. & Charnley, S. B. 2008, *MNRAS*, 385, L48.

Roueff, E., Tiné, S., Coudert, L. H., Pineau des Forêts, G., Falgarone, E., & Gerin, M. 2000, *Astron. & Astrophys.*, 354, L63.

Rybicki, G. B., & Hummer, D. G. 1991, *Astron. & Astrophys.*, 245, 171.

Salyk, C., Pontoppidan, K. M., Blake, G. A., Lahuis, F., van Dishoeck, E. F., & Evans, N. J., II. 2008, *Astrophys. J. Letters*, 676, L49.

Sandford, S. A., & Allamandola, L. J. 1993, *Astrophys. J.*, 417, 815.

Sano, T., Miyama, S. M., Umebayashi, T., & Nakano, T. 2000, *Astrophys. J.*, 543, 486.

Schutte, W. A., & Greenberg, J. M. 1991, *Astron. & Astrophys.*, 244, 190.

Sekine, Y., Sugita, S., Shido, T., Yamamoto, T., Iwasawa, Y., Kadono, T., & Matsui, T. 2005, *Icarus*, 178, 154.

Semenov, D., Wiebe, D., & Henning, T. 2004, *Astron. & Astrophys.*, 417, 93.

Shakura, N. I., & Sunyaev, R. A. 1973, *Astron. & Astrophys.*, 24, 337.

Simon, M., Dutrey, A., & Guilloteau, S. 2000, *Astrophys. J.*, 545, 1034.

Skumanich, A. 1972, *Astrophys. J.*, 171, 565.

Smith, M. D. 1998, *Icarus*, 132, 176.

Stäuber, P., Doty, S. D., van Dishoeck, E. F., & Benz, A. O. 2005, *Astron. & Astrophys.*, 440, 949.

Stevenson, D. J. 1990, *Astrophys. J.*, 348, 730.

Tanaka, H., & Koga, K. 2006, *Bull. Chem. Soc. Jpn.*, 79, 1621.

Thi, W.-F., van Zadelhoff, G.-J., & van Dishoeck, E. F. 2004, *Astron. & Astrophys.*, 425, 955.

Tielens, A. G. G. M. 1983, *Astron. & Astrophys.*, 119, 177.

———. 2005, *The Physics and Chemistry of the Interstellar Medium* (Cambridge, UK: Cambridge University Press).

Tielens, A. G. G. M., & Allamandola, L. J. 1987, in Astrophysics and Space Science Library, Vol. 134, *Interstellar Processes*, ed. D. J. Hollenbach & H. A. Thronson, Jr., 397.

Tielens, A. G. G. M., & Hollenbach, D. 1985, *Astrophys. J.*, 291, 722.

Townes, C. H., & Schawlow, A. L. 1955, *Microwave Spectroscopy* (New York: McGraw-Hill)

Tscharnuter, W. M. & Gail, H.-P. 2007, *Astron. & Astrophys.*, 463, 369.

Turner, N. J., Willacy, K., Bryden, G., & Yorke, H. W. 2006, *Astrophys. J.*, 639, 1218.

Umebayashi, T., & Nakano, T. 1981, *Pub. Astron. Soc. Japan*, 33, 617.

van der Tak, F. F. S., Schilke, P., Müller, H. S. P., Lis, D. C., Phillips, T. G., Gerin, M., & Roueff, E. 2002, *Astron. & Astrophys.*, 388, L53.

van Dishoeck, E. F. 1988, in *Rate Coefficients in Astrochemistry. Proceedings of a Conference held in UMIST, Manchester, United Kingdom, September 21–24, 1987*, ed., T. J. Millar, & D.A. Williams (Dordrecht: Kluwer Academic Publishers), 49.

van Dishoeck, E. F., & Black, J. H. 1988, *Astrophys. J.*, 334, 771.

van Dishoeck, E. F., Jonkheid, B., & van Hemert, M. C. 2006, *Faraday Discussions*, 133, 231.

van Dishoeck, E. F., Thi, W.-F., & van Zadelhoff, G.-J. 2003, *Astron. & Astrophys.*, 400, L1.

van Zadelhoff, G.-J., Aikawa, Y., Hogerheijde, M. R., & van Dishoeck, E. F. 2003, *Astron. & Astrophys.*, 397, 789.

van Zadelhoff, G.-J., van Dishoeck, E. F., Thi, W.-F., & Blake, G. A. 2001, *Astron. & Astrophys.*, 377, 566.

Vejby-Christensen, L., Andersen, L. H., Heber, O., Kella, D., Pedersen, H. B., Schmidt, H. T., & Zajfman, D. 1997, *Astrophys. J.*, 483, 531.

Viti, S., Collings, M. P., Dever, J. W., McCoustra, M. R. S., & Williams, D. A. 2004, *MNRAS*, 354, 1141.

Wadhwa, M., Amelin, Y., Davis, A. M., Lugmair, G. W., Meyer, B., Gounelle, M., & Desch, S. J. 2007, in *Protostars and Planets* V, ed. B. Reipurth, D. Jewitt, & K. Keil, 835–848.

Wagner, A. F., & Graff, M. M. 1987, *Astrophys. J.*, 317, 423.

Walmsley, C. M., Flower, D. R., & Pineau des Forêts, G. 2004, *Astron. & Astrophys.*, 418, 1035.

Wasserburg, G. J., Busso, M., Gallino, R., & Nollett, K. M. 2006, *Nuclear Physics* A, 777, 5.

Wasson, J. T., & Kallemeyn, G. W. 1988, *Royal Society of London Philosophical Transactions*, Series A, 325, 535.

Watson, W. D. 1976, Reviews of Modern Physics, 48, 513.

Webber, W. R. 1998, *Astrophys. J.*, 506, 329.

Weidenschilling, S. J. 1977, *Astrophys. Space Sci.*, 51, 153.

———. 1997, *Icarus*, 127, 290.

Weidenschilling, S. J., & Cuzzi, J. N. 1993, in *Protostars and Planets* III, ed. E. H. Levy & J. I. Lunine, 1031.

Willacy, K., Klahr, H. H., Millar, T. J., & Henning, T. 1998, *Astron. & Astrophys.*, 338, 995.

Willacy, K., Langer, W., Allen, M., & Bryden, G. 2006, *Astrophys. J.*, 644, 1202.

Willacy, K., & Langer, W. D. 2000, *Astrophys. J.*, 544, 903.

Williams, T. L., Adams, N. G., Babcock, L. M., Herd, C. R., & Geoghegan, M. 1996, *MNRAS*, 282, 413.

Wood, B. E., Linsky, J. L., Hébrard, G., Williger, G. M., Moos, H. W., & Blair, W. P. 2004, *Astrophys. J.*, 609, 838.

Woodall, J., Agúndez, M., Markwick-Kemper, A. J., & Millar, T. J. 2007a, *Astron. & Astrophys.*, 466, 1197.

Xie, T., Allen, M., & Langer, W. D. 1995, *Astrophys. J.*, 440, 674.

Yurimoto, H., Kuramoto, K., Krot, A. N., Scott, E. R. D., Cuzzi, J. N., Thiemens, M. H., & Lyons, J. R. 2007, in *Protostars and Planets* V, ed. B. Reipurth, D. Jewitt, & K. Keil, 849–862.

Zeller, R. C., & Pohl, R. O. 1971, *Phys. Rev.* B, 4, 2029.

4

THOMAS HENNING AND GWENDOLYN MEEUS

DUST PROCESSING AND MINERALOGY IN PROTOPLANETARY ACCRETION DISKS

In this chapter we discuss the different dust components a protoplanetary disk is made of, with a special emphasis on grain composition, size, and structure. The chapter will highlight the role these dust grains play in protoplanetary disks surrounding young stars, as well as observational results supporting this knowledge. First, the path dust travels from the interstellar medium into the circumstellar disk is described. Then dust-condensation sequences from the gas are introduced to determine the most likely species that occur in a disk. The characteristics of silicates are handled in detail: composition, lattice structure, magnesium-to-iron ratio, and spectral features. The other main dust-forming component of the interstellar medium, carbon, is presented in its many forms, from molecules to more complex grains. Observational evidence for polycyclic aromatic hydrocarbons (PAHs) is given for both young stars and solar system material. We show how light-scattering theory and laboratory data can be used to provide the optical properties of dust grains. From the observer's point of view, we discuss how infrared spectra can be used to derive dust properties, and we present the main spectral analysis methods currently used and their limitations. Observational results, determining the dust properties in protoplanetary disks, are given, first for the bright intermediate-mass Herbig Ae/Be stars and then for the lower-mass Tauri stars and brown dwarfs. Here we present results from the *Infrared Space Observatory* (*ISO*) and the *Spitzer Space Telescope*, as well as from the mid-infrared instrument at the *Very Large Telescope Interferometer* (*VLTI*), and summarize the main findings. We discuss observational evidence for grain growth in both Herbig Ae/Be and T Tauri stars, and its relation to spectral type and dust settling. We conclude with an outlook on future space missions that will open new windows toward longer wavelengths and even fainter objects.

1. Introduction

Dust grains dominate the opacity in protoplanetary disks whenever they are present. This implies that their radiation properties play a crucial role in determining the temperature and density structure of these disks. The initial population of (sub)micron-sized particles evolves over time toward planetesimals, eventually providing the building blocks for terrestrial planets.

The dust grains shield the interior of protoplanetary disks from energetic cosmic particles and stellar X-ray and ultraviolet (UV) radiation and provide the surface for electron recombination. The presence of dust grains regulates the ionization structure of disks, which is an important ingredient for the magnetorotational instability to operate and to drive angular momentum transport. The dust particles are equally important for disk chemistry because chemistry on grain surfaces leads to the formation of molecular ices and, possibly, complex organic molecules, which will enter the gas phase when their evaporation temperature is reached.

Infrared spectroscopy is a powerful tool to characterize the properties of protoplanetary dust. With ground-based telescopes, the *ISO* and *Spitzer Space Telescope*, an enormous amount of data has been obtained to characterize the mineralogy of disks around a variety of objects, ranging in luminosity from Herbig Ae/Be (HAeBe) stars to T Tauri stars and even brown dwarfs. These data allow us to put constraints on the chemical composition and amorphous/crystalline state of the dust particles and to address questions such as radial distribution and mixing processes. Mid-infrared long-baseline interferometry is starting to contribute as well to our understanding of the structural properties of dust in the different radial zones of protoplanetary disks.

2. Dust Components in Protoplanetary Disks

2.1. General Overview

Because the infalling disk material originates from the interstellar medium (ISM), more precisely from the parental molecular cloud core, the initial dust composition in a protoplanetary accretion disk is assumed to be similar to the molecular cloud dust composition. It can be slightly altered from this composition because volatile molecular ices could evaporate during the passage of the accretion-shock front, oxygen could convert into water, and quartz (SiO_2) could form from silicon atoms. For a detailed discussion of the various dust populations in space, including dust in molecular clouds, we refer to the review by Dorschner & Henning (1995). The most abundant cosmic dust species are compounds of O, Si, Mg, Fe, and C: silicates and carbonaceous dust.

Depending on the angular momentum of the material, molecular cloud dust will accrete onto the disk at different radial distances from the star, which may influence subsequent grain evolution (Dullemond et al. 2006). In protoplanetary disks, a wide range of modification processes are expected to occur, including thermal annealing in hot regions of the inner disk, ion irradiation by stellar flares, X-ray and UV irradiation, destruction of carbonaceous dust by oxidation close to the central star, equilibration with the gas through sublimation-condensation processes, and solid-phase reactions, as well as molecular ice formation in the outer disk. In addition, grain growth and both radial and vertical mixing processes need to be considered, and thus we are led to the expectation that the grain composition changes both over time and with radial distance from the star, so that a relatively large diversity of dust compositions can be expected in protoplanetary disks.

As a reference for the composition of protoplanetary dust in the outer disk regions, the dust model introduced by Pollack et al. (1994) is widely used. It is based on the solar elemental composition and the ISM gas-depletion pattern, the composition of primitive solar system material, including the mass-spectroscopy results of the space probes to comet Halley, and theoretical considerations. The model contains the iron-magnesium silicate minerals olivines and pyroxenes, quartz, metallic iron, troilite (FeS), and volatile and refractory organics, as well as water ice in the outer disk. Pollack et al. already noted that the silicates are certainly mixtures of amorphous and crystalline silicates, with the amorphous silicates dominating. In fact, they used the optical constants of amorphous silicates at mid-infrared wavelengths as a pragmatic choice. The dust model divides the products of the most abundant dust-forming elements O, C, Si, Mg, Fe, S, and N into three categories: gases, molecular ices, and refractory grains. On the basis of the dust composition and the optical properties of these various components, Pollack et al. (1994) also derived dust opacities for disks. These were further improved by including dust aggregates (Henning & Stognienko 1996), as well as updated optical data for the various materials (Semenov et al. 2003).

Molecular ices exist only in the cold outer disk, while FeS forms from Fe and H_2S at condensation temperatures of 680 K. Besides water ice, CO, CO_2, NH_3, CH_4, and CH_3OH ices have also been detected in the infrared spectra of disks (Pontoppidan et al. 2005; Zasowski et al. 2009).

The exact fractional abundances of the various solids remain an open question. In general, we would expect a radial variation of the dust composition, as already mentioned for Fe/FeS and molecular ices. As another example, carbonaceous dust should be destroyed in the inner disk when it is evolving toward an equilibrium dust mixture under oxygen-rich conditions.

The bulk of the ISM/molecular cloud dust consists of amorphous silicates and amorphous carbonaceous material. *ISO* observations around 10 micron have shown that most of the silicates ($\geq$ 98% by mass) in the ISM have an amorphous structure (e.g., Kemper et al. 2004). In contrast, infrared spectroscopy of HAeBe stars, T Tauri stars, and brown dwarfs has demonstrated that a significant fraction of the dust in protoplanetary disks is in a crystalline state, implying that these crystals have been formed inside the disks (see § 5). Amorphous silicates crystallize only at relatively high temperatures through thermal annealing processes (e.g., Fabian et al. 2000). This fact suggests that these materials have experienced high temperatures in the disks (typically 800–1,000 K).

In an actively accreting disk, the main accretion flow points to the star, and most of the dust will be destroyed by sublimation and subsequently incorporated into the star. Once the main accretion phase has terminated and the star has grown close to its final mass, the material will only slowly move inward. In both cases, the dust grains will experience an increase in temperature when approaching the star. This will lead to both (1) annealing of the amorphous material into a more crystalline structure and (2) chemical processing through evaporation and recondensation, thereby changing the abundance of the different species. It is important to remember that the dust species are the main source of the opacities, which determine the disk structure. When chemical processing or sublimation causes a certain dust species to disappear, a change in the opacities will occur, with consequences for the structure of the disk.

2.2. Condensation of Dust

Condensation sequences of dust from the gas phase are often based on chemical equilibrium calculations, from the early studies by Larimer (1967), Grossman (1972), and Lattimer et al. (1978) to the more recent investigations by Gail (1998) and Krot et al. (2000). Although the gas-dust mixture may often not be in a state of chemical equilibrium, these calculations are a useful tool to predict which dust species can be expected in the considered elemental gas mixture.

Gail (1998; see also Gail 2003 for a comprehensive discussion) proposed a scheme to determine the stable dust materials expected to be present in a disk. It starts from the dust mixture—vaporized into the gas phase—from Pollack et al. (1994) and is based on chemical equilibrium considerations. The most important parameters in this scheme are the temperature and the pressure, which critically depend on the location in the disk; as a consequence, the stability of a certain dust species (and hence its presence) will be radially dependent. To summarize Gail's results, we will give an overview of the condensates that are expected to be stable at a certain temperature, starting from the outer disk region and moving inward:

- At low temperatures (below 700 K), FeS (troilite) will be formed and is stable, with the remaining iron (excess Fe over S) contained in pure iron particles. Silicon will be in magnesium-rich amorphous silicates, while SiO_2 is found to be unstable in chemical equilibrium.
- At higher temperatures (around 800 K), FeS will disappear and contribute to metallic iron, while amorphous silicates will anneal into crystalline silicates.
- At even higher temperatures (1,300–1,400 K), both crystalline silicates and solid iron can no longer survive and are destroyed; the last remaining dust particles are aluminum-rich species such as corundum (Al_2O_3).
- Above 1,850 K these last remaining dust species also can no longer survive.

From a comparison with dust-condensation experiments by, e.g., Nuth and Donn (1982), it is known that non-equilibrium processes must also play a role in the formation of dust (e.g., Tielens et al. 2005). Unfortunately, condensation paths that also include kinetic considerations are not yet available (e.g., Gail 2003), mainly because of the lack of measured reaction rates. This is especially true for the first step in the dust-formation process in oxygen-rich environments, the nucleation of tiny seed particles from the gas phase.

However, the predictions based on chemical equilibrium considerations are already in relatively good agreement with observations. In a more refined model, Gail (2004) added radial mixing, which moves processed material from the inner disk to more outward regions. We should also note that several additional dust components can exist in the temperature-density regime of protoplanetary disks, including Al- and Ca-containing compounds such as hibonite ($CaAl_{12}O_{19}$) and spinel ($MgAl_2O_4$).

2.3. Silicates

Silicates form a diverse class of materials, ranging from amorphous and glasslike structures, characterized by three-dimensional disordered networks, to well-ordered crystals (e.g., Colangeli et al. 2003; Henning 2009, 2010). The different silicates are assembled by linking the corners of individual $[SiO_4]^{4-}$ tetrahedra through their oxygen atoms with different levels of complexity. The negative net charge of the ion group must be balanced by metal or hydrogen cations to produce an electrically neutral compound. The cations are dispersed between the individual tetrahedra or the tetrahedra arrays. Mineral structures that have been extensively discussed in the context of protoplanetary dust are Mg-Fe olivines and pyroxenes. Olivines with the composition $Mg_{2x}Fe_{2(1-x)}SiO_4$ can be considered a solid solution of their end members forsterite (Mg_2SiO_4) and fayalite (Fe_2SiO_4). Pyroxene with the composition $Mg_xFe_{(1-x)}SiO_3$ is a solid solution formed from enstatite ($MgSiO_3$) and ferrosilite ($FeSiO_3$). The lattice structures of olivines and pyroxenes are very different: olivines are island silicates (nesosilicates) with isolated tetrahedra, while pyroxenes are chain silicates (inosilicates) in which one oxygen atom of every tetrahedron is shared with its neighbor. We should explicitly note that the terminology of olivines and pyroxenes refers to crystal structures and should not be used for amorphous silicates of the same chemical composition. The silicate particles show a wide variety of infrared features, characteristic of their chemical composition and structure, as is shown for the wavelength range between 8 and 13 μm in Fig. 4.1, where we also show an example of a template with PAH features.

In the equilibrium calculations for the inner disk by Gail (1998), iron is seldom found to be incorporated into silicates, so only the magnesium-rich end members of the compounds are present: forsterite (Mg_2SiO_4) for the olivines and enstatite ($MgSiO_3$) for the pyroxenes. Enstatite appears to be the more stable of those two, with forsterite appearing only in a narrow temperature range, just below the stability limit. It is thus predicted that in a disk, enstatite is the dominating component of the crystalline silicates, and forsterite is important only in the more inward regions of the disk, near the region where it becomes too hot to survive (around 1,400 K). However, Gail (2004) cautions that the high enstatite abundance he predicts might need to be lowered because the forsterite-enstatite conversion might be too slow to reach complete chemical equilibrium within the relevant timescales involved. We will discuss the ratio of enstatite to forsterite, as determined from observations, in § 5.

The Mg/Fe ratio of the amorphous silicates located in the outer regions of protoplanetary disks is not well constrained. In the Pollack et al. (1994) dust model, an average value of 0.3 was assumed for the Fe/(Fe+Mg) ratio, guided by the mass-spectroscopy results of comet Halley dust (Jessberger et al. 1989), as well as results obtained for anhydrous chondritic porous interplanetary dust

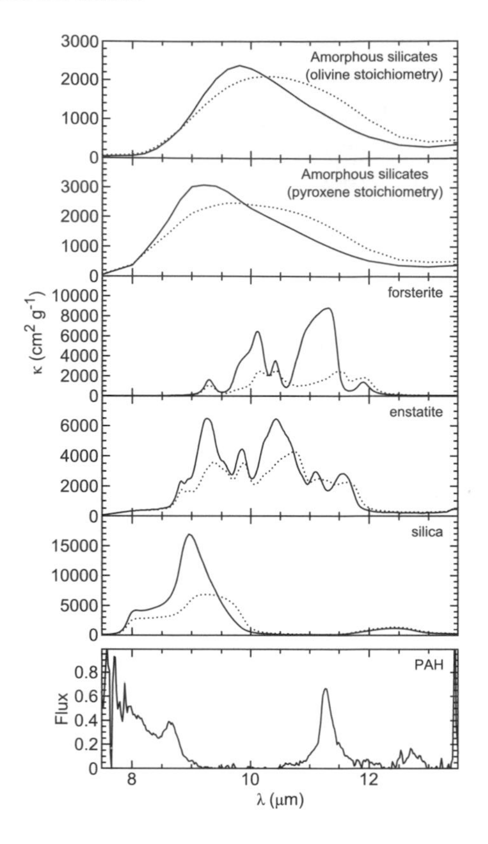

Fig. 4.1. Mass-absorption coefficients of the various silicate grains. Homogeneous spheres have been assumed for the amorphous grains and a distribution of hollow spheres for the crystalline forsterite and enstatite and the amorphous silica. Grains with volume-equivalent grain radii of 0.1 μm (solid lines) and 1.5 μm (dotted lines) have been used. In addition, a polycyclic aromatic hydrocarbon (PAH) template is shown. After van Boekel et al. (2005).

particles (CP IDPs; Bradley et al. 1988) that were collected in the stratosphere. Such IDPs consist mainly of glass with embedded metal and sulfides (GEMS), whose properties are consistent with those of interstellar amorphous silicates (Bradley et al. 1994). However, their origin is currently under debate: Bradley & Ishii (2008) argue that they are of presolar nature because some of the GEMS have a nonsolar isotopic composition, and the ISM dust is in any case characterized only for a small part by nonsolar isotopic composition, while Min et al. (2007) argue that most of them were formed in the solar system. From an analysis of the shape and position of interstellar silicate features, Min et al. (2007) concluded that interstellar silicates are predominantly Mg rich. However, the interplay between shape and composition in determining the relatively broad infrared features of amorphous silicates introduces considerable uncertainty in such an analysis (see, e.g., Chiar & Tielens 2006 for different conclusions). So far, in the analysis of protoplanetary disk spectra, mainly amorphous silicates with a mixed Fe/Mg ratio have been used (e.g., Bouwman et al. 2001;

Kessler-Silacci et al. 2006). Therefore, a detailed examination of high-quality spectra, taking into account different chemical compositions and size/shape effects, still needs to be performed.

Amorphous silicates with the same stoichiometry as pyroxenes and olivines (note again that the amorphous silicates have a different structure from pyroxenes and olivines despite the same chemical bulk composition) typically show two broad infrared bands at about 10 and 18 μm, corresponding to Si–O stretching and O–Si–O bending vibrations, respectively. These bands are frequently observed in the spectra of protoplanetary disks. The large width of the bands results from a distribution of bond lengths and angles, typical of the amorphous structure of these solids. The 18 μm band is additionally broadened and generally weaker because of the coupling of the bending mode to the metal-oxygen stretching vibrations occurring in this spectral region. The exact position of the Si–O stretching vibration depends on the level of SiO_4 polymerization. As an example, the band is shifted from 9 μm for pure (sub)micron-sized SiO_2 grains to about 10.5 μm for $Mg_{2.4}SiO_{4.4}$ (Jäger et al. 2003a).

In contrast to amorphous silicates, crystalline pyroxenes and olivines produce a wealth of narrow bands from the mid-infrared to the far-infrared wavelength range because of metal-oxygen vibrations. In crystalline pyroxenes and olivines, the majority of the infrared peaks are shifted to longer wavelengths with increasing iron content. These observed shifts are caused by an increase in bond lengths between the metal cations and the oxygen atoms when Mg^{2+} is substituted by Fe^{2+}. The wave-number shift is very closely related to the Fe content and allows a determination of the Mg/Fe ratio from infrared spectroscopy (Jäger et al. 1998). However, it is difficult to derive the Mg/Fe ratio from 10 μm observations alone because the shift there is rather small; fortunately, it is more pronounced for bands at longer wavelengths, enabling the determination of the ratio. On the basis of laboratory data, forsterite grains have strong bands at 10.0, 11.3, 16.3, 19.8, 23.5, 27.5, 33.5, and 69.7 μm, while enstatite grains have bands at 9.4, 9.9, 10.6, 11.1, 11.6, 18.2, 19.3, and 21.5 μm. The exact position of these features varies with the quality of the crystals and with temperature and will also depend on the shape distribution of the particles.

By observing the mid-infrared wavelength range, the *ISO* and *Spitzer* missions provided a wealth of information on the presence of crystalline olivines and pyroxenes in protoplanetary disks through spectroscopy (e.g., Bouwman et al. 2001, 2008; Kessler-Silacci et al. 2006; see also § 5). Strong bands were observed at 9.3, 10.1, 11.3, 19.0, 23.4, 27.8, and 33.5 μm. In Fig. 4.2 we illustrate the wealth of crystalline features observed in T Tauri stars.

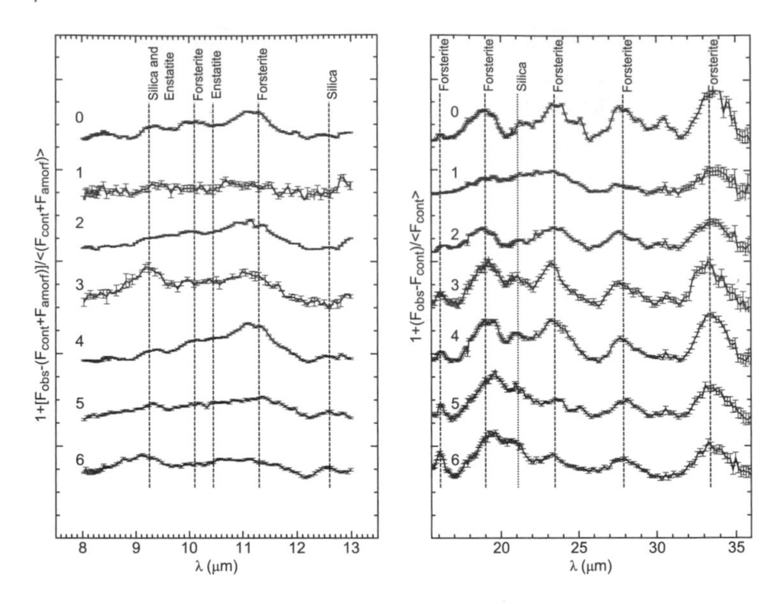

Fig. 4.2. The emission bands of crystalline silicates, as observed with the *Spitzer Space Telescope* in the spectra of seven T Tauri stars. The spectra have been normalized to a dust-model fit, thereby removing the amorphous silicate and PAH features and enhancing the crystalline features. Note that the silica identification is less secure. After Bouwman et al. (2008).

2.4. Carbonaceous Grains

Carbon is a major player in the ISM because it is a primary cooling and heating agent. It can be present in many different forms: as pure atoms, in simple molecules (e.g., CO, SiC, CN, and CH), and more complex ones (polycyclic aromatic hydrocarbons, PAHs) up to carbonaceous solids. Carbonaceous solids span a wide range of materials (Henning & Salama 1998; Henning et al. 2004) from ordered structures, such as graphite and diamond, to complicated amorphous structures, such as a variety of hydrogenated carbonaceous particles. The relevance of these materials as cosmic dust analogs and their spectroscopic properties have been summarized by Henning et al. (2004). Hydrogen-deficient amorphous carbon grains have only weak or lacking infrared modes and will be difficult to identify in the spectra of protoplanetary disks. Saturated aliphatic hydrocarbons show CH stretching and deformation modes in CH_2 and CH_3 groups at 3.4 and 6.6 μm, seen in the diffuse ISM but so far not detected in the spectra of protoplanetary disks. An exception is the Herbig Ae/Be star HD 163296, where *ISO* spectroscopy provided evidence for the presence of aliphatic carbonaceous dust (Bouwman et al. 2001). We can certainly expect more evidence for this dust component from a detailed analysis of high-quality *Spitzer* data (e.g., Juhász et al. 2010).

In the Pollack et al. (1994) model for the dust composition of the cooler outer disk, a kerogen-like material is assumed: it is a carbon-rich structure containing a significant amount of H, N, and O, typical of the matrix material in carbonaceous chondrites. This material should show C=O stretching vibrations in carbonyl groups. However, spectroscopic evidence for the presence of this material in disks is still lacking. So far, we have to conclude that the nature of the carbonaceous material in protoplanetary disks remains ill defined.

In protoplanetary disks around HAeBe stars, PAHs were convincingly detected at 3.3, 6.2, 7.7, 7.9, 8.2, 8.6, 11.2, and 12.7 μm (Peeters et al. 2002; Sloan et al. 2005; Boersma et al. 2008; Keller et al. 2008; Acke et al. 2010). The ratio of the band strengths can be used to derive the charge state of the PAHs, from which the ionization parameter can be calculated (e.g., Bakes et al. 2001). Furthermore, the band ratio can also be related to the size of the emitting PAHs (Allamandola et al. 1985). In T Tauri stars there is much less evidence for the presence of PAHs, certainly because of the different stellar UV radiation fields (Geers et al. 2006, 2007). The presence of PAHs in the disks of HAeBe stars shows that at least some carbonaceous material survives in protoplanetary disks.

Evidence for the presence of nanodiamonds has been found in a very small number of HAeBe stars (van Kerckhoven et al. 2002). Their infrared features at 3.43 and 3.53 μm have been interpreted as vibrational modes of hydrogen-terminated crystalline facets of diamond particles (Guillois et al. 1999; pure nanodiamonds have no infrared features). Alternatively, diamondoid molecules were recently discussed as the carriers of these bands (Pirali et al. 2007). There are actually only three HAeBe stars known to date that have clear diamond signatures: HD 97048 (Whittet et al. 1983), MWC 297 (Terada et al. 2001), and Elias 1 (Whittet et al. 1984). An extensive 3 μm spectroscopic survey of over 60 HAeBe stars did not add a single source with a pronounced diamond spectrum to the already-known objects (Acke & van den Ancker 2006). Goto et al. (2009) argued that the presence of diamonds in the disks around selected HAeBe stars may be related to the transformation of graphitic material into diamond under the irradiation of highly energetic particles.

Carbonaceous material was also found in solar system material: in chondritic meteorites, nanodiamonds and graphitic particles were detected (e.g., Sandford 1996; Hill et al. 1997), while PAHs were observed in comets (e.g., Moreels et al. 1994; Joblin et al. 1997).

2.5. *Iron-Containing Grains*

Iron-containing particles have an important effect on the dust opacity (Ossenkopf et al. 1992; Henning & Stognienko 1996). In amorphous silicate grains, they strongly influence the near-infrared absorptivity and the temperature of the grains (Dorschner et al. 1995). Pure iron aggregates would dramatically increase the dust opacities (Henning & Stognienko 1996). In the Pollack et al. (1994) dust model, most of the iron is in silicates and troilite (FeS), and the remaining 20% of its solar abundance is assumed to be in metallic iron. In the inner disk, most of the iron would be in metallic iron. Only at temperatures below 700 K will part of the metallic iron be incorporated into FeS.

There is strong evidence for the widespread occurrence of Fe and FeS in primitive solar system material, including the dust analyzed by the mass spectrometers on board the space probes to comet Halley and the *Stardust* samples from Comet 81P/Wild 2 (e.g., Bradley et al. 1988; Bradley 1994; Zolensky et al. 2006). Solid iron and troilite form solid solutions with the available Ni and NiS under the conditions of protoplanetary disks and standard cosmic element mixtures.

Despite the important role of iron in determining the opacities of protoplanetary dust, so far, no convincing spectroscopic identification in astronomical spectra has been presented. Small iron particles will contribute only to the general infrared continuum and will show no infrared resonance. Therefore, they cannot be identified by distinct infrared spectroscopic features. Laboratory measurements demonstrated that (sub)micron-sized FeS particles should have relatively strong infrared features between 30 and 45 μm (Begemann et al. 1994; Mutschke et al. 1994). Again, no observational evidence for the presence of these features exists in spectra of protoplanetary disks.

3. Optical Properties of Dust Particles

In order to interpret astronomical spectra and to be able to assign solid-state features to given species, the optical properties of the dust particles need to be calculated. Such calculations are based on light-scattering theory and laboratory data.

The interaction of a radiation field with a system of solid particles can be described by their absorption and scattering cross sections C_{abs} and C_{sca}, respectively. They describe what fraction of the incoming radiation is absorbed or scattered by a dust particle. The extinction cross section C_{ext} is given by

$$C_{ext} = C_{abs} + C_{sca}. \tag{1}$$

The cross sections depend on the chemical and structural properties of the solid particles, ranging from the atomic scale (chemical composition and crystal and defect structure) to the mesoscopic scale (porosity and inhomogeneities, mantles, surface states) and finally the macroscopic morphology (size and shape distribution, agglomeration, coalescence).

Instead of the extinction cross section, the mass-extinction coefficient κ_m, defined as the extinction cross section per unit of particle mass, is often used to characterize the extinction of light. This quantity is actually more appropriate in describing how much light is removed from the incoming radiation field by a fixed mass of particles. For a spherical particle, $\kappa_m = \frac{3Q_{ext}}{4a\delta}$, where a is the particle radius, δ is the material density, and Q_{ext} is the extinction efficiency (extinction cross section per geometric cross section, $C_{ext}/\pi a^2$).

The optical properties of small particles can deviate considerably from those of bulk materials because of the occurrence of surface modes. The structure of the interface of small particles, including their shape, can have strong effects on their optical behavior. A comprehensive description of the classical electrodynamics of light absorption and scattering by small particles goes far beyond the goal of this section, and we refer to the excellent textbook by Bohren & Huffman (1983) for a detailed discussion.

The qualitative features of the absorption and scattering of light strongly depend on the ratio between the wavelength of the incident light λ and the size of the particle (for a spherical particle, the radius a). We can distinguish three cases:

1. Geometrical optics (size parameter $x = 2\pi a/\lambda \gg 1$): The propagation of light is described by rays that are reflected and refracted at the surface of the scatterer and finally transmitted, according to Snell's law and the Fresnel formulae. The scattering of a wave incident on a particle can be described as a combination of a reflected and a transmitted wave. For absorbing materials, light can penetrate only within the skin depth. Scattering, therefore, is mainly a surface effect, and the absorption cross section becomes proportional to the area of the particle as the radius increases. In this case, the mass-absorption coefficient for a sphere scales roughly as $1/a$. We should note that for very large size parameters, the extinction efficiency Q_{ext} approaches the limiting value 2.
2. Wave optics ($\lambda \sim a$): The angular and wavelength dependence of the scattered radiation is dominated by interferences and resonances. For spherical particles, this is the domain of Gustav Mie's (1908) scattering theory, which is often applied in astrophysics.
3. Rayleigh limit (size parameter $x = 2\pi a/\lambda \ll 1$): If, in addition, we have $\mid m \mid x \ll 1$, where $m = n + ik$ is the (complex) refractive index of the particle, we are in the quasi-static limit. Then both the incident field and

the internal field can be regarded as static fields. In this regime, phase shifts over the particle size are negligible. For nonmagnetic particles this implies that it is generally sufficient to consider only the dipolar electric mode.

The interaction of infrared and (sub)millimeter radiation with submicron sized grains can generally be considered good examples of the quasi-static case. However, particles with high imaginary parts of the refractive index (metals, semiconductors, crystalline grains) and particles of somewhat larger sizes can easily violate the conditions for the quasi-static limit, even at infrared wavelengths.

In Fig. 4.3, we show an example of the extinction efficiencies of four different grain sizes for the case of an infrared resonance of amorphous silicates in the 10 μm range, caused by Si–O stretching vibrations. It is clear that the feature changes shape with increasing grain size: larger (micron-sized) grains show typical "flat-topped" features and eventually disappear. This behavior can be used to trace the size of the particles by infrared spectroscopy. Here we want to note that the infrared features lose their diagnostic value for grain sizes much larger than the wavelength of the feature.

For the sake of simplicity and physical insight, we will discuss only the quasi-static case (Rayleigh limit) in the following. In this case, there is a connection between electrostatics and scattering by particles. Therefore, the expressions for the scattering and absorption cross sections for (small) spherical particles can be derived by treating the particle as an ideal dipole, with the dipole moment given by electrostatic theory (see, e.g., Bohren & Huffman 1998).

According to Rayleigh's law, the scattering cross section scales with k^4, where $k = 2\pi/\lambda$ is the wave number. This means that for very small absorbing particles (compared with the wavelength of incident radiation), the extinction

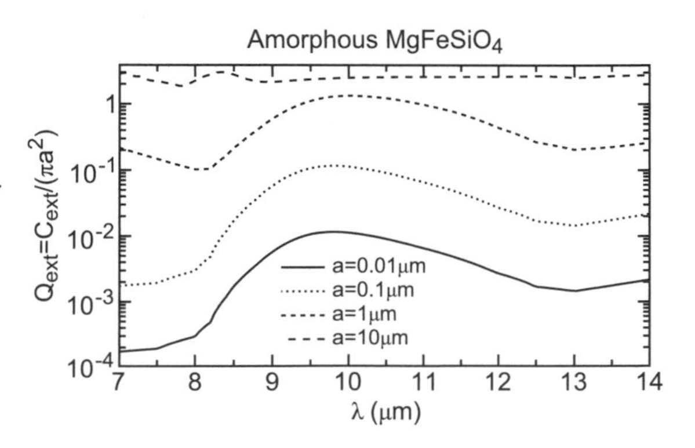

Fig. 4.3. Extinction efficiencies for spherical amorphous silicates of different sizes. Optical constants after Dorschner et al. (1995).

cross section is given by the absorption cross section. For the extinction cross section we can write

$$C_{\text{ext}} = k\text{Im}(\alpha), \quad (2)$$

where the quantity α denotes the polarizability and Im stands for the imaginary part of this quantity. The polarizability is defined as the ratio of the induced electrical dipole moment to the electric field that produces this dipole moment. This quantity depends on the complex dielectric function (dielectric permittivity) $\epsilon = \epsilon_1 + i\epsilon_2$ (for dielectrics $\epsilon = m^2$) of the particle and on the dielectric function of the embedding medium, ϵ_m, which is in most cases a wavelength-dependent real number (in vacuum $\epsilon_m = 1$).

One can treat not only spheres in the electrostatic approximation but also other particles, as long as their characteristic dimensions fulfill the same conditions as those defined for the spherical particles. Therefore, we will now consider ellipsoids as a more general particle shape, including both spheres (all axes equal) and spheroids (two axes having the same length).

For ellipsoids, the polarizability, α_i, in an electric field parallel to one of the principal axes is given by

$$\alpha_i = V(\epsilon - \epsilon_m)/(\epsilon_m + L_i(\epsilon - \epsilon_m)), \quad (3)$$

where L_i are geometric factors and V is the volume of the ellipsoid. The relation $L_1 + L_2 + L_3 = 1$ implies that only two of these three factors are independent. For a continuous distribution of ellipsoids (CDE) we get the relation (in vacuum with $\epsilon_\text{m} = 1$)

$$\alpha = V(2\epsilon/(\epsilon - 1)) \log \epsilon, \quad (4)$$

where $\log \epsilon$ denotes the principal value of the logarithm of the complex number ϵ. The CDE in this form assumes equal probability for the presence of every shape and averages over all orientations. This implies that extreme shapes are also equally weighted, although they are less likely to be present in real shape distributions. Nevertheless, the CDE can be used for a first estimate of how important shape effects for a certain resonance really are. It is important to note that the previous equations also demonstrate that the mass-extinction coefficient for such particles is independent of their size, but not of their shape.

The simplest case of light scattering in the quasi-static limit is that of a spherical particle where all axes are equal and $L_i = 1/3$. This gives $\alpha_i = \alpha$ and results in the expression

$$C_{\text{ext}} = 4x\text{Im}((\epsilon - \epsilon_m)/(\epsilon + 2\epsilon_m)). \quad (5)$$

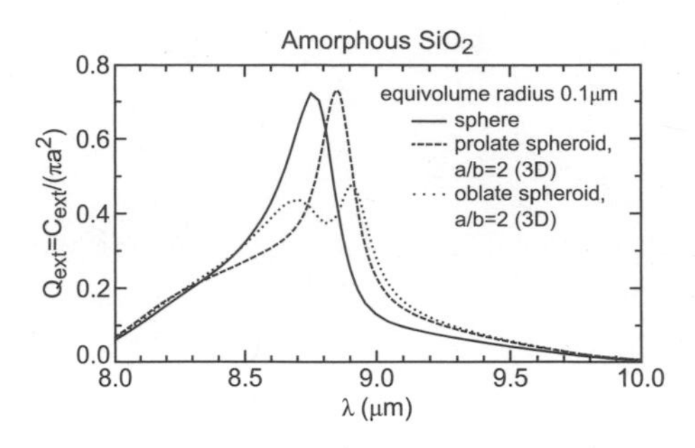

Fig. 4.4. Extinction efficiencies for silica particles of different shapes. Optical constants from Henning & Mutschke (1997).

Eq. (3) demonstrates that resonances for particles surrounded by a nonabsorbing medium occur close to the wavelength where the imaginary part of the dielectric function is close to zero and the real part fulfills the condition

$$\epsilon_1 = \epsilon_m(1 - 1/L_i). \tag{6}$$

This equation immediately implies that the resonance wavelengths depend on the shape of the particles. The resonances can occur only in regions where the real part of the dielectric function is negative (for a sphere $\epsilon_1 = -2\epsilon_m$). Examples of astronomically relevant materials that fulfill these criteria are SiC and SiO_2 in the infrared and graphite in the UV. In contrast, the lattice features of amorphous silicates do not always show this behavior. In Fig. 4.4 we show an example of the shape effects for SiO_2 particles. In the case of lattice modes, the resonances are always located between the transverse and longitudinal phonon frequencies, which makes it possible to estimate the wavelength range where the peak absorption can occur. The equations also show that the positions of the resonances depend on the surrounding medium. This is important for protoplanetary dust because it consists of core-mantle grains where the core is made of refractory material and the mantle is composed of molecular ices.

For particles of arbitrary shapes there exists no analytical solution, not even in the quasi-static limit. Numerical models frequently used for non-spherical particles are the separation-of-variables method, the T-matrix method, and the discrete dipole (or multipole) method (see, e.g., Draine 1988; Michel et al. 1996; Voshchinnikov et al. 2000, 2006; Min et al. 2008).

We should stress again that the occurrence of resonances is a property typical of small particles. For metal particles, resonances even occur at wavelengths where the bulk material does not show any absorption bands. In reality, the resonances will be modified and smeared out by a distribution of shapes, and the real difficulty is to evaluate which shape distribution would be a realistic description of the observed infrared features (Min et al. 2005, 2007; Voshchinnikov & Henning 2008). An analysis of the dust composition, based on the simple assumption that the particles are compact spheres, certainly leads to unreliable results. Min et al. (2005, 2007) recommend the application of a distribution of hollow spheres (DHS) for the calculations of the dust cross sections, which is computationally easy to use. In the Rayleigh limit, the absorption properties for a distribution of spheroidal particles and for a distribution of hollow spheres are very similar. However, the DHS method can also be used outside the Rayleigh limit, in contrast to the CDE. Absorption properties calculated by the DHS method seem to provide a reasonable representation of the shape of observed dust features (see, e.g., van Boekel et al. 2005; Kessler-Silacci et al. 2006).

In the dense regions of protoplanetary disks, we expect that particles coagulate and form larger particles (Beckwith et al. 2000; Henning et al. 2006; Natta et al. 2007). For a description of the interaction of electromagnetic radiation with fluffy aggregates composed of individual particles, two distinct approaches are possible:

1. The *Deterministic* approach: The frequency-domain Maxwell equation is solved for an individual cluster. The resulting cross sections are calculated for many clusters and then averaged over both the ensemble of clusters and their orientation. The advantage of this approach is that for special systems (e.g., clusters of spheres), exact solutions of the problem exist. For computational reasons, however, these methods are often limited to either comparatively small clusters or moderately absorbing systems. Examples of this kind of approach are the discrete-dipole and multipole approximations (DDA/DMA) and the extended Mie theory for multisphere aggregates.
2. The *Statistical* approach: The equations are formulated in terms of statistically relevant quantities (e.g., average radial density function of the clusters, density correlation function), without any explicit treatment of individual particles. Whereas a given cluster generally does not have any symmetry, statistical averages show rotational invariance unless alignment mechanisms break this symmetry. The advantage of this approach is that only the necessary information (ensemble- and orientation-averaged quantities) enter the calculations. Examples of this approach are the different effective medium theories and the strong permittivity fluctuation theory.

The dielectric function ϵ or the complex refractive index m can be determined in the laboratory for relevant astronomical materials. The term *optical constants* for the real (n) and imaginary (k) parts of the complex refractive index is somewhat misleading since the quantities strongly depend on frequency. For certain materials they also depend on temperature. The optical constants or dielectric functions are macroscopic quantities and lose their meaning for small clusters and molecules.

The dispersion with frequency is determined by resonances of the electronic system and of the ionic lattice and, at very low frequencies, by the relaxation of permanent dipoles. In the resonance regions the absorption becomes strong (high imaginary part), and the real part n shows "anomalous dispersion," i.e., a decrease with frequency. In many cases this behavior can be described by Lorentzian oscillators.

Compilations of optical constants of solid materials can be found in a number of databases that are available either in the form of books or of electronic media. The most important database in book form is the *Handbook of Optical Constants of Solids*, edited by E. D. Palik, which currently consists of three volumes that appeared in 1985, 1991, and 1997. These books are highly recommended since they comprise detailed discussions of the origin and errors of each data set. A database that is especially dedicated to cosmic dust has been developed by Henning et al. (1999) and can be found in its updated electronic form at http://www.mpia-hd.mpg.de/HJPDOC (Heidelberg-Jena–St. Petersburg Database of Optical Constants).

4. Spectral Analysis Methods

Dust properties are best studied in the infrared because it is here that the vibrational resonances of many astronomically relevant materials occur. Indeed, silicates and other oxides, sulfides, hydrogenated amorphous carbon particles, PAH molecules, and even hydrogen-terminated nanodiamonds all show characteristic features at those wavelengths (see § 2). This makes the infrared *the* fingerprint region for cosmic dust studies. Sensitivity, spectral coverage, and resolution are important observational parameters for such studies. Ground-based mid-infrared spectroscopy is mostly limited to the 8–13 μm atmospheric window, but can deliver data with very high spectral and spatial resolution, however, always with limited sensitivity because of the thermal background of the atmosphere. The *ISO* Short Wavelength Spectrometer (SWS) covered the extremely interesting wavelength range between 2 and 45 μm with a spectral resolution between 1,000 and 2,000. In addition, the *ISO* Long Wavelength Spectrometer (LWS) extended the wavelength range to wavelengths between 45 and 200 μm with a resolution from 100–200 up to 6,800–9,700 in high-resolution mode. The *Spitzer* Short-Low and Long-Low spectrometers covered a wavelength range between 5.3 and 14.5 μm and 14.2 and 38.0 μm, respectively, with a spectral resolution of 60–120. The Short-High and Long-High spectrometers had a wavelength coverage between 10.0 and 19.5 μm and 19.3 and 37.2 μm with a spectral resolution of 600. The *ISO* mission was the first infrared space observatory that delivered high-quality data for the bright HAeBe stars over a wide wavelength range. The *Spitzer Space Telescope*, with its unprecedented sensitivity, provided spectra of very high quality not only for these intermediate-mass stars but also for large and statistically significant samples of T Tauri stars and even made the first measurements of spectra of brown-dwarf disks possible.

4.1. Location of the "Observable" Dust Grains

The dust spectral features that are seen in emission arise in the optically thin warm disk atmosphere; the dust regions closer to the midplane are optically thick and do not show spectral features. This means that we can trace only a very small part of the total disk material with infrared spectroscopy. A major concern here is the differential sedimentation of particles: larger grains would reach a smaller scale height, that is an equilibrium between sedimentation and vertical turbulent mixing and this implies that through spectroscopy, we get only selective information about the uppermost disk layer. In the case of efficient mixing between the optically thick (featureless) disk midplane and the upper disk layers, the observations would provide information on the complete dust population of the disks.

The properties of the dust grains—size, shape, agglomeration state, chemical composition, and material structure—as well as the dust temperature distribution, determine the shape of the spectral features that occur on top of the continuum, arising in the optically thick part of the disk. In addition, the disk geometry also has an important influence on the observed spectrum: in disks with a flared geometry, the outer disk contributes significantly to the spectrum in the 10 μm range, whereas in disks with a flat geometry, the inner disk edge (the inner "rim") is more dominant. The inner disk edge is often strongly "crystallized," whereas the regions farther out contain much less crystalline material. Therefore, a disk with a flat geometry may appear to have a higher crystallinity than a flared disk, even though the actual composition of the dust is identical; this is purely a contrast effect.

Furthermore, it is important to realize that spectra around 10 μm trace different regions of the disk as a function of (sub)stellar luminosity: $R_{10} \propto L_*^{0.56}$. For brown dwarfs and T Tauri stars, the emission comes from regions of 0.05 and 0.5 AU distance from the brown dwarf and star, respectively, whereas the 10 μm emission from HAeBe stars traces the 10 AU range (see Kessler-Silacci et al. 2007 for a discussion of the scaling of the silicate emission region with luminosity). In the inner disk, more rapid grain growth is expected, leading to different contrast ratios between amorphous and crystalline silicate features, because the amorphous features flatten with increasing size (see Fig. 4.3). This fact also needs to be kept in mind when one is comparing spectra of objects with different luminosities and arriving at conclusions about the amount of crystalline grains.

4.2. Spectral Decomposition

The combination of disk thermal structure and dust optical properties can lead to degeneracies between these properties in the modeling results, even if sophisticated radiative transfer calculations are used to interpret the spectral energy distributions. Such degeneracies can be reduced when additional interferometric data, intensity maps, and polarization data are used to further constrain the disk properties. An additional problem is that some materials do not show any specific resonance (e.g., metallic iron particles in the infrared) and that large particles "lose" their characteristic features. Such "featureless" grains produce a continuum that is difficult to separate from the continuum of the optically thick part of the disk.

The most studied wavelength region in the context of dust is around 10 μm, where amorphous and crystalline silicates show features, as well as silica and PAHs. It is fortunate that this region is also observable from the ground, unlike most longer-wavelength regions. However, many of the features typical of crystalline grains are located at longer wavelengths and can be observed only by space missions. Because the dust emission arises in an optically thin region, the modeling of the features should be straightforward, but it is complicated by the temperature distribution of the particles and the underlying continuum. Here we should note that a certain wavelength range in the spectrum corresponds to a certain temperature range in the disk. This immediately implies that the analysis of a wider spectral range requires the application of a temperature distribution in the analysis.

In order to analyze larger data sets, especially in the 8–13 μm region, different simple spectral decomposition methods have been used in the literature to narrow down the properties of the grains that produce the features and to determine quantities such as composition, shape, size, and crystallinity. In a simple approach, the continuum below the spectral features is modeled by a polynomial and then subtracted from the measured spectrum. In this continuum-subtraction method (Bouwman et al. 2001), the continuum is fitted outside the feature, which is often hardly possible in ground-based observations because the spectrum cannot be sampled in the atmospherically opaque regions. In addition, the continuum can be associated with the bands themselves because the grains generally contribute to the continuum mass-absorption coefficients. It is further assumed that the dust grains have a single temperature. The continuum-subtracted feature is then fitted with a linear combination of mass-absorption coefficients of dust particles of different sizes, composition, and structure (see Bouwman et al. 2001); the necessary optical constants for the materials are provided by laboratory measurements.

Two other methods are very similar but differ in the modeling of the underlying continuum. In the single-temperature method (van Boekel et al. 2005) the spectrum is fitted by a linear combination of optically thin and thick emission components. In this approach it is assumed that the disk continuum is well represented by a Planck function and that the temperature of this continuum is the same as the temperature of the optically thin disk emission. The two-temperature method (Bouwman et al. 2008) is formally identical to the single-temperature method but fits the temperatures for the feature and the continuum separately. In the two-temperature method the observed monochromatic flux F_ν is given by

$$F_\nu = B_\nu(T_{\mathrm{cont}})C_0 + B_\nu(T_{\mathrm{dust}})\left(\sum_{i=1}^{3}\sum_{j=1}^{5} C_{i,j}\kappa_\nu^{i,j}\right) + C_{\mathrm{PAH}}F_\nu^{\mathrm{PAH}}, \tag{7}$$

where $B_\nu(T_{\mathrm{cont}})$ is the Planck function with a continuum temperature T_{cont}, $B_\nu(T_{\mathrm{dust}})$ is the Planck function with the characteristic dust temperature, $\kappa_\nu^{i,j}$ is the mass-absorption coefficient for species j and grain size i, and F_ν^{PAH} is the PAH template spectrum. C_0, $C_{i,j}$, and C_{PAH} are weighting factors.

The strongest limitations of these approaches are that the underlying continuum of the optically thick dusty disk is assumed to be well represented by a Planck function, which is never the case in realistic disk models (e.g., Chiang & Goldreich 1997; Men'shchikov & Henning 1997; Dullemond et al. 2001; Dullemond & Dominik 2004), and that even in the narrow spectral range from 8 to 13 μm, grains of quite different temperatures contribute to the emission.

Recently, Juhász et al. (2009) developed a more realistic but still fast approach, the two-layer temperature-distribution method, in which a continuous distribution of temperatures is applied, rather than a fixed temperature, and the correct continuum is calculated. The continuum emission below the silicate features consists of three components: a high-temperature component from the star and from the inner rim of the disk, a low-temperature component from the cold disk midplane, and the optically thin emission from featureless grains (e.g., carbon) from the disk atmosphere. This again illustrates that the real continuum is complex and cannot (and should not) be modeled by a Planck function with a single temperature.

In the two-layer temperature-distribution method, the observed monochromatic flux F_ν is given by

$$F_\nu = F_{\nu,\mathrm{atm}} + D_0 \frac{\pi R_\star^2 B_\nu(T_\star)}{d^2} + D1 \int_{T_{\mathrm{rim},0}}^{T_{\mathrm{rim,min}}} \frac{2\pi}{d^2} B_\nu(T) T^{\frac{2-q}{q}} \, dT + D2 \int_{T_{\mathrm{mid},0}}^{T_{\mathrm{mid,min}}} \frac{2\pi}{d^2} B_\nu(T) T^{\frac{2-q}{q}} \, dT. \quad (8)$$

Here $R_\star$ and $T_\star$ are the radius and temperature of the central star, $B_\nu(T)$ is the Planck function, and q is the exponent of the temperature distribution, which is assumed to be a power law. The subscripts "atm", "rim", and "mid" refer to the disk atmosphere, puffed-up inner rim, and disk midplane, respectively. Note that one fits the value of q for the disk atmosphere, inner rim, and disk midplane separately. $F_{\nu,\mathrm{atm}}$ denotes the flux of the optically thin disk atmosphere, which is given by

$$F_{\nu,\mathrm{atm}} = \sum_{i=1}^{N} \sum_{j=1}^{M} D_{i,j} \kappa_{i,j} \int_{T_{\mathrm{atm},0}}^{T_{\mathrm{atm,min}}} \frac{2\pi}{d^2} B_\nu(T) T^{\frac{2-q}{q}} \, dT, \quad (9)$$

where N and M are the number of dust species and of grain sizes used, respectively. The temperature range for the integrals is fixed by the contribution from grains of different temperatures to the emission of the actually analyzed spectral range. In other words, the outer disk does not contribute and does not need to be considered when one is analyzing the emission in the 10 μm range. In this method (and the other methods), one assumes that the dust mixture is uniform over the fitted range of disk temperatures and corresponding disk radii. This means that one has to split the wavelength interval if one wants to analyze the whole spectral region covered by the *ISO* or *Spitzer Space Telescope.*

Juhász et al. (2009) compared the robustness of the various methods mentioned above with the aid of synthetic disk spectra, calculated with a two-dimensional (2D) radiative transfer code, so that the input dust composition is known. They showed that the two-layer temperature distribution method does the best job in retrieving the input composition. Furthermore, they showed that within the interval 5 to 35 microns the wavelength region between 7 and 17 microns is best suited to derive the various dust parameters if one assumes a dust composition close to that generally obtained in the analysis of disk spectra.

We note, however, that the two-layer temperature-distribution method has its own limitations because it assumes that all dust species have the same temperature at a given disk location and does not consider the influence of the amount of disk flaring. This may lead to spurious inverse correlations between the derived mass-averaged grain size and the flaring of the disk because the radiative transfer effect, discussed at the beginning of this section, is not appropriately taken into account. In addition, the input grain model should contain the major dust species expected to be present in protoplanetary disks.

The next step in sophistication would be to use a real radiative transfer model, but this would need additional information to constrain the model. In the most recent models, it is now possible to include a realistic treatment of the irradiation of the inner disk by the central star and different flaring configurations, as well as dust sedimentation (see, e.g., Dullemond et al. 2001; Dullemond & Dominik 2008).

5. Mineralogy of Protoplanetary Dust

The derivation of the dust properties in protoplanetary disks can be tackled through the analysis of primitive material in the solar system as an analog for young disks, as well as through infrared spectroscopy of protoplanetary disks (see Henning 2003 for a comprehensive coverage of the field of astromineralogy). In this context, the *ISO* and *Spitzer* missions provided a legacy of spectroscopic data on protoplanetary disks. In the following, we will discuss the knowledge these space missions have provided about the dust properties in disks around young stars.

5.1. Intermediate-Mass Stars: The HAeBe Stars

Protoplanetary dust has most thoroughly been studied for disks around intermediate-mass (2 to 8 $M_{\odot}$) pre-main-sequence stars, the HAeBe stars. This is mainly because they are brighter than their lower-mass counterparts and provide spectra with high signal-to-noise ratios. The characterization of the dust in the disks around HAeBe stars made important steps forward with the launch of *ISO*, which provided high-quality infrared spectra, and also with ground-based data in the 8–13 μm window. Broad emission features from silicates (at $\sim$ 9.7 and 18 μm), features assigned to crystalline silicates, and PAH features (see § 2.4) have been observed (e.g., Malfait et al. 1998; Bouwman et al. 2001; Acke & van den Ancker 2004; van Boekel et al. 2005).

These observations showed that the dust composition and size distribution vary widely from object to object: some objects (e.g., AB Aur; Bouwman et al. 2000; van Boekel et al. 2005) have dust features typical of amorphous dust grains as found in molecular clouds and in the diffuse interstellar medium. Other HAeBe stars have dust features showing a large fraction of crystalline dust grains, similar to solar system bodies such as comets and interplanetary dust particles. The striking similarity between the silicate mineralogy of comet Hale-Bopp's dust and the dust around the isolated HAeBe star HD 100546 was both an exciting and surprising result, showing that some cometary material has seen partial processing at high temperatures and can serve as an analog for silicate minerals in disks (Malfait et al. 1998). An amazing observation with the *Spitzer Space Telescope* (Lisse et al. 2006) and ground-based N-band observations (Harker et al. 2007) was the detection of "crystalline" silicate features after the Deep Impact Encounter on comet 9P/Tempel 1. That encounter lifted grains from the cometary nucleus into the coma. Although the claims about the abundance of minor grain components from the analysis of the infrared *Spitzer* spectrum by Lisse et al. (2006) should be taken with some caution, the experiment clearly demonstrates the presence of crystalline silicates in the cometary nucleus. For a more detailed discussion of cometary grains and their implications for dust mineralogy and heating, as well as of radial mixing in protoplanetary disks, we refer to Hanner (2003) and Wooden et al. (2007).

The analysis of long-wavelength spectroscopy data ($\lambda > 20\ \mu m$) can provide a determination of the Mg/Fe content of the crystalline silicates, as discussed in § 2.3. Focused on the 69 μm feature, such an analysis demonstrated that crystalline olivines in the outflows of evolved stars are predominantly Mg rich (Molster et al. 2002). A similar result was obtained for the pyroxenes, based on the 40.5 μm feature. The exact determination of the chemical composition of crystalline silicates in the disks around young stars is more complicated because of a lack of relevant spectral features observed with high S/N ratio. The only source where the 69 μm feature has been detected through *ISO* observations was the bright HAeBe star HD 100546. Here the analysis also showed that the crystalline silicates are predominately Mg rich (e.g., Malfait et al. 1998; Bouwman et al. 2003). This result has recently been confirmed by high-quality data from the Heschel observatory (Sturm et al. 2010). The shorter-wavelength data on crystalline silicate bands in the spectra of disks around young stars (and comet Hale-Bopp) also point to a low iron content in crystalline silicates (e.g., Bouwman et al. 2008; Juhász et al. 2010).

Spitzer observations provided a compilation of high-signal-to-noise-ratio spectra for about 45 HAeBe disk sources (Juhász et al. 2010). These spectra provide a strong indication that forsterite grains dominate the dust composition

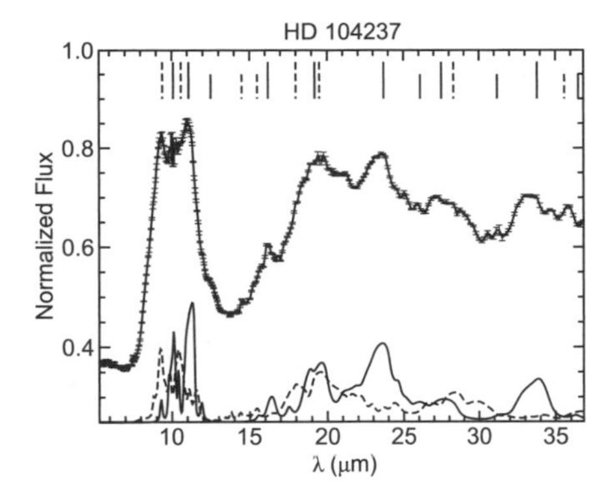

Fig. 4.5. Normalized *Spitzer* spectrum of the HAeBe star HD 104237. The positions of the forsterite (solid line) and enstatite (dashed line) bands are indicated. In addition, the mass-absorption coefficients for 0.1 μm forsterite and enstatite grains (distribution of hollow-spheres model, volume-equivalent radius) are shown. After Juhász et al. (2010).

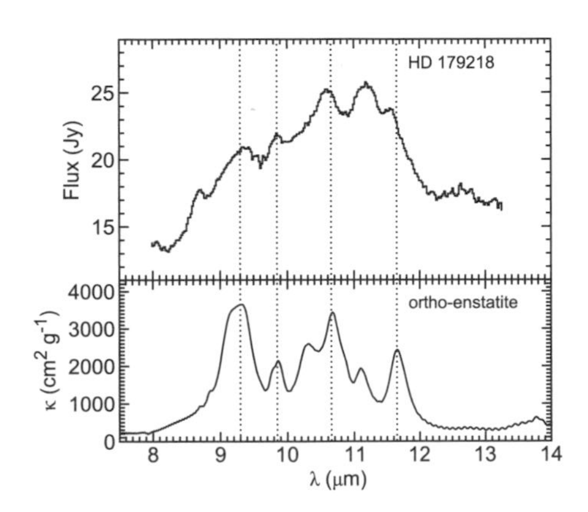

Fig. 4.6. The N-band spectrum of HD 179218 (upper panel) and the measured mass-absorption coefficients of ortho-enstatite from Chihara et al. (2002; lower panel). The wavelengths of the most prominent emission bands are indicated by the dotted lines. In this object, enstatite grains are an important constituent of the grain population that causes the 10 μm feature. After van Boekel et al. (2005).

in the outer disk, characterized by the long-wavelength data. In Fig. 4.5 we show the high-quality spectrum of the optically brightest HAeBe star, HD 104237.

Fig. 4.6 shows a high-quality ground-based spectrum of a HAeBe star that illustrates that such data can also provide important constraints on the dust composition. The spectrum shows very convincing evidence for the presence of enstatite in HD 179218.

Interferometric measurements with the Mid-infrared Instrument (*MIDI*) at the *Very Large Telescope Interferometer*, providing for the first time spatially resolved spectroscopy, demonstrate that dust processing by thermal annealing and coagulation is most efficient in the innermost parts of the disk (< 2 AU; van Boekel et al. 2004): a spatial gradient in amount of crystallinity and size distribution was found in the disks of 3 HAeBe stars. These data also indicated a radial dependence of the chemical composition of the crystalline silicate dust: olivines dominate in the inner disk, while pyroxenes dominate in the outer disk.

5.2. The Lower-Mass T Tauri Stars and Brown Dwarfs

The grain properties in the lower-mass ($\sim$ 1 solar mass) T Tauri stars (TTSs) have also been derived in the last few years, and a diversity in dust properties similar to that observed in HAeBe stars was found (e.g., Meeus et al. 2003; Przygodda et al. 2003). Thanks to the sensitivity of *Spitzer*, it soon became possible to study larger samples of TTSs, confirming the similarity in dust properties with the higher-mass stars (e.g., Forrest et al. 2004; Kessler-Silacci et al. 2006; Sicilia-Aguilar et al. 2007; Watson et al. 2009). Kessler-Silacci et al. (2006) also found that half of their sample (40 TTSs) show crystalline silicate features at longer wavelengths (33–36 μm). Sicilia-Aguilar et al. (2007) and Watson et al. (2009) found no correlation between the crystallinity and any stellar parameter but showed that, in general, the crystallinity in TTSs is relatively low (less than 20%). In Fig. 4.7 we show a comparison between the *Spitzer* spectrum of the disk around a low-mass star and the *ISO* spectrum of comet Hale-Bopp. The two spectra show remarkable similarity, indicating that comet-like material is present in the disks around TTSs.

In two different samples of TTSs, Bouwman et al. (2008) and Meeus et al. (2009) determined that the forsterite-to-enstatite ratio is low in the inner disk (1 AU), while forsterite dominates in the outer (5–15 AU), colder regions. The same results were obtained for the *Spitzer* sample of HAeBes (Juhász et al. 2010). This is in contradiction to the chemical-equilibrium calculations by Gail (2004), which predict that—if one assumes that the crystalline silicates form as high-temperature gas-phase condensates—forsterite is present only in the innermost regions where crystalline silicates can survive, while enstatite dominates in the more distant regions. Also, radial mixing, investigated in the study by Gail (2004), would not resolve this discrepancy, because this model still predicts that enstatite will dominate in the outer disk regions. Here one should keep in mind that the reaction rates for the conversion from forsterite to enstatite and vice versa are still relatively uncertain.

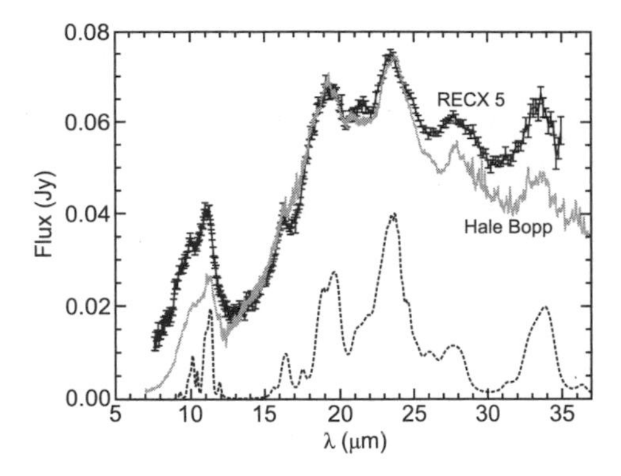

Fig. 4.7. *Spitzer* spectrum of the ~ 9 Myr-old M4-type star RECX5 compared with the *ISO* spectrum of comet Hale-Bopp (Crovisier et al. 1997). For identification purposes, we also show an emission spectrum (bottom dashed line) for a distribution of hollow forsterite spheres (volume-equivalent radius of 0.1 μm) at a temperature of 200 K.

Most likely, nonequilibrium processes contribute to the formation of enstatite (see the extensive discussion of this topic by Bouwman et al. 2008). As an alternative scenario for the formation of crystals, local heating in transient events has been discussed, either by lightning in the disks (Desch & Cuzzi 2000) or by shocks caused by gravitational instabilities (Harker & Desch 2002). Compositional and structural features of enstatite and forsterite in primitive chondrite matrices also point to formation through shock heating in the Solar nebula, at distances from 2 to 10 AU from the Sun (Scott & Krot 2005). Here we should note that potential correlations between stellar parameters and crystallinity may be erased by additional processes, such as amorphization of grains through ion irradiation associated with stellar activity (see, e.g., Jäger et al. 2003b).

In the mid-infrared spectra of a few T Tauri stars, taken with the *Spitzer Space Telescope*, Sargent et al. (2009) detected prominent narrow emission features at 9.0, 12.6, 20, and sometimes 16.0 μm, indicating the presence of crystalline SiO_2 (silica). Modeling suggests that the two polymorphs of silica, tridymite and predominantly cristobalite, which form at high temperatures and low pressures, are the dominant forms of silica responsible for the spectral features. This material is certainly largely the result of processing of primitive material in the protoplanetary disks around these stars. Tridymite and cristobalite, once formed, must be cooled quickly enough to keep their crystalline structure.

Finally, for brown dwarfs, the field of disk studies is quite new, but also here Spitzer has somewhat lifted the veil. The silicate emission feature observed in young brown dwarfs in Chamaeleon I (2 Myr, Apai et al. 2005) suggests dust processing similar to that in the young stellar disks: the 10 μm feature varies from source to source, with different degrees of disk flattening (due to larger grains). Remarkably, given their cooler temperature, a high degree of crystallinity (between 10% and 50%) is derived.

Here we note that the brown-dwarf spectra generally have a lower signal-to-noise ratio than those of the brighter stars, which makes the derivation of dust parameters less reliable. In addition, because of the low luminosity of brown dwarfs, we probe a physically much smaller region of their disks, so we see only those regions where grain growth is expected to be very fast (see also § 6). This may also explain the observational evidence for often flatter disk geometries in brown dwarfs, due to the fast settling of larger grains and the associated transition from flared to flat disk structures. We already noted in § 4.1 that a flatter disk may have a higher apparent crystallinity than a flared disk, even though the actual composition of the dust is identical; this may at least partly explain the brown dwarf results.

In this context it is interesting to note that the 10 μm feature in brown dwarfs seems to disappear quickly; in the 5 Myr-old Upper Scorpius region, the feature is either absent or very weak (Scholz et al. 2007), while the feature is completely absent in the three brown dwarfs observed in the TW Hya association (10 Myr; Morrow et al. 2008), so not much is known about dust evolution in brown dwarf disks.

5.3. PAHs and Nonsilicate Dust

PAH emission is widespread in disks around HAeBe stars (see § 2.4), with stronger features in more flaring disks and weak or absent features in geometrically flat disks (e.g., Acke & van den Ancker 2004). This can be explained by the fact that flared disks intercept a larger fraction of the UV radiation from the central star than flat disks. However, Dullemond et al. (2007) found that dust sedimentation can enhance the infrared features of PAHs. For disks with low turbulence, the sedimentation causes the thermal (larger) dust grains to sink below the photosphere, while the PAHs still stay well mixed in the surface layer. The sedimentation of the larger grains would also lead to a reduced far-infrared flux. Therefore, this investigation predicts that sources with weak far-infrared flux have stronger PAH features, which is—at least among the HAeBe stars opposite to what has been observed. This suggests that sedimentation is not the only factor responsible for the weak mid- to far-infrared excess in some disks. We also refer to Keller et al. (2008) and Acke et al. (2010) for a discussion of this topic.

PAH emission is also expected to be less strong in the cooler T Tauri stars because the PAH molecules are excited by UV photons. Geers et al. (2007) showed that at least 8% of TTSs show PAH features, but their latest spectral type is only G8. Bouwman et al. (2008) report the first detection of the 8.2 μm PAH feature in young low-mass objects and observe this feature in 5 of their 7 sources.

It is interesting to note that apart from the discussed silicates and PAHs, not much convincing spectroscopic evidence for other grain components has been found, so far.

6. Evidence for Grain Growth

Grain growth in protoplanetary disks is a complex process, driven by gas-grain dynamics that lead to collisions between the particles and finally coagulation (Beckwith et al. 2000; Henning et al. 2006; Dominik et al. 2007; Natta et al. 2007). A state-of-the art grain-growth model for the conditions of protoplanetary disks has recently been developed that takes into account radial drift and vertical sedimentation, as well as coagulation and the relevant microphysics (Brauer et al. 2008).

In the ISM, the mass-averaged grain size for silicates is smaller than 0.1 micron (Kemper et al. 2004). In young disks, the derived average grain size varies strongly between objects (e.g., Bouwman et al. 2001; van Boekel et al. 2005), but it is generally found to be much larger than in the diffuse interstellar medium. Van Boekel et al. (2003) related the shape and the strength of the 10 μm feature and showed that this relation provides proof for grain growth (see Fig. 4.8): strong and triangular 10 μm features are typical of submicron-sized grains, whereas a weaker and flattened structure indicates the presence of grains with sizes between 2 and 4 μm. Here we should again note that the shape of the feature can be influenced by other parameters than size (Min et al. 2006; Voshchinnikov et al. 2006; Voshchinnikov & Henning 2008). The actual particle size can be underestimated if one uses homogeneous spheres instead of fractal dust aggregates, as pointed out by Min et al. (2006).

Spitzer observations of 40 TTSs by Kessler-Silacci et al. (2006) reveal evidence for fast grain growth to micron-sized grains in the disk surface. They did not find a correlation either with age of the systems or with accretion rates. Kessler-Silacci et al. (2006) further found that later-type (M) stars show flatter 10 μm features (pointing to larger grain sizes) than earlier-type (A/B) stars. This finding was confirmed by Sicilia-Aguilar et al. (2007), who reported that strong silicate features are a factor of $\sim$ 4 less frequent among disks

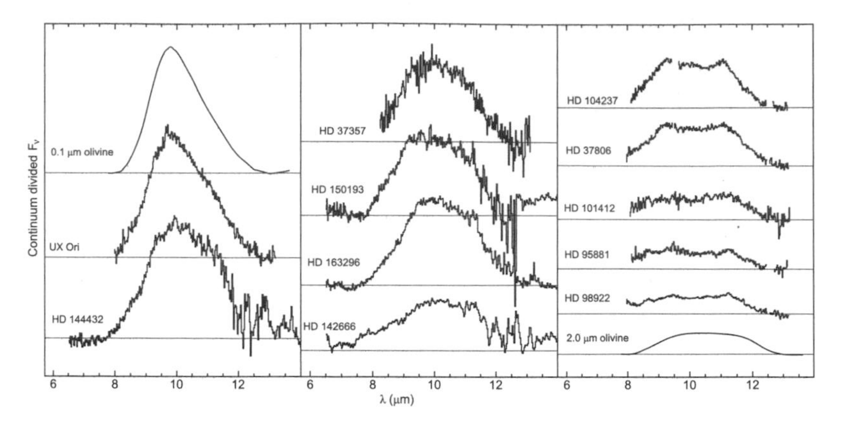

Fig. 4.8. Infrared spectra of HAeBe stars ordered by peak strength, illustrating the size effect on the shape and strength of the silicate feature. After van Boekel et al. (2003).

around M-type stars than around stars of earlier type. This inverse relation between stellar luminosity and grain size was further investigated in a study by Kessler-Silacci et al. (2007). It can easily be explained by a different location being probed at 10 μm: for later-type stars, this is more inward than for earlier types (e.g., TTSs 0.1–1 AU, versus HAeBe stars 0.5–50 AU). In the inner disk we would expect faster grain growth because of the higher collision rates of grains. This may also explain the fast disappearance of the silicate features in the spectra of brown dwarfs, discussed in the previous section.

Sicilia-Aguilar et al. (2007) found a somewhat counterintuitive relation for TTSs in the cluster Tr37. Only the youngest (0–2 Myr) objects showed evidence of large grains by flat and weak spectral features, while features typical of submicron-sized grains were seen in the oldest objects ($>$ 6 Myr), indicating that dust settling removes larger grains from the disk atmosphere. In addition, objects with a lot of turbulence (witnessed by larger accretion rates) have larger dust: it is likely that the turbulence supports large grains against settling toward the disk midplane. They also relate the observed weakness of silicate emission features in later-type objects to inner disk evolution.

Dullemond & Dominik (2008) investigated the effect of differential settling of grains on the appearance of the silicate feature. They confirmed that sedimentation can turn a "flat" feature into a "triangular" one, but only to a limited degree and for a limited range of grain sizes. Only in the case of a bimodal size distribution, i.e., a very small grain population and a bigger grain population, is the effect strong. If sedimentation were the sole cause of the feature variation, one would expect disks with weak mid- to far-infrared excess to have a stronger 10 μm silicate feature than disks with a strong excess at these wavelengths, but this is not what has been observed in a sample of 46 HAeBe stars with a wide range of mid- to far-infrared excesses (Acke & van den Ancker 2004).

In a sample of 24 HAeBe stars, van Boekel et al. (2005) derived the composition and grain size of the dust with the single-temperature method (see § 4.2). They used this decomposition to derive the mass fraction of the observable grains in the 10 μm region and found that a high crystallinity (above 10%) is observed only in those cases where the mass fraction in large grains is higher than 85%, which suggests that crystallization and grain growth are related. However, this may partly be a contrast effect, because the strength of the amorphous feature becomes weaker with increasing grain size, thus more clearly revealing crystalline features.

For grains with sizes larger than about 5 μm, 10 μm spectroscopy will no longer be able to trace such particles, which anyway may sediment below the optically thin atmosphere. Here millimeter observations of HAeBe stars and TTSs have provided evidence for the presence of even centimeter-sized grains in protoplanetary disks (see, e.g., Natta et al. 2007 for a review). In case the disks are optically thin at millimeter wavelengths—a feature that can be expected because of the low mass-absorption coefficients at these wavelengths—the slope of the spectral energy distribution ($F_\nu \propto \nu^\alpha$) can be directly related to the frequency dependence of the mass-absorption coefficient ($\kappa(\nu) \propto \nu^\beta$) via the relation $\beta = \alpha - 2$. Here the Planck function is represented by its Rayleigh-Jeans approximation. For typical submicron-sized dust grains in the diffuse ISM, the spectral index β has been found to be close to 2 (e.g., Draine & Lee 1984). For particles that are very large compared with the wavelength, which will block the radiation by virtue of their geometric cross sections, the mass-absorption coefficient becomes independent of frequency (gray behavior). Particles of about the same size as the wavelength, for instance, 7 mm, selected for studies at the *Very Large Array* (*VLA*), i.e. pebble-sized particles, are expected to have spectral indices in the intermediate range. For silicate particles, β indices ≤ 1 suggest the presence of dust particles with millimeter sizes. Such values have been found in the *VLA* studies of disks around HAeBe stars (Natta et al. 2004) and TTSs (Rodmann et al. 2006), clearly indicating the presence of large particles.

7. Conclusions and Future Directions

The last decade has seen huge progress in our knowledge of dust properties in protoplanetary disks, which was made possible mostly through observations with both the *ISO* and the *Spitzer Space Telescope*. The first surprise was the vast variety in appearance of dust features around young stars. A more detailed analysis, however, showed that those dust features, although very different in appearance, could be related to a handful of dust species with different sizes and structures.

The dust in protoplanetary disks bears the signature of processing compared with ISM dust; the grains are clearly larger, and most young sources also show evidence for crystalline silicates.

Despite the large number of spectra now available to the community, it has proved difficult to pin down the relationships between the observed dust properties, on the one hand, and the stellar and disk properties, on the other hand. The obviously expected relation of increasing grain size with age, or of increasing crystallinity with higher temperature of the central star, has proved to be incorrect.

The problem lies in the fact that there are many parameters to take into account: age, stellar luminosity, disk flaring angle, dust sedimentation, and presence of a close companion, to name just a few. Furthermore, dust features cannot be directly related to a specific dust species and size: other parameters, such as the dust temperature or the shape of the dust particles, also play an important role. In addition, good laboratory data for astronomical dust species over the full observable wavelength range remain a much-needed ingredient. The same is true for reaction rates, which determine the conversion between different grain species.

Therefore, we expect to make more progress in the coming years by comparing the wealth of data provided by *Spitzer* in a thorough statistical study, eliminating as much as possible those parameters that can be determined, so that, e.g., in studying the size distribution, stars with a similar luminosity are being compared.

In the coming years, several new observatories and instruments will become available. With the launch of the *Herschel Space Observatory* in 2009, covering the far-IR wavelength region (Photodector Array Camera and Spectrometer (PACS), 57–210 μm), we now are able to study the lattice vibrations of heavy ions or ion groups with low bond energies with a high signal-to-noise ratio. In particular, studies of the forsterite band at 69 μm will be promising in the context of determining the dust temperature and the composition of the olivines, while aqueous alteration can be studied through features of hydrous silicates at 100–110 μm.

Further on the horizon lies the launch of the *James Webb Space Telescope* (JWST), where the unprecedented sensitivity will allow us to study the disks of brown dwarfs with the same ease with which we now study T Tauri stars, opening up yet another region of the parameter space in dust-processing studies.

Last but not least, much can be learned from comparing observations of protoplanetary disks and dust with the composition of primitive material in the solar system, provided by the analysis of meteoritic material and interplanetary dust particles of cometary origin as collected by the *STARDUST* mission.

References

Acke, B., van den Ancker, M. E. 2004, A&A, 426, 151.

Acke, B., van den Ancker, M. E., Dullemond, C. P., van Boekel, R., Waters, L. B. F. M. 2004, A&A, 422, 621.

Acke, B., Bouwman, J., Juhász, A. et al. 2010, APJ, 718, 558.

Acke, B., van den Ancker, M. E. 2006, A&A, 457, 171.

Apai, D., Pascucci, I., Bouwman, J., Natta, A., Henning, Th., Dullemond, C. P. 2005, Science, 310, 834.

Allamandola, L. J., Tielens, A. G. G. M., Barker, J. R. 1985, ApJ, 290, L25.

Bakes, E. L. O., Tielens, A. G. G. M., Bauschlicher, C. et al. 2001, ApJ, 560, 261.

Beckwith, S. V. W., Henning, Th., Nakagawa, Y. 2000, in Mannings, V., Boss, A. P., Russell, S. S. (eds.), Protostars & Planets IV, University of Arizona Press, Tucson, p. 533.

Begemann, B., Dorschner, J., Henning, Th., Mutschke, H., Thamm, E. 1994, ApJ, 423, L71.

Boersma, C., Bouwman, J., Lahuis, F., et al. 2008, A&A, 484, 241.

Bohren, C. F., Huffman, D. R. 1998, Absorption and Scattering of Light by Small Particles, J. Wiley & Sons, New York.

Bouwman, J., de Koter, A., van den Ancker, M. E., Waters, L. B. F. M. 2000, A&A, 360, 213.

Bouwman, J., Meeus, G., de Koter, A. et al. 2001, A&A, 375, 950.

Bouwman, J., de Koter, A. Dominik, C., and Waters, L. B. F. M. 2003, A&A, 401, 577.

Bouwman, J., Henning, Th., Hillenbrand, L. A. et al. 2008, ApJ, 683, 479.

Bradley, J. P., Sandford, S. A., Walker, R. M. 1988, in Kerridge, J. F., Matthews, M. S. (eds.), Meteorites and the Early Solar System, University of Arizona Press, Tucson, p. 861.

Bradley, J. P. 1994, Science, 265, 925.

Bradley, J. P., Ishii, H. A. 2008, A&A 486, 781.

Brauer, F., Dullemond, C. P., Henning, Th. 2008, A&A, 480, 859.

Chiang, E. I., Goldreich, P. 1997, ApJ, 490, 368.

Chiar, J. E., Tielens, A. G. G. M. 2006, ApJ, 637, 774.

Chihara, H., Koike, C., Tsuchiyama, A., Tachibana, S., Sakamoto, D. 2002, A&A, 391, 267.

Colangeli, L., Henning, Th., Brucato, J. R., et al. 2003, A&A Rev., 11, 97

Crovisier, J., Leech, K., Bockélee-Morvan, D. 1997, Science, 275, 1904.

Desch, S. J., Cuzzi, J. N. 2000, Icarus, 143, 87.

Dominik, C., Blum, J., Cuzzi, J. N., Wurm, G. 2007, in Reipurth, B., Jewitt, D., Keil, K. (eds.), Protostars & Planets V, University of Arizona Press, Tucson, p. 783.

Dorschner, J., Henning, Th. 1995, A&A Rev., 6, 271.

Dorschner, J., Begemann, B., Henning, Th., Jäger, C., Mutschke, H. 1995, A&A, 300, 503.

Draine, B. T. 1988, ApJ, 333, 848.

Draine, B. T., Lee, H. M. 1984, ApJ, 285, 89.
Dullemond, C. P., Dominik, C., Natta, A. 2001, ApJ, 560, 957.
Dullemond, C. P., Dominik, C. 2004, A&A, 417, 159.
Dullemond, C. P., Apai, D., Walch, S. 2006, ApJ, 640, L67.
Dullemond, C. P., Henning, Th., Visser, R., et al. 2007, A&A, 473, 457.
Dullemond, C. P., Dominik, C. 2008, A&A, 487, 205.
Fabian, D., Jäger, C., Henning, Th., Dorschner, J., Mutschke, H. 2000, A&A, 364, 282.
Forrest, W. J., Sargent, B., Furlan, E., et al. 2004, ApJS, 154, 443.
Gail, H.-P. 1998, A&A, 332, 1099.
Gail, H.-P. 2003, in Henning, Th. (ed.), Astromineralogy, Springer, Berlin, p. 55.
Gail, H.-P. 2004, A&A, 413, 571.
Geers, V. C., Augereau, V.-C., Pontoppidan, K. M., et al. 2006, A&A, 459, 545.
Geers, V. C., van Dishoeck, E. F., Visser, R., et al. 2007, A&A, 476, 279.
Goto, M., Henning, Th., Kouchi, A., et al. 2009, ApJ, 693, 610.
Grossman, L. 1972, Geochimica et Cosmochimica Acta, 36, 597.
Guillois, O., Ledoux, G., Reynaud, C. 1999, ApJ, 521, L133.
Hanner, M. S. 2003, in Henning, Th. (ed.), Astromineralogy, Springer, Berlin, p. 171.
Harker, D. E., Desch, S. J. 2002, ApJ, 565, L109.
Harker, D. E., Woodward, C. E., Wooden, D. H., Fisher, R. S., Trujillo, C. A. 2007, Icarus, 191, 432.
Henning, Th., Stognienko, R. 1996, A&A, 311, 291.
Henning, Th., Mutschke, H. 1997, A&A, 327, 743.
Henning, Th., Salama, F. 1998, Science, 282, 2204.
Henning, Th., Ilin, V. B., Krivova, N. A., Michel, B., Voshchinnikov, N. V. 1999, A&A Suppl. Ser., 136, 405.
Henning, Th. (ed.) 2003, Astromineralogy, Springer, Berlin.
Henning, Th., Jäger, C., Mutschke, H. 2004, in Witt, A. N., Clayton, G. C., Draine, B. T. (eds.), Astrophysics of Dust, ASP Conf. Ser. 309, p. 603.
Henning, Th., Dullemond, C. P., Wolf, S., Dominik, C. 2006, in Klahr, H., Brandner, W. (eds.), Planet Formation, Cambridge University Press, Cambridge, p. 112.
Henning, Th. 2009, in Boulanger, F., Joblin, C., Jones, A., Madden, S. (eds.), Interstellar Dust: From Astronomical Observations to Fundamental Studies, EDP Sciences, EAS Publ. Ser. 35, p. 103.
Henning, Th. 2010, Ann. Rev. Astron. Astrophys., 48, 21.
Hill, H. G. M., d'Hendecourt, L. B., Perron, C., Jones, A. P. 1997, M&PS 32, 713.
Jäger, C., Molster, F. J., Dorschner, J., et al. 1998, A&A, 339, 904.
Jäger, C., Dorschner, J., Mutschke, H., Posch, Th., Henning, Th. 2003a, A&A, 408, 193.
Jäger, C., Fabian, D., Schrempel, F., Dorschner, J., Henning, Th., Wesch, W. 2003b, A&A, 401, 57.
Jessberger, E. K., Kissel, J., Rahe, J. 1989, in Atreya, S. K., Pollack, J. B., Matthews, M. S. (eds.), Origin and Evolution of Planetary and Satellite Atmospheres, University of Arizona Press, Tucson, p. 167.
Joblin, C., Boissel, P., de Parseval, P. 1997, P&SS, 45, 1539.
Juhász, A., Henning Th., Bouwman J., et al. 2009, ApJ, 695, 1024.
Juhász, A., Bouwman, J., Henning Th., et al. 2010, ApJ, 721, 431.
Keller, L. D., Sloan, G. C., Forrest, W. J., et al. 2008, ApJ, 684, 411.
Kemper, F., Vriend, W. J., Tielens, A. G. G. M. 2004, ApJ, 609, 826.
Kessler-Silacci, J., Augereau, J.-C., Dullemond, C. P., et al. 2006, ApJ, 639, 275.

Kessler-Silacci, J., Dullemond, C. P., Augereau, J.-C., et al. 2007, ApJ, 659, 680.
Krot, A. N., Fegley, B., Jr., Lodders, K., Palme, H. 2000, in Mannings, V., Boss, A. P., Russel, S. S. (eds.), Protostars & Planets IV, University of Arizona Press, Tucson, p. 1019.
Larimer, J. W. 1967, Geochimica et Cosmochimica Acta, 31, 1215.
Lattimer, J. M., Schramm, D. N., Grossman, L. 1978, ApJ, 219, 230.
Lisse, C. M., VanCleve, J., Adams, A. C. et al. 2006, Science, 313, 635.
Malfait, K., Waelkens, C., Waters, L. B. F. M. 1998, A&A, 332, L25.
Meeus G., Waters, L. B. F. M., Bouwman, J., et al. 2001, A&A, 365, 476.
Meeus G., Sterzik M., Bouwman J., et al. 2003, A&A, 409, L25.
Meeus G., Juhász, A., Henning Th., et al. 2009, A&A, 497, 379.
Men'shchikov, A. B., Henning, Th. 1997, A&A, 318, 879.
Michel, B., Henning, Th., Stognienko, R., Rouleau, F. 1996, ApJ, 468, 834.
Mie, G. 1908, Ann. Phys., 25, 377.
Min, M., Hovenier, J. W., de Koter, A. 2005, A&A, 432, 909.
Min, M., Dominik, C., Hovenier, J. W., Koter, A., Waters, L. B. F. M. 2006, A&A, 445, 1005.
Min, M., Waters, L. B. F. M., de Koter, A., et al. 2007, A&A, 462, 667.
Min, M., Hovenier, J. W., Waters, L. B. F. M., de Koter, A. 2008, A&A, 489, 135.
Molster, F. J., Waters, L. B. F. M., Tielens, A. G. G. M., Koike, C., Chihara, H. 2002, A&A 382, 241.
Moreels, G., Clairemidi, J., Hermine, P., et al. 1994, A&A, 282, 643.
Morrow, A. L., Luhman, K. L., Espaillat, C., et al. 2008, ApJ, 676, 143.
Mutschke, H., Begemann, B., Dorschner, J., Henning, Th. 1994, Infrared Physics & Technology, 35, 361.
Natta, A., Testi, L., Neri, R., Shepherd, D. S., Wilner, D. J. 2004, A&A, 416, 179.
Natta, A., Testi, L., Calvet, N., Henning, Th., Waters, R., Wilner, D. 2007, in Reipurth, B., Jewitt, D., Keil, K. (eds.), Protostars & Planets V, University of Arizona Press, Tucson, p. 767.
Nuth, J. A., Donn, B. 1982, J. Chem. Phys., 77, 2639.
Ossenkopf, V., Henning, Th., Mathis, J. S. 1992, A&A, 261, 567.
Palik, E. (ed.) 1985, Handbook of Optical Constants of Solids, New York Academic Press, New York.
Palik, E. (ed.) 1991, Handbook of Optical Constants of Solids II, Boston Academic Press, Boston.
Palik, E, (ed.) 1997, Handbook of Optical Coustatus of Solids, Elsevier.
Peeters, E., Hony, S., van Kerckhoven, C., et al. 2002, A&A, 390, 1089.
Pirali, O., Vervloet, M., Dahl, J. E., et al. 2007, ApJ, 661, 919.
Pollack, J. B., Hollenbach, D., Beckwith, S., et al. 1994, ApJ, 421, 615.
Pontoppidan, K. M., Dullemond, C. P., van Dishoeck, E. F., et al. 2005, ApJ, 622, 463.
Przygodda, F., van Boekel, R., Abraham, P., et al. 2003, A&A, 412, L43.
Rodmann, J., Henning, Th., Chandler, C. J., Mundy, L. G., Wilner, D. J. 2006, A&A, 446, 211.
Sandford, S. A. 1996, M&PS, 31, 449.
Sargent, B. A., Forrest, W. J., Tayrien, C., et al. 2009, ApJ, 690, 1193.
Scholz, A., Jayawardhana R., Wood, K., et al. 2007, ApJ, 660, 1517.
Scott, E. R. D., Krot, A. N. 2005, ApJ, 623, 571.
Semenov, D., Henning, Th., Helling, C., et al. 2003, A&A, 410, 611.
Sicilia-Aguilar, A., Hartmann, L. W., Watson, D., et al. 2007, ApJ, 659, 1637.
Sloan, G. C., Keller, L. D., Forrest, W. J., et al. 2005, ApJ, 632, 956.

Sturm, B., Bouwman, J., Henning, Th., et al. 2010, A&A, 518, L129.
Terada, H., Imanishi, M., Goto, M., Maihara, T. 2001, A&A, 377, 994.
Tielens, A. G. G. M., Waters, L. B. F. M., Bernatowicz, T. J. 2005, ASPC, 341, 605.
van Boekel, R., Waters, L. B. F. M., Dominik, C. et al. 2003, A&A, 400, L21.
van Boekel, R., Min, M., Leinert, Ch., et al. 2004, Nature, 432, 479.
van Boekel, R., Min, M., Waters, L. B. F. M., et al. 2005, A&A, 437, 189.
van Kerckhoven, C., Tielens, A. G. G. M., Waelkens, C. 2002, A&A, 384, 568.
Voshchinnikov, N. V., Henning, Th. 2008, A&A, 483, L9.
Voshchinnikov, N. V., Il'in, V. B., Henning, Th., Michel, B., Farafonov, V. G. 2000, JQRST, 65, 877.
Voshchinnikov, N. V., Il'in, V. B., Henning, Th., Dubkova, D. N. 2006, A&A, 445, 167.
Watson, D. M., Leisenring, J. M., Furlan, E., et al. 2009, ApJS, 180, 84.
Whittet, D. C. B., Williams, P. M., Zealey, W. J., Bode, M. F., Davies, J. K. 1983, A&A, 123, 301.
Whittet, D. C. B., McFadzean, A. D., Geballe, T. R. 1984, MNRAS, 211, 29.
Wooden, D., Desch, S., Harker, D., Gail, H.-P., Keller, L. 2007, in Reipurth, B., Jewitt, D., Keil, K. (eds.), Protostars & Planets V, University of Arizona Press, Tucson, p. 815.
Zasowski, G., Kemper, F., Watson, D. M., et al. 2009, ApJ, 694, 459.
Zolensky, M. E., Zega, Th. J., Yano, H., et al. 2006, Science, 314, 1735.

5

RICHARD H. DURISEN

DISK HYDRODYNAMICS

1. Introduction

I assert, among all things of nature whose first cause one investigates the origin of the world system and the formation of celestial bodies together with the causes of their motions is the one which one may hope to grasp first in a fundamental and satisfactory way. . . . Give me matter, I will show you how a world might arise from it. For if there is matter available which is endowed with an essential attractive force, then it is not difficult to determine the causes which can contribute to the arrangement of the world, considered at large.—(Kant, 1755; translated by S. L. Jaki)

Kant and Laplace are credited with being the earliest proponents of the so-called nebular hypothesis, the idea that our own planetary system evolved into its current state from a swirling disk of dust and gas around the young Sun. The above quote from Kant's "Opening Discourse" to *Universal Natural History and Theory of the Heavens* (Kant, 1755) may be viewed by modern readers as naively optimistic, but the spirit of Kant's remarks remains with us today as we attempt to understand from simple physical principles how matter distributed in space can pull itself together into the elegant and cozy architecture of a solar system. Although details of their 18th- and 19th-century cosmogonics were wrong, the genius of these men was to be among the first to envision the origin of planetary systems as a natural evolutionary process guided by physical laws. Although free energy, deterministic chaos, nonlinearity, and environmental effects, among others, endow the problem with a complexity that Kant and Laplace could hardly have imagined, we can in fact learn a great deal about planetary system formation and disk evolution through arguments from first principles.

This chapter focuses on the theory of protoplanetary gas disks as mechanical systems, specifically hydrodynamic systems. We will find, of course, that we are quickly impelled to consider thermodynamics and radiative physics in order to understand more fully how disks behave. By design, this chapter will avoid extensive treatment of magnetic-field effects (see the chapters by Balbus and by Königl and Salmeron), except as one possible source of turbulence. The big questions guiding our discussion will be the following:

- What hydrodynamic processes affect the evolution of disks?
  - Which can and do alter their structure?
  - Which, if any, play a role in planet formation?
- Which of these hydrodynamic processes can transport mass and angular momentum in disks?
  - How efficient are they?
  - Are they local or global?

Because of the focus on transport and global structure, there are a number of interesting disk phenomena that I will not discuss, such as bending waves (see Ogilvie, 2000) and migration of embedded objects (see Papaloizou and Terquem, 2006; Papaloizou et al., 2007).

I strive to develop each topic clearly from its basics so that this chapter can serve as a primer on the subject for future researchers. However, the confines of a chapter length do not permit me to provide many derivations. Instead, I will attempt to introduce results with clear explanations and offer references where the derivations can be found. I have come to disks from the direction and background of work on star formation and the structure and stability of rotating stars, and so my perspective has been shaped over the years by several books and review articles in that area (notably Chandrasekhar, 1969; Fricke and Kippenhahn, 1972; Tassoul, 1978; Cox, 1980; Tohline, 1982; Durisen and Tohline, 1985; Bodenheimer, 1995). For specific processes, texts on hydrodynamics and galactic dynamics are helpful (e.g., Landau and Lifshitz, 1959; Chandrasekhar, 1961; Binney and Tremaine, 1987), and I draw heavily from reviews on disk transport mechanisms by Pringle (1981), Papaloizou and Lin (1995), Balbus and Hawley (1998), Balbus (2003), and Gammie and Johnson (2005). Of course, the Hartmann (2009) book on disks around young stars is indispensable and strongly influences the structure and content of this chapter. The reader will also find two very accessible introductions to astrophysical fluid dynamics with contemporary applications in Shu (1992) and Clarke and Carswell (2007).

The chapter begins in § 2 with the basics of disk equilibrium and dynamics, classic views about disk origin, and an introduction to the evolution of disks due to internal stresses. The quest for the origin of stresses that cause transport leads in § 3 to consideration of various instabilities that might produce turbulence in disks. I first describe each mechanism in its simplest form and then consider how it may or may not apply to disks. Although some other interesting contenders remain, such as baroclinic instabilities and instabilities driven by structures in disks, there are two mechanisms, namely, gravitational and magnetorotational instabilities, that are considered most likely to produce strong transport of mass and angular momentum. The rest of this chapter focuses on what is currently known about gas-phase gravitational instabilities (GIs) in protoplanetary disks. Section 4 presents areas of agreement about how GIs behave under idealized conditions, when the disk physics is kept relatively simple, such as isothermal or polytropic equations of state and simple cooling prescriptions. Efforts to include more realistic radiative physics, as presented in § 5, have not produced consensus results at the time of writing. Because of its importance for the question of giant-planet formation, § 5 describes this contentious area in some detail and lays out paths to a resolution. Section 5 ends with a smorgasbord of special topics that tie GIs to other aspects of disk evolution, and I permit myself to indulge in some pet speculations about scenarios that may integrate a range of phenomena. Section 6 offers some concluding remarks.

2. Hydrodynamic Theory of Gas Disks

A chapter on the hydrodynamics of disks must begin with the partial differential equations of hydrodynamics:

$$\frac{\partial \mathbf{v}}{\partial t} + \mathbf{v} \cdot \nabla \mathbf{v} = -\frac{1}{\rho} \nabla P - \nabla \Phi + \frac{1}{\rho} \nabla \cdot \tau_\nu, \tag{1}$$

$$\frac{\partial \rho}{\partial t} + \nabla \cdot \rho \mathbf{v} = 0, \tag{2}$$

$$\rho \left(\frac{\partial e}{\partial t} + \mathbf{v} \cdot \nabla e \right) = -P \nabla \cdot \mathbf{v} + D_\nu + \Gamma - \Lambda - \nabla \cdot \mathbf{F}, \tag{3}$$

and

$$\nabla^2 \Phi_d = 4\pi G \rho. \tag{4}$$

Eqs. (1)–(4) are the inertial frame momentum equation (or equation of motion), the mass continuity equation, the internal energy equation as derived from the first law of thermodynamics, and Poisson's equation for disk self-gravity. The variables are defined as follows; ρ is the gas mass density, $\mathbf{v}$ the gas bulk velocity, P the gas pressure, Φ the total gravitational potential, τ_ν the viscous stress tensor, e the specific internal energy, D_ν the viscous dissipation function, Γ the local heating rate per unit volume (*volumetric heating*), Λ the cooling rate per unit volume (*volumetric cooling*), $\mathbf{F}$ the net energy flux due to thermal transport mechanisms like radiation and conduction, and Φ_d the contribution to the gravitational potential from the disk itself. The full gravitational potential can be a sum over various components,

$$\Phi = \Phi_s + \Phi_d + \Phi_{ext}, \tag{5}$$

where $\Phi_s = -GM_s/r$ is the gravitational potential of the central star referenced to infinity, with the radial coordinate r centered on the star, and Φ_{ext} is a possible contribution from an external potential, e.g., a binary companion or massive objects embedded in the disk. In the latter case, an indirect potential can be added to cast the equations into the accelerated reference frame of the star (e.g., R. P. Nelson et al., 2000b).

The two viscous terms involve shear and divergence in the flow in all directions. They can be found written out fully in cylindrical, spherical, and Cartesian coordinates in Appendix B of Tassoul (1978). In this chapter I will mainly consider the shear terms in a nearly axisymmetric Keplerian disk, and I will typically write the fluid equations in cylindrical coordinates ($\varpi-, \phi-, z$) centered on the star. Hereafter, the terms *radial*, *azimuthal*, and *vertical* will be used to refer to the $\varpi-$, $\phi-$, and z-directions, respectively. The important viscous term in (1) for our purposes is

$$\tau_{\varpi\phi} = \tau_{\phi\varpi} = \mu\varpi \frac{\partial \Omega}{\partial \varpi}, \tag{6}$$

and the heating rate due to viscous friction is the stress multiplied by the rate of strain, or

$$D_\nu = \tau_{\varpi\phi}\left(\varpi \frac{\partial \Omega}{\partial \varpi}\right) = \mu\left(\varpi \frac{\partial \Omega}{\partial \varpi}\right)^2. \tag{7}$$

In (6) and (7), $\mu = \rho\nu$ is the dynamic shear viscosity, ν is the kinematic shear viscosity, and $\Omega = v_\phi/\varpi$ is the angular rotation rate of the fluid about the z-axis. Eqs. (6) and (7) characterize the force and energy dissipation due to the rotational shear in a nearly Keplerian disk with viscosity ν.

Eqs. (1)–(7) must be supplemented by gas characteristic equations, such as the ideal gas equation of state,

$$P = \rho kT/m_{gas}, \tag{8}$$

where T is the gas temperature and m_{gas} is the mean molecular weight of the gas. The appropriate heating and cooling rates in (3) and the flow of energy by radiation must then be specified, plus an addition relation for $e(\rho, T)$. For the latter, it is sufficient to use

$$P = (\gamma - 1)\rho e \tag{9}$$

for the special case of an ideal gas with a constant value of $\gamma =$ the ratio of specific heats and no phase changes. Thermal convection can either be separately modeled in **F** or be allowed to develop naturally in dynamic calculations from solution of the above equations (e.g., Boley et al., 2007a). Inclusion of magnetic forces and the additional equations required to describe the magnetic field for ideal and nonideal magnetohydrodynamics (MHD) are discussed in the chapters by Balbus and by Königl and Salmeron.

2.1. The Basics of a Circumstellar Gas Disk

In standard accretion-disk theory, it is common to think about circumstellar gas disks as quasi-equilibrium axisymmetric states in pure rotation that evolve because of friction and dissipation of energy. The timescale for this evolution t_{ev} is long compared with the dynamic timescale t_{dyn} on which the disk changes because of force imbalance. For now, let us set $\Phi_{ext} = 0$. *Pure rotation* means that $\mathbf{v} = v_\phi \mathbf{i}_\phi = \varpi \Omega \mathbf{i}_\phi = (2\pi\varpi/t_{rot})\mathbf{i}_\phi$, where $\mathbf{i}_\phi$ is the unit vector in the ϕ-direction; *axisymmetry* means that all ϕ-derivatives are zero. The slow evolution involves additional flow, primarily radial, with $v_\varpi/v_\phi \sim t_{dyn}/t_{ev}$. Equilibrium, satisfied to order t_{dyn}/t_{ev}, is found by setting the left-hand side of (1) to zero. In the ϖ-direction, this gives a balance of pressure-gradient, centrifugal, and gravitational forces when

$$-\frac{1}{\rho}\frac{\partial P}{\partial \varpi} + \varpi\Omega^2 - \frac{GM_s\varpi}{r^3} - \frac{\partial \Phi_d}{\partial \varpi} = 0. \tag{10}$$

In the z-direction, pressure-gradient forces balance the vertical components of gravity, i.e.,

$$-\frac{1}{\rho}\frac{\partial P}{\partial z} - \frac{GM_s z}{r^3} - \frac{\partial \Phi_d}{\partial z} = 0. \tag{11}$$

It is easy to show that in the special case where $P = P(\rho)$ only, (10) and (11) imply $\partial\Omega/\partial z = 0$; in other words, "rotation is constant on cylinders" (the *Poincaré-Wavre theorem*; see chapter 4 of Tassoul, 1978).

DISKS AS THIN. Suppose that the pressure-gradient force in (10) is negligible compared with the other terms. Because $P \sim \rho {c_s}^2$, this implies by dimensional analysis of the first two terms in (10) that the gas sound speed $c_s << \varpi\Omega$. If the disk mass $M_d << M_s$, then (10) gives the rotation law for a Keplerian disk,

$$\Omega(\varpi) = \Omega_K(\varpi) = (GM_s/\varpi^3)^{1/2}. \tag{12}$$

When M_d is not negligible, $\Omega = (GM_\varpi/\varpi^3)^{1/2}$ is a fairly good approximation, where M_ϖ is the mass of the star plus disk inside the cylindrical radius ϖ. When disk self-gravity is negligible in the z-direction, dimensional analysis of (11) tells us that $c_s \approx \Omega H$, where the disk vertical half-thickness H can be defined by

$$\Sigma = \int_{-\infty}^{\infty} \rho dz = 2\rho_0 H \tag{13}$$

and ρ_0 is the midplane density. Note that H is determined by the vertical distribution of the bulk of the disk mass. The bulk of the mass is therefore vertically compact because $c_s << \varpi\Omega$ means that $H/\varpi << 1$. The extreme flaring observed in some Class 2 disks traces a hot disk atmosphere that contains relatively little mass (see the chapter by Calvet and D'Alessio). An interesting consequence of disk thinness is that the thermal internal energy of the disk gas is much smaller than the rotational kinetic energy of the disk. In fact, $e/v_\phi^2 \sim (c_s/\varpi\Omega)^2 \sim (H/\varpi)^2$. If $H/\varpi < 0.1$, then $e/v_\phi^2 < 0.01$.

When disks are thin ($H/\varpi << 1$), the viscous disk evolution equation can be derived from the time-dependent versions of eqs. (1) and (2) by the following simplifications. Ignore all ϕ derivatives by axisymmetry, assume that $v_\phi = \varpi\Omega_K$ with high accuracy because of the conditions $t_{dyn}/t_{ev} << 1$ and $H/\varpi << 1$, and integrate over the z-direction. As shown in Pringle (1981) and Hartmann (2009), the result is a viscous diffusion equation for disk surface density,

$$\frac{\partial\Sigma}{\partial t} = \frac{3\partial}{\varpi\partial\varpi}\left[\varpi^{1/2}\frac{\partial}{\partial\varpi}(\varpi^{1/2}\nu\Sigma)\right]. \tag{14}$$

If ν were known as a function of Σ, say, this equation alone could be solved for $\Sigma(\varpi, t)$. For constant ν, there is an analytic solution for a δ-function mass ring spreading radially outward and inward with time (Lynden-Bell and Pringle, 1974; Pringle, 1981). Asymptotically, all the mass accretes to the central star, while the disk angular momentum is carried away to infinity by a vanishingly small amount of the mass.

In an α-disk (Shakura and Sunyaev, 1973), it is assumed that the viscosity can be written as $\nu = \alpha c_{s0} H$, where α is a dimensionless parameter that specifies the strength of an effective "turbulent" viscosity and c_{s0} is the midplane sound speed. Dimensionally, eq. (14) tells us that the viscous evolution time for a disk is $t_{ev} \approx \varpi^2/\nu \approx (\varpi/H)^2(t_{rot}/2\pi\alpha)$, or

$$t_{ev} \approx 6 \cdot 10^6 \text{ yr} \left(\frac{0.01}{\alpha}\right) \left(\frac{0.05}{H/\varpi}\right)^2 \left(\frac{\varpi}{100 \text{ AU}}\right)^{3/2} \left(\frac{\text{M}_\odot}{M_s}\right)^{1/2}. \quad (15)$$

If the disk is in a steady state, then there is a constant mass inflow rate far from the boundaries given by $\dot{M} = 3\pi \nu \Sigma$, or

$$\dot{M} \approx 10^{-7} \text{ M}_\odot \text{ yr}^{-1} \left(\frac{\alpha}{0.01}\right) \left(\frac{H/\varpi}{0.05}\right)^2 \left(\frac{\varpi}{5 \text{ AU}}\right)^{1/2} \left(\frac{M_s}{\text{M}_\odot}\right)^{1/2} \left(\frac{\Sigma}{300 \text{ g cm}^{-2}}\right). \quad (16)$$

$\dot{M}$ is sustained by viscous torques that transport angular momentum outward. As discussed in § 3.3, outward transport of angular momentum in a Keplerian disk releases energy. In a steady-state disk, the energy dissipated by viscous friction must be radiated away and is given by

$$\int_{-\infty}^{+\infty} D_\nu dz = \nu \Sigma \left(\varpi \frac{d\Omega}{d\varpi}\right)^2. \quad (17)$$

Far from a disk boundary in a Keplerian disk, (14) and (17) give the energy radiated per unit disk area as $3GM_s\dot{M}/4\pi\varpi^3$. Half of it is radiated upward and half downward.

The α prescription for ν brings thermal physics into the problem through c_{s0}. If we use quantities integrated over the vertical direction, as in thin-disk modeling (e.g., Gammie, 2001), (9) becomes $\mathcal{P} = (\gamma_2 - 1)\mathcal{U}$, where $\mathcal{P} = \int_{-\infty}^{\infty} P dz$, $\mathcal{U} = \int_{-\infty}^{\infty} \rho e dz$, and γ_2 is the two-dimensional analog of the three-dimensional γ. The relation between γ_2 and γ is not straightforward. Using a plane-parallel sheet approximation, one can show that

$$\gamma_2 = (3\gamma - 1)/(\gamma + 1) \quad (18)$$

when the star dominates the vertical gravity (second term on the left-hand side of [11]) and

$$\gamma_2 = 3 - 2/\gamma \quad (19)$$

when disk self-gravity (third term) dominates (Gammie, 2001). For the special case of an isothermal gas, $\gamma_2 = \gamma = 1$, eq. (9) is invalid, and we may use c_s instead of c_{s0} in the definition of α without ambiguity.

Fig. 5.1 illustrates the evolution of a simple solar nebula with $\alpha = 0.01$ and an assumed $T(\varpi)$ to fix c_{s0}. An initial ring of material with 0.1 $\text{M}_\odot$ at 10 AU evolves into a disk in only 10^4 yr and is depleted by over a factor of 10 in 10^6 yr. Even in this simplistic case, the disk is only truly a steady-state disk over a restricted range of radii for a moderate interval of time. Nevertheless, as described in the chapter by Calvet and D'Alessio, steady-state models are

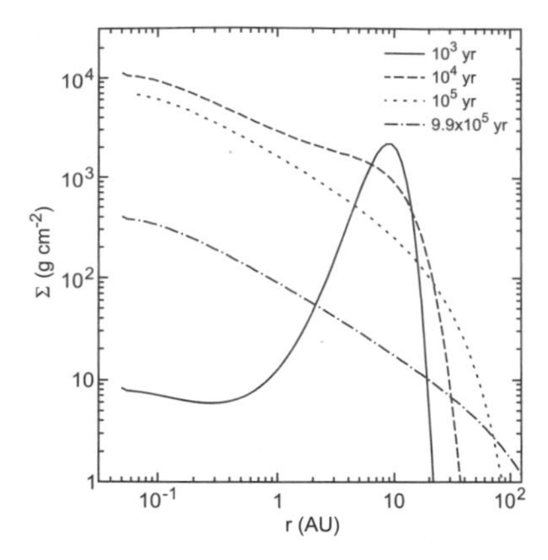

Fig. 5.1. Evolution of a simple α-disk model around a 1 $M_\odot$ star that assumes $\alpha = 10^{-2}$ and $T = 300$ K $(r/\text{AU})^{-3/4}$ and starts with a narrow ring of material at 10 AU. Adapted from a figure in Gammie and Johnson (2005).

the norm for modeling observed disks. The evolution of real disks depends strongly on how energy transport affects $T(\varpi)$.

DISKS AS THICK. Considering disks to be thin and integrating over the vertical structure obscure interesting features that can dominate disk appearance (see the chapter by Calvet and D'Alessio). In fact, writing the α-disk equations in full r and z, Takeuchi and Lin (2002) find that the drift velocities of gas responsible for the net transport of mass actually vary with height above the disk midplane and can even reverse sign. The radial flow tends to be inward at high disk altitudes and outward in the midplane, with significant implications for mixing and dust transport (see also Ciesla, 2009). Whether these "channel flows" become even more complex in three dimensions is not currently known. Nevertheless, if disks really do evolve slowly because of an effective viscosity, as envisioned above, then the thin-disk approximation is probably accurate enough to describe the overall mass transport of the gas, and vertical hydrostatic equilibrium probably remains valid over the bulk of each disk column, even in the presence of accretion onto the disk surface or wind outflows.

What about dynamic processes? When global disk dynamics are modeled in a thin-disk approximation, as in A. F. Nelson et al. (2000a) or Gammie (2001), it is effectively assumed that the vertical structure adjusts to the hydrostatic equilibrium of eq. (11) instantaneously. How realistic is that? The dynamic timescale, i.e., the time in which the disk responds to global force imbalances, is usually considered to be set by consideration of radial equilibrium, and so t_{dyn} is the disk rotation period t_{rot}. The timescale t_z associated with violations of vertical force balance is the vertical sound crossing time $2H/c_s$, the time it takes a sound wave to propagate from the disk midplane to the surface and back. Because $c_s/H \approx \Omega$ in a thin disk, we get $t_z \approx t_{dyn}/\pi$. In other words, the dynamic timescales in the radial and vertical directions are essentially the same. In this sense, for processes that happen on a dynamic timescale, no disk is thin, because the vertical structure does not adjust to hydrostatic equilibrium on a timescale much shorter than t_{rot}. Dramatic examples of this are the shock bores described in § 4.3.

The same conclusion can be reached from a different direction by examining ordered sequences of configurations whose members in one limit are in fact geometrically thick and have non-Keplerian rotation but whose other end members are "thin" and nearly Keplerian. Rotating, self-gravitating fluid configurations have been studied systematically since the late 18th century, when Maclaurin discovered the equilibrium states for a uniformly rotating incompressible fluid with an oblate spheroidal surface (see Chandrasekhar, 1969). With proper normalization (constant mass and density), the Maclaurin spheroids are a one-parameter family where the parameter can be taken to be total angular momentum, the eccentricity of the meridional cross section (in a plane including the rotation axis), or the ratio of the total rotational kinetic energy $\mathcal{T}$ to the absolute value of the total gravitational energy $|\mathcal{W}|$. It was shown in the 19th century that when $\mathcal{T}/|\mathcal{W}| \geq 0.274$, and equivalently eccentricity ≥ 0.953 (or $z_p/\varpi_{eq} \leq 0.303$, where z_p and ϖ_{eq} are the polar and equatorial radii, respectively), the so-called bar mode, which distorts the axisymmetric spheroid into an ellipsoid tumbling around the rotation axis, becomes dynamically unstable and grows exponentially.

What do these 18th- and 19th-century results have to do with disks? Since the 1960s researchers have been able to construct analogs of the Maclaurin spheroids for differentially rotating and compressible fluids and have found that they exhibit a similar dynamic barlike instability at a remarkably similar value of $\mathcal{T}/|\mathcal{W}|$, independent of details (see Tassoul, 1978; Durisen and Tohline, 1985). In compressible fluids, the central part of the barlike mode is indeed a radially coherent bar, but the eigenfunction breaks into a trailing spiral in the regions toward the equator that are more strongly rotationally supported. The analog Maclaurin sequences depend on the assumed distribution of angular momentum in the equilibrium configuration. Pickett et al. (1996) considered a series of such analog families where the angular momentum distribution was itself varied in an ordered way by a single parameter. When the angular momentum distribution puts more of the angular momentum into a disklike, nearly Keplerian equatorial region of the configuration, the character of the barlike instability changes from having a pronounced central bar to having an extensive region of tightly wrapped spirals. In extreme cases, multiple spiral modes can be simultaneously unstable, a behavior that will sound very familiar when we discuss gravitational instabilities in § 4. This tells us something about spiral instabilities that is missed in the vertically integrated treatments common in spiral wave analysis of galactic disks (e.g., Binney and Tremaine, 1987) or of gas disks before the 1990s.

The normal modes of rotating stars identified by linearizing eqs. (1)–(4) for adiabatic perturbations of an ideal gas can be classified as *p*-, *g*-, *r*-, and *f*-modes (Tassoul, 1978; Cox, 1980). The principal restoring forces for *p*-, *g*-, and *r*-modes are the *pressure*-gradient forces, *gravity*, and inertial forces due to *rotation*, respectively. The *f*- or *fundamental* modes, by contrast, do not have a single dominant restoring force. The low-order nonspherical *f*-modes are sometimes considered to be surface distortion modes because the *f*-mode eigenfunctions have large relative amplitudes near the surface and characteristically induce oscillations of the star's surface shape. The bar modes that become unstable in the Maclaurin spheroids can be identified as the $l = |m| = 2$ *f*-modes, where l and m are the poloidal and azimuthal mode numbers in a standard spherical harmonic expansion. The work of Pickett et al. (1996) suggests that the spiral waves appearing in gravitationally unstable disks are *f*-modes, i.e., they are surface distortions of the disk as much as they are "density" waves. The Pickett et al. models are isentropic fluids and have temperature stratification in the vertical direction. Three-dimensional (3D) hydrodynamics simulations of spiral waves in these models (Pickett et al., 1996, 2000, 2003) show strong vertical distortion of the disks and enhanced dissipation due to shocks at high disk altitudes. Analytic studies of external forcing in stratified disks (Lubow and Ogilvie, 1998) confirm that the disk response predominantly has the character of an *f*-mode with mode amplitude concentrated toward the disk surface. This is in contrast to a disk that is vertically isothermal, where the modes do not distort the surface and can more properly be called density waves. The situation for nonlinear waves is even more complex, as discussed in § 4.3.

So, even when a disk is geometrically thin, i.e., H is significantly $< \varpi$, the dynamics of the vertical direction are not simple, and disks changing on a dynamic timescale should be treated three-dimensionally. This becomes all the more evident when we learn that disk dynamics are sensitive to the energy flow in the vertical direction, a central theme of § 5.

2.2. How and Why Do Disks Form?

The fundamental reason that disks exist around young stars is that the typical specific angular momentum j of molecular cloud cores, about 10^{21} cm^2 s^{-1}, is orders of magnitude larger than the maximum j for a star in radial force balance, about 10^{17} cm^2 s^{-1} (see Table 1 of Bodenheimer, 1995). On the other hand, for Jupiter's and Neptune's orbits, $j \approx 10^{20}$ and $3 \cdot 10^{20}$ cm^2 s^{-1}, respectively. So a cloud that collapses to form a single star rather than a multiple star is likely to form a circumstellar disk of solar system size.

ILLUSTRATIVE SPECIAL CASE. The collapse of a uniformly rotating, singular isothermal sphere has been worked out analytically in great detail (Shu, 1977; Cassen and Moosman, 1981; Cassen and Summers, 1983; Terebey et al., 1984). Although it is debatable whether this applies in detail to any real case of star formation, it provides some insight into disk formation.

Define the free-fall time t_{ff} for a shell of radius r in a nonrotating spherical cloud as the time it would take the shell to collapse to the cloud center through gravity alone. As long as $d\rho/dr \leq 0$, shells will not cross during free fall, and the collapse of each shell proceeds as if the mass interior to the shell M_r initially were a central point mass. If each fluid element starts from rest, then its trajectory is a degenerate ellipse with an eccentricity of 1, a semimajor axis $r/2$, and a period $2t_{ff}$. Application of Kepler's third law then gives

$$t_{ff} = \left(\frac{3\pi}{32G\bar{\rho}}\right)^{1/2}, \tag{20}$$

where $\bar{\rho}$ is the initial average density inside the shell at the start of collapse.

Consider a sphere of ideal gas with an isothermal sound speed $c_i = (kT/m_{gas})^{1/2}$. In hydrostatic equilibrium, where only the pressure gradient supports the sphere against gravity, we deduce from eq. (1) that

$$\frac{dP}{dr} = -\frac{\rho G M_r}{r^2}. \tag{21}$$

It is easy to verify the singular isothermal sphere solution

$$\rho(r) = \frac{c_i{}^2}{2\pi G r^2} \tag{22}$$

by substitution of (22) into (21) together with (8). Because $\rho \sim r^{-2}$, $\bar{\rho}$ also $\sim r^{-2}$, which gives $t_{ff} \sim r$. Note that M_r also $\sim r$. The singular isothermal sphere is unstable, and collapse proceeds inside out because the inner shells have the shortest t_{ff}. Shu (1977) described a similarity solution for the collapse where an expansion wave moves outward at speed c_i. Inside the wave front the cloud is collapsing, while outside it is still hydrostatic. In a similarity solution, the central accretion rate $\dot{M}$ has to be some fraction of the constant $M_r/t_{ff}(r) = 8c_i{}^3/\pi G$. Shu's solution gives

$$\dot{M} = 0.975\,\frac{c_i^{\,3}}{G} = 1.5 \cdot 10^{-6}\ \mathrm{M}_{\odot}\ \mathrm{yr}^{-1}\left(\frac{T}{10\ K}\right)^{3/2}, \tag{23}$$

where we assume that $m_{gas} = 2.4$ amu, the value appropriate for gas with a solar composition that has hydrogen in its molecular form.

If the cloud is in slow uniform rotation at angular speed Ω_0, then along the cloud's equator $j = \varpi^2 \Omega_0$. At any moment, the maximum j collapsing to the center $j_{max} \sim t^2$ as the expansion wave moves outward. In a truly singular cloud, there is always enough low-j matter, for reasonable parameters, to form a low-mass star initially, but once j_{max} becomes large enough to put gas into orbit at the star's equator, a disk will form and grow rapidly in outer radius r_d. Without angular momentum transport, $r_d \approx {j_{max}}^2/GM \sim t^3$, where $M = \dot{M}t$ is the total collapsed mass. The actual growth rate of the disk radius and the fate of its accreted mass depend on the competition between accretion onto the disk and accretion through it due to transport.

As illustrated in Fig. 5.2, once the bulk of the infalling material has high j, what happens next depends on how the timescale for accretion onto the disk $t_{in} \approx M_d/\dot{M}$ compares with $t_{ev}(\varpi)$ from eq. (15) (Cassen, 1994). As shown on the left-hand side of Fig. 5.2, when $t_{in} > t_{ev}$ at all ϖ, the disk will empty out onto the star as fast as mass arrives, and the radius of the disk will expand even faster than t^3 because of outward transport of angular momentum by viscosity. On the other hand, if $t_{in} < t_{ev}$ everywhere, the disk accumulates mass and grows as $r_d \sim t^3$ (Pickett et al., 1997). Suppose that the transport is characterized by an α-viscosity. Because t_{ev} in (15) is a function of ϖ, if the collapse proceeds long enough and α is constant, the outer regions of any disk should eventually accumulate mass faster than α-viscosity can accommodate. This results in the onset of GIs (Laughlin and Bodenheimer, 1994; see § 4.2). Whether the disk anywhere enters such an accumulation phase depends on

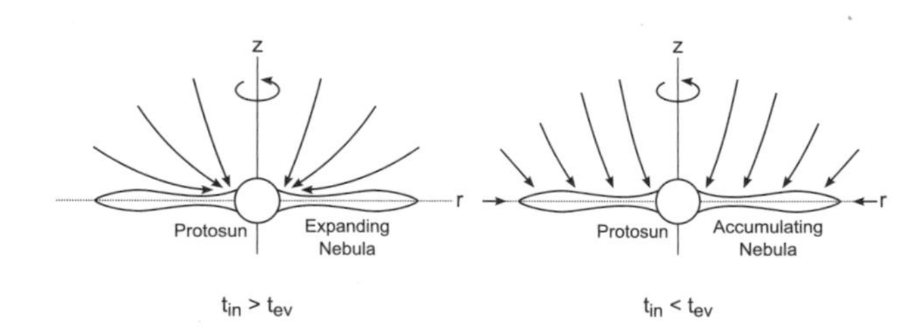

Fig. 5.2. Left panel: Mass empties out onto the star through the disk, and the disk outer radius expands rapidly because of outward angular momentum transport. Right panel: Mass accumulates in place, and the disk radius expands only as the angular momentum of the infalling material increases. In both panels, the size of the central star is exaggerated. Adapted from Cassen and Summers (1983).

whatever process terminates infall, for instance, the finite size of the initial cloud determined by the star-forming environment.

"REALISTIC" COLLAPSE? Although the collapse of a singular isothermal sphere is just a toy mathematical problem, it already illustrates the effect on disk formation of several parameters: the initial cloud conditions through c_i and Ω_0, the cloud's environment through whatever process terminates accretion, and internal disk transport mechanisms through α (see, e.g., Dullemond et al., 2006; Visser et al., 2009). Uncertainties magnify once the broad set of initial collapse conditions advocated by various authors is considered. For an "isolated" cloud, disk formation will be affected by the velocity, density, and temperature structure of the initial cloud and the relative importance of magnetic fields, all of which can affect how mass arrives when and where carrying how much angular momentum. Even if we keep everything the same as in the previous section but allow the initial cloud to have a central region of roughly constant density, the picture changes considerably. For reasonable values of Ω_0, the first mass to arrive near the center of the collapse typically has too large a j to form a star directly. Then the so-called first-equilibrium core, which forms when the collapsing molecular gas becomes opaque enough to behave adiabatically rather than isothermally, will be rapidly rotating. Bate (1998) and Machida et al. (2007) illustrate how this additional complexity affects the collapse for nonmagnetic and magnetic cases, respectively. As described in the chapter by Königl and Salmeron, disk collapse and accretion for low-mass stars may in fact be completely regulated by magnetic fields, except perhaps in dead zones (see § 4.2), and then disk masses need never become very large. For high-enough magnetic flux in the initial cloud, a rotationally supported disk may not even form (Hennebelle and Fromang, 2008). On the other hand, using sheetlike collapses induced by magnetic channeling at the earliest stages of cloud collapse, Vorobyov and Basu (2005, 2006) find that GIs may occur in repetitive outbursts and dominate accretion onto the star.

There is no consensus at present about how stars "really" form, except that at some stage formation involves gravitational collapse of cores in environments where many stars are forming at once (see articles in Reipurth et al., 2007). How messy this can be for disks around solar-type stars is illustrated by decaying turbulence simulations of star-forming regions (e.g., Bate and Bonnell, 2005), where mass accretes onto multiple interacting systems of protostars along sheets and filaments (see Plate 6). Disks that form can be truncated by collisions and tidal interactions. Accretion may stop when stars drift out of, are ejected from, or use up reservoirs of dense gas, and accretion can resume when subsystems merge or reenter dense regions. Even for the relatively quiescent virial equilibrium initial conditions assumed by Krumholz et al. (2007a) for their three-dimensional (3D) radiative hydrodynamics collapse simulation of a 100 $M_\odot$ molecular cloud core, they find that a large (100s AU), massive ($M_d \sim M_s$) disk forms. This disk is also fed asymmetrically by streams and mergers, experiences bursts of GIs, and fragments into stellar-mass companions.

It should be clear from these few examples that theoretically, there is no well-accepted, canonical set of initial conditions for disks. In fact, it is very likely that disk formation is a chaotic and highly diverse process in nature. The remaining discussion takes the presence of a disk around a star as a starting point.

2.3. *What Causes Disks to Evolve?*

As detailed in the chapter by Calvet and D'Alessio, we have ample evidence that disks exist around young stars of low and intermediate masses for times $\sim$ few Myr, corresponding to $\alpha \sim 10^{-3}$ to 10^{-2} in eq. (15), and that they are actively accreting matter onto their central stars. Accretion of the disk onto the star requires some process that lowers the j of fluid elements and allows them to drift to orbits of smaller ϖ. For the simple thin-disk theory of § 2.1 to be valid, there must be a physical basis for the α-viscosity that transports the angular momentum.

EVOLUTION DUE TO TURBULENCE. Under the assumptions leading to eq. (14), $t_{ev} >> t_{rot}$, and the disk evolves diffusively while remaining in an axisymmetric equilibrium. In a protoplanetary disk, molecular viscosity $\nu_{mole} \sim c_s l$, where l is the gas particle mean free path. For typical parameters, $\nu_{mole} \sim 10^4$ to 10^5 cm^2/s, yielding $t_{ev} >>$ Hubble time. Real disks evolve much more quickly, with an effective ν that appears to be a significant fraction ($\alpha \sim 10^{-3}$ to 10^{-2}) of $c_s H$. This suggests a turbulent process where the gas churns with perturbed velocities on the order of a fraction of c_s on scales on the order of a fraction of H.

Consider small fluctuations δf about the instantaneous axisymmetric equilibrium in a fluid quantity f. We can put $f + \delta f$ into eqs. (1) and (2), cancel terms of order f, integrate over z under the usual thin-disk assumption, and retain only the lowest-order terms necessary to describe the secular evolution due to transport of angular momentum. Allowing for finite radial pressure support in the zero-order equilibrium state, Balbus (2003) shows that

$$\frac{\partial \Sigma}{\partial t} = \frac{\partial}{\varpi \partial \varpi} \left[\frac{1}{(\varpi^2 \Omega)'} \frac{\partial}{\partial \varpi} (\varpi^2 \Sigma W_{\varpi\phi}) \right], \tag{24}$$

where the prime denotes a derivative with respect to ϖ and $W_{\varpi\phi}$ is the vertically integrated ϖ,ϕ-component of the specific (per unit mass) stress tensor. Note that generally,

$$W_{\varpi\phi} = -\frac{1}{\Sigma}\int \tau_{\varpi\phi} dz. \tag{25}$$

For viscous (or Navier-Stokes) stresses, in the special case where ν is independent of z,

$$W_{\varpi\phi}^{NS} = -\nu\varpi \frac{\partial\Omega}{\partial\varpi}, \tag{26}$$

and we recover eq. (14) for a strictly Keplerian disk. Thus, to within a factor of order unity, we can define an effective α_{eff} that relates *any* $W_{\varpi\phi}$ to an equivalent α-viscosity by

$$\alpha_{eff} = W_{\varpi\phi}/c_s^2. \tag{27}$$

Eq. (27) effectively normalizes the torque-producing nonaxisymmetric stresses ($W_{\varpi\phi}$) to the gas pressure in the disk ($\sim c_s^2$).

THE BIG THREE SOURCES OF STRESS. There are three major contributions to $W_{\varpi\phi}$ generally considered in disks, those due to fluctuations in the velocity, the gravitational field (**g**), and the magnetic field (**B**), as given, respectively, by

$$W_{\varpi\phi}^{R} = \left\langle \int \delta v_{\varpi}\, \delta v_{\phi} dz \right\rangle, \tag{28}$$

$$W_{\varpi\phi}^{N} = \left\langle \int \frac{\delta g_{\varpi}\, \delta g_{\phi}}{4\pi G\rho}\, dz \right\rangle, \tag{29}$$

and

$$W_{\varpi\phi}^{M} = \left\langle \int \frac{\delta B_{\varpi}\, \delta B_{\phi}}{4\pi\rho}\, dz \right\rangle. \tag{30}$$

They are referred to as the *hydrodynamic* (or Reynolds) stress, the *gravitational* (or Newton) stress, and the *magnetic* (or Maxwell) stress, respectively. Even though, in some cases, the fluctuating quantities may have no zero-order counterpart, I do not suppress the δ to emphasize in principle that I am talking about fluctuating first-order quantities. The derivation of eq. (29) can be found in Lynden-Bell and Kalnajs (1972).

Net transport of mass and angular momentum will occur when, over some suitable spatial and temporal average, these stresses are nonvanishing. For application to eq. (24), the spatial average must include integration over all ϕ. Net torques result from correlations in the ϖ- and ϕ-components of the fluctuating vector fields. For example, in (28), if the sign of δv_{ϖ} tends to correlate with the sign of δv_{ϕ}, which is proportional to δj, then the turbulence causes a net outward flux of angular momentum. Obviously, a macroscopic ordered **B**-field with nonzero $B_{\varpi} B_{\phi}$ will also induce torques and disk evolution, as discussed in the chapter by Königl and Salmeron.

To compute disk evolution over long times like an α-disk using eq. (24), one would need to be able to compute $W_{\varpi\phi}$ from local zero-order disk properties without recourse to a detailed simulation of the fluctuations. This implies that one should not do ϖ averages in (28) to (30) over more than a few scale heights, which in turn requires that the correlated fluctuations leading to a significant net $W_{\varpi\phi}$ should be characterized by radial wavelengths less than a few H. I will refer to a process that satisfies this condition as being *structurally* local.

So far, I have emphasized mass and angular momentum transport (eqs. [14] and [24]). In the evolution of an α-viscosity disk, there is energy dissipated locally at a rate given by eq. (17). To recover a valid α-disk evolutionary model from a turbulent process, the local dissipation of energy needs to agree with (17) when (26) and (27) are used. Balbus and Papaloizou (1999) argue that such a self-consistent local α-disk picture is possible for turbulent hydrodynamic and magnetic stresses but is probably not possible, except perhaps near the corotation radius of a spiral wave (see §§ 3.2 and 4.1), for gravitational stresses. The problem is that self-gravitating waves can transport energy globally rather than just dissipating it locally. The validity of an α-disk picture requires self-consistency of local energy dissipation, and I will refer to this condition as being *dissipatively* local. An important consideration in the case of any transport mechanism in a disk is whether and in what sense these locality conditions are satisfied.

3. The Wonderland of Instabilities

3.1. Classic Results

Until the late 1990s, most researchers looked to Reynolds stresses as the way to justify use of α-disk models. Turbulence generally occurs when perturbations arising from instability at some driving scale length cause a nonlinear *cascade* of energy down to the scale on which it is dissipated by molecular viscosity (see Fig. 5.3). This produces fluctuations with spatial and temporal correlations on all scales in between and leads to nonzero Reynolds stresses. So the search for real α-like behavior in disks became a search for instability mechanisms that lead to turbulence, followed by an analysis of the resulting fluctuations

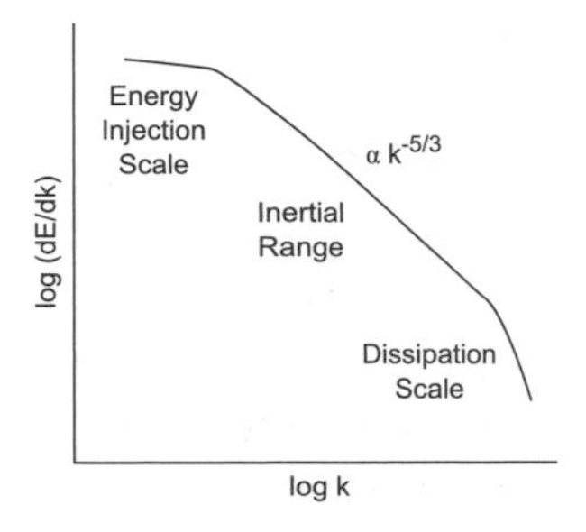

Fig. 5.3. A Komolgorov energy spectrum for steady-state turbulence versus wave number k of the turbulent motions. If energy is injected at large scales (small k), presumably by an instability, a self-similar region of intermediate k, labeled the *inertial range*, carries energy down to small-enough scales for molecular viscosity to dissipate the energy.

to determine the magnitude and sign of $W^R_{\varpi\phi}$. Even decades before α-disk models were proposed, consideration of classic hydrodynamic instabilities led some researchers to conclude that Keplerian disks must be turbulent (von Weizsäcker, 1948). In this section I review some classic ideas for sources of turbulence. Although some controversy has continued to the present day, modern results generally show that many of the "obvious" sources of turbulence probably do not work or, at least, may not provide the large positive values of α implied by the observed timescale for disk evolution.

HYDRODYNAMIC PROCESSES. On small scales, where curvature can be ignored, a Keplerian disk superficially resembles a planar shearing flow; i.e., $dv_x/dy \neq 0$ in local Cartesian coordinates where y is analogous to ϖ and v_x to v_ϕ. A simple energy argument helps explain why a planar shearing flow should be unstable. Consider an incompressible fluid with a linear shear, i.e., where $v_x(y) = v_0 + v'(y - y_0)$ with $v' =$ constant. Now mix the fluid elements while conserving linear momentum so that the new $v_x(y) =$ constant $= v_0$ over $y = y_0 \pm \epsilon$. The fractional change in specific kinetic energy of the flow is $-(1/3)(\Delta v/v_0)^2$, where $\Delta v = \epsilon v'$ is the difference in velocity over a change in y of ϵ. In other words, there is "free" energy in the shear. If there is some process maintaining the shear, then an instability could generate and sustain turbulence that feeds on this energy. A finite molecular viscosity can stabilize a shear flow, but only up to a certain point. Typically, in nature, ν can keep a shear flow laminar only for a Reynolds number $= Re < Re_{crit} =$ about a few $\cdot 10^2$ to 10^4, depending on the geometry of the flow, where $Re = v'L^2/\nu$ and L is the size of the shearing region. Applying this to a Keplerian disk around a 1 $M_\odot$ star at 1 AU using $\nu_{mole} = 10^5$ cm^2/s, we get $Re = 5{\cdot}10^{14} >> 10^4$.

But disks are in fact a rotating flow. Shear is then defined as nonuniform rotation, $d\Omega/d\varpi \neq 0$. If we let $q = d\log\Omega/d\log\varpi$, a Keplerian disk (eq.[12]) has $q = -1.5$. Let us reexamine our energy argument. If we conserve angular rather than linear momentum as we exchange mass elements, then kinetic energy is liberated if and only if $q < -2$; in other words, $dj/d\varpi = d(\varpi^2\Omega)/d\varpi < 0$. It can in fact be proved rigorously (see Tassoul, 1978) that for axisymmetric perturbations in an inviscid fluid where the j for the fluid elements will be strictly conserved, the fluid is stable if and only if $dj/d\varpi \geq 0$, the so-called Rayleigh criterion. The instability for $dj/d\varpi < 0$ is linear and grows on a dynamic timescale $\sim t_{rot}$. So, even at infinite *Re*, a Keplerian disk is linearly stable to axisymmetric perturbations, but what about nonaxisymmetric instabilities and nonlinear perturbations?

Consider Taylor-Couette flow, the flow of an incompressible fluid between two fixed cylindrical surfaces with radii $\varpi_1 < \varpi_2$ forced to rotate at angular speeds Ω_1 and Ω_2, respectively. Laboratory experiments and analyses available at the time α-disk models were proposed in the 1970s indicated that Taylor-Couette flows that satisfy the Rayleigh stability criterion are in fact linearly stable for all accessible *Re*, as illustrated using a historic diagram by Taylor in Fig. 5.4. However, it was also known that nonlinear, nonaxisymmetric perturbations could produce turbulence in flows with q marginally greater than -2. These incursions of nonlinear instability into the Rayleigh stable region are deeper as *Re* increases, leading some to argue that real Keplerian disks must be unstable in this nonlinear sense (see Richard and Zahn, 1999).

There is at least one other nonlinear effect of possible interest for disk transport, namely, the growth of large stable vortices. Vorticity is defined as

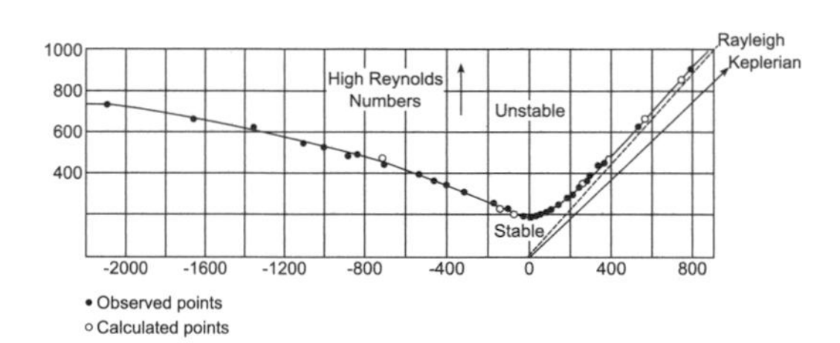

Fig. 5.4. Stability diagram for Taylor-Couette flow. The two axes represent the angular rotation speeds of the outer (vertical) and inner (horizontal) cylinders normalized to the molecular viscosity of the fluid. Effective Reynolds numbers increase away from the origin. The unstable region is above the curve traced by dots and open circles. The Rayleigh criterion for axisymmetric stability is the dashed curve. The approximate Ω_2/Ω_1 for Keplerian rotation is shown by the arrow. Adapted from Brenner and Stone (2000), who based their figure on an original by Taylor from 1923.

$\nabla \times \mathbf{v}$. For a Keplerian disk, the vorticity of the mean flow is $\Omega \mathbf{i_z}/2$, and, for uniform rotation, it is just $\Omega \mathbf{i_z}$, where $\mathbf{i_z}$ is the unit vector in the z-direction. Baroclinic instabilities, as well as other instabilities associated with structures in the disk, can produce localized vorticity perturbations. In shallow shearing flows, like Jupiter's atmosphere (Marcus, 1993), a large vortex, like the Great Red Spot, can grow at the expense of smaller vortices. This phenomenon is referred to as *negative eddy viscosity* because the effect of smaller vortices in this case is to enhance rather than dissipate the energy in the large vortex. An asymmetric vortex with correlated velocity fluctuations can produce a net Reynolds stress.

THERMODYNAMIC PROCESSES. Many of these results carry over to compressible fluids as long as motions are substantially subsonic and the fluid remains isentropic, so that no net work is done or entropy generated when fluid elements are exchanged. Entropy gradients, however, add a new dimension. Let s' be the specific entropy gradient in the y-direction, and let g be the magnitude of the gravitational field in the negative y-direction. If $s' > 0$, the fluid has a negative buoyancy that will stabilize a linear shear v' when the Richardson number

$$Ri = \frac{gs'}{C_P(v')^2} > \text{about } \frac{1}{4}, \tag{31}$$

where C_P is the specific heat at constant pressure. Ri is a ratio that effectively measures the relative importance of the energies associated with the negative buoyancy and with the velocity shear. If we apply this naively to the radial direction in a disk, $|Ri| \sim 10^{-2}$ for a typical radial temperature distribution $T(\varpi)$. Such a low Ri seems insufficient in itself to stabilize Keplerian shear.

Now consider a plane-parallel case where $v' = 0$. When $s' < 0$, positive buoyancy of the fluid elements will cause a Rayleigh-Taylor instability that overturns the fluid. Basically, because an upward-displaced fluid element has lower entropy than its surroundings, it has lower density and experiences a net upward gravitational force, like a hot-air balloon. The fluid stabilizes only if the elements become rearranged so that $s' \geq 0$. If a thermal energy flux is being driven through the fluid in the positive y-direction and if the temperature gradient required for thermal conduction to carry this flux sustains $s' < 0$, then the instability leads to *thermal convection*, where the fluid turns over continuously and the advective fluid motions carry part of the thermal flux. To test whether a region in which there is thermal energy flux in the $+y$-direction is convective or not, compute the s' necessary for conduction to carry all the flux. Call this s'_{rad} because "conduction" in a real disk is due to radiation. Then the Schwarzschild criterion for the occurrence of thermal convection is $s'_{rad} < 0$. So far, in this discussion, it is assumed that the stratified fluid is in hydrostatic equilibrium in the y-direction. If the same arguments about heat flow are applied instead to the vertical or z-direction, then there will be turnover on scales comparable to H when the Schwarzschild criterion for convective instability is satisfied in that direction. As shown in Lin and Papaloizou (1980) and Ruden and Pollack (1991), this can indeed occur for reasonable disk parameters in the absence of strong heating of the upper disk surface by external irradiation. Irradiation, if strong enough, imposes a positive s'_{rad}. The convective motions introduce δv_{ϖ} and δv_{ϕ} fluctuations that may become correlated to produce Reynolds stresses.

Another classic instability in a rotating fluid that may lead to angular momentum transport in a disk is *baroclinic instability*. A *baroclinic* fluid is one in which surfaces of constant entropy and surfaces of constant effective potential (gravity plus rotation) are not the same. The *Hoiland criterion* states that an axisymmetric rotating fluid configuration will be unstable when j decreases outward along a surface of constant s. This is basically just the Rayleigh criterion applied along a constant-s surface. Baroclinic instabilities cause r-modes to grow, and, in the Earth's atmosphere, they generate large-scale Rossby waves and contribute to the development of cyclonic storms (see references in Petersen et al., 2007a, 2007b). Effects of heating and cooling will almost inevitably make disks baroclinic to some degree.

3.2. Applications to Disks

HYDRODYNAMIC INSTABILITIES. The study of "purely" hydrodynamic effects has required highly refined techniques because it is difficult to achieve effective values of *Re* much greater than 10^6 in either laboratory experiments or numerical codes. On the laboratory side, the most exquisite Taylor-Couette experiment to date has been reported by Ji et al. (2006). A serious difficulty with classic Taylor-Couette experiments for applications to disks is the rotation of the end caps on the cylinder. Ji et al. approximate Keplerian shear by breaking the end caps into radially inner and outer pieces that can be rotated independently of the vertical cylindrical walls. They also directly measure Reynolds stresses in so-called β-viscosity units where the effective $\nu = \beta \Delta v \Delta \varpi$. Here Δv is the velocity difference between the cylindrical walls, and $\Delta \varpi$ is the radial gap between the cylinders. They find that for $Re \approx 10^6$, $\beta < 10^{-5}$ and is indistinguishable from zero. This is definitely at odds with Richard and Zahn (1999), who argued from earlier Taylor-Couette experiments that β would be at least on the order of 10^{-5}. Ji et al. assert that α in disk geometry should be similar in magnitude to the β of Taylor-Couette geometry. The critical Reynolds number Re_{crit} at which nonlinear nonaxisymmetric instabilities set in to produce a nonzero effective α thus appears to be $> 10^6$ for Keplerian shear. Well-accepted scaling arguments suggest that for $Re >> Re_{crit}$, $\alpha \approx 1/Re$. So, even if nonlinear instabilities cause Keplerian disks to become turbulent at very high *Re*, the αs are unlikely to be large enough to cause disk evolution at the observed rate, which seems to require $\alpha \sim 10^{-3}$ to 10^{-2}.

Two numerical studies (Hawley et al., 1999; Lesur and Longaretti, 2005) support these conclusions. In both cases, shearing box techniques are used to simulate a local piece of a rotating flow in cylindrical coordinates. Only a limited range in azimuth is modeled by assuming periodic boundary conditions in ϕ. The radially inner and outer boundary conditions at ϖ_1 and $\varpi_2 = \varpi_1 + \Delta\varpi$ are also assumed to be periodic, but, to account for shear, fluid quantities at ϕ for $\varpi = \varpi_1$ are set equal to those at $\phi + \Delta\Omega t$ for $\varpi = \varpi_2$, where $\Delta\Omega = \Delta\varpi\,\Omega'$ and Ω' is the assumed rotational shear. Shearing boxes can be constructed in both 2D (z-integrated) and 3D. Such simulations are perforce local and generally will miss instabilities with radial scales $> \Delta\varpi$. Both Hawley et al. and Lesur and Longaretti use more than one hydrodynamics code at several resolutions to increase confidence in their results. Hawley et al. employ two grid-based codes, while Lesur and Longaretti include one code based on spectral methods, i.e., where the numerical calculation is discretized in wave number space, not real space. The *Re* for simulations in these articles approaches 10^5.

Hawley et al. vary the q of the shear and test nonlinear stability by imposing nonlinear initial fluctuations. Near the Rayleigh stability limit, $q = -2$, a rotating flow behaves like a linear shearing flow. To understand how this can be, let us consider the epicyclic frequency κ, the frequency at which a fluid element perturbed from pure rotation will oscillate about its unperturbed position in an elliptical epicycle because of Coriolis forces. In a Keplerian flow, these oscillations are retrograde, i.e., opposite the sense of rotation of the mean flow. As shown in many references (e.g., Binney and Tremaine, 1987), $\kappa^2 = dj^2/r^3dr = 2(2+q)\Omega^2$, and hence $\kappa = 0$ for $q = -2$. Note also that $\kappa^2 < 0$ when $q < -2$, which is just a manifestation of the instability that sets in when the Rayleigh stability criterion is violated. In other words, for $q < -2$, instead of oscillating, fluid elements deviate exponentially from their equilibrium positions after a small perturbation.

For $Re \sim 10^5$, Hawley et al. find that all linear perturbations damp for $q \geq -1.99$, while initial velocity noise of a few percent amplitude grows for $q < -1.95$ and damps for $q = -1.94$. The first result confirms the expectation that the flow is linearly stable for $q > -2$, and the second suggests nonlinear instability for q slightly greater than -2. Tests with the Keplerian value $q = -1.5$ show damping of nonlinear perturbations for all codes and resolutions. They conclude that Re_{crit} for Keplerian shear, if it exists, is probably $>> 10^6$. Lesur and Longaretti verify the $\alpha \approx 1/Re_{crit}$ scaling for accessible values of Re_{crit} and then show that even the most optimistic extrapolation to $q = -1.5$ yields $\alpha < 3 \cdot 10^{-6}$. A conservative interpretation of these numerical and laboratory experiments is that even if Keplerian shear is nonlinearly unstable and produces turbulence, the resulting α is so low that the timescale in eq. (15) will be much longer than the observed lifetimes of protoplanetary disks. Simple shear flow instability does not seem viable as an evolutionary mechanism for disks. Recent efforts to treat this problem analytically have been inconclusive (Rincon et al., 2007).

An important exception has been known since the mid-1980s (Papaloizou and Pringle, 1984, 1985; Goldreich et al., 1986). Define the corotation radius (CR) ϖ_{CR} of a wave as the place where the wave's azimuthal *pattern speed* Ω_{pat} is the same as the local gas angular rotation rate Ω. A global, purely hydrodynamic instability is possible in an orbiting fluid flow when nonaxisymmetric waves on opposite sides of their common ϖ_{CR} are able to communicate with one another. The reason is succinctly described in Balbus (2003). On either side of their common CR, waves have different signs of energy perturbation. If communication by some global effect can move energy from the negative to the positive energy wave, then the wave amplitudes will grow. Edges are one way to introduce the needed communication by wave reflection. For the special case of a narrow torus with a power law $\Omega(r)$, cross talk between waves through distortions of the edges of the torus leads to instability for $q < -\sqrt{3} \approx -1.732$. Although this is an interesting phenomenon, the unstable q-values are considerably more negative than for a Keplerian shear ($q = -1.5$). A related idea is that inwardly propagating waves stimulated by forcing distortions at an outer edge can carry negative angular momentum inward (Vishniac and Diamond, 1989).

CONVECTIVE INSTABILITIES. The current assessment regarding convection as a source of useful Reynolds stress is even less encouraging than that for hydrodynamic instabilities. Local simulations similar to those discussed above (Cabot, 1996; Stone and Balbus, 1996) find that convective instability in the z-direction results in small negative values of $\alpha \sim -\text{few} \cdot 10^{-4}$ to -10^{-5}; i.e., thermal convection in a rotating disk leads to very weak inward transport of angular momentum, contrary to the modestly strong outward transport required to sustain the mass inflow observed in real disks. Stone and Balbus show that this conclusion is independent of shear; convection behaves the same way in a uniformly rotating disk. Balbus (2000) explains this in terms of the retrograde character of the perturbed epicyclic motions of fluid elements in disks with $q > -2$. If we imagine inner and outer perturbed fluid elements interacting at ϖ near the ends of epicyclic excursions in an energetically plausible way, i.e., where their total kinetic energy decreases because of the interaction, then the angular momentum exchange must go in the direction of equalizing the js of the fluid elements. The fluid that moves back inward will have a higher j than it did before. Balbus argument breaks down when the perturbations are so strong that the epicyclic approximation becomes invalid or when wave transport becomes important. With negative Reynolds stress, transport of mass by convective turbulence cannot sustain itself through release of heat, because net release of heat by turbulent dissipation requires outward angular momentum transport. So, in the absence of strong irradiation, convection can be sustained by a dissipative release of heat in the disk interior due to other mechanisms (as in Klahr et al., 1999), but it cannot, by itself, produce self-consistent α-disk behavior.

BAROCLINIC INSTABILITIES, VORTICES, AND FINITE STRUCTURES. If large-scale vortices can grow by negative eddy viscosity in shallow shearing flows (§ 3.1), one might expect them to occur in thin, nearly Keplerian disks. In 2D simulations (e.g., the shearing-box computations of Johnson and Gammie, 2005), when nearly Keplerian disks are perturbed initially by introducing small-scale, random vorticity fluctuations $\nabla \times \delta\mathbf{v}$, large and long-lived vortices organize themselves within a few to 10s t_{rot}. The vortices tend to be elongated, with the long axis oriented along the azimuthal (ϕ) direction. All of them are retrograde (anticyclonic) and grow through merger of the small negative vorticity perturbations initially imposed. Although, in principle, the vorticity of a fluid element is conserved in the absence of entropy variations or dissipation (see chapter 3 of Tassoul, 1978), the initially prograde, cyclonic vorticity perturbations become sheared out rapidly in physical space and lose their coherence. Anticyclones are preferred in Keplerian disks because the epicyclic motions of perturbed fluid elements are also retrograde, so that a collective macroscopic nonlinear structure can be readily constructed from nonlinear oscillations of fluid elements. Another equivalent but perhaps simpler way to understand this is to note that a retrograde vortex "rolls" with the shear, while a prograde vortex rolls counter to the shear (Adams and Watkins, 1995).

Anticyclones in Keplerian disks are analogs of high-pressure systems in the Earth's atmosphere. Compared with the ambient flow, negative vorticity causes fluid elements to move around the star somewhat more slowly than the mean flow in the radially outer part of the vortex and somewhat more rapidly in the radially inner part. As a result, a perturbed pressure gradient in the vortex, with a maximum at the center, is necessary to maintain the overall steady-state equilibrium of the flow. The central pressure maximum has interesting consequences for planetesimal formation (see § 5.4).

Large-scale vortices may actually form in disks as the result of r-modes produced by baroclinic instabilities or through instabilities generated by narrow axisymmetric structures in an equilibrium disk, such as edges (Papaloizou and Pringle, 1985), vortensity extrema (Papaloizou and Lin, 1989), and other local extrema in physical properties (Lovelace et al., 1999; Li et al., 2000, 2001). Here *vortensity* is the ratio of the disk vorticity to surface density. Lovelace et al. (1999) effectively combine a number of these instabilities in 2D into the occurrence of a radial extremum in a single disk variable $\mathcal{L}(\varpi) = (\Sigma\Omega/\kappa^2)s^{2/\gamma_2}$. Through the incorporation of s, extrema in $\mathcal{L}$ effectively include some baroclinic instabilities. Linear analyses and 2D hydrodynamics simulations by Li et al. illustrate growth of r-modes due to narrow jumps and bumps in $\Sigma(\varpi)$ and $P(\varpi)$ for equilibrium disks. Large-scale and long-lived anticyclonic vortices are the nonlinear outcome. This is sometimes referred to as Rossby wave instability (RWI). Fig. 5.5 illustrates an unstable r-mode and the resultant nonlinear vortices that can occur.

The problem with instabilities that result from extrema as a path to an effective α is that they require significant preexisting radial nonuniformity in a disk (for examples of ways to produce such structures, see Durisen et al., 2005;

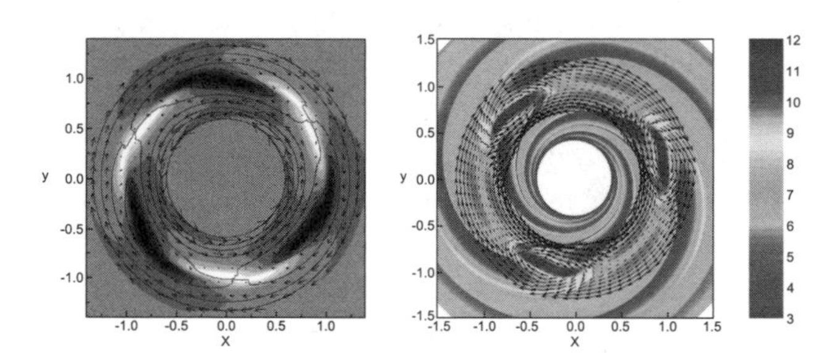

Fig. 5.5. Vortex formation due to narrow radial structure. Both these disks have bumps in their otherwise smooth radial entropy distributions at radii near $\varpi = 1$. The left panel, from Li et al. (2000), shows the linearly unstable r-mode that results from the bump. The gray scale in this plot represents the surface-density perturbation, with black being maximum. The right panel from Li et al. (2001), is the nonlinear development of such an unstable mode. The units in these plots are arbitrary. The pressure gray scale on the right goes with the right plot. The velocity vectors show perturbed velocities relative to the mean flow. Notice that the vortices are anticyclonic and have a pressure maximum at their centers.

Kretke and Lin, 2007; Lyra et al., 2008, 2009). Whether baroclinic instabilities due to more gradual, smooth entropy gradients in disks can produce vortices remains somewhat controversial, with different studies producing results that seem discrepant. For example, a 3D hydrodynamics simulation by Klahr and Bodenheimer (2003) of a disk with radiative cooling shows production of a very large-scale vortex, while Keplerian disks with radial entropy gradients are found to be stable regardless of the value of Ri (see eq. [31]) in 2D shearing-box simulations by Johnson and Gammie (2006). Perhaps the best work on this topic at the time of writing is the study by Petersen et al. (2007a,b), who consider baroclinic instabilities in 2D simulations that include cooling and thermal conductivity in a vertically integrated approximation. To allow integration over hundreds of orbit periods, these authors use an *anelastic* approximation. In effect, compressibility is ignored except insofar as it affects the buoyancy of fluid elements. This preserves the essential physics of the baroclinic instabilities without requiring that a code resolve and follow sound waves. For $\Sigma \sim \varpi^{-3/2}$, large-scale vortices are readily produced and sustained in simulations where the radial temperature gradient is steeper than $T(\varpi) \sim \varpi^{-1/2}$ and the cooling time t_{cool} is $<$ about 10 t_{rot}. All else being equal, higher Re and higher thermal conductivity are both conducive to sustaining the baroclinic conditions that produce vortices. For the $\Sigma(\varpi)$ chosen, $T \sim \varpi^{-1/2}$ corresponds to $Ri = 0$. So the simulations suggest that fast-cooling disks with negative values of Ri are susceptible to sustained baroclinic instabilities and vortex production.

Vortices are of interest for angular momentum transport because nonlinear effects, such as shock waves at the leading and trailing edges of a vortex, can cause their elongated shapes to be asymmetrical and not exactly aligned along the ϕ-direction, as in the right panel of Fig. 5.5. The correlated δv_{ϖ} and δv_{ϕ} in the distorted vortical flow may then lead to net radial transfer of angular momentum. Although the transport is due to a macroscopic structure, not turbulence, this still results in a net Reynolds stress with an associated effective α (Petersen et al., 2007b). Li et al. (2001) report effective αs for their vortices of $\sim 10^{-4}$ to 10^{-2} for some choices of parameters, while Petersen et al. find the Reynolds stresses to be highly time variable, with average values of uncertain magnitude and sign. This is perhaps not surprising because the anelastic approximation would interfere with shock formation.

Most of the work on vortices discussed so far has been 2D, but, as with other aspects of disk physics (e.g., §§ 2.1 and 4.3), the behavior of vortices can be strongly influenced by the 3D structure of a stratified disk, even if it is geometrically thin. As pointed out by Barranco and Marcus (2005), there is a fundamental difference between a 2D and a 3D picture. In 2D, vorticity

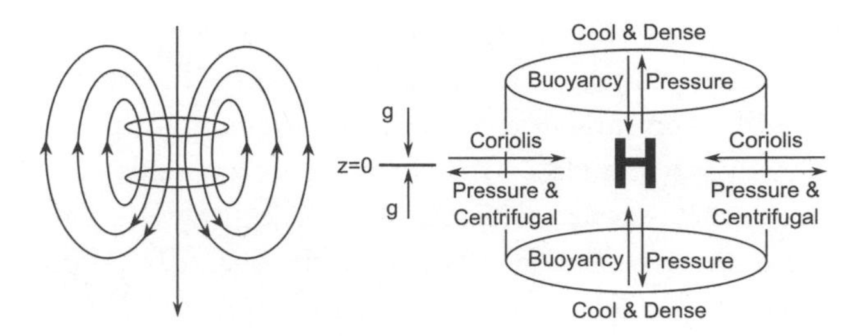

Fig. 5.6. Anatomy of a 3D vortex with symmetry about the midplane. The left panel illustrates lines of constant vorticity, which must be divergence free. A vortex that is anticyclonic in its center must have a cyclonic halo. The right panel shows the perturbed forces about the axisymmetric disk equilibrium necessary to sustain the vortex. Adapted from figures in Barranco and Marcus (2005).

perturbations are quantized to be either positive or negative. In 3D, vorticity in an inviscid flow is divergence free, like a magnetic field (see chapter 3 of Tassoul, 1978), and lines of constant vorticity are closed loops. Any 3D vortex with a retrograde core necessarily has an outer prograde halo. Fig. 5.6 illustrates the structure of a 3D vortex centered on the midplane with an anticyclonic core. The center of the vortex has to have an excess pressure, as indicated by the "H" for "high pressure," to counteract the Coriolis force. When Barranco and Marcus simulate thin, vertically stratified disks, midplane vortices are not long lived but disappear within a few rotation periods, to be replaced by weaker vortices that are centered several scale heights above the midplane.

Obviously, the occurrence and nature of baroclinic instabilities and large-scale vortices in real disks deserve a great deal of further study, and it is difficult at the present time to assess whether and how they may contribute to angular momentum transport. The 2D results of Petersen et al. underscore the complexity of the problem, because the thermal physics is also critically important. Simulations with radiative transport and high intrinsic *Re* followed over many orbits in full 3D seem required.

3.3. The Two Big Instabilities

TAPPING THE ENERGY IN KEPLERIAN SHEAR. Nature tends to increase the entropy in a system by turning reservoirs of ordered energy into heat. A Keplerian shear ($q = -1.5$) is sustained in a disk by the gravitational field of the central star. The hydrodynamic and convective instabilities we have considered so far involve only local fluid stresses that are unable to tap the "free" energy in the shear because noncontiguous fluid elements do not act on each other directly as they shear apart. In these cases, angular momentum is not

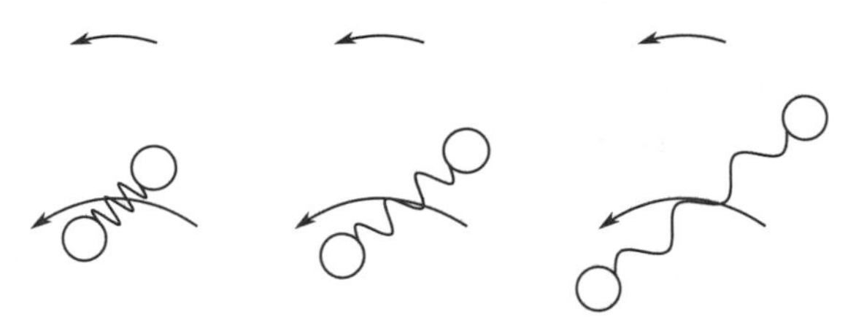

Fig. 5.7. Coupling of fluid elements in a shearing flow. The arrows indicate the outwardly decreasing Ω with the star toward the bottom. The three panels are sequential from left to right. The circles are two fluid elements that interact because of some perturbing force represented by the spring. As the shear causes the fluid elements to separate in ϕ, the tension in the spring produces a negative torque on the inner fluid element and a positive torque on the outer fluid element. As a result, the fluid elements drift even farther apart radially, which accentuates the effects of the shear. Adapted from Gammie and Johnson (2005).

transported by the linear instabilities themselves but only by the turbulence or by large-scale structures, like vortices, that result from nonlinear development of the instabilities. Conservation of angular momentum constrains the effect of the linear instability. On the other hand, if there is a physical force that links two fluid elements together like a rubber band or spring along the line between them as they move apart with the shearing flow (see Fig. 5.7), then the angular momenta of the fluid elements are no longer conserved. The energy in the flow can then be tapped directly, and the linear instability itself can feed on the energy released.

Consider, as we have before, two unit-mass fluid elements at radii ϖ_1 and ϖ_2 with $\varpi_2 > \varpi_1$ in a purely Keplerian disk (eq. [12]). The total energy per gram of the two fluid elements is then

$$E = \frac{1}{2}\left(\frac{j_1{}^2}{\varpi_1{}^2} + \frac{j_2{}^2}{\varpi_2{}^2}\right) - GM_s\left(\frac{1}{\varpi_1} + \frac{1}{\varpi_2}\right). \tag{32}$$

Exchange a small amount of angular momentum δj from 1 to 2 and move the fluid elements to the new radii corresponding to their altered values of j. Then (see Hartmann, 2009; pp. 130–131), the change in the total mechanical energy ΔE of the fluid elements is

$$\Delta E = -\frac{j_2 \delta j}{\varpi_2{}^2}\left[\left(\frac{\varpi_2}{\varpi_1}\right)^{3/2} - 1\right]. \tag{33}$$

Energy is released when angular momentum is moved outward ($\delta j > 0$). In addition to being the source of heat in a steady-state accretion disk (§ 2.1), this energy can spur the further growth of an instability and sustain turbulence and large-scale disturbances in the disk.

GRAVITATIONAL AND MAGNETOROTATIONAL INSTABILITIES. There are two known mechanisms that can act to exchange angular momentum outward between fluid elements shearing apart in a Keplerian flow: a magnetic-field line threading both elements and the gravitational interaction between the elements. The so-called magnetorotational instability (MRI) occurs in a gas disk with $q < 0$ and a weak magnetic field if the gas is a sufficiently good conductor. The MRI is described in detail in the chapter by Balbus. The focus of the remainder of this chapter is gas-phase *gravitational instabilities* (GIs), which become effective when disk self-gravity is important and couples fluid elements through long-range gravitational interactions. Both MRIs and GIs have been demonstrated to occur in disks, and both have the potential to produce interesting values of effective α, although the situation with MRIs has proved to be somewhat more complicated than originally expected (see, e.g., Fromang et al., 2007).

4. Gravitational Instabilities: Idealized Conditions

4.1. Basics

LINEAR REGIME: GENERAL. Consider a razor-thin disk, i.e., one that is geometrically very thin and where hydrostatic equilibrium (eq. [11]) is assumed in the vertical direction so that the disk can be described by vertically integrated quantities. One can then study the stability of the disk by linearizing the hydrodynamic equations (eqs. [1] and [2]) plus Poisson's equation (eq. [4]) about the zero-order equilibrium described by eq. (10). Assume that the equation of state has the form $\mathcal{P} \sim \Sigma^{\gamma_2}$ with corresponding 2D adiabatic sound speed $c_s = \sqrt{\gamma_2 \mathcal{P}/\Sigma}$. Restrict the analysis to a local region and to perturbations in physical variable f of the form

$$\delta f = f_1 e^{i(m\phi - \omega t + k\varpi)}, \quad (34)$$

where k is the radial wave number. Because f_1 can be complex, which introduces a ϖ-dependent phase shift in ϕ, such perturbations can represent coherent m-armed spiral waves propagating in the ϕ-direction with pattern speed $\Omega_{pat} = \omega/m$ and pattern period $P_{pat} = 2\pi/\Omega_{pat}$. The frequency ω represents the frequency of oscillation of the fluid elements in the wave, and Ω_{pat} is the angular speed with which the entire pattern rotates about the disk in an inertial frame.

For spiral waves in a disk, there are usually three radii where the motions associated with the disturbance resonate with fluid motions. One is the corotation radius ϖ_{CR} defined by $\Omega = \Omega_{pat}$, where fluid elements in the zero-order flow comove with the pattern. The others are the inner and outer Lindblad resonances, defined by the plus and minus signs, respectively, in

$$\Omega = \Omega_{pat} \pm \kappa/m. \tag{35}$$

At the Lindblad resonances, freely oscillating fluid elements execute exactly one epicyclic motion between the arrival of spiral arms, and so free oscillations of the fluid resonate with the oscillations associated with the wave. In the linear regime, this leads to divergence in the amplitude of a spiral wave with pattern period Ω_{pat}. As a result, the Lindblad resonances represent places where interactions between the disk and the wave can significantly alter the angular momentum of the fluid (see Goldreich and Tremaine, 1979). No Lindblad resonances exist for $m = 0$, and they also do not exist for $m = 1$ in a strictly Keplerian disk where $\kappa = \Omega$. Generally, for $m \geq 2$, the existence of Lindblad resonances depends in detail on $\Omega(\varpi)$ and on the radial extent of a disk, and they always exist in nearly Keplerian disks if ϖ_{CR} is far from any disk boundary.

As derived, for example, in Toomre (1981) and in chapter 6 of Binney and Tremaine (1987), a Wentzel-Kramers-Brillouin (WKB) approximation applied to the linearized equations yields the following dispersion relation for the waves represented by eq. (34):

$$(m\Omega - \omega)^2 = \kappa^2 - 2\pi G\Sigma|k| + k^2 {c_s}^2. \tag{36}$$

LINEAR REGIME: AXISYMMETRIC WAVES. Analysis of the dispersion relation (36) for the special case of axisymmetric or ringlike perturbations ($m = 0$) is instructive (Toomre, 1964). The left-hand side becomes simply ω^2. When $\omega^2 > 0$, the perturbations are stable axisymmetric waves, but when $\omega^2 < 0$, the waves are unstable and grow exponentially. The first term on the right-hand-side represents epicyclic oscillations of perturbed fluid elements and is due mostly to orbit motion in the central potential of the star. Because it is positive, it is a stabilizing term independent of k and so is most effective relative to the other two terms on the right-hand side as $k \to 0$. Another way of saying the same thing is that the rotation of the disk stabilizes long-wavelength waves. The last term on the right-hand side, which represents sound waves, is also positive and dominates as $k \to \infty$. In other words, short-wavelength disturbances are stabilized by gas pressure. The middle term on the right-hand side is due to the self-gravity of the disk itself and has a purely destabilizing influence. Because it is linear in k, it is clear, qualitatively, that it will be most effective relative to the other two terms at intermediate wave numbers provided $2\pi G\Sigma$ is large enough.

If we replace k by $2\pi/\lambda$ in eq. (36), where λ is the radial wavelength, and set $m = 0$, then one can show that the unstable intermediate wavelengths fall in the range bounded by

$$\lambda_{\pm} = \frac{2\pi^2 G\Sigma}{\kappa^2}\,[1 \pm (1 - Q^2)^{1/2}]. \tag{37}$$

Here Q is the Toomre stability parameter defined by

$$Q = \frac{c_s\kappa}{\pi G\Sigma} \sim \frac{{t_{ff,z}}^2}{t_z t_{rot}} \sim \frac{2g_z(\text{star})}{g_z(\text{self})}, \tag{38}$$

where $t_{ff,z}$ is the timescale for gravitational collapse in the z-direction for a plane-parallel sheet with surface density Σ, $g_z(\text{star}) = GM_sH/\varpi^3$ is the vertical component of gravity due to the star at $z = H$, and $g_z(\text{self}) = 2\pi G\Sigma$ is the z-component of gravity due to the gas disk itself at its surface. Notice in (37) that instability occurs when $Q < 1$, i.e., when a disk is sufficiently cool (low c_s) or massive (high Σ) or, equivalently, when the disk z-collapse time due to self-gravity is less than the harmonic mean of the timescales associated with the stabilizing pressure and rotation. $Q < 1$ is usually referred to as the Toomre criterion for the instability of ring modes in disks. Notice that for $Q < 2$, disk self-gravity dominates, and so, as one might expect, disk self-gravity leads to axisymmetric instability once it exceeds the gravity of the central star in the z-direction by a sufficient amount.

The *most unstable* $m = 0$ mode, i.e., the one that grows first as Q drops across the stability boundary at $Q = 1$, is

$$\lambda_{most} = \frac{2\pi^2 G\Sigma}{\kappa^2} \approx \frac{2\pi H}{Q}, \tag{39}$$

the λ in the middle of the range given in (37). For fixed κ and Σ, the fastest-growing mode at a given value of Q is

$$\lambda_f = Q^2\lambda_{most} \tag{40}$$

and has growth rate

$$\kappa(1/Q^2 - 1)^{1/2} \sim \Omega. \tag{41}$$

The upshot of all this analysis can be summarized as follows: for a disk that is sufficiently cool (low c_s) and/or massive (high Σ), as determined by $Q < 1$, gravitational instabilities with characteristic wavelengths of order λ_{most} will grow exponentially on a dynamic timescale from small amplitude noise.

LINEAR REGIME: NONAXISYMMETRIC WAVES. Unfortunately, the local analysis leading to eq. (36) does not offer a great deal of insight into the occurrence

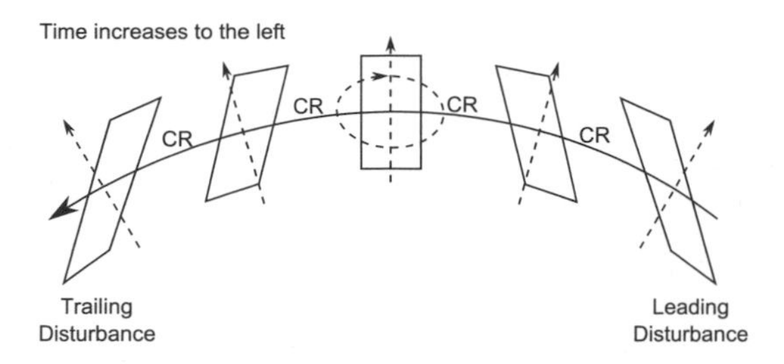

Fig. 5.8. The parallelograms represent a leading density disturbance that is being sheared from leading to trailing by the differential rotation of the disk near corotation (CR) at five different times increasing to the left. The ellipse in the central image shows the direction of epicyclic motion of the fluid. The arrow indicates the radial outward direction. Adapted from Figure 6-18 in Binney and Tremaine (1987).

of instabilities with $m \neq 0$. A large body of 2D and 3D numerical simulations, including those referenced in this chapter and dating back at least to Papaloizou and Savonije (1991), has shown that nonaxisymmetric spiral waves become unstable for $Q <$ about 1.5 to 1.7. The instability is again linear and dynamic, which means that small perturbations grow exponentially on a timescale $\sim t_{rot}$ (e.g., Laughlin et al., 1998; A. F. Nelson et al., 1998; Pickett et al., 1998). The precise Q-limit for instability depends on the structure of the disk. Simulations show that the growing multiarmed spirals have a predominantly trailing pattern. If a disk cools (decreasing the 2D or 3D c_s) and/or increases its mass (increasing Σ) from stable to unstable conditions (lower Q), then the nonaxisymmetric GIs will occur first. Being dynamic, the GIs will dominate further evolution unless the cooling or accretion of mass occurs on an even shorter timescale.

As confirmed by simulations (Durisen et al., 2008; Cossins et al., 2009), the most unstable m-armed spiral has a radial wave number well characterized by $k = 2\pi bm/\lambda_{most}$, where b is a factor of order unity. As shown in Binney and Tremaine (1987), the pitch angle i of the spiral, i.e., the angle between the azimuthal direction and a tangent to the spiral arm, is given by

$$\cot i = \left| \frac{k\varpi}{m} \right| \approx \frac{bQ\varpi}{H} . \quad (42)$$

For typical disk parameters, the unstable spiral modes in the linear regime are expected to be moderately tight, $i \approx 5^\circ$ to 20°, and this seems to apply as well to the nonlinear regime even in strong bursts of GIs (Boley and Durisen, 2008).

The mechanism for instability is most likely swing amplification. As shown in Fig. 5.8, consider a disturbance near its own CR and suppose that it has a leading spiral character. Differential rotation will shear out the disturbance from leading to trailing. In a nearly Keplerian disk, where $\kappa = \Omega$, the fluid elements in the shearing disturbance, following retrograde epicyclic motions, will move with the disturbance as it shears. If self-gravity in the disk is strong enough, gravity has time to collapse the disturbance on itself while it shears. This will cause a strong pulse of amplification provided

$$X = \frac{\pi \varpi}{m \lambda_{most}} < \text{about } 3, \quad (43)$$

or

$$m > \text{about } 5 \left(\frac{0.05}{H/\varpi} \right) \left(\frac{Q}{1.5} \right) \quad (44)$$

(see Binney and Tremaine, 1987). If X is too large, the wave moves away from the region of effective amplification before it can amplify. Eq. (43) suggests that strong growth is favored when modes are global (large λ_{most}), and eq. (44) suggests that for thin disks, the initially growing modes of low order are likely to have four or five arms, as seen in many simulations (e.g., Mejía et al., 2005; Durisen et al., 2008). Thicker disks are likely to be dominated by modes with smaller numbers of arms (see, e.g., Lodato and Rice, 2005).

In gas disk simulations where nonaxisymmetric GIs occur and where X has been measured, X is typically $\sim$ few or less over some region of the disk (e.g., Pickett et al., 1998; Mayer et al., 2004), and in cases where pattern periods and wave amplitudes have been measured, the dominant growing spiral waves usually straddle their CR with power in the wave concentrated between the Lindblad resonances (e.g., Mejía et al., 2005; Boley et al., 2006, 2007a; Cossins et al., 2009). These behaviors are consistent with swing amplification. Typically, when $Q < 1.5$, the instability results in simultaneous growth of multiple spiral waves with different ms (e.g., A. F. Nelson et al., 1998; Pickett et al., 1998). Sustained growth through swing is usually attributed to a feedback mechanism that returns leading spiral waves to the CR in phase with the growing wave, but the feedback loop in 3D gas disk simulations has not been rigorously traced.

Another mechanism for growth of spiral waves in a self-gravitating disk is SLING (stimulation by the long-range interaction of Newtonian gravity; Adams et al., 1989; Shu et al., 1990), where spiral disturbances that displace the star from the center of mass (COM) of the star/disk system grow through a feedback loop involving reflections from the distorted edge of the disk. One-armed spirals ($m = 1$) are especially susceptible to SLING because their propagation through the disk is not hindered by Lindblad resonances in a Keplerian disk. A similar phenomenon has been seen in simulations of rapidly rotating polytropic stars, where a one-armed spiral in the Keplerian equatorial regions can slosh to the opposite side of the COM from the more slowly rotating core (Pickett et al., 1997). The growth rate for SLING is dynamic, but because it involves reflection from the outer boundary, the e-folding time is typically a few times t_{rot} for the outer disk edge. In an extended disk, this is much longer than the growth times of $m > 1$ waves by swing, which are $\sim t_{rot}$ for their CRs. Also, Shu et al. estimate that SLING instability requires $M_d/M_s >$ about 1/3. Such large disk masses may occur during the early embedded phase of disk accretion, but even then, if $Q < 1.5$, multiarmed GIs should still grow by swing much more quickly than by SLING. To date, although one-armed spirals do appear in some simulations (e.g., Laughlin and Rozyczka, 1996), sustained SLING amplification does not appear to play an important role in disk simulations.

NONLINEAR REGIME. Once GIs initiate, there are several key questions about their nonlinear outcome: What mechanisms can limit the nonlinear growth? What happens to a disk when limiting mechanisms fail and under what conditions do they fail? Three processes turn out to be critical to the nonlinear dynamics: thermal physics, mode coupling, and fragmentation. This subsection introduces these basic ideas as a prelude to the extended discussions of §§ 4.2 and 4.4.

Thermal Physics. A suggestion made by several authors (e.g., Paczynski, 1978; Lin and Pringle, 1987) and foreshadowed with remarkable prescience by Goldreich and Lynden-Bell (1965) is that the amplitude of GIs will self-limit at a point where heating by the instability is balanced by radiative cooling. Heating of the gas due to GIs occurs for several reasons. The gravitational torques in the trailing spirals induce outward transport of angular momentum, and this releases gravitational energy because of net inward transport of mass into the central potential well, as discussed in § 3.3. Larson (1984) and Boss (1984) were among the first to propose that GI torques may operate in this way as a transport mechanism for young stellar disks. How this works is easy to discern (e.g., Durisen et al., 1986) for the polytropic equilibrium states that resemble rapidly rotating stars (see § 2.1), because these objects produce a single, well-defined, global spiral mode when they become dynamically unstable. Fluid elements outside the CR of the wave gain angular momentum and expand outward, while those inside the CR lose angular momentum and contract. Detailed analysis of disk simulations demonstrates that this special role of CR generalizes to the dominant spiral modes in GI-active disks (e.g., Mejía et al., 2005; Boley et al., 2006, 2007a).

Another source of gas heating occurs as the spiral waves steepen into shocks at nonlinear amplitudes. As discussed in § 4.3, the Mach number $\mathcal{M}$ is the ratio, measured in the shock frame, of the normal component of the gas velocity $|u_1|$ to the sound speed. For spiral waves with pitch angle i propagating azimuthally with pattern speed Ω_{pat},

$$\mathcal{M} \approx \varpi|\Omega - \Omega_{pat}| \frac{\sin i}{c_s} \tag{45}$$

These shocks produce entropy generation in relatively thin shock fronts. Ultimately, this localized heating also comes from tapping the energy in the central potential through net inward transport, although some additional energy comes from compression of the gas into the thin, partially self-gravitating structures. Eq. (45) is only approximate because it assumes that the gas is flowing along the ϕ-direction as it enters the shocks. In the nonlinear regime for a well-established wave, the preshock streamlines are not purely azimuthal. Notice that the strength of the shock depends on location relative to the CR of the wave.

Heating by GIs tends to increase c_s and hence Q. If the disk heats indefinitely, the instability will shut off once Q becomes too large, but if the disk cools radiatively, heating and cooling can reach a balance at nonlinear wave amplitude. The result can be sustained GI activity at a relatively constant but unstable value of Q (Pringle, 1981). Although their simulations were rather primitive by contemporary standards, Tomley et al. (1991, 1994) first convincingly demonstrated that thermal physics can control GI amplitudes by a balance of heating and cooling, and, as described below, this has been confirmed by many subsequent studies (e.g., Pickett et al., 1998, 2000, 2003; Gammie, 2001; Boss, 2002a; Rice et al., 2003; Mejía et al., 2005; Boley et al., 2006, 2007a; Stamatellos and Whitworth, 2008; Cossins et al., 2009).

Mode Coupling. Using second- and third-order governing equations, which are basically just high-order members of the equation sets obtained by expanding in small perturbations about the disk equilibrium state, Laughlin et al. (1997, 1998) showed that two-armed GI spiral waves in disks spawn structures with different numbers of arms when they become nonlinear. This is easy to see from the elementary trigonometric identity $\cos^2 2\phi = (1 - \cos 4\phi)/2$. In the second-order equations, $\cos 2\phi$ terms will multiply each other and hence produce driving terms for $m = 0$ and $m = 4$ disturbances. The $m = 0$ terms correspond to reorganization of the basic axisymmetric structure through transport by torques, while the $m = 4$ terms represent the development of power in four-armed structures as the two-armed mode becomes nonlinear.

At first, $m = 4$ is not in itself a distinct mode but a harmonic of the $m = 2$ mode that grows in phase with it (Imamura et al. 2000). Nevertheless, independent $m = 4$ modes can be forced or stimulated by this driving term, and independent unstable modes may be triggered. Similarly, Laughlin and Korchagin (1996) showed that the nonlinear interaction of independent $m = 2$ and $m = 3$ modes can give rise to an $m = 1$ wave.

In fully nonlinear hydrodynamic simulations of low-Q disks, multiple modes can become unstable initially in the linear regime (e.g., A. F. Nelson et al., 1998; Pickett et al., 1998), and even if the initial growth is dominated by one mode, numerous modes usually appear in the nonlinear regime. The result is that if a disk achieves a quasi-steady balance of heating and cooling, what has been called the *asymptotic state,* then there tends to be power at all m-values resolved by a given numerical method (e.g., Mejía et al., 2005; Boley et al., 2006). Gammie (2001) refers to this state as *gravitoturbulence,* but it is still unclear what relationship it may have to traditional notions of turbulence. Despite mode coupling, the asymptotic amplitude of GIs and the resultant torques and net quasi-steady mass transport increase as the cooling time t_{cool} decreases (e.g., Mejía et al., 2005; Cai et al., 2006, 2008). Mode coupling makes the asymptotic state complicated, but it is thermal physics that determines the limiting nonlinear amplitude. Although mode coupling is an important phenomenon that deserves further study, it does not seem to play a primary controlling role and is not further elaborated in this chapter.

Fragmentation. When the stresses induced by GIs become too large, or, equivalently, when t_{cool} becomes short enough (Gammie, 2001; Rice et al., 2003, 2005), the spiral arms become extremely thin and fragment gravitationally into dense self-gravitating clumps, which may be either transient or permanent. This behavior is alluded to in Tomley et al. (1991), but, as shown in Plate 7, was first convincingly demonstrated by Boss (2000) and by Mayer et al. (2002) for isothermal disks, where t_{cool} effectively $\rightarrow 0$. It is the phenomenon of fragmentation that led to the revival by Boss (1997) of the disk instability (DI) theory for gas-giant planet formation (see § 5.2). The conditions under which fragmentation may occur in real disks are a topic of ongoing debate. There are two aspects to this problem, numerical and physical; when is fragmentation induced or suppressed by purely numerical effects (see Nelson, 2006; Durisen et al., 2007; for recent discussions), and under what conditions does fragmentation really occur as a physical process in disks? An in-depth discussion of numerical issues is beyond the scope of this chapter, and I here emphasize fragmentation results that are thought to be independent of numerics.

ISSUES. Among the key questions about GIs in disks are the following:

- When and where do we expect GIs to occur in protoplanetary disks?
- How do GIs affect the structure of the disk?
- What controls their behavior in the nonlinear regime, particularly steady mass transport versus fragmentation?
  - What can we learn from simple treatments of the equation of state (EOS)?
  - What can we learn from simple cooling laws?
- What happens when GIs occur in disks with realistic treatments of radiative cooling?
- What, if anything, do GIs have to do with planet formation and other disk phenomena?

The sections and subsections that follow discuss what is currently known about these issues. Although there is growing consensus about generic conditions for mass transport and fragmentation due to GIs, applications to real disks and to planet formation remain controversial.

4.2. Occurrence of GIs

Examination of eq. (38) in terms of typical parameters provides insight into when GIs might occur in protoplanetary disks. I will assume that we can replace c_s by c_{s0} when required without loss of accuracy. For a Keplerian disk, where $\kappa = \Omega = \Omega_K$, instability of nonaxisymmetric modes is expected when

$$Q < Q_{crit} \approx 1.7 \left(\frac{H/\varpi}{0.05}\right)\left(\frac{5\ \mathrm{AU}}{\varpi}\right)^2\left(\frac{M_s}{\mathrm{M}_\odot}\right)\left(\frac{3\cdot 10^3\ \mathrm{g\ cm^{-2}}}{\Sigma}\right). \quad (46)$$

Notice that $\pi\Sigma\varpi^2 \sim M_d$, and so GIs will occur under two conditions: if M_d/M_s is large or if the disk is cold and thin in a region of significant surface-density enhancement. Rewriting Q reveals a third interesting possibility. If we again set $\kappa = \Omega = \Omega_K$,

$$Q \sim \frac{T^{1/2}}{\Sigma\varpi^{3/2}} \sim \varpi^{(p-3-2r)/2}, \quad (47)$$

where we have assumed $T \sim \varpi^p$ and $\Sigma \sim \varpi^r$. Eq. (47) tells us that any disk with $p - 3 - 2r < 0$ extending to large-enough radii will have GIs in the outer disk. We now examine each of the three cases in more detail.

LARGE DISK MASS. A Hayashi et al. (1985) minimum-mass solar nebula (MMSN) has a total M_d of about 0.01 $M_\odot$, which is typical of what is cited for T Tauri stars (see the chapter by Calvet and D'Alessio). The surface density of $3 \cdot 10^3$ g cm^{-2} at 5 AU used in eq. (46) is about 20 times that of the Hayashi et al. MMSN but is only about twice as large as the more recently proposed compact nice MMSN (Desch, 2007). So, for ϖ < a few 10s of AUs, a protoplanetary disk is unlikely to be globally unstable to GIs unless $M_d/M_s >$ about 0.1 $M_\odot$, a disk mass at the upper bound of T Tauri values. Many recent numerical studies of GIs in protoplanetary disks have concentrated on cases where $M_d/M_s \sim 0.1$ because these would seem to be the highest relative disk masses one might expect in disks that are forming planets. On the other hand, disks in embedded (Class 1) systems, which are still in the early cloud-accretion phase, can have masses comparable to those of their central stars (e.g., Osorio et al., 2003). This makes sense because early accretion rates from the collapsing interstellar cloud are likely to be high. As discussed in § 2.2, if [illegible] < [illegible], then mass accretes into the disk faster than α-viscosity can transfer it onto the star. The steady increase of Σ should inevitably lead to GIs. Plugging $\varpi = 5$ AU into eq. (15) gives about 10^5 yr, and so GIs should occur at these radii for typical parameters if $\dot{M} >$ about 10^{-5} $M_\odot$ yr^{-1}.

Laughlin and Bodenheimer (1994) were the first to demonstrate, using 2D and 3D simulations, that GIs would occur in a circumstellar disk around a young solar-type star for the particular case of a collapsing singular isothermal cloud. With $\dot{M} \approx$ few $\cdot 10^{-4}$ $M_\odot$ yr^{-1} in their simulations, GIs erupt almost as soon as the disk forms, as shown in Fig. 5.9. Because these authors assumed isothermal disk behavior in the GI-active region (see §§ 4.4 and the discussion near eq. [69]), the GIs in their disks become extremely strong and transport mass on a dynamic timescale. This work was extended to massive star formation by Yorke and Bodenheimer (1999). A somewhat more realistic treatment by Rice and Armitage (2009), which accounts for the effect of radiative cooling on the strength of GIs, also leads to a scenario where GIs can dominate disk evolution. Assuming instead a magnetically dominated cloud-collapse model, Vorobyov and Basu (2006, 2008) nevertheless confirm early onset of GIs and suggest that GIs may control the entire accretion history of the disk. Simulations show that disks with masses comparable to those of their central stars can be GI unstable and that gravitational instability of global modes results both for low- and high-mass central stars (e.g., Lodato and Rice, 2005; Krumholz et al., 2007a; Stamatellos and Whitworth, 2009). Recently, Cai et al. (2008) simulated a massive circumstellar disk based closely on the disk parameters deduced observationally by Osorio et al. (2003) for the disks in the Class 1 L1551 IRS5

Fig. 5.9. Surface overdensity gray scales illustrate the eruption and maintenance of global GIs in a disk accreting from a rotating singular isothermal sphere. The time markers are in units of 477 years, and the outer radius of the disk is 230 AU. This is a low-resolution (25,000 particles) 3D SPH simulation. Adapted from Laughlin and Bodenheimer (1994).

multiple system. Even with strong irradiation from the surrounding infalling envelope, the disk is dynamically unstable to a global two-armed mode.

Despite these simulation results, it is important to remember that the initial conditions for star formation are still far from being well understood. They probably depend on the star-formation environment and vary considerably from case to case. With different but plausible assumptions, it is possible to imagine disk formation where, even in the earliest phases, t_{in} is always and everywhere $> t_{ev}$ (Dullemond et al., 2006).

DEAD ZONES. The MRI can occur in disks only where the gas is ionized enough to couple to a weak magnetic field (see the chapter by Balbus). Thermal ionization of gas-phase alkali metals occurs in LTE for $T >$ about 1,000 K. When dust is included, however, recombination on the dust grains is very efficient, so thermal ionization may not actually occur until T exceeds the temperature at which all the dust is completely sublimated, or about 1,700 K (Desch, 1999; Sano et al., 2000). Temperatures this high occur only in the innermost regions of disks ($<$ few $\cdot$ 0.1 AU) and in very low-density irradiated regions high in the disk atmosphere (see the chapter by Calvet and D'Alessio). Elsewhere, disk gas must be ionized by other means, such as ultraviolet (UV) rays, X-rays, or energetic particles due to activity of the central star, cosmic rays, or nearby stars. It typically takes a column density of about 100 g cm^{-2} to attenuate these sources of ionizing radiation (see the chapters by Clarke and by Calvet and D'Alessio). Column densities in the disk from a few tenths out to about 10 AU may be high enough that the bulk of the gas disk mass is not ionized. For these radii, the MRI can occur only in layers with a surface density of about 100 g cm^{-2} at the top and bottom surfaces of the disk. Gammie (1996) suggested that this results in *layered accretion* with mass transport concentrated in the ionized layers, and he referred to the region between the ionized layers as the *dead zone*. A key issue (see the chapters by Balbus and by Königl and Salmeron) is how much turbulence the ionized layers can induce in the neutral layers. Even if turbulent MHD torques and hence mass accretion are not entirely quenched in the dead zone (e.g., Oishi et al., 2007), leading to what has been dubbed an *undead zone*, they may be considerably reduced, resulting in both inefficient heating and a buildup of mass, both of which are conducive to episodic outbursts of GIs (e.g., Armitage et al., 2001).

OUTER DISKS. Disk observations and associated modeling discussed in the chapter by Calvet and D'Alessio suggest that the effective temperature of the thermal radiation from disks $T_{eff} \sim \varpi^{-3/4}$ to $\varpi^{-1/2}$, where the former corresponds to a geometrically flat disk dominated by viscous dissipation and the latter to a flared disk dominated by stellar irradation. How this translates to a midplane $T(\varpi)$ depends on optical depth and on the relative contributions of internal heating by turbulent viscosity and external irradiation. Let us suppose that $T(\varpi)$ has a similar range of power laws $p = -3/4$ to $-1/2$. Then, for a Keplerian disk, eqs. (46) and (47) imply that any large-enough disk ($\sim$ 100 AU or more) will be GI unstable in its outer regions if Σ falls off less steeply than $\varpi^{-7/4}$ for $p = -1/2$ or less steeply than $\varpi^{-15/8}$ for $p = -3/4$. This possibility has been pointed out by many researchers over the years, and at least some disks may satisfy these conditions, especially during the early accretion phase.

4.3. Spiral Shock Bores

GIs infect a disk with large-amplitude, prograde spiral waves that have a predominantly trailing character. Fluid elements outside the CR of the wave have slower angular speed and are overtaken by the wave, while those inside the CR overtake the wave and enter the dense arms from behind. Because c_s is typically much less than the disk orbital speed $\varpi \Omega$, shocks are a natural consequence as the waves steepen in the nonlinear regime. As mentioned in § 2.1, it is common for researchers to think of these spirals as *density* waves, where the main effects are compression and heating of the gas. This is a partial truth that can be terribly misleading but has become set in people's minds because so much excellent work on spiral waves before the 1990s was done using the vertically integrated thin-disk approximation and/or using an isothermal EOS. As first emphasized by Pickett et al. (1998, 2000, 2003), the surfaces of GI-active disks develop strong spiral surface corrugations in 3D simulations. They demonstrated the phenomenon but did not precisely identify the underlying mechanism. Following up insights from MHD studies of galactic gas disks by Martos and Cox (1998), Boley and Durisen (2006) used analytic arguments combined with simulations to show that spiral shock waves in vertically stratified disks are also hydraulic jumps, and they introduced the term *spiral shock bore* to capture both the shock and jump characteristics. Vertical jumps behind shocks are intrinsic to the dynamics of strong spiral waves in realistic disks, and they produce complex flows and large disk-surface distortions, which may have direct bearing on the observational and physical consequences of GIs. Henceforth, it is hoped that readers of this chapter will think of a strong spiral wave in a thermally stratified gas disk as something more than simply a density wave.

SHOCK WAVES. To understand how spiral shocks are also hydraulic jumps, it helps to review the simple picture of a steady plane-parallel shock. A shock is effectively a nonlinear sound wave in which dissipative mechanisms convert some of the kinetic energy of the preshock flow into thermal energy. Let subscripts 1 and 2 refer to preshock (upstream) and postshock (downstream) gas properties, respectively. I will ignore the details of how energy changes form and assume that it happens in a thin region represented as an abrupt discontinuity. Pre- and postshock quantities can be related by using constraints imposed by conservation laws. Let u be the component of the fluid velocity perpendicular to the shock front in the shock reference frame. For a steady, plane-parallel flow, if we ignore gravity and assume that viscosity acts only in the thin shock front, eqs. (1) and (2) express momentum and mass conservation across the shock, respectively, and yield the shock jump conditions

$$P_1 + \rho_1 {u_1}^2 = P_2 + \rho_2 {u_2}^2 \tag{48}$$

and

$$\rho_1 u_1 = \rho_2 u_2 \tag{49}$$

Any velocity component parallel to the shock front is conserved across the shock.

To determine postshock quantities from preshock quantities requires a third jump condition specifying what happens to the energy in the gas. A glance at eq. (3) indicates how tricky the energy budget of the gas can be. There are three simple limiting cases for the third jump condition that are instructive. It we assume that the gas is ideal with a constant γ and no phase changes across the shock,

$$\frac{1}{2}\,{u_1}^2 + \left(\frac{\gamma}{\gamma+1}\right)\frac{P_1}{\rho_1} = \frac{1}{2}\,{u_2}^2 + \left(\frac{\gamma}{\gamma+1}\right)\frac{P_2}{\rho_2} \quad \text{(adiabatic)}, \tag{50}$$

$$P_1{\rho_1}^{-\gamma} = P_2{\rho_2}^{-\gamma} \quad \text{(isentropic)}, \tag{51}$$

and

$$P_1{\rho_1}^{-1} = P_2{\rho_2}^{-1} \quad \text{(isothermal)}. \tag{52}$$

In these simple cases, we refer to the shocks as *adiabatic, isentropic,* or *isothermal* when the specific energy [eq. [50]), specific entropy [eq. [51]), or temperature [eq. [52]), respectively, is the same on both sides of the shock. The latter two cases imply a loss of energy in the shock, while the first case involves an increase in the specific entropy of the gas. The isothermal limit is the most extreme because the internal energy per gram does not change through the shock even though the bulk kinetic energy is substantially diminished.

The jump conditions for an adiabatic shock yield the following results of interest to us:

$$\frac{u_2}{u_1} = \frac{\rho_1}{\rho_2} = \frac{\gamma-1}{\gamma+1} + \left(\frac{2}{\gamma+1}\right)\frac{1}{\mathcal{M}^2} \tag{53}$$

and

$$\frac{P_2}{P_1} = \frac{2\gamma\mathcal{M}^2}{\gamma+1} + \frac{\gamma-1}{\gamma+1}, \tag{54}$$

where the Mach number $\mathcal{M}$ is defined as $|u_1|/c_{s1}$. Eqs. (53) and (54) illustrate that the postshock pressure jump increases as $\mathcal{M}^2$, while $\rho_2/\rho_1 = u_1/u_2 \to (\gamma+1)/\gamma-1)$ as $\mathcal{M} \to \infty$. For an isothermal shock, eq. (53) is replaced by

$$\frac{\rho_2}{\rho_1} = \gamma\mathcal{M}^2, \tag{55}$$

where γ becomes unity if the Mach number is defined in terms of the isothermal sound speed c_i rather than the adiabatic sound speed c_s. In other words, the compression across a strong, i.e., high-$\mathcal{M}$, isothermal shock can be very large. This is one of the reasons that spiral waves in isothermal evolutions may validly be characterized as density waves.

SPIRAL SHOCK WAVES IN DISKS ARE ALSO HYDRAULIC JUMPS. Consider the flow of water in a shallow channel on the surface of the Earth. *Hydraulic jumps* occur in such flows at subsonic speeds when there is an abrupt decrease in the flow speed. Because water is nearly incompressible, conservation of mass, momentum, and energy then requires that the bulk kinetic energy of the pre-jump flow be converted across the jump into a gravitational potential energy and also, under some conditions, disordered turbulent motion. The result is that the vertical height h of the flow increases.

Even though the gas in a protoplanetary disk is compressible, the formation of a shock wave produces a similar phenomenon. Suppose that the disk is in vertical hydrostatic equilibrium in the preshock region. For an adiabatic shock in the strong shock limit ($\mathcal{M} \to \infty$), eqs. (53) and (54) tell us that the vertical pressure-gradient force per gram will increase by the factor

$$\frac{1}{\rho_2}\frac{dP_2}{dz}\left(\frac{1}{\rho_1}\frac{dP_1}{dz}\right)^{-1} = \frac{2\gamma(\gamma-1)\mathcal{M}^2}{(\gamma+1)^2} \tag{56}$$

as the gas moves through a thin spiral shock front. On the other hand, if the disk is not self-gravitating, the ratio of the vertical gravitational field is unity between the pre- and postshock regions because it is then determined entirely by the central force field of the star. When gas self-gravity is included, eqs. (10) and (11) of Boley and Durisen (2006) show that the postshock gravitational field increases only by a factor of order unity through the shock even as $\mathcal{M} \to \infty$. The ratio of the pressure-gradient jump to the jump in gravitational field is called the *jump factor J* by Boley and Durisen. Clearly, passing through a strong adiabatic shock disrupts vertical hydrostatic equilibrium. Because $J > 1$, material expands upward at the sound speed on a timescale of order $H/c_s \approx \Omega^{-1} = t_{rot}/2\pi$. In the limit of a strong adiabatic shock where the disk is non-self-gravitating, J is just the right-hand side of eq. (56).

We would, of course, like to know how high the gas will jump behind the shock. In a classic hydraulic jump at the Earth's surface, where $g_z =$ constant, this is determined by the Froude number, defined by $|u_1|/\sqrt{g_z h_1}$. Then, in a nondissipative jump, $h_2/h_1 = (1/4 + 2F^2)^{-1/2} - 1/2$ (Massey, 1970); in other words, for a strong jump, $h_2/h_1 \sim F$. Using an approximate energy argument, Boley and Durisen show that for a strong adiabatic shock in a non-self-gravitating disk, $h_2/h_1 \approx \sqrt{J} \sim \mathcal{M}$. In other words, to within a factor of order unity, the Froude number is the same as the Mach number. This illustrates mathematically the intimate relationship between adiabatic spiral shocks and hydraulic jumps in a stratified disk. For more details concerning the general case of shocks with any $\mathcal{M}$ or γ in partially self-gravitating disks, see Boley and Durisen. Even if the g_z due to disk self-gravity is comparable to the g_z due to the central star ($Q \approx 2$ in eq. [38]), shocks are accompanied by jumps for all $\mathcal{M} >$ about 2.5 when $\gamma = 7/5$ or higher.

By contrast, for an isothermal shock, the right-hand side of eq. (56) is unity. Then, if disk self-gravity is negligible, $J = 1$, and the disk height does not change. Moreover, if disk self-gravity is not negligible, $J < 1$, and the disk will actually compress vertically behind the shock. This is the second reason that spiral waves in isothermally evolved disks can properly be called density waves; strong compression of the gas is the main effect of the shock.

ANATOMY OF A SHOCK BORE IN A DISK. The morphology of the flow in shock bores is made rather complex by the rotational shear. Let us focus on a region inside the CR of a spiral wave, where the gas catches up to the spiral arm and the shock front lies along the inner edge of the arm. As the gas jumps in the postshock region, it expands upward. Because the preshock disk has a smaller height, the jumping gas also expands horizontally and moves out over the preshock gas. At the same time, the component of the gas motion tangential to the shock is unchanged. For a trailing spiral, this leads to streaming of the gas along the spiral arm in the postshock region. The jumping gas does the same thing. After reaching a height characterized by $\sqrt{J}$, the gas falls back onto the disk. The result inside the CR of the spiral pattern is a breaking wave that curls over and crashes down onto the preshock gas at a radius that is approximately one or two disk scale heights interior to the radius at which it jumped, as illustrated in Fig. 5.10.

4.4. Simple Equations of State

METHODOLOGY. A proper treatment of how nonlinear GIs affect disk evolution requires a numerical approach, and a variety of techniques have been employed. Some researchers solve 2D vertically integrated hydrodynamics equations, either globally (e.g., the vertically isentropic simulations of A. F. Nelson et al., 2000a) or locally (e.g., the shearing-box simulations of Gammie, 2001). These do not capture the real complexities of the vertical direction (see §§ 2.1 and 4.3) but allow high-resolution simulations at low computational cost. Most contemporary work on GIs is now fully 3D, and a wide variety of 3D techniques have been brought to bear. They fall into three general categories:

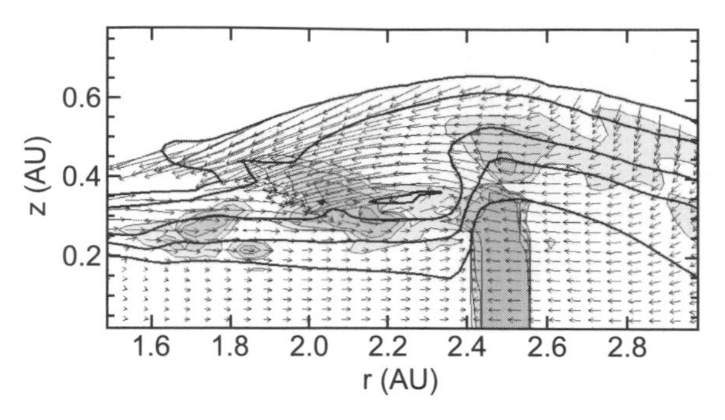

Fig. 5.10. A slice in the ϖ,z-plane of a shock bore. The solid lines are density contours, and the arrows indicate the direction and magnitude of the fluid velocity. The code uses artificial bulk viscosity (AV) to mediate shocks, and the gray scale regions, indicating strong AV dissipation, mark the locations of strong shocks. The vertical gray region at 2.5 AU is the intersection of this plane by a trailing spiral shock with a CR at 5 AU. The wave is moving into the paper. Gas jumping behind the shock at $\varpi > 2.5$ AU is curling over the shock front and crashing back down onto the disk between 1.8 and 2.2 AU in the form of a large breaking wave. Adapted from Boley and Durisen (2006).

fixed-grid (FG) calculations using finite-difference equations (e.g., Boss, 1997, 2000; Pickett et al., 1998, 2000, 2003), smoothed-particle hydrodynamics (SPH) (e.g., Mayer et al., 2002; Rice et al., 2003), and, only recently, adaptive mesh refinement (AMR) schemes using Riemann solvers (e.g., Krumholz et al., 2007a; see also Plate 8). The focus of this chapter is physical processes, not techniques. Except for the important but still controversial issues of radiative cooling, locality of GIs, and longevity of clumps in fragmenting disks, this chapter emphasizes consensus results that are independent of method. More discussion of the nature and relative merits of different approaches can be found in Durisen et al. (2007).

ISOTHERMAL, ISENTROPIC, AND ADIABATIC. Toward the end of the 1990s and early in the 21st century, papers appeared by various authors describing simulations of GIs using no explicit cooling and simple assumptions about the EOS. The terminology in these papers often differed or was unclear, and to avoid further confusion, Durisen et al. (2007) suggested the following standardized terminology:

An *isothermal* disk simulation is one where the temperature of the gas is held constant spatially at the fixed-grid cells (Eulerian) of FG calculations or held constant for the moving fluid elements (Lagrangian) of SPH calculations. This does not mean that the initial disk structure is itself isothermal; usually it is not, and usually it does not make much difference in the results whether the isothermality is Eulerian or Lagrangian.

An *isentropic* simulation holds the specific entropy s of the gas fixed instead of the temperature. Again, the initial structure of the disk itself need not also be isentropic; in general, it is not, but this varies from study to study. The isentropic condition can again be enforced in either an Eulerian or Lagrangian manner. Sometimes, isentropic evolution is achieved by adopting a *polytropic* EOS, where

57 $$P = K\rho^{1+1/n} = K\rho^{\gamma},$$

for constant K and polytropic index n. The second equality corresponds to isentropic behaviors for an ideal gas with ratio of specific heats γ where log K is proportional to s. GIs lead to compression and shocks in the nonlinear regime, and so both isothermal and isentropic simulations result in net energy loss by the disk because, according to eqs. (51) and (52), all or part of the work done by compression $P\nabla \cdot \mathbf{v}$ or heating by viscous dissipation in the shock fronts is not conserved but simply disappears (Pickett et al., 1998, 2000).

An *adiabatic* calculation is one where no thermal energy is lost as the fluid evolves. Then s increases as fluid elements go through shock fronts, as a consequence of eq. (50). To model this in numerical calculations requires the inclusion of artificial viscosity or other shock-capturing techniques. Then, unless cooling is introduced, GIs shock-heat the gas, raise Q, and shut themselves off (Pickett et al., 2000; Boss, 2002a, 2003). An adiabatic calculation can employ eq. (9) with a fixed γ or a more realistic EOS that accounts for excitation of H_2 and other compositional changes (Boss, 2001; Boley et al., 2007b).

Global 3D FG calculations using simple EOSs by Pickett et al. (1998, 2000, 2003) and Boss (2002a) demonstrate that as the EOS allows more energy to be lost in strong compressions, the resultant GIs have higher amplitude; in other words, the thermal physics of the disk controls the amplitude and nonlinear behavior of GIs. This is perhaps the most important consensus conclusion to emerge from work near the turn of the millennium.

THE ISOTHERMAL LIMIT. Investigations of simple EOSs without radiative cooling have been largely superseded by those that include cooling and will not be discussed further here except for the isothermal case. Because all thermal energy produced by compressional heating or friction is, in effect, radiated away immediately by a fluid element, the effective cooling time is essentially zero. As an extreme limiting case without complicated gas physics or radiative transport, isothermal simulations, although unrealistic, can be used to explore the most severe effects possible from GIs, such as fragmentation of a disk into compact bound pieces. They also test the relative performance of different types of simulation codes. It is in the context of isothermal simulations (see Plate 7) that dense clumps persisting for multiple orbits were first reliably demonstrated in fragmenting disks (Boss, 2000; Mayer et al., 2002). An important consensus result is that with sufficient resolution in any reasonable hydrodynamics code, an isothermally evolved disk with low-enough Q will fragment into dense protoplanet-like clumps. For isothermal disks, fragmentation seems to set in when $Q_i <$ about 1.4, where Q_i is Q evaluated using the midplane c_i in place of c_s (Johnson and Gammie, 2003; Mayer et al., 2004; Durisen et al., 2008). The state of the art in isothermal calculations is the so-called Wengen collaboration, where a large number of codes are used to evolve the same low-Q disk up to the point of clump formation. As shown in Plate 8, with sufficient resolution, codes of all types can agree about fragmentation for the same initial conditions in an isothermally evolved disk.

As a specific example of the continued utility of isothermal disk simulations beyond code testing, it is possible to understand analytically why clump formation in low-Q isothermal calculations occurs at corotation of the fastest-growing mode (Durisen et al., 2008). The dense isothermal sheet behind the spiral shock front is more likely to collapse because of self-gravity where it is least compressed because the pressure-gradient force in a thin sheet increases faster with isothermal compression than gravity does. As a result, a spiral wave is most likely to fragment where the preshock speeds for gas entering the shock are relatively low, namely, in the vicinity of the mode's CR. This insight into the process of fragmentation can be gleaned because the isothermal case permits a highly simplified treatment of the postshock gas.

LONGEVITY OF FRAGMENTS. Even for the simplest assumptions about disk physics, one area that lack consensus at present is whether the clumps that form in a fragmenting disk become permanent bound objects. The Wengen collaboration has shown that the detailed behavior of clumps in isothermal disks is tricky to resolve properly even with AMR schemes. How one refines or derefines a grid can affect the apparent growth of a clump. Many authors try to isolate clumps in their simulations and compute their total thermal and gravitational energies. Usually, by this measure, clumps are bound, i.e., have negative total energies. Actually, however, possibly significant surface-energy terms are usually ignored. The environment of a fragmenting disk is violent, and clumps are subject to collisions, mergers, external pressure, shear stresses, and tidal stresses. Moreover, many schemes introduce explicit artificial shear and/or bulk viscosity to mediate shocks (FG schemes) and/or to enforce fluidlike behavior (SPH), and all numerical schemes have some additional level of uncontrolled friction introduced by truncation error. Although too high an artificial viscosity can prevent clump formation altogether (Mayer et al., 2004), a moderate amount of artificial viscosity can prolong clump lifetimes once slumps form because the violent environment of a fragmenting disk is calmed by the damping of velocity differences (Pickett and Durisen, 2007). Consequently, a very cautious scientist might say that the formation of permanent bound gas-giant protoplanets by disk fragmentation has not yet been reliably demonstrated. Efforts beyond the current Wengen collaboration (Plate 8) will be necessary to resolve this thorny question.

4.5. Idealized Cooling Laws

In order to characterize how thermal physics affects the amplitude and behavior of GIs, one needs to introduce quantitative control of disk energy loss. An approach similar to that in the classic experiments by Tomley et al. (1991, 1994) is to include explicit volumetric cooling in eq. (3), where

$$\Lambda = \rho e / t_{cool}. \tag{58}$$

In other words, over the time interval t_{cool}, every parcel of gas loses an amount of energy, presumably through radiation, equal to its internal energy. In this approximation, cooling is treated as a local energy loss, not as transport. Different choices for how t_{cool} varies spatially or temporally yield different insights.

CONSTANT $t_{cool}\Omega$. The first published studies using eq. (58) assumed that $t_{cool}\Omega$ is constant over the disk. In effect, t_{cool} is then a fixed multiple of the local t_{rot}. This would be exactly the case in a steady-state accretion disk with a constant α, but it may not be appropriate for a globally evolving disk with realistic opacities and self-adjusting radiative cooling. A signature study by Gammie (2001) used local 2D (thin-disk) shearing-box simulations. If the disk reaches an asymptotic state where heating in the disk is balanced locally by cooling, Gammie showed analytically that one expects (see also Pringle, 1981) that

$$\alpha = \frac{4}{9}\left[\gamma_2(\gamma_2 - 1)t_{cool}\Omega\right]^{-1}, \tag{59}$$

where α is a composite of contributions from the Reynolds and Newton stresses (eqs. [28] and [29]) and γ_2 is the effective 2D γ for a vertically integrated thin disk (see § 2.1). The derivation of eq. (59) is straightforward. As long as there is no wave transport of energy (Balbus and Papaloizou, 1999), the heating of the disk is due entirely to a local dissipative term analogous to eq. (7), namely,

$$D = \tau_{\varpi\phi}^{R+N} \left(\varpi \frac{\partial \Omega}{\partial \varpi} \right), \tag{60}$$

where $\tau_{\varpi\phi}^{R}$ and $\tau_{\varpi\phi}^{N}$ are defined by eqs. (25), (28), and (29). Eq. (59) is then derived by setting eq. (60) equal to eq. (58) after integrating over z.

Gammie found α_{eff}s consistent with (59), with the Reynolds and Newton stresses contributing about equally, once his simulations relaxed to a quasi-steady asymptotic state of sustained GIs at a roughly constant Q. In his simulations, this state is achieved within a few dynamic times. He also found that for sufficiently small $t_{cool}\Omega$, the disk no longer reaches an asymptotic state but instead fragments into dense pieces. Specifically,

$$t_{cool}\Omega < (t\Omega)_{crit} \Rightarrow \text{disk fragmentation.} \tag{61}$$

For $\gamma_2 = 2$, $(t\Omega)_{crit} \approx 3$. Note that $(t\Omega)_{crit} = 3$ in (61) is roughly equivalent to $t_{cool} < t_{rot}/2$. Although eq. (61) is commonly referred to as the *Gammie criterion* for disk fragmentation, the idea that fragmentation might set in when the cooling time is comparable to the dynamic time was anticipated by others (e.g., Shlosman and Begelman, 1989) and demonstrated, albeit crudely, by the Tomley et al. (1991, 1994) numerical experiments.

Global 3D SPH simulations by Rice et al. (2003) and by Lodato and Rice (2004, 2005) using $t_{cool}\Omega$ = constant with a $\gamma = 5/3$ EOS generally confirm these results. Lodato and Rice insert γ for γ_2 in eq. (59), which is correct only when the vertical gravitational field is somewhere between dominance by the star ($\gamma_2 = 1.5$) and dominance by disk self-gravity ($\gamma_2 = 1.8$), as determined by eqs. (18) and (19), respectively. This gravitational-field condition occurs for $Q \approx 2$, as shown by eq. (38), so their substitution of γ for γ_2 ends up being a decent approximation in a GI-active disk with $\gamma = 5/3$. They find that the α_{eff}s they measure in their simulations agree to within several 10s of percent with eq. (59). Curiously, however, unlike Gammie, they report that the Reynolds stresses tend to be small and that α is dominated by the gravitational stresses. By bracketing the $t_{cool}\Omega$ for fragmenting and nonfragmenting cases, Rice et al. (2003) find that for $\gamma = 5/3$, $(t\Omega)_{crit}$ is between 3 and 5, with a tendency for more massive disks to have a higher $(t\Omega)_{crit}$.

Cossins et al. (2009) present a detailed analysis of GI structure in disks where $\gamma = 5/3$ with constant values of t_{cool}/t_{rot} from 4 to 10. Their αs are also in good agreement with Gammie's formula, eq. (59). They find that the number of arms in the waves are consistent with WKB estimates, that the density amplitudes go inversely as the square root of t_{cool}/t_{rot} in a manner consistent with shock heating, and that the waves extend about corotation so that the Mach number of the flow is of order unity.

CONSTANT t_{cool}. It is by no means obvious that a real GI-active disk will have $t_{cool}\Omega$ = constant even in the asymptotic state. Another limiting case that makes sense to consider is t_{cool} = constant applied to the entire disk, usually chosen to be some multiple or fraction of the outer rotation period (orp) of the disk. For modest disk masses, $M_d/M_s \leq 0.25$, the whole disk in $t_{cool}\Omega$ = constant SPH simulations tends to erupt smoothly into nearly asymptotic GI activity. By contrast, as shown in Plate 9, for nonfragmenting t_{cool} = constant global simulations with a grid-based FD code, an initially marginally unstable disk with low-amplitude random-density noise goes through several phases of evolution (Pickett et al., 2003; Mejía et al., 2005): first an *axisymmetric cooling* that lasts several orps, a strong *burst* of GIs in one or two discrete global modes that produces a significant rearrangement of the disk surface density in one or two orps, a *transition* lasting several orps where heating temporarily washes out some of the nonaxisymmetry, and finally a quasi-steady, long-lived *asymptotic* state of sustained GI activity with a nearly constant unstable $Q \approx 1.5$ over a large part of the disk and with an overall balance of heating and cooling.

The t_{cool} = 2 orp simulation in Mejía et al. has an initial 40 AU disk with $M_d = 0.07$ $M_\odot$, $M_s = 0.5$ $M_\odot$, and $\Sigma \sim \varpi^{-1/2}$ and covers almost 24 orps (6,000 yrs). About half this time is spent in the asymptotic phase. The burst in this simulation is a dramatic event with ordered global spiral patterns that produce prodigious mass inflow rates of about 10^{-5} $M_\odot$ yr^{-1} in the 10 to 30 AU region of the disk. This corresponds to $\alpha_{eff} \sim 0.1$ to 0.2 near the peak of the burst, about an order of magnitude higher than predicted by local energy balance in eq. (59). Wave transport of energy must be extremely large. The presence of only a few discrete global modes and an α_{eff} inconsistent with local thermal balance means that bursts are nonlocal in both the structural and dissipative senses (see § 2.3). The occurrence of bursts depends on a disk's initial conditions, environment, and competition with other transport mechanisms.

By measuring $\dot{M}$ in the asymptotic phase of the t_{cool} = 2 orp simulation and using the steady-state α-disk result $\dot{M} = 3\pi\nu\Sigma$ to estimate α_{eff}, Mejía et al. report values at least several times larger than given by eq. (59). In work currently under way by my own IU Hydro Group (Michael et al., *private communication*), we are performing experiments to determine how numerical resolution affects the measured value of α_{eff}. The α_{eff}s are now computed properly by using eqs. (27) and (29). Apparently, 128 azimuthal zones, as used by Mejía et al., are too few, and low resolution tends to overestimate α_{eff} by a factor of about 2. The value of α_{eff}, averaged over radius and over many outer rotations, converges for 256 and 512 azimuthal zones at a value of about 0.024, close to but slightly lower than the value given by (59) when the self-gravitating disk $\gamma_2(\gamma)$ relation (18) is used. As shown by eq. (38), with $Q \sim 1.5$ in the asymptotic state, γ_2 should be between the values given by (18) and (19). So the measured converged value of the spatially and temporally averaged α_{eff} is roughly consistent with Gammie's simple formula within the uncertainties of the $\gamma_2(\gamma)$ relation. However, even in the asymptotic state, local balance of heating and cooling is probably violated because there are significant radial and temporal variations in α_{eff} over shorter times and over length scales comparable to the wavelengths of low-order spirals. The IU Hydro Group also finds, like Lodato and Rice, that α_{eff} is almost entirely due to the gravitational stresses. The Reynolds stress, though more difficult to measure, seems relatively small in the region of the simulation where most of the asymptotic-phase mass transport happens.

By varying the value of t_{cool} in their simulations, Mejía et al. find that $\dot{M} \sim t_{cool}^{-1}$ in nonfragmenting disks, as expected from eq. (59), and that the Gammie criterion for fragmentation with $(t\Omega)_{crit} \approx 3$ to 5 is valid in their $\gamma = 5/3$ simulations. The latter is confirmed by restarting the evolution of a nonfragmenting disk in the asymptotic phase with an abruptly lowered t_{cool}, noting the radii at which fragments form, and computing the $t_{cool}\Omega$ for these regions.

LOCAL VERSUS GLOBAL BEHAVIOR. The α-disk picture described in §§ 2.1 and 2.3 is intrinsically local, in the sense that only local properties are needed to compute an effective viscosity and evolve a disk in a smooth manner. Balbus and Papaloizou (1999) argued that the Newton stresses due to GIs are likely to violate both the structural and dissipative locality conditions necessary to treat GI disk evolution with confidence using an α-disk description. The global simulations discussed above give mixed results on this issue.

In $t_{cool}\Omega =$ constant simulations, the cooling time is by assumption determined locally. In fact, this cooling prescription is a necessary, though not sufficient, condition for a steady-state accretion disk. Perhaps it is then not so surprising that with this constraint, the Lodato and Rice (2004) and Cossins et al. (2009) global disk simulations behave in a manner consistent with Gammie's local simulations when M_d/M_s is not too large, namely, there is no burst, GIs erupt uniformly everywhere, α_{eff} agrees with eq. (59), the radial surface-density distribution remains smooth, and the spiral waves tend to be tightly wrapped and confined near their corotation radii. For $M_d/M_s \geq 0.5$, however, Lodato and Rice (2005) find that disks can behave in an eruptive manner with more open, larger-scale, lower-order spirals that come and go. Their $M_d/M_s = 0.5$ disk experiences an initial burst, as in the the lower-mass

$t_{cool} = 2$ orp Mejía et al. disk, with an α that exceeds eq. (59) by a factor of up to 5. The latter behavior is both structurally and dissipatively nonlocal.

For $t_{cool} =$ constant simulations, it is implicitly assumed that the physics imposes a globally applicable cooling time. Now, the GIs initiate with a global burst even for relatively modest disk mass, and these bursts are structurally and dissipatively nonlocal. Although one could argue that the bursting behavior is just a transient introduced by unrealistic starting conditions in the simulations, Mejía et al. present strong evidence that even in the asymptotic phase, global modes, particularly $m = 2$, dominate the mass and angular momentum transport. Fourier analysis of ρ in the azimuthal direction shows not only strong power in $m = 2$ but coherent patterns that extend between the Lindblad resonances (LRs) for a small number of discrete modes. Two of these dominant two-armed spirals have their CR roughly aligned with a sign change in the time-averaged $\dot{M}$, as expected for angular momentum transport dominated by a single trailing spiral (see § 4.1). Moreover, the several discrete $m = 2$ modes present tend to have alignments between the CRs and LRs of other modes, and the CRs and LRs of strong modes often align with enhancements in the azimuthally averaged $\Sigma(\varpi)$. Note, from eq. (35), that $m = 2$ modes have a long reach between their LRs. Altogether, this implies that GI transport is at least structurally nonlocal in the asymptotic state.

Nevertheless, for $t_{cool} =$ constant, it is true that there is considerable power present at all resolved m-values in the asymptotic state. Strong, discrete modes with three on four spiral arms are identified in Fourier analyses. For $m \geq 5$, there is power almost everywhere in very many modes with relatively localized reach. Coherent power tends to be concentrated between the LRs for any m or Ω_{pat}. The high-order behavior is what Gammie calls gravitoturbulence, but its relationship to ordinary hydrodynamic turbulence is unclear at present. The break point between local and global modes can be quantified roughly as the m-value at which the distance $\Delta\varpi_{LR}$ between the LRs is only a few times H. For $H/\varpi \approx 0.1$, one gets $\Delta\varpi_{LR} = 2H$ for $m \approx 7$, the point where the amplitude spectrum does begin a significant drop-off (see Figure 12 of Boley et al., 2006). In simulations with real radiative cooling, Cai et al. (2008) decompose $W^N_{\varpi\phi}$ into contributions by different m. Typically, as shown in Fig. 5.11, for mildly irradiated disks, $m = 2$ contributes 50% or more of the net torque that goes into the gravitational α_{eff}.

Dissipative locality has not yet been as rigorously tested. In the $t_{cool}\Omega =$ constant simulations with low disk mass, Lodato and Rice (2004) compare local cooling with local energy dissipation (see their Figure 6). There is general agreement in the trends of these two quantities with ϖ, but the deviations

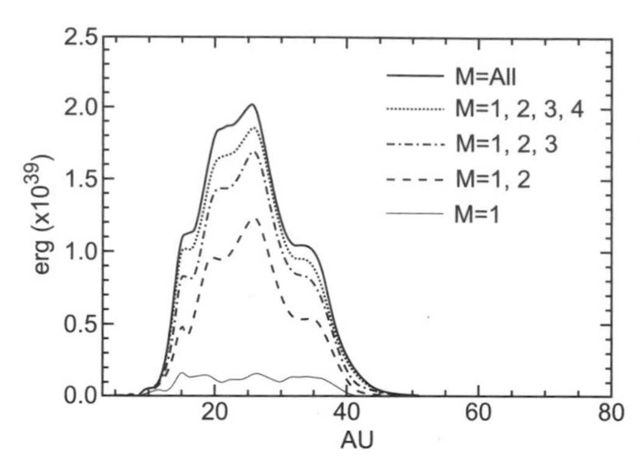

Fig. 5.11. Gravitational torques due to GIs in the Mejía et al. (2005) disk from Plate 9 for the case of realistic radiative cooling and mild external irradiation in the asymptotic phase. Fourier analysis is used to separate contributions from different numbers of arms. The curves show first $m = 1$ torques, then $m = 1 + 2$, and so on. Note that about half of the internal gravitational stress is contributed by two-armed spirals and that the structural features in the total torque profile are imposed mostly by $m = 2$. Adapted from Cai et al. (2008).

can be as large as a factor of two. In our own convergence testing, although $\alpha_{eff} \rightarrow$ eq. (59) to within 10s of percent for $t_{cool} = 2$ orp, this is only with heavy spatial and temporal averaging. The asymptotic-phase stresses show large fluctuations, as already demonstrated in Mejía et al. through analysis of the mass transport. More refined analyses are certainly warranted. Even if waves carry only 10s of percent of the energy over the radial distances spanned by the $\Delta\varpi_{LR}$ for $m = 2$, that could have an impact on disk appearance. Another hint that purely local treatment of GIs misrepresents the physics is that although Reynolds and Newton stresses contribute equally to α_{eff} in Gammie's local simulations, both $t_{cool}\Omega$ and $t_{cool} =$ constant global simulations show that α_{eff} is dominated by the gravitational stresses in the regions of active mass transport. Despite these concerns, eq. (59) is probably a useful estimate of the effective transport in a nonfragmenting GI-active disk for the purpose of long-term disk evolution, as in Rice and Armitage (2009).

ADDITIONAL EFFECTS. γ, $\Sigma(r)$, $t_{cool}(t)$, and $\dot{M}_d$. In a more refined set of $t_{cool}\Omega =$ constant global SPH simulations designed to examine the fragmentation criterion, Rice et al. (2005) find that $(t\Omega)_{crit} \approx 6$ to 7 for $\gamma = 5/3$ and ≈ 12 to 13 for $\gamma = 7/5$. These two values of γ are of particular interest in disks, because $\gamma = 5/3$ corresponds either to atomic or ionized hydrogen or to molecular hydrogen so cold that the rotation states are not excited, while $\gamma = 7/5$ corresponds to molecular hydrogen with excited rotation states. In real disks, the former value is expected in the inner and outer disks, while the latter value is more appropriate for intermediate radii. Exactly how to treat the rotation states of molecular hydrogen is a tricky issue. A detailed discussion goes beyond the scope of this chapter, but the reader is referred to Boley et al. (2007b) for more details. According to the Rice et al. results, as the gas EOS "softens" ($\gamma \rightarrow 1$), a disk becomes more susceptible to fragmentation; i.e., fragmentation sets in at larger $t_{cool}\Omega$. Grid-based FD global simulations with t_{cool} = constant confirm this result to within 10s of percent (Michael et al., private communication). By relating $(t\Omega)_{crit}$ to α using eq. (59) with γ_2 set equal to γ, Rice et al. find that for both γ values, $\alpha_{crit} \approx 0.06$ for fragmentation. They conclude that there is a maximum gravitational stress that a disk can sustain without fragmenting. In t_{cool} = constant simulations, this does not seem to be true during bursts.

Unpublished work by my own group (Michael et al., private communication) examines how much the Mejía et al. (2005) results are affected by varying the initial $\Sigma(\varpi) \sim \varpi^r$, by changing the initial perturbation, and by relaxing the condition that the star remains fixed at the center of the inertial frame. One interesting feature is that as the exponent r decreases from $-1/2$ to $-3/2$, the bursts get weaker, but the asymptotic r in the outer part of the disk after the burst is always roughly $-5/2$. It can be shown analytically (see Boley and Durisen, 2008) that this is the expected $\Sigma(\varpi)$ for a radially and vertically isentropic Keplerian disk with roughly constant Q and $\gamma = 5/3$. It makes sense that Q would settle into a constant unstable value in the asymptotic state, but it is not clear why the disk would tend to be isentropic unless waves transport energy. This is probably worth further study. My group also finds that when the initial random perturbation is nonlinear ($\geq 5\%$ cell-to-cell perturbations in the initial equilibrium density distribution), GIs erupt more smoothly, without a burst. Bursts are thus sensitive to initial conditions and are not a necessary consequence of t_{cool} = constant.

Clarke et al. (2007) show that $(t\Omega)_{crit}$ depends on how quickly the unstable cooling condition is achieved. If $t_{cool}\Omega$ decreases from a stable value more slowly than the dynamic time, fragmentation does not set in until a $(t\Omega)_{crit}$ is reached that is about half the value quoted above by Rice et al.. However, this lower value of $(t\Omega)_{crit}$ does not seem to decrease further when the cooling time is decreased even more slowly. Clarke et al. again argue for a maximum gravitational stress in a nonfragmenting disk, but with a value increased to $\alpha_{crit} \approx 0.12$. Whether the latter claim applies to bursts remains to be tested. The interesting implication is that the propensity of a disk to fragment depends not just on its instantaneous thermodynamic state but also on its thermodynamic history, even in simulations with highly simplified thermal physics. Again, the message is clear: Disk thermal physics controls the strength and behavior of GIs.

Although some simulations have included growth in disk mass ($\dot{M}_d > 0$) (e.g., Mayer et al., 2004; Boley, 2009), little has been done to model infall onto GI-active disks with any detail over significant lengths of time. When $\dot{M}_d$ is large during the early accretion phase, it probably has an impact on whether a disk will fragment (Boley, 2009; Clarke, 2009; Rafikov, 2009). So far, it is unknown whether strong nonaxisymmetry in the inflow can affect the fragmentation criterion.

4.6. Conclusions

The following list summarizes both consensus results and remaining unsettled questions from studies with simplified EOSs and cooling laws.

- Thermal physics
    - Thermal physics controls the nonlinear behavior of GIs.
    - Heating mechanisms tend to damp GIs, while cooling mechanisms tend to destabilize the disk and increase GI amplitudes.
    - For large t_{cool}, disks achieve an asymptotic state where heating by GIs balances cooling at a roughly constant unstable value of Q.
    - When t_{cool} becomes small enough compared with t_{rot}, disks fragment into dense clumps.
- Local versus global
    - GIs are always dominated by low-order, usually two-armed global waves when $M_d >$ about 0.5 M_s.
    - For lower-mass disks, there is no clear consensus, but results to date suggest the following:
        - GIs act more locally when $t_{cool}\Omega$, i.e., t_{cool}/t_{rot}, is constant.
        - GIs are always dominated by low-order global modes when $t_{cool} =$ constant.
- Mass and angular momentum transport
    - In cases where GIs are local, α_{eff} seems well described by a simple analytic formula related to the local $t_{cool}\Omega$.
    - In cases where GIs are global, α_{eff} can be much larger during bursts. During the asymptotic phase, it fluctuates but agrees roughly with the simple analytic formula in an average sense.
- Fragmentation
    - Fragmentation occurs in isothermally evolved disks when Q is low enough ($<$ about 1.4).

- Fragmentation occurs in disks with other EOSs when $t_{cool}\Omega$ is small enough, with more compressible EOSs exhibiting fragmentation at higher values of $t_{cool}\Omega$.
- There may be a maximum gravitational stress $\alpha \approx 0.06$ to 0.12 sustainable by nonfragmenting disks in the asymptotic phase.
- It remains unclear whether clumps formed by fragmentation are always bound and permanent.
- It is not known how infall onto the disk might affect the fragmentation criterion.

5. Gravitational Instabilities: Realistic Applications

Of course, what we want to know is what real disks will do. In reality, protoplanetary disks cool radiatively, which makes the story of their behavior more intricate (see the chapter by Calvet and D'Alessio), and as a result, there is no consensus at the time of writing. Probably the most important unsettled question is whether a real disk can fragment directly into planet-sized pieces in the few to 40 AU region, where most gas-giant exoplanets are expected to form (but see § 5.4). In the spirit of emphasizing fundamentals, this section will lay out general principles and techniques but will not present too much detail about the disparate numerical results (see Durisen et al., 2007). Developments will undoubtedly be fast paced in the coming decade, and much that I could report in this chapter may soon become obsolete.

5.1. Radiative Cooling

A full understanding of why radiative cooling is difficult requires beginning at a basic level. Although this subsection may be somewhat redundant with other chapters, I here emphasize principles that underlie models for radiative cooling in dynamic codes.

THE RADIATIVE TRANSFER EQUATION. The radiative transfer equation, which describes how the specific intensity I_ν changes along a ray through an emitting and absorbing medium, is given by

$$\frac{dI_\nu}{d\tau_\nu} = I_\nu - S_\nu, \qquad (62)$$

where I_ν is the radiant energy per unit time per unit area per unit solid angle flowing in the x-direction per unit frequency interval, $d\tau_\nu = -\kappa_\nu \rho dx$, and κ_ν is the mass-absorption coefficient, the cross section per unit mass for absorption of radiation at frequency ν (Chandrasekhar, 1960; Mihalas and Weibel Mihalas, 1984). For simplicity, I do not make a distinction here between true absorption and scattering. As indicated by the minus sign in the $d\tau_\nu$ relation, the optical depth τ_ν is conventionally measured in a direction opposite to the direction which the photons are flowing, as if one is looking into the medium backward along the ray. Consequently, the first term on the right-hand side of (62), although positive, represents the absorption of radiation as it moves along dx. The second term on the right-hand side, although negative, contains the source function S_ν and represents radiation added to the beam by emission along dx. The exact nature of S_ν depends on the application, but when the emission is dominated by processes in local thermodynamic equilibrium (LTE) at the local gas temperature, $S_\nu = B_\nu(T)$, where T is the local gas temperature and B_ν is the Planck function. So far, all radiative treatments of GIs in disks have assumed LTE. This is likely to be fine for deep layers of the disk but can be inappropriate for the upper disk atmosphere (see the chapters by Calvet and D'Alessio and by Clarke).

Eq. (62) is deceptively simple in appearance. It has to be solved at each frequency everywhere in space for every direction through the emitting and absorbing medium, and, in general, all directions, positions, and frequencies are coupled through the absorption and reradiation of energy. In a hydrodynamics calculation, time dependence is added, and the transfer of radiation can affect the dynamics through radiation and gas-pressure forces. Moreover, as one looks into a source of radiation along a ray, the medium can go from optically thin ($\tau_\nu << 1$) to optically thick ($\tau_\nu >> 1$). Energy transport is non-local in optically thin regions. This condition necessarily pertains somewhere, because at the edge of any radiation source there has to be an optically thin region where the photon mean free path ($= 1/\kappa_\nu \rho$) is as large as the system. In the outer parts of this region, photons are well approximated as free streaming away from the source. Nevertheless, emission and absorption of light may still control the local gas T, and regions can be optically thin at some frequencies and not others. By its nature, radiative transfer is poorly treated as a simple diffusion problem, but let us first examine this limiting case.

THE DIFFUSION LIMIT. Ignore the frequency dependence for simplicity, i.e., adopt a gray opacity approximation, and define radiation energy density $\epsilon_{rad} = c^{-1} \int_{4\pi} I d\omega$, radiative flux in direction $\mathbf{z}$ by $\mathbf{F} = \int_{4\pi} I \cos\,\theta d\omega$, and the radiation pressure across the surface perpendicular to direction $\mathbf{z}$ as $P_{rad} = c^{-1} \int_{4\pi} I \cos^2\theta d\omega$. Here θ is the polar angle measured from direction $\mathbf{z}$, ω represents the solid angle, and c is the speed of light. For LTE and high optical depths, it is usually the case that, locally, $P_{rad} = \epsilon_{rad}/3 = aT^4/3$, where a is the radiation constant. Recognizing that the position vector **x** implicit in eq. (62) can be at an angle θ to direction **z**, we can multiply (62) by cos θ and integrate over all directions to get

$$\mathrm{F}_{rad} = c\,\frac{dP_{rad}}{d\tau_z} = -\,\frac{4acT^3}{3\kappa\rho}\,\frac{dT}{dz}\,, \tag{63}$$

where F_{rad} is the flux in direction **z** and $\tau_z = \int_z^\infty \kappa\rho dz$ is the optical depth measured downward along the z-direction. In 3D, this can be done in all directions to obtain a heat-conduction equation

$$\mathbf{F}_{rad} = -\mathcal{K}_{rad}\nabla T, \tag{64}$$

where $\mathcal{K}_{rad} = 4acT^3/3\kappa\rho$ is the radiative conductivity. When $\mathbf{F}_{rad}$ is inserted for **F** in the energy eq. (3), $\nabla \cdot \mathbf{F}_{rad}$ when positive represents net radiative cooling. An implicit assumption of such an approach is that the photon mean free path is small everywhere, but this breaks down in optically thin regions. If we assume LTE and the limit of high τ_ν at all ν, eqs. (63) and (64) are valid in a nongray case provided one replaces κ by the Rosseland mean opacity κ_R given by

$$\frac{1}{\kappa_R} = \frac{\int_0^\infty \frac{1}{\kappa_\nu}\frac{dB_\nu}{dT}\,d\nu}{\int_0^\infty \frac{dB_\nu}{dT}\,d\nu}\,. \tag{65}$$

PLANE-PARALLEL ATMOSPHERE. A particularly simple solution of (62) can be obtained for a gray, LTE, plane-parallel atmosphere that is in radiative equilibrium, i.e., where radiative absorption and emission are in a steady state with a constant and steady upward flux of radiation σT_{eff}^4 carried through the atmosphere. Here σ is the Stefan-Boltzmann constant and T_{eff} is the effective temperature. If one makes the Eddington approximation ($\epsilon_{rad} = P_{rad}/3$ everywhere) and assumes no incoming radiation at $\tau_z = 0$, then

$$T^4 = \frac{3}{4}\,T_{eff}^4\,(\tau_z + 2/3)\,, \tag{66}$$

and the mean intensity $J = 1/4\pi \int_{4\pi} Id\omega$ is just the frequency-integrated Planck intensity $B(T) = acT^4/4\pi$. At $\tau_z = 2/3$, the actual gas temperature $T = T_{eff}$. In applications to stars, $\tau \approx 2/3$ is taken to define the photosphere, the "surface" layer of a star from which most of the light escapes. However, it should also be noticed that $T/T_{eff} \to (0.5)^{1/4} \approx 0.841$ as $z \to \infty$. An Eddignton gray atmosphere that is also in hydrostatic equilibrium in a constant vertical gravitational field thus tends to a uniform temperature and hence an exponential mass-density drop-off as $z \to \infty$. For disks, a useful generalization of this result is to add a downward influx of radiation at $\tau_z = 0$ of σT_{irr}^4, where T_{irr} is the effective irradiation temperature (Cai et al., 2008). Then eq. (66) becomes

67
$$T^4 = \frac{3}{4} T_{eff}^4 (\tau_z + 2/3) + T_{irr}^4.$$

Thus in radiative equilibrium, a disk irradiated from above will, of course, be hotter everywhere. This also allows solutions where $T_{eff}^4 < 0$, i.e., the net heat flow is downward into the high-τ_z region.

DISK OPACITY. All simulations of GIs in disks with radiative cooling use a gray or monochromatic approximation, where the opacity is treated as a frequency-weighted average. In optically thick regions, where diffusion is a good approximation, κ_R is the sensible choice. However, much of the volume of even a massive protoplanetary disk can be optically thin (Cai et al., 2006). For $\tau << 1$, the first term on the right-hand side of (62) can be ignored. In LTE, with $S_\nu = B_\nu(T)$, the frequency-weighted opacity that makes sense to characterize emission and hence energy loss is then the Planck mean opacity κ_P, given by

68
$$\kappa_P = \frac{\int_0^\infty \kappa_\nu B_\nu d\nu}{\int_0^\infty B_\nu d\nu}.$$

Even when κ_ν varies smoothly with ν, the difference between (65) and (68) can be a factor of two or more.

As discussed in the chapter by Calvet and D'Alessio, the continuous opacity in disks outside the radially innermost regions is dominated by the dust. For $\varpi >$ about 1 AU, line opacity and emissivity tend to be important only in the upper layers of disks heated by starlight or energetic events and possibly in thin layers produced by strong shocks. The vertically outermost disk layers, which can be photodissociation or photoionization regions with unequal gas and dust temperatures, have so far not been modeled in GI simulations. For GIs, the interesting process is the cooling by transport of radiation at near to far-infrared and millimeter wavelengths in the bulk of the gas mass close to the midplane. There dust is dominant.

Numerical simulations of disks typically adopt either the Pollack et al. (1994) or the D'Alessio et al. (1998, 2001) opacities. Pollack et al. opacities assume that the dust grains are spherical and have sizes appropriate for the interstellar medium. The dust material is divided into water ice, organics, troilite, and silicates with evaporation edges near 150 K, 400 K, 700 K, and 1,200 K, respectively. As these thresholds are crossed from below, κ_R and κ_P tend to drop by factors of up to about 2, and by orders of magnitude in the case of silicates. The dust in D'Alessio et al. opacities is similar except that the grain size distribution can be varied. If a is the grain radius, then the fraction of grains in interval da between some minimum a_{min} and maximum a_{max} radius is assumed to be a power law, $a^g da$. In D'Alessio et al., a_{min} is usually set to 0.005 microns, while a_{max} and g can be varied, with $g = -3.5$ being similar to the choice implicit in Pollack et al. opacities over most of the range of sizes they consider. The values of κ_R and κ_P are independent of ρ and tend to be $\sim$ several $cm^2\ gm^{-1}$ over 100 K to 1,000 K for $a_{max} = 1$ micron and decrease to ~ 1 or 2 $cm^2\ gm^{-1}$ or less as a_{max} is increased to 1 mm. Both mean opacities $\to 0$ as $T \to 0$. See Figure 16 of Boley et al. (2006) for examples of $\kappa(T)$.

Given the dependence of radiative cooling on the dust opacity, which in turn depends on the composition, size, and spatial distributions of dust, an important message is that the disk dynamics is coupled to the evolution of solids (see the chapter by Henning and Meeus) just by the opacity alone. Disk metallicity and the growth and settling of dust are major considerations for the behavior and consequences of GIs in disks.

DISK RADIATIVE COOLING TIMES. When one is considering a new physical process, it is useful to estimate the associated timescales. First, the light travel time across a solar-system-sized disk is of order a fraction of a day and is the time on which photons can leave optically thin regions. Somewhat more interesting is the photon-diffusion time in the z-direction for high optical depths. The mean free path of a photon is about $1/\kappa\rho$, and so, for $\tau_z >> 1$, the time it takes a photon to randomly walk vertically over a scale height is

$$t_{phot} = \frac{H\tau}{c} \approx 0.04\ \mathrm{yr} \left(\frac{H/\varpi}{0.05}\right) \left(\frac{\varpi}{5\ \mathrm{AU}}\right) \left(\frac{\tau}{10^4}\right). \quad (69)$$

The shortness of this timescale caused some researchers (e.g., Laughlin and Bodenheimer, 1994) to claim that disks cool in a time $<< t_{rot}$ and so evolve isothermally. As we now know, this would also cause them to fragment for $Q <$ about 1.4. However, the photon-diffusion time for the Sun ($\sim 10^4$ yr) is much smaller than its heat-diffusion timescale ($\sim$ few$\cdot 10^7$ yr) because photons degrade in energy as they diffuse, and anyhow the energy-loss rate has to be compared with the thermal energy reservoir. So t_{phot} is not the cooling time.

Realistic estimates for the vertical radiative cooling time can be obtained most easily in the optically thick and thin limiting cases. For $\tau_z >> 1$, as an estimate, we apply the radiative equilibrium result eq. (66) with T as the midplane temperature T_{mid} and $\tau_z = \tau_z(z=0) = \tau_{mid}$. Then σT_{eff}^4 represents the flux leaving the top of a vertical column. So, for a $\gamma = 5/3$ ideal gas,

$$t_{cool}(\text{thick}) \approx \frac{\Sigma e(T_{mid})}{2\sigma T_{eff}^4} = \frac{9k\Sigma(\tau_{mid}+2/3)}{16 m_{gas}\sigma T_{mid}^3}, \quad (70)$$

where I have approximated the internal energy in a unit area column as $\Sigma e(T_{mid})$ and the 2 in the denominator accounts for the disk having two radiative surfaces. Eq. (70) suffices as an estimate, but the reader should consult Hubeny (1990) for a proper treatment of a plane-parallel slab of finite thickness. For $\tau_{mid} \approx \kappa_R \Sigma/2 >> 2/3$, we get

$$t_{cool}(\text{thick}) \approx 10^5 \text{ yr} \left(\frac{\Sigma}{3 \cdot 10^3 \text{ g cm}^{-2}}\right)^2 \left(\frac{\kappa_R}{1 \text{ cm}^2 \text{ gm}^{-1}}\right) \left(\frac{75 \ K}{T_{mid}}\right)^3, \quad (71)$$

where I have chosen m_{gas} to be 2.4 amu.

For $\tau_{mid} << 1$, the second term on the right-hand side of eq. (62) tells us that in LTE, the energy-loss rate per unit mass by emission integrated over all frequencies and directions is $4\kappa_P B(T)$, and the cooling time of a fluid element is just

$$t_{cool}(\text{thin}) \approx \frac{\pi e(T)}{4\kappa_P \sigma T^4} \approx 200 \text{ yr} \left(\frac{0.1 \text{ cm}^2 \text{ gm}^{-1}}{\kappa_P}\right) \left(\frac{10 \ K}{T}\right)^3, \quad (72)$$

where parameters appropriate for $\varpi \sim 100$ AU are used, a region that is likely to be optically thin.

In general, radiant energy can be flowing in any direction. Given $\mathbf{F}_{rad}$, a local radiative heating or cooling time for any fluid element can be computed via

$$t_{cool}(\text{local}) = \frac{\rho e(T)}{\nabla \cdot \mathbf{F}_{rad}}, \quad (73)$$

where a negative value of t_{cool} implies net heating by radiative transport. By integrating the numerator and the denominator over z before the division, this can be turned into a columnwise t_{cool}(column) that measures the timescale for net radiative losses (or gains) in a unit column of the disk. Net heating of a volume element or even a whole column by radiation is possible because of strong radiative shocks and irradiation.

RAFIKOV'S CRITIQUE OF PLANET FORMATION BY DISK INSTABILITY. The disk instability theory for gas-giant planet formation (Boss, 1997) relies on GIs to fragment disks into bound gas-giant protoplanets. To test this idea, the parameters chosen for eq. (71) are roughly compatible with those in eq. (46)

for a marginally unstable disk with $H/\varpi = 0.05$ around a 1 $M_\odot$ star at 5 AU. The corresponding τ_{mid} is about 1,500. At 5 AU, $t_{rot} = 12$ yr, and so $t_{cool}(\text{thick})\Omega = 5 \cdot 10^4$. Having derived results similar to (46) and (71), Rafikov (2005) offered the following critique of DI: In the region where one would like to form gas-giant planets, say a few to 40 AU, it is not possible for a disk simultaneously to be cool enough to have gas-phase GIs (46) and to cool fast enough to fragment (61). Matzner and Levin (2005), who include irradiation (see below), present similar arguments, and Rafikov (2007) also finds that the inclusion of energy transport by convection does not substantively change the result.

One can try to minimize $t_{cool}(\text{thick})$ by decreasing disk metallicity (and hence dust content) or by allowing dust to grow and settle to the midplane. No matter how this is done, the most efficient radiative cooling occurs when $\tau_{mid} \approx 2/3$. Lowering the opacity further makes cooling optically thin and therefore less efficient again. This can be seen in (71) and (72) because $t_{cool}(\text{thick}) \sim \kappa$ for $\tau_{mid} >> 2/3$, while $t_{cool}(\text{thin}) \sim 1/\kappa$ for $\tau_{mid} << 2/3$. Keeping all other parameters the same and letting $\tau_{mid} \to 2/3$ in eq. (70) give $t_{cool}(\text{shortest}) \approx 100$ yr at 5 AU or $t_{cool}(\text{shortest})\Omega \approx 50$, still far from the values of 6 to 12 required for fragmentation (Rice et al., 2005). Boss (2005) critiqued Rafikov's original analysis for using $Q_{crit} \approx 1$ as the GI stability limit. Here I have adopted a more generous value $Q_{crit} \approx 1.7$ in eq. (46) and obtain essentially the same result. Even if factors of order unity in the analysis or the assumed parameters soften the conclusion, Rafikov's argument sounds a loud cautionary note about expecting real disks to fragment near 5 AU. He also points out that t_{cool} in (70) increases as T_{mid} decreases. Therefore, if a disk cools to the threshold of instability, it becomes less likely to fragment because t_{cool} increases.

On the other hand, when applied to $\varpi \sim 100$ AU, eq. (72) tells us that $t_{cool}(\text{thin}) \sim t_{rot}$. Therefore, as pointed out by Rafikov and others, marginally unstable massive disks may very well fragment if they extend to large-enough radii (see § 4.2).

IRRADIATION. Disk radiative cooling is complicated by the possibility of external radiation shining onto the disk from its central star, an infalling envelope, or neighboring stars (see the chapter by Calvet and D'Alessio for a more complete discussion). Matzner and Levin (2005) argued analytically that irradiation plays a role in suppressing fragmentation by GIs in the early stages of disk evolution around solar-type stars. A simple estimate is instructive. Solving eq. (38) for the temperature and using eq. (46) to represent the marginal instability limit for nonaxisymmetric modes, one gets that

$$T > T_{crit} \approx 75 \text{ K} \left(\frac{Q_{crit}}{1.7}\right) \left(\frac{\varpi}{5 \text{ AU}}\right)^3 \left(\frac{M_\odot}{M_s}\right) \left(\frac{\Sigma}{3 \cdot 10^3 \text{ gm cm}^{-2}}\right)^2 \quad (74)$$

will stabilize a disk. In a constant-Q disk hovering near instability, T_{crit} decreases outward only if $\Sigma(\varpi)$ falls off more steeply than $\varpi^{-3/2}$.

External irradiation affects GIs in the optically thick part of the disk near the midplane through a flux of infrared radiation coming down onto the disk from above. For irradiation by an infalling envelope, the incoming radiation is already at mid- to far-infrared wavelengths. For irradiation caused by the central star or nearby stars, shorter-wavelength optical radiation is absorbed by and heats the uppermost layers of the disk atmosphere. This superheated region reradiates some of its energy downward toward the midplane in the infrared. Let σT_{irr}^4 represent the downward-irradiating IR flux. Then irradiation is likely to stabilize the disk at ϖ if $T_{irr} > T_{crit}$, because the disk temperature should be at least T_{irr} in the absence of other heating. In fact, values of T_{irr} similar to eq. (74) can be achieved by stellar and envelope irradiation. For a uniformly flared disk dominated by stellar irradiation, we expect $T_{irr} \sim \varpi^{-1/2}$. Envelope irradiation may be due to reprocessing of starlight at large distances from the disk and so could be more uniform with ϖ, as assumed in Cai et al. (2008). What part of a real disk can be kept stable by irradiation then depends on $\Sigma(\varpi)$. Real disks could behave in complicated ways if parts of the disk are shaded by a raised inner rim or by time-dependent spiral ridges due to shock bores. The interaction of irradiation with GIs is only beginning to be explored in simulations (e.g., Cai et al., 2008; Stamatellos and Whitworth, 2008).

SPECIAL ISSUES FOR GI-ACTIVE DISKS. Even without irradiation, there are aspects of disks that make treatment of radiative cooling challenging. Several of these harken back to issues raised in §§ 2.1 and 4.3. Although disks are usually geometrically thin ($H/\varpi \leq 0.1$), they are not dynamically thin ($t_z \sim t_{rot}$). The disk vertical and horizontal structure is strongly variable once GIs erupt, with shocks at all altitudes plus breaking waves and some forms of shock dissipation that are concentrated in the upper layers (Pickett et al., 2000; Boley and Durisen, 2006). Geometric thinness implies that the vertical temperature gradients are, on average, largest in the z-direction and that heat will mostly diffuse or be radiated vertically. As a result, some radiative cooling schemes implicitly assume that all energy is lost in the z-direction (A. F. Nelson et al., 2000a; Johnson and Gammie, 2003). Nevertheless, strong spiral shocks do produce large horizontal temperature gradients, and energy diffusion in the ϖ- and ϕ-directions can also be important for modeling energy loss, at least at the 10s of percent level (Boley et al., 2007a). Dynamic variability also means that deviations from radiative equilibrium could be pandemic in both optically thick and thin regions. Furthermore, a substantial fraction of the net cooling can come from the optically thin region above the disk photosphere and from the radially outermost regions where $\tau_{mid} < 2/3$ (Cai et al., 2006). Nonequilibrium processes probably should be treated in these regions, and the cooling scheme needs to be able to handle both $\tau_{mid} > 2/3$ and $\tau_{mid} < 2/3$. This can tax a code's ability to step in time because the heating and cooling times in low-τ regions of low density are sometimes rather short.

Despite all these difficulties, the thermal physics, including energy transport, should be dealt with as carefully as the dynamics. As we have seen, one of the fundamental insights now generally accepted is that the rate of energy loss by the disk controls the amplitude and behavior of GIs. For $H/\varpi \sim 0.05$ to 0.1, the internal energy of the disk is 1% or less of the rotational or gravitational energy of the disk (see § 2.1). All the energy liberated by GIs that escapes the disk must go through this internal energy reservoir. This is similar to the case of a main-sequence star, where energy from the very large nuclear reservoir is first dumped into a relatively small internal energy reservoir before being radiated; it is the thermal transport of energy that controls the rate of evolution of the star and regulates the release of nuclear energy. In a disk, radiative losses from a mostly flat rather than spherical surface regulate the rate at which the dynamic GI process can release gravitational energy. Just as for a star, the transport of energy through the disk's small thermal reservoir has to be computed accurately.

TWO- AND THREE-DIMENSIONAL TREATMENTS FOR DISK SIMULATIONS. This section reviews the principal techniques used so far to model realistic radiative cooling in disk simulations. The presentation emphasizes the philosophy of each method and is given in rough chronological order according to the first major paper that adopted it. I try to give enough detail for the reader to judge the relative merits of the schemes, but the original papers need to be consulted by anyone who intends to implement them.

Several GI simulation efforts approximate realistic radiative cooling by assuming or fitting a 1D radiative equilibrium solution in z, either the one in eq. (66) or the Hubeny variant, to estimate loss of energy by the disk in the vertical direction (A. F. Nelson et al., 2000a; Johnson and Gammie, 2003; Mejía et al., 2003; Stamatellos and Whitworth, 2008). The hydrodynamics in the first two papers is 2D and deals only with vertically integrated quantities. The vertical direction is assumed to be hydrostatic and isentropic over z. From τ_{mid} and T_{mid}, an effective flux σT_{eff}^4 from the top of each column is computed and used to cool the column. The last two papers employ 3D hydrodynamics, so the z-structure is resolved and dynamic. In their Case 2, Mejía et al. again estimate σT_{eff}^4 from τ_{mid} and T_{mid} and distribute the heat loss over the column. The scheme in Stamatellos & Whitworth is somewhat harder to characterize. Each SPH particle in their simulation is cooled by using an approach similar to (66) applied by estimating, in an average sense, how far the particle is from the photosphere. All these approaches treat the radiative physics as a local cooling process, without explicit transport of energy. They are "realistic" in that they estimate the cooling by assuming radiative equilibrium with τ calculated from a realistic κ_R, either from Pollack et al. (1994) or D'Alessio et al. (1998, 2001).

Boss (2001, 2002a, 2002b) makes an important step forward by including 3D diffusive transport (eq. [64]) with the Pollack et al. κ_R in his grid-based FD code. Energy is now actually transported and can move in all directions. Because his grid is in spherical coordinates, he measures the optical depth τ_r inward along spherical coordinate radial spokes to determine a disk photosphere ($\tau_r = \tau_c$). Diffusion is done only for $\tau_r > \tau_c$. Boss does not estimate or fit radiative fluxes at the photosphere but instead enforces a constant temperature in low-τ_r regions. He finds little difference between using $\tau_c = 1$ or 10. Mayer et al. (2007b) use an SPH version of 3D diffusion developed by Cleary and Monaghan (1999). To locate the radiative surface of the disk, they look for edge particles with no neighbors within some cone angle of the vertical and/or cylindrically radial directions. Edge particles in z are allowed to radiate like a blackbody at their own temperature over a hemisphere with a radius equal to the particle's smoothing length. There is no separate treatment of optically thin regions. The Boss and Mayer et al. schemes are similar in the following ways: 3D diffusion in optically thick regions, no explicit radiative treatment of optically thin regions, and boundary conditions (BCs) that are not easily characterized in terms of classic radiative transfer.

As argued in the preceding section, the z-direction may not be in radiative or hydrostatic equilibrium in a GI-active disk. Also, much of the disk volume can be optically thin, and shocks or external radiation fields can cause significant heating in optically thin regions. In an effort to deal with all these issues, Mejía (2004) and Cai (2006) develop a hybrid scheme with the following features: full 3D diffusion in optically thick regions ($\tau_z > 2/3$), explicit treatment of optically thin regions to allow for nonequilibrium conditions, and coupling of the optically thin atmosphere for $\tau_z < 2/3$ to the optically thick region when $\tau_{mid} > 2/3$. A flux limiter (Bodenheimer et al., 1990) is applied to the diffusion calculation so that the flux cannot exceed the physical limit where all photons are streaming in one direction. This is done to prevent excessive transport in the horizontal directions (ϖ and ϕ) if cells that are vertically optically thick become horizontally optically thin. However, even this limiter can fail under extreme conditions with large horizontal gradients (Cai et al., 2008). When $\tau_{mid} < 2/3$, usually in the outer disk, direct cooling to space is assumed at the rate $4\kappa_P \sigma T^4$ used in eq. (72). This produces a radial temperature discontinuity of 10s of percent near where $\tau_{mid} \approx 2/3$.

A key aspect of the Cai/Mejía scheme is that the BC for the diffusion problem is determined by using eq. (67) to set the flux at the upper face of the cell in which the cell-centered τ_z first becomes $> 2/3$ as one goes into the disk along a vertical column. Specifically, the τ_z and T of this cell are used in (67) to calculate σT_{eff}^4 for the upper cell face. An additional term σT_{atm}^4 is added to (67) to account for energy radiated down onto the disk by the atmosphere, and some of the upward-moving radiative flux is assumed to be absorbed by the atmosphere. The cell-to-cell coupling in the atmosphere is not perfect and results in a temperature drop of 10s of percent across the photosphere (Boley et al., 2006). Nevertheless, the scheme computes the correct flux out of the optically thick region in tests against an analytic solution based on Hubeny's work.

A simple modification to the Cai/Mejía scheme (Zhu et al., 2009; Cai et al., 2010) improves the high- and low-τ coupling, eliminates the vertical temperature drop at the photosphere, and reduces the radial jump near $\tau_{mid} \approx 2/3$. For $\tau_z < 2/3$, the radiative cooling (or heating when negative) is taken to be

$$\Lambda = 4\kappa_P(\sigma T^4 - \pi J), \tag{75}$$

where the mean radiation field J is the one that would obtain in radiative equilibrium by setting $J(\tau_z) = B(T_{fit})$ and using $T_{fit}(\tau_z)$ as the temperature expected from eq. (67) after the photospheric fitting described above. This accounts for the possibility that the atmosphere is slightly out of radiative equilibrium but also allows it to relax accurately to equilibrium when conditions warrant. Calculating Λ relative to radiative equilibrium makes t_{cool}(local) longer and more manageable numerically in low-τ_z regions. This scheme works best when the optically thin layers are close to radiative equilibrium but can produce numerical difficulties when it is implemented in a time-explicit form for situations where optically thin regions are far from radiative equilibrium.

Boley (2007) and Boley et al. (2007a) introduce a hybrid scheme that employs true radiative transfer in the vertical direction. Flux-limited diffusion is still used to calculate the ϖ and ϕ contributions to $\nabla \cdot \mathbf{F}_{rad}$ in optically thick regions, but the discrete-ordinate method (Chandrasekhar, 1960; Mihalas and

Weibel Mihalas, 1984) is employed in the z-direction as if each vertical column were a plane-parallel atmosphere. No vertical BC is now needed, except an incoming intensity at the top of the grid, because the vertical direction is a complete, though approximate, monochromatic radiative transfer calculation using either κ_R or a weighted average of κ_R and κ_P depending on τ_{mid}. The current version of this code, called CHYMERA, uses only one upward and one downward ray to compute the vertical radiative flux at horizontal cell faces. The vertical fluxes at the top and bottom cell faces are then used to determine the z-derivative part of $\nabla \cdot \mathbf{F}_{rad}$. Results for a GI-active disk in the asymptotic phase agree reasonably well with the Cai/Mejía scheme, but there are no artificial discontinuities in $T(z)$ at the photosphere, and the scheme is more numerically stable in some respects (Boley et al., 2007a). A different kind of hybrid scheme for SPH has recently been developed by Forgan et al. (2009) that efficiently handles optically thin emission without recourse to cumbersome photospheric fitting. It uses the SPH diffusion approximation in high-τ regions, as in Mayer et al. (2007b), but blended with the Stamatellos et al. (2007b) cooling scheme to treat regions with $\tau <$ about unity.

All the radiative cooling and transport schemes for disks described above assume a particular limiting case of radiative hydrodynamics. As discussed in Krumholz et al. (2007b), when $\tau >> 1$ and $\tau v/c > 1$, one enters the dynamic diffusion regime where the radiation field has significant energy content and exerts significant forces. These effects have to be included in the hydrodynamics equations. Fortunately, for protoplanetary disks orbiting a 1 $M_\odot$ star, $v/c < 10^{-4}$ for $\varpi > 1$ AU, while optical depths are typically $< 10^4$. A particularly impressive application of disk radiative hydrodynamics is the simulation of a massive rotating molecular cloud core by Krumholz et al. (2007a) in the $\tau v/c > 1$ limit with full 3D flux-limited diffusion and a high-order adaptive mesh-refinement scheme.

The time steps that can be taken in explicit radiative hydrodynamics calculations, regardless of scheme, cannot be longer than some fraction of t_{cool}(local). In explicit codes, one has to limit the allowed cooling (or heating) to keep the step sizes large enough to be useful, but these limitations on heating and cooling rates can be monitored to be sure that they do not affect significant regions of the disk. In the experience of the IU Hydro Group, overly short shock heating or radiative cooling timescales most commonly arise in low-density optically thin regions that do not contain much of the disk mass or energy content. Performance of radiative schemes for disks can be substantially improved by using implicit or semi-implicit approaches (e.g., Forgan et al., 2009).

5.2. Simulations of GIs with Radiative Cooling

GI fragmentation of spatially large disks (> 100 AU) with M_d/M_s substantially > 0.1 has been argued analytically or found in simulations by several groups (e.g., Vorobyov and Basu, 2005, 2006; Kratter and Matzner, 2006; Krumholz et al., 2007a; Stamatellos et al., 2007a; Boley, 2009; Clarke, 2009; Rafikov, 2009; Stamatellos and Whitworth, 2009; Boley et al., 2010). In these cases, the fragments can have gas-giant, brown-dwarf, or even stellar masses. At the time this chapter is being written, computational results on fragmentation in outer disks are somewhat sparse but are not generally considered controversial (see, however, Boss, 2006b), because the discussions attendant on eqs. (47) and (72) make fragmentation plausible in large, massive disks. Current disagreements in applications to outer disks focus more on the masses and ultimate fate of the fragments as they migrate and as infall onto these massive young disks continues.

The more serious debate about the DI planet-formation mechanism concerns whether GIs can produce gas-giant protoplanets in the traditional planet-forming region (a few to 10s AU), where most known exoplanets are believed to have formed. Here DI envisions fragmentation of disks around young solar-type stars with more modest masses $M_d/M_s \sim 0.1$ and spatial extents. The importance of this debate is that DI is currently the only serious alternative to formation of gas giants by core accretion plus gas capture in this region (see Durisen et al., 2007; Lissauer and Stevenson, 2007). The rest of the discussion in this subsection will be confined to simulations with realistic radiative cooling that test DI for protoplanetary disks with the sizes and masses appropriate for gas-giant planet formation inside 10s of AU around stars of roughly solar mass or below.

THE CONTROVERSY. The situation at the time of writing is as follows. On the one hand, simulations by A. F. Nelson et al. (2000a), Mejía et al. (2003), Boley et al. (2006, 2007a), Cai et al. (2006, 2008, 2010), Boley and Durisen (2008), Stamatellos and Whitworth (2008), and Forgan et al. (2009) support the analytic arguments of Rafikov (2005, 2007), namely, that with realistic radiative cooling, the cooling times (eq. [71]) in GI-unstable disks (eq. [46]) are too long for fragmentation to occur, and disks do not fragment. On the other hand, in a long series of influential papers, Boss finds that protoplanetary disks do fragment readily into dense clumps. Boss's position is supported, but for more restrictive conditions, by simulations in Mayer et al. (2007b), where fragmentation occurs when disks are somewhat more massive than the Boss disks ($\geq 0.12\ M_\odot$) and have higher mean molecular weight (> 2.4 amu).

Examples of both fragmenting and nonfragmenting radiatively cooled disks are shown in Plate 10.

The disagreements extend to how disk behavior varies with physical conditions. Simulations in Boss (2002b) indicate that variations in metallicity, which are assumed to cause proportionate changes in the dust opacity, make little difference in the occurrence of fragmentation even over the broad range of 0.1 to 10 times solar metallicity values, representing a factor of 100 in dust opacity and radiative cooling times. On the other hand, simulations by Cai et al. (2006) indicate that variations in metallicity do matter. Although GIs do not produce fragmentation in any of their simulations, the GIs become stronger in amplitude because of shorter cooling times as metallicity decreases even over the narrow range of 0.25 to 2 times the solar value. Cai et al. (2006) also find that increasing a_{max} weakens GIs for $\varpi > 10$ AU at fixed metallicity because the opacity gets larger with a_{max} at these radii.

On the assumption that the nonfragmenting results are correct, it is worth mentioning that in the asymptotic phase of nonfragmenting disks (Boley et al., 2006, 2007a; Cai et al., 2008), global gravitational torques are dominated by low-order modes, especially $m = 2$ (see Fig. 5.11), with spatially and temporally averaged $\alpha_{eff} \sim 10^{-2}$, roughly consistent with eq. (59), but with large fluctuations on the dynamic timescale. Radial density concentrations correlate with the CRs and LRs of large-scale modes. It would be difficult to model these disks as simple α-disks, at least structurally.

Boss (2002a) asserts that his use of a boundary temperature mimics irradiation. In his simulations, rather high boundary temperatures (≥ 150 K) are required to suppress fragmentation, whereas Boley et al. (2006) find that even nonirradiated disks do not fragment, and Cai et al. (2008) see GIs weakened ($T_{irr} \sim 15$ to 25 K) or suppressed ($T_{irr} \approx 50$ K, in rough agreement with eq. [74]) by modest amounts of irradiation. $T_{irr} = 25\ K$ applied to a disk extending from about 2 to 50 AU corresponds to intercepting about 2% of a solar luminosity, an amount typical for an irradiated T Tauri disk (see the chapter by Calvet and D'Alessio). Weakening and suppression of GIs by irradiation are also supported by analytic arguments (Matzner and Levin, 2005) and other simulations (Stamatellos and Whitworth, 2008).

POSSIBLE RESOLUTIONS. I am optimistic that by the time this chapter becomes readily available, the controversy will be resolved. Nevertheless, I think that it is instructive to discuss possible causes for the disagreement and to predict how a resolution will be reached.

The Usual Suspects. One obstacle at present is that only a few of the disks in the various works use the same initial conditions. Adopted disks vary in inner and outer radii, M_d/M_s, $\Sigma(\varpi)$, $Q(\varpi)$, and initial perturbations. The local 2D work of Johnson & Gammie (2003) raises a caution in this regard by showing that one cannot formulate a useful generalization of the Gammie fragmentation condition (61) by using the radiative t_{cool} of the initial state. One has to use a spatially averaged t_{cool}(column) after the GIs have reached an asymptotic behavior to recover something resembling (61). This probably does not obviate the Rafikov (2005) argument, because he finds that the typical $t_{cool}\Omega$ for the disks of interest is quite large, but it does mean that Gammie's criterion alone is not reliable when applied to a specific disk with arbitrary initial conditions. This problem is compounded by the fact that most initial disks used by researchers are not close to radiative equilibrium and perforce have a strong initial radiative transient. The only practical recourse is to run simulations and see whether the disks fragment. Boss does not calculate cooling times explicitly, but he infers short cooling times because he sees fragmentation. Researchers who track cooling times (e.g., Boley et al., 2006, 2007a; Boley and Durisen, 2008) find that t_{cool}(column) increases with time and is large enough to be consistent with the lack of fragmentation.

The EOSs in the various simulations can also be different, ranging from use of a constant $\gamma = 5/3$ (e.g., Boley et al., 2006, 2007a) or a constant $\gamma = 7/5$ (e.g., Mayer et al., 2007b) to an EOS that includes some treatment of the rotation states of H_2 (e.g., most Boss radiative hydrodynamics simulations; Boley and Durisen, 2008; Stamatellos and Whitworth, 2008). The EOS for H_2 can be somewhat controversial itself. Compare, for instance, Boley et al. (2007b) and Boss (2007b). A related issue is whether artificial viscosity is used to mediate shocks. Boss usually does not include artificial viscosity, which means that his shocks are not constrained to conserve energy.

Boss (2004a) and Mayer et al. (2007b) actually agree with Rafikov (2005) that radiative cooling alone cannot cause fragmentation. They attribute the fast cooling in their simulations, as required for the fragmentation they see, to convection associated with shock heating. Boley et al. (2006, 2007a) examine their own simulations for vertical entropy inversions and thermal convective flows in the optically thick regions and find none. In addition, Boley et al. (2007a) set up toy problems where convection should occur in disks according to the analyses of Lin and Papaloizou (1980) and Ruden and Pollack (1991). Their code produces convection in these toy problems when expected, but, as argued analytically by Rafikov (2007), the convection carries no more than a few 10s of percent of the vertical flux in the optically thick region. To produce fragmentation in the disks at issue here, convection would have to reduce the cooling times by one or more orders of magnitude, not just 10s of percent. Boley et al. (2007a) suggest that the vertical upwellings associated with spiral shocks reported by Boss and Mayer et al. are not convection but shock bores (§ 4.3). Shock bores by themselves should not produce much enhanced cooling, and in fact they do not do so in the bursting dead-zone models of Boley and Durisen (2008).

In early work by the IU Hydro Group (Pickett et al., 2003), the gravitational force holding dense clumps together was underestimated in the azimuthal direction because of an inadequate number of azimuthal grid cells L_ϕ (typically only 128) and possibly also because of a limited number of terms in computing the boundary potential for use in the Poisson solver. Boss (2007b) pointed to these as major concerns. Recently, as part of the Wengen Project, the IU boundary-potential solver has been tested against other solvers and found to be fully adequate. However, it is true that a larger L_ϕ, at least 512, is needed to get good agreement with SPH and AMR schemes on fragment formation. Boley et al. (2006, 2007a) performed long radiative runs with $L_\phi = 128$ but reran stretches at 512 to verify, as strongly suggested by the 128 results, that the disks do not fragment. The more recent Boley and Durisen (2008) simulations are very telling in this regard. The disks in these simulations are chosen to be unrealistically massive and small (0.17 $M_\odot$ from 2 to 10 AU) to bias them as much as possible to having strong GIs in the Jupiter-formation region. Some of these calculations have extremely high resolution in all directions, up to $512 \times 1,024 \times 128$ above the midplane in (ϖ, ϕ, z). Some simulations include large reductions in opacity, by factors up to 10^4, to mimic extreme dust growth and settling. At this level of opacity reduction, $\tau \approx 1$, which provides the most efficient radiative cooling possible. Even with these extreme physical conditions and very high numerical resolution, no fragmentation and no fast cooling due to convection occur.

In my opinion, the most likely culprit for the disagreement is treatment of the radiative boundary conditions (BCs). Consider again the analogy to stellar evolution mentioned in § 5.1. The pace of evolution of a star is controlled by how quickly it can lose energy through transport. Under some conditions, particularly for cool stars, the surface BCs must be treated carefully to compute accurate luminosities (Kippenhahn and Weigert, 1990). These are obtained for stellar evolution by fitting stellar atmosphere-like calculations with various levels of sophistication, sometimes as simple as eq. (66), to the photospheric layers of the star. Naturally, I am biased toward results from my own IU Hydro Group, where we apply this same approach to disks. Philosophically, one could also argue that radiative heat diffusion in the optically thick disk interior requires a boundary condition on the flux of energy, as in the Cai/Mejía analog to a stellar-evolution scheme, not on the temperature, as in the Boss scheme. The Mayer et al. edge algorithm for defining which SPH particles are part of the photosphere is difficult to characterize physically and requires testing against known results. Because the Boley et al. (2007a) scheme actually solves the gray radiative transfer equation in the z-direction, it can be argued to be the most sophisticated radiative cooling scheme so far applied to the problem. The code has been demonstrated to produce fragmentation when conditions are artificially contrived to cause it (Boley and Durisen, 2008). It should detect fragmentation under realistic conditions if fragmentation is really going to occur.

Paths to Glory? As we have seen, most of the simulations performed by different groups make different physical assumptions and use different initial disks. Resolution of the controversy will, in part, require that groups with different schemes compute as close to the same problem as possible. Only a little of this has been done so far. Notably, Stamatellos and Whitworth (2008) and Forgan et al. (2009) do a reasonably accurate job of matching the initial conditions and opacities of Boley et al. (2006, 2007a) and obtain similar results, namely, no fragmentation, even though they employ very different radiative cooling and hydrodynamics techniques. A collaboration has been under way for some time between the IU Hydro Group and Boss to make similar comparisons. In fact, no fragmentation occurs when the IU Cai/Mejía code (Cai et al., 2010) is used to simulate one of the same fragmenting $\gamma = 5/3$ disks reported in Boss (2007b). Unfortunately, the reason for this difference is still unclear. More simulation groups using different numerical treatments of radiative transport and hydrodynamics need to collaborate on such code comparisons and isolate the reasons for different outcomes.

Research groups with 3D radiative hydrodynamics codes are coming on line quickly, and some are beginning to direct efforts toward aspects of the DI problem. It is hoped that this will clarify the situation by a growing weight of numerical evidence on one side or the other. Along these lines, it is of critical importance that all researchers applying radiative cooling schemes to disks verify their algorithms against disk-specific analytic results. Because most energy is lost vertically, Boley et al. (2007a) propose three tests based on toy plane-parallel slab models with distributed internal heat sources using a modified version of the radiative disk solution of Hubeny (1990): a static test where the slab is relaxed to a hydrostatic radiative equilibrium, a dynamic test where it undergoes a self-similar quasi-static vertical contraction, and a convection test in which thermal convection is modeled for simple opacity laws. In addition to inspiring confidence in codes that perform well, the tests are also useful for defining code limitations. Modifications can be devised by other researchers that may be more suitable for their own schemes (e.g., Stamatellos and Whitworth, 2008), but testing of radiative routines on disklike problems is necessary.

5.3. Conclusions about GIs in Radiatively Cooled Disks

A list of consensus results, like that in § 4.6, is not yet possible, but here are some important points a reader should take away:

- Radiative transport
  - Radiative cooling adds considerable complexity to GI simulations, and a variety of approaches have been adopted.
  - Both optically thick and thin regions should be treated. Transitions between them and the radiative BCs must be properly handled.
  - Radiative cooling couples gas dynamics to the evolution and fate of the dust.
- Fragmentation controversy
  - Simulation results disagree about whether GIs can fragment protoplanetary disks into gas-giant protoplanets in the critical planet-forming region (a few to 40 AU) around solar-type stars.
  - Researchers agree that radiative cooling times in unstable disks are too long to result in fragmentation; those who see fragmentation attribute the required fast cooling to convection.
  - Massive disks that extend to large radii probably can and do fragment.
- Resolution of the controversy
  - More rigorous testing of radiative cooling routines against analytic problems is required.
  - Groups with a variety of radiation hydrodynamics schemes should use the same input physics and initial disk conditions, compare results, and isolate the reasons for significant differences.
  - The treatment of radiative BCs is probably the main concern, but differences in EOS, resolution, disk properties, and artificial viscosity may also play a role.
- Other results
  - Irradiation tends to weaken or suppress GIs.
  - Simulation results disagree on how much difference metallicity makes in the strength and consequences of GIs.

- Mass and angular momentum transport in nonfragmenting disks is at least structurally global and is dominated by low-order modes; spatially and temporally averaged α_{eff}s are $\sim 10^{-2}$ in the asymptotic phase, roughly consistent with eq. (59).

5.4. Special Effects and Planet Formation

I include this section to highlight directions where work on disk gas dynamics may have significant implications beyond mass transport and fragmentation. Until there is general agreement about the effects of radiative physics, all work on these topics should be considered tentative, and so they are not discussed in great depth. I mention them mainly to entice future young researchers and to point the reader toward a few formative papers.

SPECIAL EFFECTS

Mixing. The shock bores, associated with GI spirals can rapidly mix disk constituents, spreading abundance or isotopic anomalies and entrained dust (Boley et al., 2005; Boley and Durisen, 2006). For this particular application, whether a disk fragments is irrelevant, and, in fact, mixing is nicely investigated in nonfragmenting simulations by Boss (2004b, 2007a). He adds a "color" equation to his 3D hydrodynamics code that resembles eq. (2) but, instead, evolves a fictitious "color density" representing some unspecified difference in composition. When color is painted at the vertical surface of his disks, the color mixes to the midplane within a few dynamic times and then also mixes radially over regions spanned by strong spiral arms. Boss attributes the mixing to convection, but Boley and Durisen (2006) interpret the mixing as due to the shock bores associated with the spiral arms, as in Fig. 5.10. By tracing fluid elements in shock bores, Boley and Durisen confirm mixing to the midplane and mixing radially over several vertical scale heights on a dynamic timescale. Unfortunately, it is difficult to separate oscillations and stirring from true mixing at present, but there is a possibility that over many dynamic timescales, asymptotic-phase GIs can mix material over large distances. This may be relevant for understanding the presence of refractory materials in comet dust (Brownlee et al., 2006).

GIs and MRIs Combined. Simulations of MRIs and GIs are difficult enough when they are done separately, but it will ultimately be important and necessary to consider disks in which both are occurring. Fromang et al. (2004) and Fromang (2005) have performed global 3D MHD simulations of massive disks with simple EOSs. Even these early efforts show interesting results. Not only can MRI turbulence and GIs coexist, but they tend to interact on dynamic timescales with trade-offs of amplitude. The presence of MRI turbulence also has a stabilizing influence on GI-induced fragmentation. The interplay of separate MRI- and GI-active regions could lead to outburst phenomena in disks like FU Orionis (e.g., Armitage et al., 2001; Zhu et al., 2007, 2009). Current modeling in 2D (Zhu et al., 2009) suggests that FU Ori outbursts may be a cascade of instabilities where GIs trigger MRIs that then trigger a thermal instability (Bell and Lin, 1994) that finally produces rapid accretion onto the star. The 2D modeling does not treat GIs directly and requires assumptions about the nature of a GI/MRI boundary. However, it has been shown that GI outbursts in a dead zone can transport mass inward on timescales consistent with FU Ori events and thereby trigger the cascade (Boley and Durisen, 2008). When infall onto the disk $\dot{M}_d$ is large enough, a GI-active disk can become hot enough in the inner few AU to sustain MRI without requiring an outburst (Rice and Armitage, 2009).

Visible Consequences of GIs. Given the disagreements about the effects of radiative cooling, not too much detailed work has been done on predictions about the appearance of GI-active disks. Boley et al. (2006) find that the spectral energy distribution (SED) produced by the main GI-active part of their disk is reasonably consistent with what would be expected from a T Tauri disk with $T_{eff} \sim \varpi^{-0.6}$. However, their simulation does not cover a large range of disk radii and does not include mechanisms of grain growth or destruction, which A. F. Nelson et al. (2000a) show can strongly affect the SED. During the burst phase, the SED of the Boley et al. simulation does exhibit some FU Ori–like characteristics, but it is not a true FU Ori outburst because they do not model the disk inside 2.3 AU, where the outburst is concentrated (Zhu et al., 2007). GIs may be indirectly related to FU Ori and disk eruptions by creating dense clumps in the outer disk that are accreted by the inner disk in repeated outbursts (Vorobyov and Basu, 2005, 2006; Boley, 2009) or by episodic triggering of MRI in a dead zone (Armitage et al., 2001) that leads to thermal instability in the innermost disk, as discussed in Zhu et al. (2007, 2009) and Boley and Durisen (2008). As better interferometric methods and the Atacama Large Millimeter Array (ALMA) become available, one can, of course, look for the structural signature of GIs. Maps of surface flux, effective temperature, specific intensity, and brightness temperature for face-on GI-active disks have been computed from simulations in various papers, e.g., Figure 3 of Mejía et al. (2003), Figure 8 of Mejía et al. (2005), and Figures 11 and 12 of Boley and Durisen (2008). In all these cases, the appearance of the disk tends to be dominated by global two- or three-arm spiral patterns. A possibility that has not been thoroughly explored is that the corrugated surface of a GI-active disk could lead to dynamic timescale variability, especially when the disk is viewed more nearly edge on (Pickett et al., 2003). One possible consequence in the case of extremely large disks around massive stars is that masers might preferentially occur on lines of sight along the spiral arcs (Durisen et al., 2001).

PLANET FORMATION.

Effects of Binary Companions and Stellar Encounters. Obviously, a stellar or brown-dwarf companion or intruder can tidally perturb a disk that is marginally GI unstable. The enhanced amplitude of the spiral waves due to tidal forces can have two countervailing effects. They could either induce fragmentation when it would not have otherwise occurred or lead to stronger heating of the disk, which suppresses fragmentation that might have otherwise occurred. Because planets are found in binary systems, it is important to understand whether these might be more or less favorable environments for the DI theory. There have been several studies (e.g., A. F. Nelson, 2000; Boss, 2006a; Lodato et al., 2007; Mayer et al., 2007a). In all cases, the disks are of modest mass ($M_d/M_s \sim 0.1$) and $\sim$ 10 to 10s of AU in size around solar-type stars, and the perturbers are binary companions in eccentric orbits or parabolic interlopers. Just as with the problem of fragmentation in an isolated disk, there is no general agreement about the direction of the effects. Given my biases, I suspect that the simulations by Nelson (2000) and Lodato et al. (2007) give the correct answer, namely, that fragmentation tends to be strongly suppressed by the heating associated with shocks caused by tidal disturbances. However, this problem needs to be revisited with better radiative codes, and a wider range of parameter space needs to be explored. For instance, even with good radiative cooling schemes, disk behavior may be sensitive to assumptions about opacity and irradiation. Also, few results have been reported on how binary companions affect GIs as a transport mechanism.

Interaction of GIs with Solids. A particularly exciting area for future research will be the interaction of GIs with solid material. As discussed in more detail in the chapters by Calvet and D'Alessio and by Henning and Meeus, dust in disks is likely to grow and settle to the midplane on fairly short timescales, and the rocky subdisk may itself become susceptible to its own instabilities (e.g., Johansen et al., 2007); this could lead to planetesimal production. Gas-phase GIs could also play some role in growing large bodies. While the smallest dust particles remain entrained with the gas, are stirred by turbulence, and may be mixed over large distances by shock bores, particles of millimeter sizes and larger experience secular drifts relative to the gas. In particular, particles tend to drift opposite to the direction of a pressure gradient, with the largest drift speeds $\sim$ 100s m s^{-1} attained for particles with radii of about 1 meter. These drift speeds are large compared with orbit speeds of 30 km s^{-1} at 1 AU and can result in rapid loss of solids into the star if the pressure drops monotonically outward. Above about 100s meter sizes, the drift speeds for particles due to gas drag become small again. GIs can produce axisymmetric concentrations of mass (Haghighipour and Boss, 2003) and pressure ridges along dense spiral waves where particles of roughly meter size may accumulate rapidly, on timescales not much longer than the dynamic time on which the GIs themselves act (Rice et al., 2004), as shown in Plate 11. It is even possible that if there is enough mass present in particles of the optimal size, concentration in spiral arms may lead to gravitational instability of the concentrated solids themselves (Rice et al., 2006). The result could be enhanced rates of planetesimal production, which, in turn, could accelerate the accretion of the solid cores required by the core-accretion theory for forming gas-giant planets.

Hybrid Planet-Formation Theories. The discussion of the previous paragraph suggests that core accretion of gas-giant planet formation may not occur in a laminar disk. A variety of dynamic structures could be present that enhance the rate of planetesimal and planetary embryo production and core growth. These include rings at the boundaries between GI-active and inactive regions (Pickett and Lim, 2004; Durisen et al., 2005), the spiral arms discussed above, and pressure maxima at the centers of vortices produced by baroclinic and other instabilities (Adams and Watkins, 1995; Klahr and Henning, 1997). These are exciting possibilities worthy of further study. I point especially to Klahr and Bodenheimer (2006), where gas-giant formation by core accretion is computed for the environment of a large-scale vortex in a disk. Similar effects might occur at the snow line (Kretke and Lin, 2007) or at the edge of a dead zone (Lyra et al., 2008, 2009). I refer to these as *hybrid* theories because they incorporate elements of standard core accretion but invoke additional disk dynamics to facilitate and accelerate the process.

Formation of Gas Giants at Large Distances. The emphasis in the earlier part of this section is on the region inside about 40 AU, but, as discussed in § 4.2, GIs can occur at large disk radii if massive disks are spatially extended. Beyond about 50 AU, Rafikov's (2005) arguments against fragmentation no longer apply, and in fact fragmentation is expected if Q becomes low enough. Simulations of large disks around A-type and solar-type stars with realistic treatments of radiative cooling show that the resulting bound objects tend to have the masses of gas giants and brown dwarfs (Stamatellos and Whitworth, 2009; Boley, 2009; Boley et al., 2010). The recent detection of super-Jupiters orbiting their stars at large radial distances (many 10s to 100s of AUs) is difficult to understand in terms of core accretion plus gas capture because of the long formation times required at those distances and because of the difficulty of migrating or scattering planets so far outward (Dodson-Robinson et al., 2009). This raises the interesting possibility, worthy of further investigation, that planet formation is actually a bimodal process, with formation by core accretion effective in inner disks (< about 50 AU) and formation by DI in outer disks (> about 100 AU) (Boley, 2009; Clarke, 2009; Rafikov, 2009). The major uncertainty now is the fate of these distant gas-giant protoplanets. How many migrate toward the star before they contract to small size and get tidally disrupted (Boley et al., 2010)? What consequences would that have for disk evolution and planet formation? How many will survive as true planets with continued rapid mass inflow onto the disk (Kratter et al., 2010)? Krumholz et al. (2007a) have also shown that spatially extended massive disks formed during massive star formation can fragment into stellar-mass clumps.

A Unified Theory? As discussed in Boss and Durisen (2005) and Boley and Durisen (2008), which borrow ideas from papers like Gammie (1996), Wood (1996), Armitage et al. (2001), and Vorobyov and Basu (2005), it is at least conceivable that several young-star and early solar system phenomena are intimately connected, namely GIs, planetesimal production, gas-giant planet formation, thermal processing of solids by shocks, and FU Ori outbursts. To test aspects of this idea, Boley and Durisen (2008) bias their simulations of bursting dead zones at 4 to 5 AU heavily in favor of producing the strongest possible GIs. Although some thermal processing of solids, such as annealing of silicates, would occur in their simulations, it proves difficult, if not impossible, to get shocks strong enough to make chondrules (see the chapter by Henning and Meeus). The main problem is that the pitch angle of the spiral arms (eq. [42]) never becomes larger than about 10° even in the nonlinear regime, so that the Mach numbers of the spiral shocks in eq. (45) remain relatively low (see also Cossins et al., 2009). The more successful result reported in Boss and Durisen was dependent on GIs producing Jupiter-mass fragments accompanied by transient high-pitch angle wakes. Boley and Durisen note that outbursts centered closer to 1 AU (Zhu et al., 2007, 2009) can work in principle, and this may be a fruitful area for further study. Another criticism of connecting GI bursts with chondrule formation is that GIs require a relatively massive disk, most likely early in disk evolution, while chondrules apparently formed 1 or 2 million years after the Solar Nebula first took shape. The connection between GI bursts and planet formation is also unclear. Boss would argue that Jupiter formed in a GI burst, but Boley and Durisen see no evidence of disk fragmentation even under extreme assumptions. If GI bursts are related to planet formation, it may be through a phase of rapid planetesimal and embryo production induced by the GIs, as discussed in the previous paragraph, or through bursts of planet formation by DI in an outer disk followed by rapid inward migration (e.g., Vorobyov and Basu, 2006; Boley, 2009; Boley et al., 2010). Whether the notion of a unified theory has any real merit remains to be seen.

6. Conclusion

The intricate uncertainties surrounding some issues raised in this chapter might at first seem discouraging. Kant's philosophical project, to derive the formation of planetary systems from first principles, seems to become more elusive the more deeply we delve into it. In reality, however, the complexity that emerges from a modest number of physical ingredients is one of the wondrous features of the universe in which we live. Isolating important processes from unimportant processes in real complex systems is challenging enough in the laboratory. Astronomers and planetary scientists are brash enough to try to understand complex phenomena that are as remote in time as our solar system's origin and as remote in space as the young circumstellar disks in star-forming regions. Observations tell us that planetary systems are not only common but also exhibit a wild diversity of architectures. So, while the processes of formation must be robust enough to be widespread, they must also be highly nonlinear and chaotic in nature. Complexity is no surprise, given the variety of effects that can strongly influence the outcome of disk evolution—initial protostellar cloud conditions, strength of magnetic fields, interactions among protostars, irradiation, tidal disturbances, self-gravity, radiative cooling, and the production of turbulence and global structure through the nonlinear development of various instabilities, to name a few. Moreover, gas dynamics is coupled to the growth and migration of solids through opacity, ionization equilibrium, and multifluid behavior. All this complexity means that we have more work to do, which is not in itself a bad thing. It is very likely that Kant would have been fascinated by what we now know, and that Laplace, if he lived today, would have written several chapters of this book himself.

References

Adams, F. C., Ruden, S. P., and Shu, F. H. (1989). Eccentric Gravitational Instabilities in Nearly Keplerian Disks. *Astrophys. J.*, 347:959–976.

Adams, F. C., and Watkins, R. (1995). Vortices in Circumstellar Disks. *Astrophys. J.*, 451: 314–+.

Armitage, P. J., Livio, M., and Pringle, J. E. (2001). Episodic Accretion in Magnetically Layered Protoplanetary Discs. *MNRAS*, 324:705–711.

Balbus, S. A. (2000). Stability, Instability, and "Backward" Transport in Stratified Fluids. *Astrophys. J.*, 534:420–427.

Balbus, S. A. (2003). Enhanced Angular Momentum Transport in Accretion Disks. *Ann. Rev. Astron. Astrophys.*, 41:555–597.

Balbus, S. A., and Hawley, J. F. (1998). Instability, Turbulence, and Enhanced Transport in Accretion Disks. *Reviews of Modern Physics*, 70:1–53.

Balbus, S. A., and Papaloizou, J. C. B. (1999). On the Dynamical Foundations of Alpha Disks. *Astrophys. J.*, 521:650–658.

Barranco, J. A., and Marcus, P. S. (2005). Three-Dimensional Vortices in Stratified Protoplanetary Disks. *Astrophys. J.*, 623:1157–1170.

Bate, M. R., (1998). Collapse of a Molecular Cloud Core to Stellar Densities: The First Three-Dimensional Calculations. *Astrophys. J. Letters*, 508:L95–L98.

Bate, M. R., and Bonnell, I. A. (2005). The Origin of the Initial Mass Function and Its Dependence on the Mean Jeans Mass in Molecular Clouds. *MNRAS*, 356:1201–1221.

Bell, K. R., and Lin, D. N. C. (1994). Using FU Orionis Outbursts to Constrain Self-Regulated Protostellar Disk Models. *Astrophys. J.*, 427:987–1004.

Binney, J., and Tremaine, S. (1987). *Galactic Dynamics*. Princeton, NJ: Princeton University Press.

Bodenheimer, P. (1995). Angular Momentum Evolution of Young Stars and Disks. *Ann. Rev. Astron. Astrophys.*, 33:199–238.

Bodenheimer, P., Yorke, H. W., Rozyczka, M., and Tohline, J. E. (1990). The Formation Phase of the Solar Nebula. *Astrophys. J.*, 355:651–660.

Boley, A. C. (2007). *The Three-Dimensional Behavior of Spiral Shocks in Protoplanetary Disks*. PhD thesis, Indiana University.

Boley, A. C. (2009). The Two Modes of Gas Giant Planet Formation. *Astrophs. J. Letters*, 695:L53–L57.

Boley, A. C., and Durisen, R. H. (2006). Hydraulic/Shock Jumps in Protoplanetary Disks. *Astrophys. J.*, 641:534–546.

Boley, A. C., and Durisen, R. H. (2008). Gravitational Instabilities, Chondrule Formation, and the FU Orionis Phenomenon. *Astrophys. J.*, 685:1193–1209.

Boley, A. C., Durisen, R. H., Nordlund, Å., and Lord, J. (2007). Three-Dimensional Radiative Hydrodynamics for Disk Stability Simulations: A Proposed Testing Standard and New Results. *Astrophys. J.*, 665:1254–1267.

Boley, A. C., Durisen, R. H., and Pickett, M. K. (2005). The Three-Dimensionality of Spiral Shocks: Did Chondrules Catch a Breaking Wave? In Krot, A. N., Scott, E. R. D., and Reipurth, B., editors, *Chondrites and the Protoplanetary Disk*, volume 341 of *Astronomical Society of the Pacific Conference Series*, pages 839–848.

Boley, A. C., Hartquist, T. W., Durisen, R. H., and Michael, S. (2007). The Internal Energy for Molecular Hydrogen in Gravitationally Unstable Protoplanetary Disks. *Astrophys. J. Letters*, 656:L89–L92.

Boley, A. C., Hayfield, T., Mayer, L., and Durisen, R. H. (2010). Clumps in the Outer Disk by Disk Instability: Why They Are Initially Gas Giants and the Legacy of Disruption. *Icorus,* 207:509–516.

Boley, A. C., Mejía, A. C., Durisen, R. H., Cai, K., Pickett, M. K., and D'Alessio, P. (2006). The Thermal Regulation of Gravitational Instabilities in Protoplanetary Disks. III. Simulations with Radiative Cooling and Realistic Opacities. *Astrophys. J.,* 651: 517–534.

Boss, A. P. (1984). Protostellar Formation in Rotating Interstellar Clouds. IV Nonisothermal Collapse. *Astrophys. J.,* 277:768–782.

Boss, A. P. (1997). Giant Planet Formation by Gravitational Instability. *Science,* 276:1836–1839.

Boss, A. P. (2000). Possible Rapid Gas Giant Planet Formation in the Solar Nebula and Other Protoplanetary Disks. *Astrophys. J. Letters,* 536:L101–L104.

Boss, A. P. (2001). Gas Giant Protoplanet Formation: Disk Instability Models with Thermodynamics and Radiative Transfer. *Astrophys. J.,* 563:367–373.

Boss, A. P. (2002a). Evolution of the Solar Nebula. V. Disk Instabilities with Varied Thermodynamics. *Astrophys. J.,* 576:462–472.

Boss, A. P. (2002b). Stellar Metallicity and the Formation of Extrasolar Gas Giant Planets. *Astrophys. J. Letters,* 567:L149–L153.

Boss, A. P. (2003). Gas Giant Protoplanet Formation: Disk Instability Models with Detailed Thermodynamics and Varied Artificial Viscosity. In Mackwell, S., and Stansbery, E., editors, *Lunar and Planetary Institute Conference Abstracts,* volume 34 of *Lunar and Planetary Institute Conference Abstracts,* page 1075.

Boss, A. P. (2004a). Convective Cooling of Protoplanetary Disks and Rapid Giant Planet Formation. *Astrophys. J.,* 610:456–463.

Boss, A. P. (2004b). Evolution of the Solar Nebula. VI. Mixing and Transport of Isotopic Heterogeneity. *Astrophys. J.,* 616:1265–1277.

Boss, A. P. (2005). Evolution of the Solar Nebula. VII. Formation and Survival of Protoplanets Formed by Disk Instability. *Astrophys. J.,* 629:535–548.

Boss, A. P. (2006a). Gas Giant Protoplanets Formed by Disk Instability in Binary Star Systems. *Astrophys. J.,* 641:1148–1161.

Boss, A. P. (2006b). On the Formation of Gas Giant Planets on Wide Orbits. *Astrophys. J. Letters,* 637:L137–L140.

Boss, A. P. (2007a). Evolution of the Solar Nebula. VIII. Spatial and Temporal Heterogeneity of Short-Lived Radioisotopes and Stable Oxygen Isotopes. *Astrophys. J.,* 660:1707–1714.

Boss, A. P. (2007b). Testing Disk Instability Models for Giant Planet Formation. *Astrophys. J. Letters,* 661:L73–L76.

Boss, A. P., and Durisen, R. H. (2005). Chondrule-Forming Shock Fronts in the Solar Nebula: A Possible Unified Scenario for Planet and Chondrite Formation. *Astrophys. J. Letters,* 621: L137–L140.

Brenner, M. P., and Stone, H. A. (2000). Modern Classical Physics through the Work of G. I. Taylor. *Physics Today,* 53:30–35.

Brownlee, D., et al. (2006). Comet 81P/Wild 2 under a Microscope. *Science,* 314:1711–1716.

Cabot, W. (1996). Numerical Simulations of Circumstellar Disk Convection. *Astrophys. J.,* 465:874–+.

Cai, K. (2006). *Three-Dimensional Hydrodynamics Simulations of Gravitational Instabilities in Embedded Protoplanetary Disks.* PhD thesis, Indiana University.

Cai, K., Durisen, R. H., Boley, A. C., Pickett, M. K., and Mejía, A. C. (2008). The Thermal Regulation of Gravitational Instabilities in Protoplanetary Disks. IV. Simulations with Envelope Irradiation. *Astrophys. J.*, 673:1138–1153.

Cai, K., Durisen, R. H., Michael, S., Boley, A. C., Mejía, A. C., Pickett, M. K., and D'Alessio, P. (2006). The Effects of Metallicity and Grain Size on Gravitational Instabilities in Protoplanetary Disks. *Astrophys. J. Letters*, 636:L149–L152.

Cai, K., Pickett, M. K., Durisen, R. H., and Milne, A. M. (2010). Giant Planet Formation by Disk Instability: A Comparison Simulation with an Improved Radiative Scheme. *Astrophys. J. Letters*, 716:L176–L180.

Cassen, P. (1994). Utilitarian Models of the Solar Nebula. *Icarus*, 112:405–429.

Cassen, P., and Moosman, A. (1981). On the Formation of Protostellar Disks. *Icarus*, 48: 353–376.

Cassen, P., and Summers, A. (1983). Models of the Formation of the Solar Nebula. *Icarus*, 53:26–40.

Chandrasekhar, S. (1960). *Radiative Transfer*. New York: Dover, 1960

Chandrasekhar, S. (1961). *Hydrodynamic and Hydromagnetic Stability*. International Series of Monographs on Physics. Oxford: Clarendon Press.

Chandrasekhar, S. (1969). *Ellipsoidal Figures of equilibrium*. Silliman Foundation Lectures. New Haven, CT: Yale University Press.

Ciesla, F. J. (2009). Two-Dimensional Transport of Solids in Viscous Protoplanetary Disks. *Icarus*, 200:655–671.

Clarke, C. J. (2009). Pseudo-viscous Modelling of Self-Gravitating Discs and the Formation of Low Mass Ratio Binaries. *MNRAS*, pages 612–+.

Clarke, C. J., and Carswell, R. F. (2007). *Principles of Astrophysical Fluid Dynamics*. Cambridge, UK: Cambridge University Press.

Clarke, C. J., Harper-Clark, E., and Lodato, G. (2007). The Response of Self-Gravitating Protostellar Discs to Slow Reduction in Cooling Time-Scale: The Fragmentation Boundary Revisited. *MNRAS*, 381:1543–1547.

Cleary, P. W., and Monaghan, J. J. (1999). Conduction Modelling Using Smoothed Particle Hydrodynamics. *Journal of Computational Physics*, 148:227–264.

Cossins, P., Lodato, G., and Clarke, C. J. (2009). Characterizing the Gravitational Instability in Cooling Accretion Discs. *MNRAS*, 393:1157–1173.

Cox, J. P. (1980). *Theory of Stellar Pulsation*. Princeton, NJ: Princeton University Press.

D'Alessio, P., Calvet, N., and Hartmann, L. (2001). Accretion Disks around Young Objects. III. Grain Growth. *Astrophys. J.*, 553:321–334.

D'Alessio, P., Cantó, J., Calvet, N., and Lizano, S. (1998). Accretion Disks around Young Objects. I. The Detailed Vertical Structure. *Astrophys. J.*, 500:411–427.

Desch, S. J. (1999). Generation of Lightning in the Solar Nebula. In *Lunar and Planetary Institute Conference Abstracts*, volume 30 of *Lunar and Planetary Inst. Technical Report*, abstract 1962.

Desch, S. J. (2007). Mass Distribution and Planet Formation in the Solar Nebula. *Astrophys. J.*, 671:878–893.

Dodson-Robinson, S. E., Veras, D., Ford, E. B., and Beichman, C. A. (2009). The Formation Mechanism of Gas Giants on Wide Orbits. *Astrophys. J.*, 707:79–88.

Dullemond, C. P., Natta, A., and Testi, L. (2006). Accretion in Protoplanetary Disks: The Imprint of Core Properties. *Astrophys. J. Letters*, 645:L69–L72.

Durisen, R. H., Boss, A. P., Mayer, L., Nelson, A. F., Quinn, T., and Rice, W. K. M. (2007). Gravitational Instabilities in Gaseous Protoplanetary Disks and Implications for Giant Planet Formation. In Reipurth, B., Jewitt, D., and Keil, K., editors, *Protostars and Planets V*, pages 607–622.

Durisen, R. H., Cai, K., Mejía, A. C., and Pickett, M. K. (2005). A Hybrid Scenario for Gas Giant Planet Formation in Rings. *Icarus*, 173:417–424.

Durisen, R. H., Gingold, R. A., Tohline, J. E., and Boss, A. P. (1986). Dynamic Fission Instabilities in Rapidly Rotating N = 3/2 Polytropes—A Comparison of Results from Finite-Difference and Smoothed Particle Hydrodynamics Codes. *Astrophys. J.*, 305:281–308.

Durisen, R. H., Hartquist, T. W., and Pickett, M. K. (2008). The Formation of Fragments at Corotation in Isothermal Protoplanetary Disks. *Astrophys. Sp. Sci.*, 317:3–8.

Durisen, R. H., Mejía, A. C., Pickett, B. K., and Hartquist, T. W. (2001). Gravitational Instabilities in the Disks of Massive Protostars as an Explanation for Linear Distributions of Methanol Masers. *Astrophys. J. Letters*, 563:L157–L160.

Durisen, R. H., and Tohline, J. E. (1985). Fission of Rapidly Rotating Fluid Systems. In Black, D. C., and Matthews, M. S., editors, *Protostars and Planets II*, pages 534–575.

Forgan, D., Rice, K., Stamatellos, D., and Whitworth, A. (2009). Introducing a Hybrid Radiative Transfer Method for Smoothed Particle Hydrodynamics. *MNRAS*, 394:882–891.

Fricke, K. J., and Kippenhahn, R. (1972). Evolution of Rotating Stars. *Ann. Rev. Astron. Astrophys.*, 10:45–72.

Fromang, S. (2005). The Effect of MHD Turbulence on Massive Protoplanetary Disk Fragmentation. *Astron. Astrophys.*, 441:1–8.

Fromang, S., Balbus, S. A., Terquem, C., and De Villiers, J.-P. (2004). Evolution of Self-Gravitating Magnetized Disks. II. Interaction between Magnetohydrodynamic Turbulence and Gravitational Instabilities. *Astrophys. J.*, 616:364–375.

Fromang, S., Papaloizou, J., Lesur, G., and Heinemann, T. (2007). MHD Simulations of the Magnetorotational Instability in a Shearing Box with Zero Net Flux. II. The Effect of Transport Coefficients. *Astron. Astrophys.*, 476:1123–1132.

Gammie, C. F. (1996). Layered Accretion in T Tauri Disks. *Astrophys. J.*, 457:355–362.

Gammie, C. F. (2001). Nonlinear Outcome of Gravitational Instability in Cooling, Gaseous Disks. *Astrophys. J.*, 553:174–183.

Gammie, C. F., and Johnson, B. M. (2005). Theoretical Studies of Gaseous Disk Evolution around Solar Mass Stars. In Krot, A. N., Scott, E. R. D., and Reipurth, B., editors, *Chondrites and the Protoplanetary Disk*, volume 341 of *Astronomical Society of the Pacific Conference Series*, pages 145–+.

Goldreich, P., Goodman, J., and Narayan, R. (1986). The Stability of Accretion Tori. I—Long-Wavelength Modes of Slender Tori. *MNRAS*, 221:339–364.

Goldreich, P., and Lynden-Bell, D. (1965). II. Spiral Arms as Sheared Gravitational Instabilities. *MNRAS*, 130:125–158.

Goldreich, P., and Tremaine, S. (1979). The Excitation of Density Waves at the Lindblad and Corotation Resonances by an External Potential. *Astrophys. J.*, 233:857–871.

Haghighipour, N., and Boss, A. P. (2003). On Pressure Gradients and Rapid Migration of Solids in a Nonuniform Solar Nebula. *Astrophys. J.*, 583:996–1003.

Hartmann, L. (2009). *Accretion Processes in Star Formation*. Cambridge, UK: Cambridge University Press.

Hawley, J. F., Balbus, S. A., and Winters, W. F. (1999). Local Hydrodynamic Stability of Accretion Disks. *Astrophys. J.*, 518:394–404.

Hayashi, C., Nakazawa, K., and Nakagawa, Y. (1985). Formation of the Solar System. In Black, D. C., and Matthews, M. S., editors, *Protostars and Planets II*, pages 1100–1153.

Hennebelle, P., and Fromang, S. (2008). Magnetic Processes in a Collapsing Dense Core. I. Accretion and Ejection. *Astron. Astrophys.*, 477:9–24.

Hubeny, I. (1990). Vertical Structure of Accretion Disks—A Simplified Analytical Model. *Astrophys. J.*, 351:632–641.

Imamura, J. N., Durisen, R. H., and Pickett, B. K. (2000). Nonaxisymmetric Dynamic Instabilities of Rotating Polytropes. II. Torques, Bars, and Mode Saturation with Applications to Protostars and Fizzlers. *Astrophys. J.*, 528:946–964.

Ji, H., Burin, M., Schartman, E., and Goodman, J. (2006). Hydrodynamic Turbulence Cannot Transport Angular Momentum Effectively in Astrophysical Disks. *Nature*, 444: 343–346.

Johansen, A., Oishi, J. S., Mac Low, M.-M., Klahr, H., Henning, T., and Youdin, A. (2007). Rapid Planetesimal Formation in Turbulent Circumstellar Disks. *Nature*, 448:1022–1025.

Johnson, B. M., and Gammie, C. F. (2003). Nonlinear Outcome of Gravitational Instability in Disks with Realistic Cooling. *Astrophys. J.*, 597:131–141.

Johnson, B. M., and Gammie, C. F. (2005). Vortices in Thin, Compressible, Unmagnetized Disks. *Astrophys. J.*, 635:149–156.

Johnson, B. M., and Gammie, C. F. (2006). Nonlinear Stability of Thin, Radially Stratified Disks. *Astrophys. J.*, 636:63–74.

Kant, I. (1755). *Universal Natural History and Theory of the Heavens.* Translated by S. L. Jaki 1981. Edinburgh: Scottish Academy Press.

Kippenhahn, R., and Weigert, A. (1990). *Stellar Structure and Evolution.* Berlin: Springer-Verlag.

Klahr, H., and Bodenheimer, P. (2006). Formation of Giant Planets by Concurrent Accretion of Solids and Gas inside an Anticyclonic Vortex. *Astrophys. J.*, 639:432–440.

Klahr, H. H., and Bodenheimer, P. (2003). Turbulence in Accretion Disks: Vorticity Generation and Angular Momentum Transport via the Global Baroclinic Instability. *Astrophys. J.*, 582:869–892.

Klahr, H. H., and Henning, T. (1997). Particle-Trapping Eddies in Protoplanetary Accretion Disks. *Icarus*, 128:213–229.

Klahr, H. H., Henning, T., and Kley, W. (1999). On the Azimuthal Structure of Thermal Convection in Circumstellar Disks. *Astrophys. J.*, 514:325–343.

Kratter, K. M., and Matzner, C. D. (2006). Fragmentation of Massive Protostellar Discs. *MNRAS*, 373:1563–1576.

Kratter, K. M., Murray-Clay, R. A., and Youdin, A. N. (2010). The Runts of the Litter: Why Planets Formed through Gravitational Instability Can Only Be Failed Binary Stars. *Astrophys. J.*, 710:1375–1386.

Kretke, K. A., and Lin, D. N. C. (2007). Grain Retention and Formation of Planetesimals near the Snow Line in MRI-Driven Turbulent Protoplanetary Disks. *Astrophys. J. Letters*, 664:L55–L58.

Krumholz, M. R., Klein, R. I., and McKee, C. F. (2007a). Radiation-Hydrodynamic Simulations of Collapse and Fragmentation in Massive Protostellar Cores. *Astrophys. J.*, 656: 959–979.

Krumholz, M. R., Klein, R. I., McKee, C. F., and Bolstad, J. (2007b). Equations and Algorithms for Mixed-Frame Flux-Limited Diffusion Radiation Hydrodynamics. *Astrophys. J.*, 667:626–643.

Landau, L. D., and Lifshitz, E. M. (1959). Fluid Mechanics, in volume 6 of *Course of Theoretical Physics*. Oxford: Pergamon Press.

Larson, R. B. (1984). Gravitational Torques and Star Formation. *MNRAS*, 206:197–207.

Laughlin, G., and Bodenheimer, P. (1994). Nonaxisymmetric Evolution in Protostellar Disks. *Astrophys. J.*, 436:335–354.

Laughlin, G., and Korchagin, V. (1996). Nonlinear Generation of One-Armed Spirals in Self-Gravitating Disks. *Astrophys. J.*, 460:855–868.

Laughlin, G., Korchagin, V., and Adams, F. C. (1997). Spiral Mode Saturation in Self-Gravitating Disks. *Astrophys. J.*, 477:410–423.

Laughlin, G., Korchagin, V., and Adams, F. C. (1998). The Dynamics of Heavy Gaseous Disks. *Astrophys. J.*, 504:945–966.

Laughlin, G., and Rozyczka, M. (1996). The Effect of Gravitational Instabilities on Protostellar Disks. *Astrophys. J.*, 456:279–291.

Lesur, G., and Longaretti, P.-Y. (2005). On the Relevance of Subcritical Hydrodynamic Turbulence to Accretion Disk Transport. *Astron. Astrophys.*, 444:25–44.

Li, H., Colgate, S. A., Wendroff, B., and Liska, R. (2001). Rossby Wave Instability of Thin Accretion Disks. III. Nonlinear Simulations. *Astrophys. J.*, 551:874–896.

Li, H., Finn, J. M., Lovelace, R. V. E., and Colgate, S. A. (2000). Rossby Wave Instability of Thin Accretion Disks. II. Detailed Linear Theory. *Astrophys. J.*, 533:1023–1034.

Lin, D. N. C., and Papaloizou, J. (1980). On the Structure and Evolution of the Primordial Solar Nebula. *MNRAS*, 191:37–48.

Lin, D. N. C., and Pringle, J. E. (1987). A Viscosity Prescription for a Self-Gravitating Accretion Disc. *MNRAS*, 225:607–613.

Lissauer, J. J., and Stevenson, D. J. (2007). Formation of Giant Planets. In Reipurth, B., Jewitt, D., and Keil, K., editors, *Protostars and Planets V*, pages 591–606.

Lodato, G., Meru, F., Clarke, C. J., and Rice, W. K. M. (2007). The Role of the Energy Equation in the Fragmentation of Protostellar Discs during Stellar Encounters. *MNRAS*, 374:590–598.

Lodato, G., and Rice, W. K. M. (2004). Testing the Locality of Transport in Self-Gravitating Accretion Discs. *MNRAS*, 351:630–642.

Lodato, G., and Rice, W. K. M. (2005). Testing the Locality of Transport in Self-Gravitating Accretion Discs—II. The Massive Disc Case. *MNRAS*, 358:1489–1500.

Lovelace, R. V. E., Li, H., Colgate, S. A., and Nelson, A. F. (1999). Rossby Wave Instability of Keplerian Accretion Disks. *Astrophys. J.*, 513:805–810.

Lubow, S. H., and Ogilvie, G. I. (1998). Three-Dimensional Waves Generated at Lindblad Resonances in Thermally Stratified Disks. *Astrophys. J.*, 504:983–995.

Lynden-Bell, D., and Kalnajs, A. J. (1972). On the Generating Mechanism of Spiral Structure. *MNRAS*, 157:1–30.

Lynden-Bell, D., and Pringle, J. E. (1974). The Evolution of Viscous Discs and the Origin of the Nebular Variables. *MNRAS*, 168:603–637.

Lyra, W., Johansen, A., Klahr, H., and Piskunov, N. (2008). Embryos Grown in the Dead Zone: Assembling the First Protoplanetary Cores in Low Mass Self-Gravitating Circumstellar Disks of Gas and Solids. *Astron. Astrophys.*, 491:L41–L44.

Lyra, W., Johansen, A., Zsom, A., Klahr, H., and Piskunov, N. (2009). Planet Formation Bursts at the Borders of the Dead Zone in 2D Numerical Simulations of Circumstellar Disks. *Astron. Astrophys.*, 497:869–888.

Machida, M. N., Inutsuka, S.-i., and Matsumoto, T. (2007). Magnetic Fields and Rotations of Protostars. *Astrophys. J.*, 670:1198–1213.

Marcus, P. S. (1993). Jupiter's Great Red Spot and Other Vortices. *Ann. Rev. Astron. Astrophys.*, 31:523–573.

Martos, M. A., and Cox, D. P. (1998). Magnetohydrodynamic Modeling of a Galactic Spiral Arm as a Combination Shock and Hydraulic Jump. *Astrophys. J.*, 509:703–716.

Massey, B. S. (1970). *Mechanics of Fluids.* London: Van Norstrand Reinhold.

Matzner, C. D., and Levin, Y. (2005). Protostellar Disks: Formation, Fragmentation, and the Brown Dwarf Desert. *Astrophys. J.*, 628:817–831.

Mayer, L., Boss, A., and Nelson, A. F. (2007a). Gravitational Instability in Binary Protoplanetary Disks. *ArXiv e-prints.*

Mayer, L., Lufkin, G., Quinn, T., and Wadsley, J. (2007b). Fragmentation of Gravitationally Unstable Gaseous Protoplanetary Disks with Radiative Transfer. *Astrophys. J. Letters*, 661:L77–L80.

Mayer, L., Quinn, T., Wadsley, J., and Stadel, J. (2002). Formation of Giant Planets by Fragmentation of Protoplanetary Disks. *Science*, 298:1756–1759.

Mayer, L., Quinn, T., Wadsley, J., and Stadel, J. (2004). The Evolution of Gravitationally Unstable Protoplanetary Disks: Fragmentation and Possible Giant Planet Formation. *Astrophys. J.*, 609:1045–1064.

Mejía, A. C. (2004). *The Thermal Regulation of Gravitational Instabilities in Disks around Young Stars.* PhD thesis, Indiana University.

Mejía, A. C., Durisen, R. H., and Pickett, B. K. (2003). Gravitational Instabilities in Disks with Radiative Cooling. In Deming, D., and Seager, S., editors, *Scientific Frontiers in Research on Extrasolar Planets*, volume 294 of *Astronomical Society of the Pacific Conference Series*, pages 287–290.

Mejía, A. C., Durisen, R. H., Pickett, M. K., and Cai, K. (2005). The Thermal Regulation of Gravitational Instabilities in Protoplanetary Disks. II. Extended Simulations with Varied Cooling Rates. *Astrophys. J.*, 619:1098–1113.

Mihalas, D., and Weibel Mihalas, B. (1984). *Foundations of Radiation Hydrodynamics.* New York: Oxford University Press.

Nelson, A. F. (2000). Planet Formation Is Unlikely in Equal-Mass Binary Systems with a $\sim$ 50 AU. *Astrophys. J. Letters*, 537:L65–L68.

Nelson, A. F. (2006). Numerical Requirements for Simulations of Self-Gravitating and Non-Self-Gravitating Discs. *MNRAS*, 373:1039–1073.

Nelson, A. F., Benz, W., Adams, F. C., and Arnett, D. (1998). Dynamics of Circumstellar Disks. *Astrophys. J.*, 502:342–371.

Nelson, A. F., Benz, W., and Ruzmaikina, T. V. (2000a). Dynamics of Circumstellar Disks. II. Heating and Cooling. *Astrophys. J.*, 529:357–390.

Nelson, R. P., Papaloizou, J. C. B., Masset, F., and Kley, W. (2000b). The Migration and Growth of Protoplanets in Protostellar Discs. *MNRAS*, 318:18–36.

Ogilvie, G. I. (2000). An Alpha Theory of Time-Dependent Warped Accretion Discs. *MNRAS*, 317:607–622.

Oishi, J. S., Mac Low, M.-M., and Menou, K. (2007). Turbulent Torques on Protoplanets in a Dead Zone. *Astrophys. J.*, 670:805–819.

Osorio, M., D'Alessio, P., Muzerolle, J., Calvet, N., and Hartmann, L. (2003). A Comprehensive Study of the L1551 IRS 5 Binary System. *Astrophys. J.*, 586:1148–1161.

Paczynski, B. (1978). A Model of Self-gravitating Accretion Disk. *Acta Astronomica*, 28: 91–109.

Papaloizou, J. C., and Savonije, G. J. (1991). Instabilities in Self-Gravitating Gaseous Discs. *MNRAS*, 248:353–369.

Papaloizou, J. C. B., and Lin, D. N. C. (1989). Nonaxisymmetric Instabilities in Thin Self-Gravitating Rings and Disks. *Astrophys. J.*, 344:645–668.

Papaloizou, J. C. B., and Lin, D. N. C. (1995). Theory of Accretion Disks I: Angular Momentum Transport Processes. *Ann. Rev. Astron. Astrophys.*, 33:505–540.

Papaloizou, J. C. B., Nelson, R. P., Kley, W., Masset, F. S., and Artymowicz, P. (2007). Disk-Planet Interactions during Planet Formation. In Reipurth, B., Jewitt, D., and Keil, K., editors, *Protostars and Planets V*, pages 655–668.

Papaloizou, J. C. B., and Pringle, J. E. (1984). The Dynamical Stability of Differentially Rotating Discs with Constant Specific Angular Momentum. *MNRAS*, 208:721–750.

Papaloizou, J. C. B., and Pringle, J. E. (1985). The Dynamical Stability of Differentially Rotating Discs. II. *MNRAS*, 213:799–820.

Papaloizou, J. C. B., and Terquem, C. (2006). Planet Formation and Migration. *Reports on Progress in Physics*, 69:119–180.

Petersen, M. R., Julien, K., and Stewart, G. R. (2007a). Baroclinic Vorticity Production in Protoplanetary Disks. I. Vortex Formation. *Astrophys. J.*, 658:1236–1251.

Petersen, M. R., Stewart, G. R., and Julien, K. (2007b). Baroclinic Vorticity Production in Protoplanetary Disks. II. Vortex Growth and Longevity. *Astrophys. J.*, 658:1252–1263.

Pickett, B. K., Cassen, P., Durisen, R. H., and Link, R. (1998). The Effects of Thermal Energetics on Three-Dimensional Hydrodynamic Instabilities in Massive Protostellar Disks. *Astrophys. J.*, 504:468–491.

Pickett, B. K., Cassen, P., Durisen, R. H., and Link, R. (2000). The Effects of Thermal Energetics on Three-Dimensional Hydrodynamic Instabilities in Massive Protostellar Disks. II. High-Resolution and Adiabatic Evolutions. *Astrophys. J.*, 529:1034–1053.

Pickett, B. K., Durisen, R. H., and Davis, G. A. (1996). The Dynamic Stability of Rotating Protostars and Protostellar Disks. I. The Effects of the Angular Momentum Distribution. *Astrophys. J.*, 458:714–738.

Pickett, B. K., Durisen, R. H., and Link, R. (1997). Rotating Protostars and Protostellar Disks. I. Equilibrium Models. *Icarus*, 126:243–260.

Pickett, B. K., Mejía, A. C., Durisen, R. H., Cassen, P. M., Berry, D. K., and Link, R. P. (2003). The Thermal Regulation of Gravitational Instabilities in Protoplanetary Disks. *Astrophys. J.*, 590:1060–1080.

Pickett, M. K. and Durisen, R. H. (2007). Numerical Viscosity and the Survival of Gas Giant Protoplanets in Disk Simulations. *Astrophys. J. Letters*, 654:L155–L158.

Pickett, M. K., and Lim, A. J. (2004). Planet Formation: The Race Is Not to the Swift. *Astronomy and Geophysics*, 45:1.12–1.17.

Pollack, J. B., Hollenbach, D., Beckwith, S., Simonelli, D. P., Roush, T., and Fong, W. (1994). Composition and Radiative Properties of Grains in Molecular Clouds and Accretion Disks. *Astrophys. J.*, 421:615–639.

Pringle, J. E. (1981). Accretion Discs in Astrophysics. *Ann. Rev. Astron. Astrophys.*, 19:137–162.

Rafikov, R. R. (2005). Can Giant Planets Form by Direct Gravitational Instability? *Astrophys. J. Letters*, 621:L69–L72.

Rafikov, R. R. (2007). Convective Cooling and Fragmentation of Gravitationally Unstable Disks. *Astrophys. J.*, 662:642–650.

Rafikov, R. R. (2009). Properties of Gravitoturbulent Accretion Disks. *ArXiv e-prints.*

Reipurth, B., Jewitt, D., and Keil, K., editors (2007). *Protostars and Planets V.* Tucson: University of Arizona Press.

Rice, W. K. M., and Armitage, P. J. (2009). Time-Dependent Models of the Structure and Stability of Self-Gravitating Protoplanetary Discs. *MNRAS*, 396:2228–2236.

Rice, W. K. M., Armitage, P. J., Bate, M. R., and Bonnell, I. A. (2003). The Effect of Cooling on the Global Stability of Self-Gravitating Protoplanetary Discs. *MNRAS*, 339:1025–1030.

Rice, W. K. M., Lodato, G., and Armitage, P. J. (2005). Investigating Fragmentation Conditions in Self-Gravitating Accretion Discs. *MNRAS*, 364:L56–L60.

Rice, W. K. M., Lodato, G., Pringle, J. E., Armitage, P. J., and Bonnell, I. A. (2004). Accelerated Planetesimal Growth in Self-Gravitating Protoplanetary Discs. *MNRAS*, 355:543–552.

Rice, W. K. M., Lodato, G., Pringle, J. E., Armitage, P. J., and Bonnell, I. A. (2006). Planetesimal Formation via Fragmentation in Self-Gravitating Protoplanetary Discs. *MNRAS*, 372:L9–L13.

Richard, D., and Zahn, J.-P. (1999). Turbulence in Differentially Rotating Flows: What Can Be Learned from the Couette-Taylor Experiment. *Astron. Astrophys.*, 347:734–738.

Rincon, F., Ogilvie, G. I., and Cossu, C. (2007). On Self-Sustaining Processes in Rayleigh-Stable Rotating Plane Couette Flows and Subcritical Transition to Turbulence in Accretion Disks. *Astron. Astrophys.*, 463:817–832.

Ruden, S. P., and Pollack, J. B. (1991). The Dynamical Evolution of the Protosolar Nebula. *Astrophys. J.*, 375:740–760.

Sano, T., Miyama, S. M., Umebayashi, T., and Nakano, T. (2000). Magnetorotational Instability in Protoplanetary Disks. II. Ionization State and Unstable Regions. *Astrophys. J.*, 543:486–501.

Shakura, N. I., and Sunyaev, R. A. (1973). Black Holes in Binary Systems: Observational Appearance. *Astron. Astrophys.*, 24:337–355.

Shlosman, I., and Begelman, M. C. (1989). Evolution of Self-Gravitating Accretion Disks in Active Galactic Nuclei. *Astrophys. J.*, 341:685–691.

Shu, F. H. (1977). Self-Similar Collapse of Isothermal Spheres and Star Formation. *Astrophys. J.*, 214:488–497.

Shu, F. H. (1992). *Physics of Astrophysics. Vol. 2.* Mill Valley, CA: University Science Books.

Shu, F. H., Tremaine, S., Adams, F. C., and Ruden, S. P. (1990). SLING Amplification and Eccentric Gravitational Instabilities in Gaseous Disks. *Astrophys. J.*, 358:495–514.

Stamatellos, D., Hubber, D. A., and Whitworth, A. P. (2007a). Brown Dwarf Formation by Gravitational Fragmentation of Massive, Extended Protostellar Discs. *MNRAS*, 382: L30–L34.

Stamatellos, D., and Whitworth, A. P. (2008). Can Giant Planets Form by Gravitational Fragmentation of Discs? *Astron. Astrophys.*, 480:879–887.

Stamatellos, D., and Whitworth, A. P. (2009). The Properties of Brown Dwarfs and Low-Mass Hydrogen-Burning Stars Formed by Disc Fragmentation. *MNRAS*, 392:413–427.

Stamatellos, D., Whitworth, A. P., Bisbas, T., and Goodwin, S. (2007b). Radiative Transfer and the Energy Equation in SPH Simulations of Star Formation. *Astron. Astrophys.*, 475: 37–49.

Stone, J. M., and Balbus, S. A. (1996). Angular Momentum Transport in Accretion Disks via Convection. *Astrophys. J.*, 464:364–372.

Takeuchi, T., and Lin, D. N. C. (2002). Radial Flow of Dust Particles in Accretion Disks. *Astrophys. J.*, 581:1344–1355.

Tassoul, J.-L. (1978). *Theory of Rotating Stars.* Princeton Series in Astrophysics. Princeton, NJ: Princeton University Press.

Terebey, S., Shu, F. H., and Cassen, P. (1984). The Collapse of the Cores of Slowly Rotating Isothermal Clouds. *Astrophys. J.*, 286:529–551.

Tohline, J. E. (1982). Hydrodynamic Collapse. *Fundamentals of Cosmic Physics*, 8:1–81.

Tomley, L., Cassen, P., and Steiman-Cameron, T. (1991). On the Evolution of Gravitationally Unstable Protostellar Disks. *Astrophys. J.*, 382:530–543.

Tomley, L., Steiman-Cameron, T. Y., and Cassen, P. (1994). Further Studies of Gravitationally Unstable Protostellar Disks. *Astrophys. J.*, 422:850–861.

Toomre, A. (1964). On the Gravitational Stability of a Disk of Stars. *Astrophys. J.*, 139:1217–1238.

Toomre, A. (1981). What Amplifies the Spirals. In Fall, S. M., and Lynden-Bell, D., editors, *Structure and Evolution of Normal Galaxies*, pages 111–136.

Vishniac, E. T., and Diamond, P. (1989). A Self-Consistent Model of Mass and Angular Momentum Transport in Accretion Disks. *Astrophys. J.*, 347:435–447.

Visser, R., van Dishoeck, E. F., Doty, S. D., and Dullemond, C. P. (2009). The Chemical History of Molecules in Circumstellar Disks. I. Ices. *Astron. Astrophys.*, 495:881–897.

von Weizsäcker, C. F. (1948). Zur Kosmogonie. *Zeitschrift für Astrophysik*, 24:181.

Vorobyov, E. I., and Basu, S. (2005). The Origin of Episodic Accretion Bursts in the Early Stages of Star Formation. *Astrophys. J. Letters*, 633:L137–L140.

Vorobyov, E. I., and Basu, S. (2006). The Burst Mode of Protostellar Accretion. *Astrophys. J.*, 650:956–969.

Vorobyov, E. I., and Basu, S. (2008). Mass Accretion Rates in Self-Regulated Disks of T Tauri Stars. *Astrophys. J. Letters*, 676:L139–L142.

Wood, J. A. (1996). Processing of Chondritic and Planetary Material in Spiral Density Waves in the Nebula. *Meteoritics and Planetary Science*, 31:641–645.

Yorke, H. W., and Bodenheimer, P. (1999). The Formation of Protostellar Disks. III. The Influence of Gravitationally Induced Angular Momentum Transport on Disk Structure and Appearance. *Astrophys. J.*, 525:330–342.

Zhu, Z., Hartmann, L., Calvet, N., Hernandez, J., Muzerolle, J., and Tannirkulam, A.-K. (2007). The Hot Inner Disk of FU Orionis. *Astrophys. J.*, 669:483–492.

Zhu, Z., Hartmann, L., Gammie, C., and McKinney, J. C. (2009). Two-Dimensional Simulations of FU Orionis Disk Outbursts. *Astrophys. J.*, 701:620–634.

6

STEVEN A. BALBUS

MAGNETOHYDRODYNAMICS OF PROTOSTELLAR DISKS

1. Introduction

Magnetic fields are at the heart of the dynamic behavior of protostellar disks; they are governed by magnetohydrodynamics (MHD). The combination of a weak (subthermal) magnetic field and Keplerian rotation is unstable to the magnetorotational instability (MRI) if the degree of ionization in the disk is sufficiently high. The MRI leads to a breakdown of laminar flow into turbulence and produces a significant outward angular momentum flux via the greatly enhanced transport.

Along with this transport, there is also a significant loss of mechanical energy via dissipation. If the turbulent energy is dissipated locally, standard α modeling should give a reasonable, if somewhat crude, estimate of the gross scalings of the disk temperature and density. Away from the central star, however, the ionization fraction of protostellar disks is small, and they are generally not in a regime of near-perfect conductivity. Nonideal MHD effects are important, and these cannot be reduced to a single, simple α parameter. Once nonideal MHD figures in the investigation, the multifluid character of the disk gas must be addressed. In the literature, this is often done without sufficient care. In this chapter, I have devoted significant space to a detailed presentation of the fundamental MHD equations appropriate for protostellar disks. I hope that it affords the reader some helpful guidance through the labyrinthine passages of nonideal MHD.

In protostellar disks on scales of 1 to 10 AU, of the various complications to ideal MHD, Ohmic dissipation processes and Hall electromotive forces are the most important to understand. The presence of dust, of course, is also critical because small interstellar-scale grains absorb free charges that are needed for good magnetic coupling. On the scales of interest there is likely to be a region near the disk midplane that is magnetically decoupled, a so-called dead zone. The growth and settling of the grains as time evolves reduce their efficiency to absorb charge. With ionization provided by coronal X-rays from the central star (and possibly also cosmic rays), protostellar disks may be sufficiently magnetized throughout most of their lives to be MRI active, especially away from the disk midplane. The devil, as always, is in the details.

An understanding of the MHD of protostellar disks is crucial for the theoretical development of these objects. There is no getting around the fact that the subject is decidedly inelegant. In principle, we are interested only in solving the equation $\boldsymbol{F} = m\boldsymbol{a}$, but the range of topics that bears on this problem is truly daunting. Molecular chemistry, dust-grain physicochemistry, photoionization physics, aerosol theory, and non-ideal MHD are all key players in this game. If the results of the theorists' efforts had been little or no progress, there would have been no shortage of excuses. ("More work is needed.") But in fact there has been substantial progress in understanding important issues, more perhaps than some practitioners may realize. A sort of consensus is beginning to take shape about the gross properties of protostellar disks that are dictated by the demands of MHD physics. In this review, I will try to put forward as strong a case as I can for what I somewhat boldly regard as the canonical protostellar disk model, but at the same time I will try not to gloss over what are genuine uncertainties or difficulties.

Protostellar disks are a class of accretion disks, and one area where there certainly has been a great deal of progress in recent years is in the development of accretion-disk theory. The realization that a combination of differential rotation and a weak magnetic field is profoundly unstable and produces turbulence has given the subject a foundation on which to build (Balbus & Hawley 1991, 1998). Indeed, the primary reason for studying MHD processes in protostellar disks in detail is that once these systems are no longer self-gravitating, magnetic fields profoundly influence their dynamic behavior. It is not possible to understand angular momentum transport in low-mass (T Tauri) disks without focusing on magnetic fields.

Magnetic fields provide a conduit for free energy to flow from the differential rotation to the disk itself, producing fluctuations, turbulent heating, and quite possibly the directly observed outflows. Most important, we shall see that MHD turbulence produces well-correlated fluctuations of the radial and azimuthal components of the magnetic field and fluctuating gas velocity. It is precisely these correlations that result in a substantial outward transport of angular momentum, allowing the disk gas to spiral inward and accrete onto the central star.

It may also be possible in principle for winds to remove angular momentum from the disk without completely disrupting it (see Königl and Salmenon, this volume). The difficulty that needs to be overcome for this mechanism is that the outflow must occur over the whole face of the disk, not just from the innermost regions where jets are launched: material near the inner disk edge has already lost almost all its angular momentum. Instead, the outer disk material, torqued by the magnetic field, must continuously slip relative to the field, or else the field tends to become very centrally concentrated (Lubow, Papaloizou, & Pringle 1994). How this might work without generating internal disk turbulence that itself transports angular momentum remains to be fully understood.

This understanding that magnetic fields play a crucial role as the source of disk turbulence developed more than 30 years after the basic instability (now known as the magnetorotational instability or MRI) first appeared in the literature (Velikhov 1959). But even if the importance of the instability had been grasped immediately, without the computational power that became available only after 1990, it is doubtful that the impact would have been the same. There is no substitute for being able to visualize on one's desktop the development of a linear instability into full-fledged MHD turbulence.

The MRI depends on the presence of electrical currents to do its job, and this means that the disk gas must be at least partially ionized. "Partially ionized" in practice could mean even a minute electron fraction. Tiny traces of electrons can magnetize the disk, an effect we will quantify in § 8. It is because even a wisp of a magnetic field and the merest trace of electrons take on such dynamic significance that protostellar disk dynamics depends so heavily on protostellar disk chemistry.

Young protostellar disks need sodium, potassium, and other trace "vitamins" to make them vigorous and active because these alkalis are important in regulating the delicate ionization balance in the gas. It is ultimately the ionization level that determines the resistivity that determines whether the magnetic field is well coupled to the gas or not. This is not an issue in accretion disks around compact objects, for which even a small fraction of the turbulent energy dissipated is enough to ensure that the gas is thermally ionized. Protostellar disks, on the other hand, are big (so the free energy of differential rotation is small), dusty (so grains absorb free electrons), and cold (so thermal ionization is unimportant except near the star). The overarching question of protostellar MHD research is to understand under what conditions the abundance of free electrons drops below the level needed to ensure good magnetic coupling. The magnetic coupling leads in turn to instability, turbulence, and enhanced transport, the essence of disk dynamics. Much of this review will be devoted directly to the question of disk-ionization balance. For readers wishing additional astronomical background material, the text of Lee Hartmann (2008) is an excellent choice. Balbus & Hawley (2000) and Stone et al. (2000) are earlier reviews of MHD transport processes in protostellar disks. More general disk reviews include Pringle (1981; a classic review but before-MRI), Papaloizou & Lin (1995), Lin & Papaloizou (1996), Balbus & Hawley (1998), and Balbus (2003). In addition, the *Protostars and Planets* series published by the University of Arizona Press contains a useful historical record of the development of the subject of protostellar disks; the latest volume is *Protostars and Planets V* (Reipurth, Jewitt, & Keil 2007).

2. On the Need for MHD

The modern view of protostellar disks is heavily influenced by accretion-disk formalism. Lin & Papaloizou (1980) is the first pioneering study to bring accretion-disk formalism to bear on the study of protostellar disks, although it is now thought that MHD turbulence, rather than convective turbulence, is the key to protostellar disk dynamics. To understand the central role of a magnetic field in our understanding of protostellar disks, it is of interest to review where we stand with respect to our knowledge of the onset of turbulence in rotating flows.

The principal problem for classical disk theory (Shakura & Sunyaev 1973) was to discover the origin of the very large stresses that were needed to transport angular momentum, a process then (and still often now) referred to as *anomalous viscosity*. The approach of Shakura & Sunyaev was to make a virtue of necessity, arguing that the small microscopic viscosity implied a very large Reynolds number (Re),[1] and a large Reynolds number meant that shear turbulence was present. This conclusion was sustained by the belief that the laboratory experiments showing a nonlinear breakdown of Cartesian shear layers at values of Re in excess of $\sim 10^3$ would naturally carry over to Keplerian disks, where its value was considerably higher. The fact that the inertial force associated with the Coriolis effect (not present in planar flow) was well in excess of the destabilizing shear does not seem to have been viewed as troublesome to proponents of this view.

1. The Reynolds number Re is the characteristic flow velocity V times a characteristic length L divided by the kinematic viscosity ν: $\mathrm{Re} = VL/\nu$.

The classical laboratory method for investigating the stability of rotating flows is a Couette apparatus. In this device, the space between two coaxial cylinders is filled with a liquid, almost always ordinary water. The two cylinders rotate at different angular velocities, let us say Ω_{in} and Ω_{out}. The gap between the cylinders is of order 1 centimeter, and a stable rotation profile will be attained in a matter of minutes by viscous diffusion, even if Re is very large. (An unstable rotation profile will, of course, never be found as such; the flow will remain permanently disturbed.) By choosing Ω_{in} and Ω_{out} appropriately, a section of a Keplerian disk can be accurately mimicked, and the effects of Coriolis forces can be studied.

According to the classic text of Landau & Lifschitz (1959), a Couette rotation profile that is found in the laboratory to be stable "does not actually mean, however, that the flow actually remains steady no matter how large [Re] becomes." The monograph by Zel'dovich, Ruzmaikin, & Sokoloff (1983) is even more explicit on the question, unambiguously stating that laboratory experiments had already shown that Keplerian rotation profiles were nonlinearly unstable at sufficiently large Re.

These are stunning claims. At the time these statements were written, there was certainly no credible laboratory evidence that Keplerian rotation profiles were nonlinearly unstable. The first explicit claim that Keplerian flow was unstable based on laboratory evidence seems to be the unpublished result of Richard (2001). However, a later experiment by Ji et al. (2006) suggests that the earlier finding of instability was probably due to spurious effects arising from the interaction between the flow and the end caps of the experiments (Ekman layers).[2] When care is taken to minimize this interaction, Keplerian flow is found to be linearly and nonlinearly stable at values of Re up to 2×10^6. Serious students of disk theory would do well to develop a deep skepticism of "large Re means disks are turbulent" arguments, which sadly are still promulgated in contemporary textbooks.

Why is the Coriolis force so harmful to the maintenance of turbulence? Moffatt, Kida, & Ohkitani (1994) have referred to stretched vortices as the "sinews of turbulence." Turbulence is maintained in shear flows by vortices that are ensnared along the axis of strain, coupling the free energy of the shear directly to the internal vortex motion. Two neighboring fluid elements in a vortex are rapidly pulled apart as its circulation rises. This is possible only

2. Ekman layers are narrow fluid layers between a solid boundary and a rotating flow. When the bulk of the flow is not at rest in the frame of the boundary, the fluid rotation changes rapidly in the Ekman layer and can produce a secondary circulation pattern invading the entire flow.

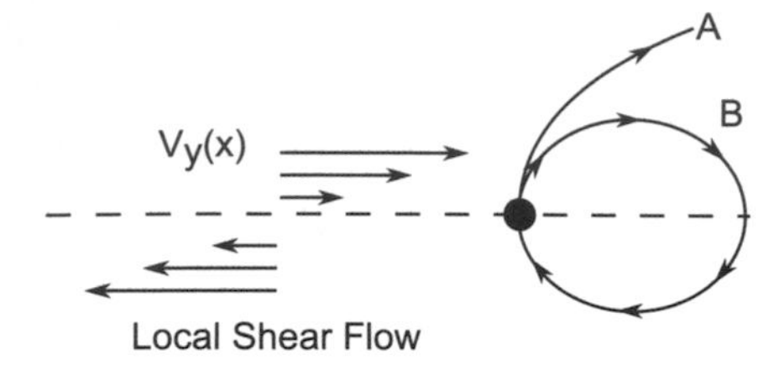

Fig. 6.1. The stabilization mechanism in a rotating disk versus Cartesian shear flow. In shear flow, displacements are unbounded, as indicated by path A. In a disk, epicyclic motions keeps the displacement tightly bound. See text for further details.

if Coriolis forces are absent. Despite the presence of shear, Coriolis forces induce epicyclic, oscillatory motions that do not allow anything resembling continuous vortex stretching. Without this, turbulence cannot be maintained. This argument breaks down if the strain rate much exceeds the formal epicyclic oscillation period, even if the latter is quite well defined. Such flows, in fact, are found to be nonlinearly unstable in numerical simulations. But for Keplerian flow, the epicyclic frequency is just the local Ω of the disk, and it exerts a strongly stabilizing influence.

Fig. 6.1 illustrates this point. Locally, Cartesian shear flow and Keplerian differential rotation appear to be similar, but perturbed fluid elements behave very differently in the two systems. Viewed from a frame moving with the undisturbed flow, a displaced fluid element in shear flow approximately follows an unbounded parabolic trajectory A, while a displaced fluid element in a Keplerian disk approximately follows a bound epicycle B. A small perturbing vortex would be stretched continuously in shear flow, tapping into the free energy source needed to maintain turbulence. By contrast, the embedded vortex would merely oscillate in a disk. It is this difference, due entirely to the presence of the rotational Coriolis force, that is ultimately responsible for the hydrodynamic stability of Keplerian disks, whereas shear flow can be nonlinearly unstable.

The title of the Ji et al. (2006) article is "Hydrodynamic Turbulence Cannot Transport Angular Momentum Effectively in Astrophysical Disks," and that seems as good a one-sentence summary of the topic as any; their experiment is probably the final word on nonlinear Keplerian hydrodynamic shear instability. It is gratifying, therefore, that all the difficulties encountered by Coriolis stabilization vanish when magnetic fields are taken into account. This is due to the fact that the magnetic field introduces new modes of response (shear waves) that are much less prone to Coriolis stabilization. One of these waves, the so-called 'slow mode,' turns into a local instability when differential rotation is present with a weak magnetic field. This is the magnetorotational instability. To understand this process in more detail, it is best to begin with the fundamentals of MHD.

3. Fundamentals

In this section, I present a detailed derivation of the fundamental MHD equations. The discussion will be more technical here than in most of the rest of the chapter, but it is very important to see how the basic governing equations of the subject arise, and much of this material is not easy to find outside specialized treatments. I hope that the reader will have the patience to read carefully through this section.

A protostellar disk is a gas of neutral particles (predominantly H_2 molecules), electrons, ions (the most important of which are generally K^+ and Na^+), and dust grains. We defer to § 8 a discussion of the complications due to the presence of dust grains and consider here the gas dynamic equations for mixtures of neutrals, ions, and atoms.

Each species (denoted by subscript s) is separately conserved and obeys the mass-conservation equation

$$\frac{\partial \rho_s}{\partial t} + \nabla \cdot (\rho_s \boldsymbol{v}_s) = 0, \tag{1}$$

where ρ_s is the mass density for species s and $\boldsymbol{v}_s$ is the velocity. The flow quantities for the dominant neutral species will henceforth be presented without subscripts.

The dynamic equation for the neutral particles is

$$\rho \frac{\partial \boldsymbol{v}}{\partial t} + \rho(\boldsymbol{v} \cdot \nabla)\boldsymbol{v} = -\nabla P - \rho \nabla \Phi - \boldsymbol{p}_{nI} - \boldsymbol{p}_{ne}. \tag{2}$$

where P is the pressure of the neutrals, Φ is the gravitational potential, and $\boldsymbol{p}_{nI}$ ($\boldsymbol{p}_{ne}$) is the momentum exchange rate between the neutrals and the ions (electrons).

Let us examine these last two important terms in more detail. The term $\boldsymbol{p}_{nI}$ takes the form

$$\boldsymbol{p}_{nI} = n\mu_{nI}(\boldsymbol{v} - \boldsymbol{v}_I)\nu_{nI}, \tag{3}$$

where n is the number density of neutrals, and μ_{nI} is the reduced mass of an ion and neutral particle,

$$\mu_{nI} \equiv \frac{m_I m_n}{m_I + m_n}, \tag{4}$$

m_I and m_n being the ion and neutral mass, respectively. The term ν_{nI} is the collision frequency of a neutral with a population of ions,

$$\nu_{nI} = n_I \langle \sigma_{nI} w_{nI} \rangle. \tag{5}$$

In eq. (5), n_I is the number density of ions, σ_{nI} is the effective cross section for neutral-ion collisions, and w_{nI} is the relative velocity between a neutral particle and an ion. The angle brackets represent an average over a Maxwellian distribution function for the relative velocity. (The mass appearing in this Maxwellian will of course be the reduced mass μ_{nI}.) For neutral-ion scattering, we may take the cross section σ_{nI} to be approximately geometric, which means that the quantity in angle brackets will be proportional to $\mu_{nI}^{-1/2}$. The order of the subscripts has no particular significance for the cross section σ_{nI}, the reduced mass μ_{nI}, or the relative velocity w_{nI}. But ν_{In} differs from ν_{nI}: the former is proportional to the neutral density n, not the ion density n_I.

Putting these definitions together gives

$$\boldsymbol{p}_{nI} = n n_I \mu_{nI} \langle \sigma_{nI} w_{nI} \rangle (\boldsymbol{v} - \boldsymbol{v}_I). \tag{6}$$

In accordance with Newton's third law, this is symmetric with respect to the interchange $n \leftrightarrow I$, except for a change in sign, $\boldsymbol{p}_{nI} = -\boldsymbol{p}_{In}$. All these considerations hold, of course, for electron-neutral scattering as well. Explicitly, we have

$$\boldsymbol{p}_{ne} = n n_e \mu_{ne} \langle \sigma_{ne} w_{ne} \rangle (\boldsymbol{v} - \boldsymbol{v}_e) \simeq n n_e m_e \langle \sigma_{ne} w_{ne} \rangle (\boldsymbol{v} - \boldsymbol{v}_e). \tag{7}$$

The gas will be locally neutral, so that $n_e = Z n_i$, where Z is the number of ionizations per ion particle. In a weakly ionized gas, $Z = 1$. The reduced mass μ_{ne} may safely be set equal to the electron mass m_e. We will use the following expressions for the collision rates (Draine, Roberge, & Dalgarno 1983):[3]

$$\langle \sigma_{nI} w_{nI} \rangle = 1.9 \times 10^{-9} \text{ cm}^3 \text{ s}^{-1} \tag{8}$$

and

$$\langle \sigma_{ne} w_{ne} \rangle = 10^{-15} \, (128 kT / 9\pi m_e)^{1/2} = 8.3 \times 10^{-10} T^{1/2} \text{ cm}^3 \text{ s}^{-1}. \tag{9}$$

The electron-neutral collision rate is essentially the ion geometric cross section times an electron thermal velocity. (The peculiar factor $(128/9\pi)^{1/2}$ comes from the way the electron velocity is projected, when averaged over the electron Maxwellian distribution function, along the direction of the mean ion flow.) But the ion-neutral collision rate is temperature independent, much more beholden to long-range induced dipole interactions, and significantly enhanced relative to a geometric cross-section assumption. Even if the ion-neutral rate were determined only by a geometric cross section, $|\boldsymbol{p}_{nI}|$ would exceed $|\boldsymbol{p}_{ne}|$ by a factor of order $(m_e/\mu_{nI})^{1/2}$. In fact, the dipole enhancement of the ion-neutral cross section makes this factor even larger.

3. A recent calculation of H-H^+ scattering by Glassgold, Krstić, & Schultz (2005) may imply a slightly higher value for the neutral-ion collision rate.

In the astrophysical literature, it is common to write the ion-neutral momentum coupling in the form

$$\boldsymbol{p}_{In} = \rho\rho_I\gamma(\boldsymbol{v}_I - \boldsymbol{v}), \tag{10}$$

where γ is the so-called drag coefficient,

$$\gamma \equiv \frac{\langle \sigma_{nI} w_{nI} \rangle}{m_I + m_n}, \tag{11}$$

and we will use this notation from here on. Numerically, $\gamma = 3 \times 10^{13}$ cm^3 s^{-1} g^{-1} for astrophysical mixtures (Draine, Roberge, & Dalgarno 1983).

The dynamic equations for the ions and electrons are

$$\rho_I \frac{\partial \boldsymbol{v}_I}{\partial t} + \rho_I \boldsymbol{v}_I \cdot \nabla \boldsymbol{v}_I = -\nabla P_I - \rho_I \nabla \Phi + Zen_I \left(\boldsymbol{E} + \frac{\boldsymbol{v}_I}{c} \times \boldsymbol{B} \right) - \boldsymbol{p}_{In} \tag{12}$$

and

$$\rho_e \frac{\partial \boldsymbol{v}_e}{\partial t} + \rho_e \boldsymbol{v}_e \cdot \nabla \boldsymbol{v}_e = -\nabla P_e - \rho_e \nabla \Phi - en_e \left(\boldsymbol{E} + \frac{\boldsymbol{v}_e}{c} \times \boldsymbol{B} \right) - \boldsymbol{p}_{en}, \tag{13}$$

respectively. Throughout this chapter, e will denote the positive charge of a proton, 4.803×10^{-10} esu. For a weakly ionized gas, the Lorentz force and collisional terms dominate in each of the latter two equations. Comparison of the magnetic and inertial forces, for example, shows that the latter are smaller than the former by the ratio of the proton or electron gyroperiod to a macroscopic flow-crossing time. Thus, to an excellent degree of approximation,

$$Zen_I \left(\boldsymbol{E} + \frac{\boldsymbol{v}_I}{c} \times \boldsymbol{B} \right) - \boldsymbol{p}_{In} = 0, \tag{14}$$

and

$$-en_e \left(\boldsymbol{E} + \frac{\boldsymbol{v}_e}{c} \times \boldsymbol{B} \right) - \boldsymbol{p}_{en} = 0. \tag{15}$$

The sum of these two equations gives

$$\frac{\boldsymbol{J}}{c} \times \boldsymbol{B} = \boldsymbol{p}_{In} + \boldsymbol{p}_{en}, \tag{16}$$

where charge neutrality $n_e = Zn_I$ has been used, and I have introduced the current density

$$\boldsymbol{J} \equiv en_e(\boldsymbol{v}_I - \boldsymbol{v}_e). \tag{17}$$

The equation for the neutrals becomes

$$\rho \frac{\partial \boldsymbol{v}}{\partial t} + \rho(\boldsymbol{v} \cdot \nabla)\boldsymbol{v} = -\nabla P - \rho \nabla \Phi + \frac{\boldsymbol{J}}{c} \times \boldsymbol{B}. \tag{18}$$

Because of collisional coupling, the neutrals are subject to the magnetic Lorentz force just as though they were a gas of charged particles. It is not the magnetic force per se that changes in a neutral gas. As we shall presently see, it is the inductive properties of the gas.

Let us return to the force-balance equation for the electrons:

$$-en_e\left(\boldsymbol{E}+\frac{\boldsymbol{v_e}}{c}\times\boldsymbol{B}\right)-\boldsymbol{p_{en}}=0. \tag{19}$$

After division by $-en_e$, this may be expanded to

$$\boldsymbol{E}+\frac{1}{c}\left[\boldsymbol{v}+(\boldsymbol{v_e}-\boldsymbol{v_I})+(\boldsymbol{v_I}-\boldsymbol{v})\right]\times\boldsymbol{B}+\frac{m_e\nu_{en}}{e}\left[(\boldsymbol{v_e}-\boldsymbol{v_I})+(\boldsymbol{v_I}-\boldsymbol{v})\right]=0, \tag{20}$$

where I have introduced the collision frequency of an electron in a population of neutrals:

$$\nu_{en}=n\langle\sigma_{ne}w_{ne}\rangle. \tag{21}$$

I have written the electron velocity $\boldsymbol{v_e}$ in terms of the dominant neutral velocity $\boldsymbol{v}$ and the key physical velocity differences of our problem. It has already been noted that in eq. (16), $\boldsymbol{p_{en}}$ is small compared with $\boldsymbol{p_{In}}$, provided that the velocity difference $|\boldsymbol{v_e}-\boldsymbol{v}|$ is not much larger than $|\boldsymbol{v_I}-\boldsymbol{v}|$. In fact, when used in (20), the $\boldsymbol{p_{en}}$ term in eq. (16) may quite generally be dropped:

$$\frac{\boldsymbol{J}}{c}\times\boldsymbol{B}\simeq\boldsymbol{p_{In}}. \tag{22}$$

It may also be shown that the final term in eq. (20),

$$\frac{m_e\nu_{en}}{e}(\boldsymbol{v_I}-\boldsymbol{v}),$$

which would then be proportional to $\boldsymbol{J}\times\boldsymbol{B}$, becomes small compared with the third term,

$$\frac{1}{c}(\boldsymbol{v_e}-\boldsymbol{v_I})\times\boldsymbol{B},$$

also proportional to $\boldsymbol{J}\times\boldsymbol{B}$. Both of these rather technical claims are justified in detail in appendix A. (In both cases, the incurred error is of order $(m_e/\mu_{In})^{1/2}$.) These simplifications allow us to write the electron force-balance equation as

$$\boldsymbol{E}+\frac{\boldsymbol{v}}{c}\times\boldsymbol{B}-\frac{\boldsymbol{J}\times\boldsymbol{B}}{en_ec}-\frac{\boldsymbol{J}}{\sigma_{cond}}+\frac{(\boldsymbol{J}\times\boldsymbol{B})\times\boldsymbol{B}}{c^2\gamma\rho\rho_I}=0, \tag{23}$$

where the electrical conductivity has been defined as

$$\sigma_{cond}\equiv\frac{e^2n_e}{m_e\nu_{en}}. \tag{24}$$

The associated resistivity η is (e.g., Jackson 1975)

$$\eta=\frac{c^2}{4\pi\sigma_{cond}}, \tag{25}$$

which has units of $\text{cm}^2\ \text{s}^{-1}$. Numerically (e.g., Blaes & Balbus 1994; Balbus & Terquem 2001),

$$\eta=234\left(\frac{n}{n_e}\right)T^{1/2}\ \text{cm}^2\ \text{s}^{-1}. \tag{26}$$

Eq. (23) is a general form of Ohm's law for a moving, multiple-fluid system.

Next, we make use of two of Maxwell's equations. The first is Faraday's induction law:

$$\nabla \times \boldsymbol{E} = -\frac{1}{c}\frac{\partial \boldsymbol{B}}{\partial t}. \tag{27}$$

We substitute $\boldsymbol{E}$ from eq. (23) to obtain an equation for the self-induction of the magnetized fluid,

$$\frac{\partial \boldsymbol{B}}{\partial t} = \nabla \times \left[\boldsymbol{v} \times \boldsymbol{B} - \frac{\boldsymbol{J} \times \boldsymbol{B}}{e n_e} + \frac{(\boldsymbol{J} \times \boldsymbol{B}) \times \boldsymbol{B}}{c \gamma \rho \rho_I} - \frac{c \boldsymbol{J}}{\sigma_{cond}} \right]. \tag{28}$$

It remains to relate the current density $\boldsymbol{J}$ to the magnetic field $\boldsymbol{B}$. This is accomplished by the second Maxwell equation,

$$\frac{4\pi}{c}\boldsymbol{J} = \nabla \times \boldsymbol{B} + \frac{1}{c}\frac{\partial \boldsymbol{E}}{\partial t} \tag{29}$$

The final term in eq. (29) is the displacement current, and it may be ignored. Indeed, since we have not, and will not, use the 'Gauss's law' equation

$$\nabla \cdot \boldsymbol{E} = 4\pi e (Z n_I - n_e), \tag{30}$$

we must not include the displacement current. In Appendix B, I show that departures from charge neutrality in $\nabla \cdot \boldsymbol{E}$ and the displacement current are both small terms that contribute at the same order: v^2/c^2. These must both be self-consistently neglected in nonrelativistic MHD. (The final Maxwell equation $\nabla \cdot \boldsymbol{B} = 0$ adds nothing new. It is automatically satisfied by eq. [27] as long as the initial magnetic field satisfies this divergence-free condition.) These considerations imply that

$$\boldsymbol{J} = \frac{c}{4\pi} \nabla \times \boldsymbol{B} \tag{31}$$

for use in eq. (28).

To summarize, the fundamental equations of a weakly ionized fluid are mass conservation of the dominant neutrals (eq. [1]),

$$\frac{\partial \rho}{\partial t} + \nabla \cdot (\rho \boldsymbol{v}) = 0, \tag{32}$$

the equation of motion (eq. [18] with [31]),

$$\rho \frac{\partial \boldsymbol{v}}{\partial t} + \rho \boldsymbol{v} \cdot \nabla \boldsymbol{v} = -\nabla P - \rho \nabla \Phi + \frac{1}{4\pi}(\nabla \times \boldsymbol{B}) \times \boldsymbol{B}, \tag{33}$$

and the induction equation (eq. [28] with [25] and [31]),

$$\frac{\partial \boldsymbol{B}}{\partial t} = \nabla \times \left[\boldsymbol{v} \times \boldsymbol{B} - \frac{c(\nabla \times \boldsymbol{B}) \times \boldsymbol{B}}{4\pi e n_e} + [(\nabla \times \boldsymbol{B}) \times \boldsymbol{B}] \frac{\times \boldsymbol{B}}{4\pi \gamma \rho \rho_I} - \eta \nabla \times \boldsymbol{B} \right]. \tag{34}$$

It is only natural that the reader should be a little taken aback by the sight of eq. (34). Be assured that it is rarely, if ever, needed in full generality: almost always one or more terms on the right side of the equation may be discarded. When only the induction term $\boldsymbol{v}\times\boldsymbol{B}$ is important, we refer to this regime as *ideal* MHD. The three remaining terms on the right are the nonideal MHD terms.

To get a better feel for the relative importance of the nonideal MHD terms in eq. (28), we denote the terms on the right side of the equation, moving from left to right, as I (induction), H (Hall), A (ambipolar diffusion), and O (ohmic resistivity). Then the scaling of the I, A, and O terms relative to the Hall term H is

$$\frac{I}{H} \sim \frac{v}{v_I - v_e}, \quad \frac{A}{H} \sim Z\frac{\omega_{cI}}{\gamma\rho}, \quad \frac{O}{H} \sim \frac{\nu_{en}}{\omega_{ce}}, \tag{35}$$

where the ion cyclotron frequency $\omega_{cI} = eB/m_I c$, and similarly for the electron frequency ω_{ce}, with m_e replacing m_I.

We will always be in a regime in which the presence of the induction term is not in question, i.e., the relative ion-electron drift velocity $v_I - v_e$ is always comparable to or in fact much less than v. More interesting is the relative importance of the nonideal terms. The explicit dependence of A/H and O/H in terms of the fluid properties of a cosmic gas is given in Balbus & Terquem (2001):

$$\frac{A}{H} = Z\left(\frac{9\times 10^{12}\ \mathrm{cm}^{-3}}{n}\right)^{1/2}\left(\frac{T}{10^3\ \mathrm{K}}\right)^{1/2}\left(\frac{v_A}{c_S}\right) \tag{36}$$

and

$$\frac{O}{H} = \left(\frac{n}{8\times 10^{17}\ \mathrm{cm}^{-3}}\right)^{1/2}\left(\frac{c_S}{v_A}\right). \tag{37}$$

Here n is the total number density of all particles, T is the kinetic temperature, v_A is the so-called Alfvén velocity,

$$\boldsymbol{v}_A = \frac{\boldsymbol{B}}{\sqrt{4\pi\rho}}, \tag{38}$$

and c_S is the isothermal speed of sound,

$$c_S^2 = 0.429\frac{kT}{m_p}, \tag{39}$$

where k is the Boltzmann constant and m_p is the mass of the proton. The coefficient 0.429 corresponds to a mean mass per particle of $2.33m_p$, appropriate for a molecular cosmic-abundance gas.

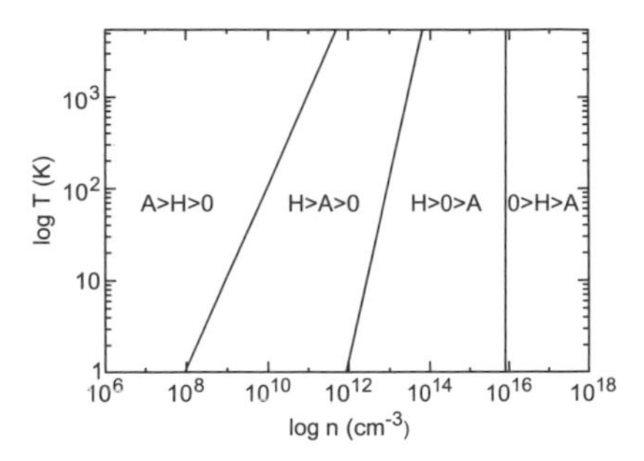

Fig. 6.2. Parameter space for nonideal MHD. The curves correspond to the case $v_A/c_S = 0.1$. From Kunz & Balbus 2004.

As reassurance that the fully general nonideal MHD induction equation is not needed for our purposes, note that eqs. (36) and (37) imply that for all three nonideal MHD terms to be comparable, $T \sim 10^8$ K. Obviously this is not an issue for protostellar disks. In Fig. 6.2, I plot the domains of relative dominance of the nonideal MHD terms in the nT plane. Protostellar disks are dominated by the resistive and Hall nonideal MHD terms, except in the innermost regions, where ohmic dissipation is largest, and the outermost regions, where ambipolar diffusion becomes important (Wardle 1999).

Our emphasis on the relative ordering of the nonideal terms in the induction equation should not obscure the fact that ideal MHD is often an excellent approximation, even when the ionization fraction $\ll 1$. For example, the ratio of the ideal inductive term to the ohmic loss term is given by the Lundquist number

$$\mathbb{L} = \frac{v_A L}{\eta}, \tag{40}$$

where L is a characteristic gradient-length scale. To orient ourselves, let us set $L = 0.1R$, where R is the radial location in the disk. Then $\mathbb{L}$ is given by

$$\mathbb{L} \simeq 2.5(n_e/n)(v_A/c_S)R_{cm},$$

where R_{cm} is the radius in centimeters. In other words, the critical ionization fraction at which $\mathbb{L} = 1$ is about

$$(n_e/n)_{crit} = 0.4(c_S/v_A)R_{cm}^{-1} \sim 10^{-13}(c_S/10v_A)$$

at $R = 1$ AU. The actual nebular ionization fraction at this location may be above or below this during the course of the solar system's evolution, but the point worth noting here is that R_{cm} is a large number for a protostellar disk, whatever the ionization source. Ionization fractions far, far below unity can render an astrophysical gas a near-perfect electrical conductor. It therefore makes a great deal of sense to begin by examining the behavior of an ideal MHD fluid.

4. Ideal MHD

The fundamental equations of ideal (single-fluid) MHD are mass conservation (e.g., [1]),

$$\frac{\partial \rho}{\partial t} + \nabla \cdot (\rho \boldsymbol{v}) = 0, \tag{41}$$

the dynamic equation of motion,

$$\frac{\partial \boldsymbol{v}}{\partial t} + (\boldsymbol{v} \cdot \nabla)\boldsymbol{v} = -\frac{1}{\rho}\nabla\left(P + \frac{B^2}{8\pi}\right) - \nabla\Phi + \left(\frac{\boldsymbol{B}}{4\pi}\cdot\right)\nabla\boldsymbol{B}, \tag{42}$$

where Φ is the (central) gravitational potential function, and our newly simplified induction equation

$$\frac{\partial \boldsymbol{B}}{\partial t} = \nabla \times (\boldsymbol{v} \times \boldsymbol{B}) \tag{43}$$

We shall work in a standard cylindrical coordinate system (R, ϕ, z), where R is the radius, ϕ is the azimuthal angle, and z the vertical coordinate. In these coordinates, the three components of the equation of motion are

$$\left[\frac{\partial}{\partial t} + \boldsymbol{v} \cdot \nabla\right] v_R - \frac{v_\phi^2}{R} = -\frac{1}{\rho}\frac{\partial}{\partial R}\left(P + \frac{B^2}{8\pi}\right) - \frac{\partial \Phi}{\partial R} + \frac{\boldsymbol{B}}{4\pi\rho}\cdot\nabla B_R - \frac{B_\phi^2}{4\pi\rho R}, \tag{44}$$

$$\left[\frac{\partial}{\partial t} + \boldsymbol{v} \cdot \nabla\right] (R v_\phi) = -\frac{1}{\rho}\frac{\partial}{\partial \phi}\left(P + \frac{B^2}{8\pi}\right) + \frac{\boldsymbol{B}}{4\pi\rho}\cdot\nabla(R B_\phi), \tag{45}$$

and

$$\left[\frac{\partial}{\partial t} + \boldsymbol{v} \cdot \nabla\right] v_z = -\frac{1}{\rho}\frac{\partial}{\partial z}\left(P + \frac{B^2}{8\pi}\right) - \frac{\partial \Phi}{\partial z} + \frac{\boldsymbol{B}}{4\pi\rho}\cdot\nabla B_z, \tag{46}$$

and the three components of the induction equation are

$$\left[\frac{\partial}{\partial t} + \boldsymbol{v} \cdot \nabla\right] B_R = -B_R \nabla \cdot \boldsymbol{v} + \boldsymbol{B} \cdot \nabla v_R, \tag{47}$$

$$\left[\frac{\partial}{\partial t} + \boldsymbol{v} \cdot \nabla\right] \frac{B_\phi}{R} = -\frac{B_\phi}{R} \nabla \cdot \boldsymbol{v} + \boldsymbol{B} \cdot \nabla\left(\frac{v_\phi}{R}\right), \tag{48}$$

and

$$\left[\frac{\partial}{\partial t} + \boldsymbol{v} \cdot \nabla\right] B_z = -B_z \nabla \cdot \boldsymbol{v} + \boldsymbol{B} \cdot \nabla v_z. \tag{49}$$

For many problems of interest, an important simplification can be made to these equations. The essence of rotational dynamics is local. Imagine going to a location R_0 in the disk. The Keplerian angular velocity Ω at $R = R_0$ is Ω_0. We hold Ω_0 fixed but allow R_0 to become arbitrarily large. Thus $v_\phi(R_0)$ is unbounded, but $v_\phi/R_0 = \Omega_0$ is finite.

Next, we introduce local coordinates (x, α, z) defined by

$$R = R_0 + x, \qquad x \ll R_0, \tag{50}$$

and

$$\alpha = \phi - \Omega_0 t \ll \pi, \tag{51}$$

and z is unchanged. We will formally introduce a new coordinate $t' = t$. This is desirable because we wish to take a time derivative at constant x, y, z, not at constant R, ϕ, z. In fact,

$$\frac{\partial}{\partial t'} = \frac{\partial}{\partial t} + \Omega_0 \frac{\partial}{\partial \alpha}. \tag{52}$$

Partial derivatives with respect to R and ϕ become partial derivatives with respect to x and α, respectively. The Lagrangian derivative

$$\frac{\partial}{\partial t} + \boldsymbol{v} \cdot \nabla \tag{53}$$

becomes

$$\frac{\partial}{\partial t'} + (\boldsymbol{v} - R\Omega_0 \boldsymbol{e_\phi}) \cdot \nabla. \tag{54}$$

(The ∇ operator now formally involves x and α derivatives.) This transformation suggests that we introduce a new azimuthal velocity,

$$w_\alpha = v_\phi - R\Omega_0, \tag{55}$$

and for notational consistency we will use w_x and w_z for the radial and vertical velocities, although they are identical to v_R and v_z.

In the local approximation, we assume that the magnitude w is much smaller than the (formally infinite) rotation velocity, and that w, the Alfvén speed v_A, and the thermal velocity $(P/\rho)^{1/2}$ are comparable in magnitude. In the limit $R_0 \to \infty$, the radial equation becomes

$$\left[\frac{\partial}{\partial t'} + \boldsymbol{w} \cdot \nabla\right] w_R - 2\Omega_0 w_\phi = R(\Omega_0^2 - \Omega_K^2) - \frac{1}{\rho}\frac{\partial}{\partial x}\left(P + \frac{B^2}{8\pi}\right) + \frac{\boldsymbol{B}}{4\pi\rho} \cdot \nabla B_R, \tag{56}$$

where the Keplerian angular velocity at R satisfies

$$R\Omega_K^2 = \frac{\partial \Phi}{\partial R}. \tag{57}$$

Expanding to first order in R, I obtain one obtains

$$R(\Omega_0^2 - \Omega_K^2) = -x\frac{d\Omega_K^2}{dR} = 3\Omega_0^2. \tag{58}$$

The final form of the equation is

$$\left[\frac{\partial}{\partial t}+\boldsymbol{w}\cdot\nabla\right]w_x-2\Omega w_y=3\Omega^2 x-\frac{1}{\rho}\frac{\partial}{\partial x}\left(P+\frac{B^2}{8\pi}\right)+\frac{\boldsymbol{B}}{4\pi\rho}\cdot\nabla B_x, \tag{59}$$

where we have dropped the subscript 0 and the prime ′. The vector components x, y, and z refer to the quasi-Cartesian system $dx = dR$, $dy = Rd\alpha$, and dz as before. (See Fig. 6.2.) When non-Keplerian disks are considered, $3\Omega^2$ should be replaced by $\frac{-d\Omega^2}{d\ln R}$.

The transformed azimuthal (y) equation is

$$\left[\frac{\partial}{\partial t}+\boldsymbol{w}\cdot\nabla\right]w_y+2\Omega w_x=-\frac{1}{\rho}\frac{\partial}{\partial y}\left(P+\frac{B^2}{8\pi}\right)+\frac{\boldsymbol{B}}{4\pi\rho}\cdot\nabla B_y, \tag{60}$$

and the z equation is

$$\left[\frac{\partial}{\partial t}+\boldsymbol{w}\cdot\nabla\right]w_z=-\frac{\partial\Phi}{\partial z}-\frac{1}{\rho}\frac{\partial}{\partial z}\left(P+\frac{B^2}{8\pi}\right)+\frac{\boldsymbol{B}}{4\pi\rho}\cdot\nabla B_z. \tag{61}$$

The mass-conservation and induction equations, like the z equation of motion, show no change of form in these corotating coordinates:

$$\frac{\partial\rho}{\partial t}+\nabla\cdot(\rho\boldsymbol{w})=0, \tag{62}$$

$$\frac{\partial\boldsymbol{B}}{\partial t}=\nabla\times(\boldsymbol{w}\times\boldsymbol{B}). \tag{63}$$

To summarize, we have transformed our general equations into a coordinate frame rotating at the Keplerian angular velocity $\Omega_K(R_0)$, restricting the spatial domain to a small neighborhood around a particular fluid element. The rotational dynamics appears in the form of a Coriolis force, $-2\boldsymbol{\Omega}\times\boldsymbol{w}$, and a centrifugal force that cancels the main gravitational force, leaving the residual tidal term $3\Omega^2 x$. These simplified local equations, which are the magnetized version of what is known as the Hill system, retain a rich dynamic content. This includes the full development of MHD turbulence.

4.1. Magnetorotational Instability

Let us apply eqs. (56)–(63) to the study of the paths of fluid elements departing from the origin $x = y = z = 0$. These *Lagrangian displacements* are in the xy orbital plane and depend only on z. I will ignore vertical stratification here, so that the equilibrium is z independent. Thus we may assume a spatial dependence of e^{ikz} for the perturbed fluid elements.

Let us assume a very simple equilibrium magnetic field: a constant vertical component B and no components in the orbital xy plane. If the displacement vector of an element of fluid in the xy plane is $\boldsymbol{\xi}$, then the perturbed magnetic field at a given location is

$$\delta \boldsymbol{B} = \nabla \times (\boldsymbol{\xi} \times \boldsymbol{B}). \tag{64}$$

This result, which is true quite generally, is lengthy to prove by a direct assault but can be intuited rather easily. The left side of the exact eq. (63) may be written $\delta \boldsymbol{B}/\delta t$, the change in $\boldsymbol{B}$ at a fixed location divided by the change in time. In the equilibrium state this is zero. Imagine perturbing this pure state by giving each fluid an additional finite velocity $\boldsymbol{U}$, but letting it act only for an infinitesimal time δt. Then

$$\delta \boldsymbol{B} = \nabla \times (\boldsymbol{w}\, \delta t \times \boldsymbol{B}) + \nabla \times (\boldsymbol{U}\, \delta t \times \boldsymbol{B}) = 0 + \nabla \times (\boldsymbol{\xi} \times \boldsymbol{B}), \tag{65}$$

where $\boldsymbol{\xi} = \boldsymbol{U}\delta t$ is the displacement of the fluid relative to its equilibrium path. This result is perhaps clearer in integral form. If we integrate over a surface A and use Stokes's theorem, we obtain

$$\int \delta \boldsymbol{B} \cdot \boldsymbol{dA} = \int (\boldsymbol{\xi} \times \boldsymbol{B}) \cdot \boldsymbol{ds} \equiv \int (\boldsymbol{ds} \times \boldsymbol{\xi}) \cdot \boldsymbol{B}, \tag{66}$$

where $\boldsymbol{ds}$ is a vector line element on the curve bounding A. This states that whatever explicit change in magnetic flux there is through A as the surface is displaced by $\boldsymbol{\xi}$, it is precisely compensated by the gain (or loss) of magnetic flux through the side of the cylinder (elemental area $\boldsymbol{ds} \times \boldsymbol{\xi}$) swept out by A. This is just a roundabout way of saying that the magnetic flux through A is conserved when it is moving with the fluid itself.

For the problem at hand,

$$\delta \boldsymbol{B} = \nabla \times (\boldsymbol{\xi} \times \boldsymbol{B}) = i\boldsymbol{k} \times (\xi \times \boldsymbol{B}) = i(\boldsymbol{k} \cdot \boldsymbol{B})\boldsymbol{\xi}, \tag{67}$$

since $\boldsymbol{k} \cdot \boldsymbol{\xi} = 0$. The magnetic term in eqs. (56) and (60) is

$$\frac{1}{4\pi\rho} \boldsymbol{B} \cdot \nabla \delta \boldsymbol{B} = -\frac{(\boldsymbol{k} \cdot \boldsymbol{B})^2}{4\pi\rho} \boldsymbol{\xi} \equiv -(\boldsymbol{k} \cdot \boldsymbol{v}_A)^2 \boldsymbol{\xi}. \tag{68}$$

We have introduced the so-called Alfvén velocity

$$\boldsymbol{v}_A = \frac{\boldsymbol{B}}{\sqrt{4\pi\rho}}, \tag{69}$$

whose physical interpretation I discuss below. For the moment, note that the magnetic force is like tension or a spring: it is always restoring and proportional to the displacement.

In eqs. (56) and (60), the derivative

$$\frac{D}{Dt} \equiv \frac{\partial}{\partial t} + \boldsymbol{v} \cdot \nabla$$

is just the time derivative following a fluid element, so we will write $D\boldsymbol{w}/Dt = \ddot{\xi}$, and of course $\boldsymbol{w} = \dot{\xi}$. Finally, we will allow for the possibility of non-Keplerian rotation. The equations of motion for the fluid element displacements ξ_x and ξ_y are then

$$\ddot{\xi}_x - 2\Omega\dot{\xi}_y = -\left[\frac{d\Omega^2}{d\ln R} + (kv_A)^2\right]\xi_x \tag{70}$$

and

$$\ddot{\xi}_y + 2\Omega\dot{\xi}_x = -(kv_A)^2\xi_y. \tag{71}$$

These are a simple set of coupled linear equations with time-dependent solutions of the form $e^{i\omega t}$. Without loss of generality, we take $\omega > 0$. In the absence of rotation, the equations decouple completely, and one finds that

$$\omega = |kv_A|. \tag{72}$$

These are waves that travel along the magnetic-field lines with group and phase velocity v_A, exactly like waves on a string. These so-called Alfvén waves are thus simple, nondispersive, incompressible disturbances.

On the other hand, in the absence of rotation, we find that

$$\omega^2 = 4\Omega^2 + \frac{d\Omega^2}{d\ln R} \equiv \kappa^2, \tag{73}$$

where κ^2 is known as the epicyclic frequency. In most astrophysical applications, Ω decreases radially outward, and $\kappa < 4\Omega^2$. The fluid dispacements for this mode execute ellipses with a major-to-minor-axis ratio of $2\Omega/\kappa$. The major axis lies along the azimuthal direction. The sense of the rotation around the ellipse is retrograde, opposite to the sense of Ω. These paths are known as epicycles. Epicyclic motion corresponds to the first-order departure from circular orbits for simple planetary motion.

The full dispersion relation resulting from eqs. (70) and (71) is more complicated than one might have expected:

$$\omega^4 - [\kappa^2 + 2(kv_A)^2]\omega^2 + (kv_A)^2\left[(kv_A)^2 + \frac{d\Omega^2}{d\ln R}\right] = 0. \tag{74}$$

This is clearly not just a matter of adding Alfvénic and epicylic motion in quadrature; something new is going on. The dispersion relation is a simple quadratic equation in ω^2, and it is straightforward to show that this quantity must be real. Stability may then be investigated by passing through the point $\omega^2 \to 0$. Therefore, marginal instability corresponds to

$$(kv_A)^2 + \frac{d\Omega^2}{d\ln R} = 0, \tag{75}$$

and instability is present when

$$\frac{d\Omega^2}{d\ln R} < 0. \tag{76}$$

Of course, this condition is nearly universally satisfied for any type of astrophysical disk. This is the simplest manifestation of the *magnetorotational instability*, or MRI.

4.2. Physical Interpretation

The equations of motion (70) and (71) have a very simple mechanical analog: they are exactly the equations of two point masses in orbit around a central body, bound together by a spring of frequency kv_A (Balbus & Hawley 1992). This mechanical analogy immediately offers a physical explanation of the MRI (see Fig. 6.3). Two masses are connected by a spring, an inner mass m_i and an outer mass m_o. The mass m_i orbits faster than the outer mass m_o, causing the spring to stretch. The spring pulls backward on m_i and forward on m_o. The negative torque on m_i causes it to lose angular momentum and sink, while the positive torque on m_o causes it to gain angular momentum and rise. Thus, even though the spring nominally supplies an attractive force, in a rotating frame this drives the masses apart. The continued stretching of the spring just makes matters worse, and the process runs away. The mass with the higher angular momentum obtains yet more, and the mass with less angular momentum loses what little it has. It is the same old story in a dynamic context: the rich get richer and the poor get poorer.

In a protostellar disk, it is the magnetic tension force that plays the role of the spring, and the two masses are any two fluid elements tethered by the field line. The linear phase of the instability is the exponentially growing separation of the displaced fluid elements, which is followed by the nonlinear mixing of gas parcels from different regions of the disk. The mixing seems to lead to

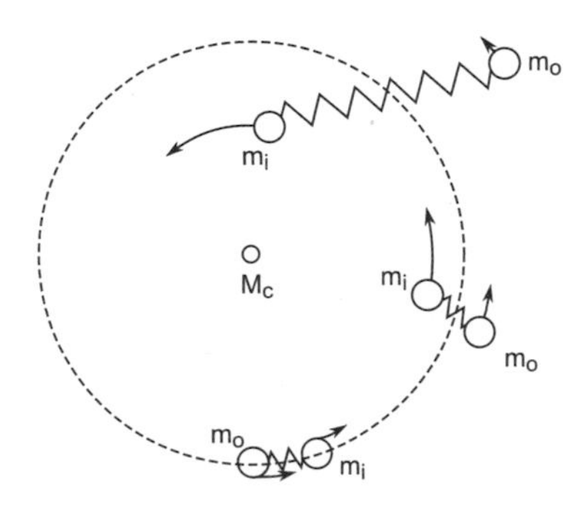

Fig. 6.3. The magnetorotational instability. Magnetic fields in a disk bind fluid elements precisely as though they were masses in orbit connected by a spring. The inner element m_i orbits faster than the outer element m_o, and the spring causes a net transfer of angular momentum from m_i to m_o. This transfer is unstable, as described in the text. The inner mass continues to sink, whereas the outer mass rises farther outward. Figure courtesy of H. Ji et al. (2006).

something resembling a classical turbulent cascade, although the details of this process, with different viscous and resistive dissipation scales, remain to be fully understood.

Notice that angular momentum transport is not something that happens as a consequence of the nonlinear development of the MRI; it is the essence of the MRI even in its linear phase. The very act of transporting angular momentum from the inner to the outer fluid elements via a magnetic couple is a spontaneously unstable process.

4.3. General Adiabatic Disturbances

If Ω is a function only of cylindrical radius R, then for general magnetic-field geometries, local incompressible WKB disturbances with space-time dependence

$$\delta X \sim \exp\left(ik_R R + ik_z z - i\omega t\right), \tag{77}$$

where δX is the Eulerian perturbation of any flow quantity, satisfy the following set of linearized dynamic equations:

$$k_R \delta v_R + k_z \delta v_z = 0, \tag{78}$$

$$-i\omega\delta v_R - 2\Omega\delta v_\phi + i\frac{k_R}{\rho}\left[\delta P + \frac{\boldsymbol{B}\cdot\nabla\boldsymbol{B}}{4\pi}\right] - i\frac{\boldsymbol{k}\cdot\boldsymbol{B}}{4\pi\rho}\delta B_R = 0, \tag{79}$$

$$-i\omega\delta v_\phi + \frac{\kappa^2}{2\Omega}\delta v_R - i\frac{\boldsymbol{k}\cdot\boldsymbol{B}}{4\pi\rho}\delta B_R = 0, \tag{80}$$

and

$$-i\omega\delta v_z + i\frac{k_z}{\rho}\left[\delta P + \frac{\boldsymbol{B}\cdot\nabla\boldsymbol{B}}{4\pi}\right] - i\frac{\boldsymbol{k}\cdot\boldsymbol{B}}{4\pi\rho}\delta B_z = 0. \tag{81}$$

The linearized induction equations are

$$-i\omega\delta B_R - i(\boldsymbol{k}\cdot\boldsymbol{B})\delta v_R = 0, \tag{82}$$

$$-i\omega\delta B_\phi - \delta B_R\frac{d\Omega}{d\ln R} - i(\boldsymbol{k}\cdot\boldsymbol{B})\delta v_\phi = 0, \tag{83}$$

and

$$-i\omega\delta B_z - i(\boldsymbol{k}\cdot\boldsymbol{B})\delta v_z = 0. \tag{84}$$

Finally, the entropy equation is

$$i\omega\gamma\frac{\delta\rho}{\rho} + \delta v_z\frac{\partial\ln P\rho^{-\gamma}}{\partial z} = 0, \tag{85}$$

where γ is the adiabatic index (not to be confused with the collision rate). We have ignored background radial gradients in the entropy and pressure but have retained their vertical gradients, in accordance with the assumption of a thin disk.

This set of linearized equations leads to the dispersion relation (Balbus & Hawley 1991)

$$\left(\frac{k^2}{k_z^2}\right)\tilde{\omega}^4 - \left[\kappa^2 + \left(\frac{k_R}{k_z}\right)^2 N^2\right]\tilde{\omega}^2 - 4\Omega^2(\boldsymbol{k}\cdot\boldsymbol{v}_A)^2 = 0. \tag{86}$$

Here

$$\tilde{\omega}^2 = \omega^2 - (\boldsymbol{k}\cdot\boldsymbol{v}_A)^2, \quad k^2 = k_z^2 + k_R^2, \tag{87}$$

and

$$N^2 = -\frac{1}{\gamma\rho}\frac{\partial P}{\partial z}\frac{\partial \ln P\rho^{-\gamma}}{\partial z}. \tag{88}$$

N is known as the Brunt-Väisälä frequency. It is the natural frequency at which a vertically displaced fluid element would oscillate in the disk because of buoyancy forces. In general, there is also a contribution due to radial gradients as well (Balbus & Hawley 1991), but this usually may be ignored in a rotationally supported (supersonic orbital speed) disk. Without any magnetic field, the dispersion relation becomes

$$\omega^2 = (k_z/k)^2\kappa^2 + (k_R/k)^2 N^2. \tag{89}$$

Since the displacement of the fluid element is incompressible, the wave number ratio k_z/k is a measure of the *radial* displacement, whereas k_R/k is a measure of the *vertical* displacement. There is no instability in this case, only a wavelike response due to the restoring Coriolis and buoyant forces.

Notice that because N vanishes in the disk plane by symmetry, it affects only the behavior of disturbances at least one scale height or so above $z = 0$. Moreover, the actual value of N is likely to be determined by radiative diffusion processes in the vertical direction. The radiation requires a source, which in disks is the turbulence we are trying to explain in the first place. (In protostellar disks, external heating of the upper layers is also a source of departure from adiabatic behavior.)

The other limit of interest is $k_R = 0$, which returns us to the dispersion relation of the previous section. (The expression kv_A should, of course, be replaced by $\boldsymbol{k}\cdot\boldsymbol{v}_A$.) This is fortunate because it can be shown (Balbus & Hawley 1992) that the most rapidly growing modes are displacements in the plane of the disk with $k_R = 0$. Since it is the differential rotation that ultimately destabilizes, it is very sensible that these displacements, which most effectively sample the differential rotation, are the most unstable. For a Keplerian rotation profile, it may be directly computed from the dispersion relation that the wave number of maximum growth is given by

$$\boldsymbol{k} \cdot \boldsymbol{v}_A = (\sqrt{15}/4)\Omega, \tag{90}$$

or 0.97Ω. By comparison, the largest unstable wavelength (i.e., smallest unstable wave number) has a value of 1.73Ω for $\boldsymbol{k} \cdot \boldsymbol{v}_A$. The maximum growth rate is

$$|\omega| = (3/4)\Omega. \tag{91}$$

This is a very large growth rate, with amplitudes growing at a rate of $\exp(3\pi/2) \sim 111$ per orbit.

4.4. Angular Momentum Transport

The azimuthal equation of motion (45) may be written as

$$\frac{\partial(\rho R v_\phi)}{\partial t} + \nabla \cdot \left[R \left(\rho \boldsymbol{v} v_\phi - \rho \boldsymbol{v}_A v_{A\phi} + \left(P + \frac{B^2}{8\pi} \right) \boldsymbol{e}_\phi \right) \right] = 0, \tag{92}$$

where $v_{A\phi}$ is the azimuthal component of the Alfvén velocity. This is an equation for angular momentum conservation, with angular momentum density $\rho R v_\phi$ and an azimuthally averaged flux of

$$\boldsymbol{F}_J = \langle R \left(\rho \boldsymbol{v} v_\phi - \rho \boldsymbol{v}_A v_{A\phi} \right) \rangle. \tag{93}$$

In protostellar disks, we are interested in the radial transport of angular momentum,

$$F_{JR} = R\rho(v_R v_\phi - v_{AR} v_{A\phi}). \tag{94}$$

It is most instructive to begin with the simple case of linear instability in a uniform, vertical magnetic field. What is the lowest-order flux that results from these exponentially growing modes? In equilibrium, v_R and the Alfvén velocities vanish. The linear perturbation of the radial velocity at a given spatial location (a so-called Eulerian perturbation) will be denoted by δv_R. Let the radial displacement ξ_R of a fluid element from its equilibrium location be given by

$$\xi_R = a e^{\gamma t} \cos(kz), \tag{95}$$

where γ and k correspond to the maximum growth rate and its associated wave number, and a is a slowly varying amplitude. Then

$$\delta v_R = \dot{\xi}_R = \gamma \xi_R. \tag{96}$$

The azimuthal velocity v_ϕ consists of an unperturbed Keplerian velocity $R\Omega$ plus a linear perturbation δv_ϕ. The product of δv_R and Ω contributes zero when a height integration is performed because of the cosine factor, so we must consider the direct product $\delta v_R \times \delta v_\phi$ to obtain the lowest-order contribution. In calculating δv_ϕ, note that

$$\delta v_\phi = \dot{\xi}_\phi - \xi_R \frac{d\Omega}{d \ln R}. \tag{97}$$

The subtraction is needed to eliminate the change in velocity a displaced fluid element would make even if there were no physical change in the rotation velocity at the new radial location: the actual change in velocity δv_ϕ is due to the change in $\dot{\xi}_\phi$ the displaced fluid element makes in excess of $\xi_R d\Omega/d \ln R$. Eq. (70) then gives

$$\delta v_\phi = \dot{\xi}_\phi - \xi_R \frac{d\Omega}{d \ln R} = \left(\frac{\gamma^2 + k^2 v_A^2}{2\Omega} \right) \xi_R, \tag{98}$$

and therefore the velocity contribution to the angular momentum flux is

$$\langle \delta v_R \delta v_\phi \rangle = \frac{\gamma}{2\Omega} (\gamma^2 + k^2 v_A^2) \langle \xi_R^2 \rangle. \tag{99}$$

Not surprisingly, it is positive (outward).

To calculate the magnetic-field correlation, we use eq. (67) to go between magnetic-field fluctuations and displacements. Then, using (71), we find that

$$\xi_\phi = -\frac{2\omega\gamma}{\gamma^2 + k^2 v_A^2} \xi_R, \tag{100}$$

from which it follows, among other things, that $\dot{\xi}_\phi$ and δv_ϕ have opposite signs. Hence

$$\frac{\langle \delta B_R \delta B_\phi \rangle}{4\pi\rho} = -\frac{2\Omega k^2 v_A^2 \gamma}{\gamma^2 + k^2 v_A^2} \langle \xi_R^2 \rangle. \tag{101}$$

Therefore,

$$\langle \delta v_R \delta v_\phi - \frac{\delta B_R \delta B_\phi}{4\pi\rho} \rangle = \frac{\gamma}{2\Omega} \left[\frac{(\gamma^2 + k^2 v_A^2)^2 + 4\Omega^2 k^2 v_A^2}{\gamma^2 + k^2 v_A^2} \right] \langle \xi_R^2 \rangle. \tag{102}$$

If we use the dispersion relation (74) to simplify this a bit, the expression for the radial angular momentum flux is

$$F_{JR} = \rho R \frac{\gamma}{2\Omega} \left[\frac{8\Omega^2 k^2 v_A^2}{\gamma^2 + k^2 v_A^2} - \kappa^2 \right] \langle \xi_R^2 \rangle. \tag{103}$$

Finally, integrating over height and defining the effective surface density Σ by

$$\int \rho \langle \xi_R^2 \rangle \, dz = \Sigma a^2 e^{2\gamma t}/2, \tag{104}$$

the angular momentum flux for the most rapidly growing mode is found to be

$$F_{JR} = \frac{3R\Sigma}{4}\Omega^2 a^2 e^{2\gamma t}. \tag{105}$$

This result bears some commentary. First, the radial flux is always positive for any unstable mode, not just the most rapidly growing one. This, as we shall see, is a direct consequence of energy conservation: energy is extracted from the differential rotation to the fluctuations only if the radial angular momentum flux is positive. Next, note that there is an outward angular momentum flux only to the extent that there is a correlation in the velocity fluctuations. This is also true in turbulent flows, and everything that we have done in this section holds in much the same way if the fluctuations are turbulent, as opposed to wavelike. The important physical point is that an instability that leads to turbulence need not lead to enhanced angular momentum transport. Only turbulence with strongly correlated velocity fields does this. Strong correlations, in turn, are necessary to extract energy from differential rotation. This is indeed the free energy source of shear-driven turbulence, but it need not be the free energy source of other forms of turbulence.

This point was made in Stone & Balbus (1996), in which the angular momentum transport resulting from convective instability was studied. The angular momentum was tiny in magnitude and *inwardly* directed: it had the "wrong" sign. Disk instabilities based on adverse thermal gradients do not, in general, lead to systematically large outward transfers of angular momentum. In angular momentum transport, not all turbulences are equivalent.

4.5. Diffusion of a Scalar

A classic problem in protostellar disk theory is to understand how dust particles are mixed with the gas. The fact that the MRI leads to vigorous radial angular momentum transport suggests that other quantities may be transported as well. In this section, we will give an argument that shows that the radial diffusion of angular momentum is indeed closely related to the radial transport of a conserved scalar quantity, say, Q. Assume that Q is a fluid-element label and satisfies an equation of the form

$$\left(\frac{\partial}{\partial t} + \boldsymbol{v}\cdot\nabla\right) Q = 0. \tag{106}$$

Then

$$\frac{\partial Q}{\partial t} + \nabla\cdot(\boldsymbol{v}Q) - Q\nabla\cdot\boldsymbol{v} = 0. \tag{107}$$

Because of mass conservation, this implies that

$$\frac{\partial Q}{\partial t} + \nabla \cdot (\boldsymbol{v}Q) + Q\frac{D \ln \rho}{Dt} = 0, \tag{108}$$

where

$$\frac{D}{Dt} = \left(\frac{\partial}{\partial t} + \boldsymbol{v} \cdot \nabla\right). \tag{109}$$

Simplifying the equation for Q, we obtain

$$\frac{\partial(\rho Q)}{\partial t} + \nabla \cdot (\rho \boldsymbol{v} Q) = 0, \tag{110}$$

which looks very much like the angular momentum conservation eq. (92). In a similar way, there will be a turbulent Q-flux given by

$$F_Q = \rho\langle \delta \boldsymbol{v} \delta Q\rangle. \tag{111}$$

Since Q is conserved as we follow a fluid element, the so-called Lagrangian perturbation ΔQ, defined by

$$\Delta Q = \delta Q + \boldsymbol{\xi} \cdot \nabla Q, \tag{112}$$

must vanish because ΔQ is constructed to be the change in Q following a fluid element as it is displaced by a small distance $\boldsymbol{\xi}$, and such changes must vanish since by definition Q is does not change along fluid-element paths. Hence the expression for the ith component of F_Q is

$$F_{Qi} = -\rho\langle \delta v_i \xi_j\rangle \partial_j Q. \tag{113}$$

This defines the diffusion tensor $\mathcal{D}_{ij}$,

$$\mathcal{D}_{ij} = \langle \delta v_i \xi_j\rangle. \tag{114}$$

In a height-integrated calculation, it is the RR component of $\mathcal{D}$ that is of importance. In the linear regime $\delta v_R = \gamma \xi_R$, so

$$\mathcal{D}_{RR} = \gamma\langle \xi_R^2\rangle. \tag{115}$$

In this case, eq. (103) gives a relationship between the flux of angular momentum and the diffusion coefficient of a passive scalar,

$$F_{JR} = \frac{\rho R}{2\Omega}\left[\frac{8\Omega^2 k^2 v_A^2}{\gamma^2 + k^2 v_A^2} - \kappa^2\right]\mathcal{D}_{RR}. \tag{116}$$

For the fastest-growing linear mode, this gives $F_{JR} = 2\Omega \mathcal{D}_{RR}$.

There is a sense in which the linear theory might find its way into nonlinear turbulent diffusion. In simulations, the MRI appears to stretch field lines locally and exponentiate velocity growth over a limited duration of time, before one fluid element becomes mixed with another and the process starts anew. If this picture is reasonably accurate, then $F_{JR} \sim \Omega\mathcal{D}_{RR}$ may well be valid in turbulent flow. (It is also, of course, what we might expect on the basis of simple dimensional analysis alone.) The important astrophysical point is that if protostellar disks are MHD turbulent, they ought to be well mixed.

5. Energetics of MHD Turbulence

5.1. Hydrodynamic Considerations

The quantity

$$\left(\rho \boldsymbol{v} v_\phi - \rho \boldsymbol{v}_A v_{A\phi}\right) \tag{117}$$

plays two conceptually different roles in the theory of accretion disks. We have seen that it is intimately linked to the direct transport of angular momentum, and in this section we shall study in detail how it extracts free energy from the large-scale differential rotation.

Let us begin with the relatively simple case of an adiabatic nonmagnetized gas. The azimuthal velocity is decomposed into a time-steady large rotational component $R\Omega$ plus a fluctuating component u_ϕ:

$$v_\phi = R\Omega + u_\phi. \tag{118}$$

I will assume neither that u_ϕ is small compared with $R\Omega$ nor that the mean value of u_ϕ vanishes, although in practice both might well be the case. We will write u_R and u_z for the radial and vertical velocity components, respectively, and $\boldsymbol{v}$ for the full velocity vector. Once again, we think of these as fluctuations, but our treatment will in fact be exact.

The adiabatic radial equation of motion is

$$\rho\left[\frac{\partial}{\partial t} + \boldsymbol{v}\cdot\nabla\right]u_R - \frac{\rho}{R}(R\Omega + u_\phi)^2 = -\frac{\partial P}{\partial R} - \rho\frac{\partial \Phi}{\partial R}. \tag{119}$$

Multiplying by u_R and regrouping, one obtains

$$\rho\left[\frac{\partial}{\partial t} + \boldsymbol{v}\cdot\nabla\right]\left(\frac{u_R^2}{2}\right) - 2\rho\Omega u_\phi u_R - \rho\frac{u_R u_\phi^2}{R} = -u_R\left(\frac{\partial P}{\partial R} + \rho\frac{\partial \Phi_{eff}}{\partial R}\right), \tag{120}$$

where

$$\Phi_{eff} = \Phi - \int^R s\Omega^2(s)\, ds. \tag{121}$$

Exactly the same manipulations with the ϕ and z equations produce

$$\rho\left[\frac{\partial}{\partial t} + \boldsymbol{v}\cdot\nabla\right]\left(\frac{u_\phi^2}{2}\right) + \rho u_\phi u_R\frac{\kappa^2}{2\Omega} + \rho\frac{u_R u_\phi^2}{R} = -u_\phi\frac{\partial P}{R\partial\phi} \tag{122}$$

and

$$\rho\left[\frac{\partial}{\partial t}+\boldsymbol{v}\cdot\nabla\right]\left(\frac{u_z^2}{2}\right)=-u_z\left(\frac{\partial P}{\partial z}+\rho\frac{\partial\Phi_{\mathit{eff}}}{\partial z}\right), \tag{123}$$

where we have assumed that Φ is independent of ϕ. Adding the three dynamic equations and using mass conservation lead to

$$\frac{\partial}{\partial t}\left(\frac{1}{2}\rho u^2\right)+\nabla\cdot(\rho u^2\boldsymbol{v}/2)+\rho u_R u_\phi\frac{d\Omega}{d\ln R}=-\boldsymbol{u}\cdot\nabla P-\rho\boldsymbol{u}\cdot\nabla\Phi_{\mathit{eff}}, \tag{124}$$

where $u^2=u_R^2+u_\phi^2+u_z^2$. Yet another use of mass conservation and a regrouping of the pressure term give us

$$\frac{\partial}{\partial t}\left(\frac{1}{2}\rho u^2+\rho\Phi_{\mathit{eff}}\right)+\nabla\cdot\left(\boldsymbol{v}(\frac{1}{2}\rho u^2+\rho\Phi_{\mathit{eff}})+P\boldsymbol{u}\right) \\ =P\nabla\cdot\boldsymbol{u}-\rho u_R u_\phi\frac{d\Omega}{d\ln R}. \tag{125}$$

We have already at hand the main structure of our energy equation. The left side is in conservation form with a well-defined energy density and energy flux, with the fluctuations isolated. (Notice that the azimuthal average of the energy equation, which we ultimately will be working with, has no rotational terms in the energy-flux divergence.) Sources of energy fluctuations are work done by pressure (which may cause heating or cooling) and the all-important $R\phi$ stress coupling to the differential rotation.

5.2. The Effects of Magnetic Fields

When magnetic fields are included, everything proceeds as before, except that the right side of eq. (124) contains the terms

$$-\boldsymbol{u}\cdot\nabla P_{tot}-\rho\boldsymbol{u}\cdot\nabla\Phi_{\mathit{eff}}+\frac{\boldsymbol{u}\cdot[\boldsymbol{B}\cdot\nabla]\boldsymbol{B}}{4\pi}. \tag{126}$$

The final magnetic term in expression (126) may be written in an index notation as

$$\frac{u_j B_i \partial_i B_j}{4\pi}, \tag{127}$$

where i, j, and k take the values x, y, and z, and repeated indices are summed over. The term ∂_i denotes the partial derivative with respect to the ith spatial variable.

To make further progress, we need the induction equation:

$$\left(\frac{\partial}{\partial t}+\boldsymbol{v}\cdot\nabla\right)\boldsymbol{B}=-\boldsymbol{B}\nabla\cdot\boldsymbol{v}+(\boldsymbol{B}\cdot\nabla)\boldsymbol{v}. \tag{128}$$

Take the dot product of this with $\boldsymbol{B}$ and write the last term in component form:

$$\left(\frac{\partial}{\partial t}+\boldsymbol{v}\cdot\nabla\right)\frac{B^2}{2}=-B^2\nabla\cdot\boldsymbol{v}+B_j B_i\partial_i v_j. \tag{129}$$

Now

$$B_j B_i \partial_i v_j = \partial_i (B_i B_j v_j) - v_j \partial_i (B_j B_i) = \nabla \cdot (\boldsymbol{v} \cdot \boldsymbol{BB}) - R\Omega [\boldsymbol{B} \cdot \nabla \boldsymbol{B}]_\phi - u_j B_i \partial_i B_j, \tag{130}$$

where we have used $\partial_i B_i = 0$. Note that with the last term, we make contact with the energy equation terms (126). The subscript ϕ in the penultimate term denotes a vector component:

$$R\Omega [\boldsymbol{B} \cdot \nabla \boldsymbol{B}]_\phi = \Omega [\boldsymbol{B} \cdot \nabla (RB_\phi)] = \nabla \cdot (R\Omega \boldsymbol{B} B_\phi) - B_\phi B_R \frac{d\Omega}{d \ln R}. \tag{131}$$

Therefore, the right side of eq. (129) becomes

$$-B^2 \nabla \cdot \boldsymbol{v} + \nabla \cdot [(\boldsymbol{u} \cdot \boldsymbol{B})\boldsymbol{B}] + B_\phi B_R \frac{d\Omega}{d \ln R} - u_j B_i \partial_i B_j. \tag{132}$$

Combining this result with the left side of eq. (129), after some cancellation of terms, gives an expression for $u_j B_i \partial_i B_j$:

$$-u_j B_i \partial_i B_j = \left(\frac{\partial}{\partial t} + \Omega \frac{\partial}{\partial \phi} \right) \frac{B^2}{2} + \nabla \cdot \left(\boldsymbol{u} \frac{B^2}{2} - (\boldsymbol{u} \cdot \boldsymbol{B}) \boldsymbol{B} \right) + \frac{B^2}{2} \nabla \cdot \boldsymbol{u} - B_\phi B_R \frac{d\Omega}{d \ln R}. \tag{133}$$

Armed with this result, we return to (126) and make a substitution for the final term. The resulting energy equation can be simplified to

$$\frac{\partial}{\partial t} \left(\frac{1}{2} \rho u^2 + \rho \Phi_{\mathit{eff}} + \frac{B^2}{8\pi} \right) + \nabla \cdot \left[\boldsymbol{v} \left(\frac{1}{2} \rho u^2 + \rho \Phi_{\mathit{eff}} \right) + P\boldsymbol{u} + \frac{\boldsymbol{B} \times (\boldsymbol{u} \times \boldsymbol{B})}{4\pi} \right] + \nabla \cdot \left(\boldsymbol{e_\phi} R\Omega \frac{B^2}{8\pi} \right) = P \nabla \cdot \boldsymbol{u} - \rho (u_R u_\phi - v_{AR} v_{A\phi}) \frac{d\Omega}{d \ln R}. \tag{134}$$

The pressure term on the right side of eq. (134) can be eliminated by using the thermal energy equation

$$\frac{3\rho}{2} \left(\frac{\partial}{\partial t} + \boldsymbol{v} \cdot \nabla \right) \frac{P}{\rho} = -P \nabla \cdot \boldsymbol{u} - \rho \mathcal{L}, \tag{135}$$

(since $\nabla \cdot \boldsymbol{v} = \nabla \cdot \boldsymbol{u}$). We have introduced the radiative energy loss term per unit volume $\rho \mathcal{L}$. Carrying through the elimination of the pressure term and averaging over ϕ lead to the total energy equation:

$$\frac{\partial \mathcal{E}}{\partial t} + \nabla \cdot \mathcal{F} = -\rho (u_R u_\phi - v_{AR} v_{A\phi}) \frac{d\Omega}{d \ln R} - \rho \mathcal{L} \qquad (\phi \text{ averaged}), \tag{136}$$

where $\mathcal{E}$ is the energy density in fluctuations,

$$\mathcal{E} = \frac{1}{2} \rho u^2 + \rho \Phi_{\mathit{eff}} + \frac{B^2}{8\pi} + \frac{3P}{2}, \tag{137}$$

and $\mathcal{F}$ is the corresponding flux,

$$\mathcal{F} = \boldsymbol{u}\left(\frac{1}{2}\rho u^2 + \rho\Phi_{\mathit{eff}}\right) + \frac{5P}{2}\boldsymbol{u} + \frac{\boldsymbol{B}\times(\boldsymbol{u}\times\boldsymbol{B})}{4\pi}. \tag{138}$$

We have completely ignored dissipation effects (viscosity and resistivity). What effect would these have on our final eq. (136)? The answer is essentially none. Although it is true that new dissipation terms would appear on the right side of eq. (134), they would also appear with the opposite sign in eq. (135). They would completely cancel on the right side of the final eq. (136): dissipation is not a loss of energy but a conversion of mechanical or magnetic-field energy into heat. Additional small energy-flux terms would appear (within the divergence operator) in connection with viscosity and resistivity, but these are generally negligible compared with the dynamic terms. The essential physical point is that total energy is conserved in the presence of dissipation even if mechanical energy is lost. Only radiative processes can remove energy from the disk.

We interpret eq. (136) as saying that energy is exchanged with the differential rotation at a volumetric rate $-T_{R\phi}d\Omega/d\ln R$, where

$$T_{R\phi} = \rho(u_R u_\phi - v_{AR}v_{A\phi}) \tag{139}$$

is the dominant component of the stress tensor. The energy made available from the differential rotation may in principle remain in velocity fluctuations in the form of waves, but in a thin, cool disk it is more likely that this energy will be locally dissipated and subsequently radiated. This is because if the energy flux varies radially over a scale of order R itself, then the $T_{R\phi}$ source term on the right will be larger than any term on the left (by an amount of order $u/(R\Omega)$). The stress can be counterbalanced only by the radiative loss term;

$$-T_{R\phi}d\Omega/d\ln R = \rho\mathcal{L}. \tag{140}$$

Therefore, although $-T_{R\phi}d\Omega/d\ln R$ is itself a nondissipative source term, in these so-called local models, it works out to be the rate at which energy is dissipated (and ultimately lost by radiation) as well. Bear in mind that the identification of the stress term with dissipation follows from a reasonable assumption about the behavior of thin disks: local dissipation of the fluctuations. Dissipation is in no way a fundamental property of the left side of eq. (140), which under different conditions could just as well be an energy source or reserve for purely adiabatic waves. There are many papers in which this point is misunderstood; caveat lector.

6. Resistive and Hall Terms

6.1. Local Dispersion Relation

Extensive regions in real protostellar disks are in a regime far from ideal MHD, in which ohmic decay and Hall electromotive forces are important (e.g., Fig. 6.2). It is important to understand how the MRI behaves under these conditions.

The fastest-growing modes, as usual, correspond to axial wave numbers $\boldsymbol{k} = k\boldsymbol{e_z}$. Fortunately, these are also the simplest to treat analytically, and we will limit my discussion to this class of disturbance.

The linearly perturbed Hall term in the induction equation is to leading order in a WKB expansion

$$-\nabla\times\left[\frac{\delta \boldsymbol{J}\times \boldsymbol{B}}{en_e}\right] = \nabla\times\left[\frac{c\boldsymbol{B}}{4\pi\, en_e}\times(i\boldsymbol{k}\times\delta\boldsymbol{B})\right], \tag{141}$$

since $\delta\boldsymbol{J} = (c/4\pi)(i\boldsymbol{k}\times\delta\boldsymbol{B})$. Taking the curl and simplifying allows us to write the right side as

$$\frac{c}{4\pi\, en_e}(\boldsymbol{k}\cdot\boldsymbol{B})(\boldsymbol{k}\times\delta\boldsymbol{B}). \tag{142}$$

(In a fully ionized disk, the presence of n_e in the denominator renders these terms negligibly small.) For our particular problem, $\boldsymbol{k}$ is axial and $\delta\boldsymbol{B}$ lies in the disk plane, and expression (142) reduces to

$$\frac{ck^2 B}{4\pi\, en_e}(\delta B_R\boldsymbol{e_\phi} - \delta B_\phi\boldsymbol{e_R}). \tag{143}$$

Including the effects of resistivity, the linearized induction eqs. (82–83) become

$$(-i\omega + \eta k^2)\delta B_R + \frac{ck^2 B_z}{4\pi\, en_e}\,\delta B_\phi - i(\boldsymbol{k}\cdot\boldsymbol{B})\delta v_R = 0 \tag{144}$$

and

$$(-i\omega + \eta k^2)\delta B_\phi - \delta B_R\left(\frac{d\Omega}{d\ln R} + \frac{ck^2 B_z}{4\pi\, en_e}\right) - i(\boldsymbol{k}\cdot\boldsymbol{B})\delta v_R = 0, \tag{145}$$

and the z equation is not needed (it is trivially satisfied). The linearized dynamic and energy equations from § 4.3 remain unchanged.

The Hall electromotive force terms introduce a phase shift in the induction that leads to a circularly polarized component of what would otherwise be an Alfvén or slow mode. When an instability is present because of differential rotation, this phase shift can either destabilize or stabilize depending on the sign of B_z.

To see this more quantitatively, we must study the dispersion relation that follows from our linearized set of equations. To this end, we define a Hall parameter v_H that has the dimensions of a velocity (Balbus & Terquem 2001),

146
$$v_H^2 = \frac{\Omega B_z c}{2\pi e n_e}.$$

Notice that if we take the sense of rotation to be positive, v_H^2 can have either sign depending on whether B_z is positive or negative. With $\sigma = -i\omega$, the dispersion relation is

147
$$\sigma^4 + 2\eta k^2 + C_2\sigma^2 + 2\eta k^2(\kappa^2 + k^2 v_A^2)\sigma + C_0 = 0,$$

where the constants C_2 and C_0 are given by

148
$$C_2 = \kappa^2 + 2k^2 v_A^2 + \eta^2 k^4 + \frac{k^2 v_H^2}{4\Omega^2}\left(\frac{d\Omega^2}{d\ln R} + k^2 v_H^2\right)$$

and

149
$$C_0 = k^2\left(v_A^2 + \frac{\kappa^2 v_H^2}{4\Omega^2}\right)\left(\frac{d\Omega^2}{d\ln R} + k^2 v_A^2 + k^2 v_H^2\right) + \kappa^2\eta^2 k^4.$$

6.2. Stability

A sufficient condition for instability is that $C_0 < 0$. Compared with the purely ohmic requirement

150
$$k^2 v_A^2\left(k^2 v_A^2 + \frac{d\Omega^2}{d\ln R}\right) + \kappa^2\eta^2 k^4 < 0,$$

the Hall condition $C_0 < 0$ is much more easily satisfied, since v_H^2 can change sign. Perhaps most striking is the possibility that the intuitive result that large wave numbers are stabilized can be lost when $v_H^2 < 0$, and that all wave numbers can be unstable, even in the presence of finite resistivity (see Fig. 6.4).

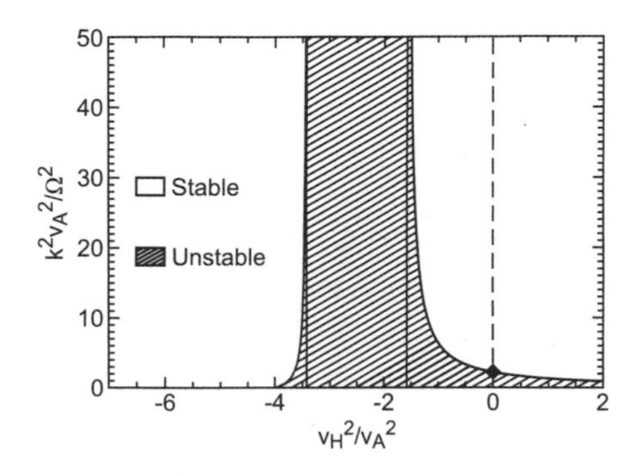

Fig. 6.4. Range of unstable wave numbers for a representative disk in the Hall-Ohm regime. The black diamond corresponds to the MRI without the Hall effect. The resistivity is chosen to be $\kappa^2\eta^2/v_A^4 = 0.35$. From Balbus & Terquem (2001).

This may be shown straightforwardly from the definition of C_0; the reader may wish to consult Balbus & Terquem (2001) for explicit details. In addition to requiring counteraligned magnetic and rotation axes, this state also demands a magnitude ratio of v_A^2 to v_H^2 that lies between about 2 and 4 (for a Keplerian disk). When the field is aligned with the rotation axis, the range of unstable wave numbers is more restricted than that found for the ideal-MHD MRI.

6.3. Numerical Simulations of the Hall-Ohm MRI

What does the preceding section imply for the MHD transport properties of protostellar disks? The first local simulations of the nonlinear development of the MRI in the presence of ohmic resistivity are those of Fleming, Stone, & Hawley (2000). These authors found the important result that the nonlinear turbulent state is easier to maintain when a mean vertical magnetic field is present. This is sometimes misunderstood as a requirement that a disk must have a global vertical magnetic field to maintain MRI turbulence, but in fact all that the very local calculations of Fleming et al. (2000) really tell us is that on this scale, working with a precisely zero mean field is probably too restrictive an approximation for a real disk. This study found a difference of two orders of magnitude in the critical magnetic Reynolds number (defined here as the product of the isothermal sound speed times the box size divided by the resistivity) needed to sustain turbulence, depending on whether a vertical field was present or not; the vertical field runs were easier to sustain. It is now understood that very high-resolution grids are needed for the local study of the MRI in a shearing box with zero mean field, so these early results should not be used quantitatively. With this precaution understood, it is interesting to note that for the vertical field runs, the Fleming et al. finding seems to be not very different from the $\mathbb{L} = 1$ criterion discussed earlier.

This nature of the nonlinear behavior of a Hall fluid was examined in a numerical investigation by Sano & Stone (2002a, 2002b). These authors carried out an extensive study of the local properties of Hall-MRI turbulence in the shearing-box formalism. The results are somewhat surprising.

In linear theory, counteraligned magnetic and rotation axes produce a very broad wave-number spectrum of instability and should be more unstable than the aligned case. What Sano and Stone found, however, was just the opposite: the aligned case showed greater levels of field coherence and higher rms fluctuations than the counteraligned case. The reason for this is interesting and illustrative of the hazards of extrapolating directly from linear theory.

In the case of counteraligned axes, the extended wave-number spectrum of linear MRI instability results, in its nonlinear resolution, in a highly efficient turbulent cascade to smaller and smaller scales. In a finite-difference numerical simulation, this cascade terminates at the grid scale, where energy is ultimately lost. By contrast, the aligned case shows instability only at longer wavelengths (small wave numbers). At larger wave numbers, the response of the fluid is wavelike, and this is likely to make a local small-scale turbulent cascade considerably less efficient. The result is a less lossy system and greater large-scale field coherence.

Finally, when simulations were done with half the field aligned with the rotation axis and half counteraligned, the Hall effect tended to wash out, and the results were similar to resistive MHD turbulence without the additional electromotive forces. The criterion for the onset of turbulence depended only on the relative strength of the resistivity, as measured by the dimensionless number $v_A^2/\eta\Omega$. This Lundquist number must be greater than unity, or MRI-induced turbulence is not maintained, whether Hall electromotive forces are present or not, according to Sano & Stone. The presence of the Hall terms in the induction equation does, however, promote a slightly elevated rms level when turbulence can be maintained.

The Hall numerical studies are fascinating and suggestive, but caution is necessary before taking the final leap from simulation to real disks. Note, for example, the role played by the numerical grid scale losses in our discussion of the nonlinear resolution of the counteraligned case. Nature works without a grid, however, and it is not clear what scale ultimately intervenes at high wave numbers in Hall protostellar disks. Moreover, it is now understood that there are important resolution questions for local shearing-box simulations with no mean field that have only recently been brought to light (Fromang et al. 2007; Lesur & Longaretti 2007). The important pioneering studies of Sano & Stone (2002a, 2002b) merit a thoroughgoing follow-up on the much larger numerical grids that are currently available to numericists.

7. Alpha Models of Protostellar Disks

7.1. Basic Principles

We are now in a position to assemble a basic "α model" (Shakura & Sunyaev 1973) of a protostellar disk. An important assumption in such models is that angular momentum and mechanical energy are transported radially through the disk. We therefore work with height- and azimuthally integrated quantities. Under these conditions, Balbus & Papaloizou (1999) have shown how the fundamental equations of MHD lead directly to the α formalism.

The velocities consist of a mean part plus a fluctuation of zero mean. For the azimuthal velocity $v\Phi$, the mean will be taken to be the Keplerian velocity

$$(R\Omega)^2 = \frac{GM}{R}, \quad (151)$$

where M is the central mass and G is the Newtonian constant. For the radial velocity v_R, the mean motion corresponds to the very slow inward accretion drift, v_2. The azimuthal fluctuation velocity u_ϕ is much less than $R\Omega$, whereas the radial fluctuation velocity u_R is much larger than the inward drift. These claims will shortly be quantified. The magnetic-field Alfvén velocities are assumed to be of the same order as the u velocities, namely, less than or of order of the isothermal thermal sound speed c_S. To summarize,

$$v_2 \ll u, v_A, c_S \ll R\Omega. \quad (152)$$

Under steady-state conditions, the local disk-accretion rate has a time-averaged value of

$$\mu \equiv 2\pi R\langle \rho v_2 + \delta\rho\, \delta u_R \rangle. \quad (153)$$

The height-integrated form of this is $-\dot{M}$, the mass-accretion rate:

$$\int \mu\, dz = -\dot{M}, \quad (154)$$

defined so that $\dot{M}$ is a positive constant. A useful average of v_2 is the drift velocity defined by

$$v_d \equiv -\frac{\dot{M}}{2\pi R\Sigma}, \quad (155)$$

where Σ is the disk surface density.

In steady state, the height-integrated average radial angular momentum flux is proportional to $1/R$. In other words,

$$-\dot{M}R^2\Omega + 2\pi \Sigma R W_{R\phi} = C/R, \quad (156)$$

where $W_{R\Phi}$ is defined by

$$\int \langle \rho u_R u_\phi - \rho v_{AR} v_{A\phi} \rangle\, dz = \Sigma W_{R\phi}, \quad (157)$$

and C is a constant to be determined. Traditionally, this has been done by asserting that $W_{R\phi}$ is proportional to the radial gradient of Ω, and at some point before the surface of the star is reached this must vanish. The constant C is then determined at this point (Pringle 1981).

This approach now seems dated and especially inappropriate for a protostellar disk in which the magnetic interactions between the disk and the star may become quite complex at small radii. Instead, let us note that at small R, C/R must be very small if $\Sigma W_{R\phi}$ does not blow up, which seems quite reasonable. If C/R is very small at the inner edge of the disk, it is clearly negligible in the body of the disk, and we shall set $C = 0$ with the understanding that this solution should not be taken to the stellar surface. Then (156) reduces to

158 $$v_d = -W_{R\phi}/R\Omega,$$

which shows that v_d (and v_2, which is of the same order) is very small compared with the u velocities. Alternatively,

159 $$\Sigma W_{R\phi} = \dot{M}R\Omega/2\pi.$$

This is a particularly useful result because $W_{R\phi}$ appears in the energy-balance eq. (140):

160 $$-\Sigma W_{R\phi}\frac{d\Omega}{d\ln R} = -\frac{d\Omega}{d\ln R}\int T_{R\phi}\,dz = \int \rho\mathcal{L}\,dz = 2\sigma T_{eff}^4.$$

In the last equality, we have equated the radiated energy per unit surface of the disk to twice the blackbody emissivity; the factor of 2 represents two radiating surfaces. Here σ is the Stefan-Boltzmann constant, and T_{eff} is the effective blackbody surface temperature. Combining (159) and (160) then gives

161 $$T_{eff}^4 = \frac{3GM\dot{M}}{8\sigma\pi R^3},$$

relating the potentially observable disk surface temperature to the central mass, accretion rate, and radial location. The unknown turbulent-stress parameter $W_{R\phi}$ has conveniently vanished.

Eq. (161) is in a restricted sense "exact." It is based on the assumption of time-steady conditions and thermal radiation, but it is independent of the explicit nature of the turbulence as long as it is local. If these conditions are all met, eq. (161) is a simple matter of energy conservation. To go beyond this result, something has to be said directly about $W_{R\phi}$. In α disk theory, that condition is

162 $$W_{R\phi} = \alpha c_S^2,$$

where α is assumed to be constant but otherwise unknown. In contrast to the simple and plausible assumption that turbulence is locally dissipated, this "α assumption" is far from obvious and in a strict mathematical sense almost certainly wrong.

The original justification for this form of the stress tensor was based on the notion of hydrodynamic turbulence and the idea that the velocity fluctuations would be restricted to some fraction of the sound speed (fluctuations in excess would cause shocks). Shakura & Sunyaev (1973) consider magnetic stresses, however, and argue that they fit within the α formalism as well since Alfvén velocities in excess of c_S are also dynamically unlikely.

The real problem with prescription (162) is that turbulence is just too complicated. Not only are long-term averages very hard to define (a problem even for a result like eq. [161]), but it is also entirely possible that α could vary in a complex nonsystematic way by an order of magnitude or more from one part of the disk to another. The primary justification for (161) is that many scaling results are very insensitive to α. On dimensional grounds c_S is is certainly an important local characteristic velocity, but it is not yet clear under what conditions a "background" magnetic field might also be providing a mean Alfvén velocity that limits or guides the turbulence.

Continuing with our α model, to link the surface temperature T_{eff} with the midplane temperature T (used in quantities like c_S), we need to introduce explicit vertical structure into the problem. The hydrostatic-equilibrium equation is

$$-\frac{\partial P}{\partial z} = \frac{GM\rho z}{R^3} = \rho z \Omega^2. \tag{163}$$

This may be solved in conjunction with a detailed energy equation, but it seems best for illustrative purposes to note that this equation serves to provide a decent and very simple estimate of the disk half-thickness, $H = c_S/\Omega$. The energy equation for simple vertical radiative diffusion defines the local radiative energy flux F_γ as

$$F_\gamma = \frac{4\sigma}{3}\frac{dT^4}{d\tau}. \tag{164}$$

The optical depth τ is given by

$$d\tau = -\rho\kappa\, dz, \tag{165}$$

where κ is the opacity of the disk (units: cross-sectional area per unit mass). In the simplest possible model, F_γ is a constant given by σT_{eff}^4, and the temperature at the midplane T is then

$$T^4 = \frac{3\tau T_{eff}^4}{4}, \tag{166}$$

where τ is the optical depth from the outer disk surface to the midplane. (In this simplest of all possible models, we set $\tau = \kappa H$, where κ is evaluated at the midplane temperature.)

The surface temperature of a disk with a mass-accretion rate of $10^{-8} M_\odot$ per year around a $1 M_\odot$ star is $85 R_{AU}^{-0.75}$ K, where R_{AU} is the radius in astronomical units. The midplane temperature T is a factor $\sim\tau^{1/4}$ larger, typically a factor of 5 or so larger. At what value of T would we expect the ionization fraction to reach the critical value of 10^{-13} we found earlier, corresponding to a Lundquist number of unity? Put in somewhat different terms, is our model of self-sustaining MHD turbulence self-consistent?

To answer this, we need to address the physics of thermal ionization. In the low-ionization regime in which we are working, thermal electrons are supplied by trace alkali elements, notably potassium, with an ionization potential of only 4.341 eV. Even this modest value corresponds to an effective temperature of 50,375 K, well above the range of 10^2 to 10^3K we expect, and potassium will barely be ionized.

The equation governing the ionization fraction x is the Saha equation.[4] In this barely ionized regime, it may be written (Stone et al. 2000) as

167
$$x^2 = as\left(\frac{2.4 \times 10^{15}}{n}\right) T^{3/2} \exp\left(-50,375/T\right),$$

where a is the abundance of potassium (10^{-7}), n is the ratio of the dominant H_2 molecules, and s is a ratio of statistical weights, expected to be near unity. We may rewrite this equation as

168
$$T = -\frac{50,375}{\ln X},$$

where

169
$$X = x^2 T^{-3/2}(n/2.4 \times 10^{15} as).$$

When $x = 10^{-13}$ and the final density factor is anything between 0.01 and 1 ($0.1 M_\odot$ of gas spread out over a region of 10 AU yields a density of about 5×10^{13} per cc), values for T are close to 10^3 K, which we will adopt as a working number.

Therefore, a surface temperature of about 200 K is required to attain a midplane temperature of 10^3 K and thereby ensure a critical level of ionization. This is the key issue for MHD theories of protostellar disks: beyond $0.2 - 0.3$ AU, or $\sim 3 \times 10^{12}$ cm, the heat generated by the dissipation associated with local turbulence is insufficient to keep the disk magnetically well coupled. Where and how do protostellar disks maintain good magnetic coupling?

4. The discussion presented here presumes that the ionization reaction rates are sufficiently rapid to keep up with the dynamic timescales of any turbulence present. At low-ionization fractions this breaks down, and a single temperature may not be enough to characterize the ionization state (Pneuman & Mitchell 1965).

8. Ionization Models of Protostellar Disks

8.1. Layered Accretion

The results of the previous section suggest that dissipative heating from MHD turbulence self-consistently generates enough heat to maintain the minimum thermal levels of requisite ionization only within a few 0.1 AU. Nonthermal sources of ionization are therefore of great interest to protostellar disk theorists. The principal ionizing agents that have been studied are cosmic rays, X-rays, and radioactivity.

Gammie (1996) argued that the low-density extended vertical layers of a protostellar disk would be exposed to an ionizing flux of interstellar cosmic rays. Just as in models of molecular cloud ionization, cosmic rays would maintain a minimal level of ionization in the upper disk layers. The range of the low-energy galactic cosmic rays is about 100 g/cm^2. This is much less than the disk column density at $\sim$ 1 AU in generic solar nebula models but can easily exceed the disk column at larger distances. If the level of ionization is high enough—and the Alfvén velocity v_A is small enough—the gas within the range of the cosmic rays will be MRI active. (The Alfvén velocity cannot be too large if disturbances are not to be stabilized by magnetic tension.) If these criteria are met, Gammie argued that turbulent accretion would proceed in the outer layers of protostellar disks, but that the midplane regions would remain laminar—in effect, a "dead zone." This is the basis of the concept of layered accretion, which has become an important idea in protostellar disk modeling (Sano et al. 2000).

Gammie's original construction was based on the assumptions that the accretion occurs in a layer of fixed columns, and that α is constant. Taken together, these assumptions preclude a steady solution; the mass flux rate is not independent of position. Instead, matter is deposited from the outer regions into the disk's inner regions, where it builds up. At some point in this scenario the disk becomes gravitationally unstable, and it was speculated that an accretion outburst might occur, which was tentatively identified with FU Orionis behavior. On the other hand, there is no compelling argument (beyond mathematical convenience) to prevent the α parameter from adjusting with position if this allows a relaxation to a time-steady solution. The overarching layered-accretion picture would, however, remain intact with this modification.

Young Stellar Objects (YSOs) are almost universally X-ray sources. Glassgold, Feigelson, & Montmerle (2000) noted that X-rays are potentially a far more powerful ionization source, even they are if attenuated, than galactic cosmic rays. These authors were thus the first to draw the link between X-ray observations and MRI activity of accretion disks. In the Glassgold et al. formulation, X-rays from a locally extended corona centered on the YSO irradiate the disk and, depending on the chosen parameters, could provide the requisite ionization levels to lessen the extent of the dead zone or eliminate it altogether.

Whether the dead zone persists in the presence of X-ray irradiation depends, among other things, on the model adopted for the disk structure. Generally, the so-called minimum-mass solar nebula model (Weidenschilling 1977) is used. This is a reconstruction of the surface-density distribution of the Sun's protostellar disk based on current planetary masses and compositions and results in an $R^{-3/2}$ radial distribution. Fromang, Terquem, & Balbus (2002) suggested that ionization fractions should be calculated self-consistently within the framework of accretion-disk theory, which in general leads to a more shallow dependence of surface density on radius. If were introduce accretion parameters whose values are free (α and the mass-accretion rate $\dot{M}$), the range of parameter space increases, and the presence and extent of a dead zone become yet more model dependent.

The ionization fraction of a protostellar disk is a chemical problem and in principle can involve a very complex, uncertain, molecular reaction network. Ilgner & Nelson (2004a) investigated the consequences for the dead zone of a considerably richer chemical network. Although the quantitative details were sensitive to the chemistry, the qualitative structure was not. A dead zone is likely to be present in any plausible chemical network that has been studied up to the present, but there are conditions in which its extent can be very small or possibly even zero. Ilgner & Nelson (2004b), for example, made the interesting comment that flaring activity by the central star can have a significant effect on the extent of the dead zone.

8.2. Activity in the Dead Zone

Just because the dead zone is unable to host the MRI does not mean that it is well and truly dead. Fleming & Stone (2003) carried out numerical simulations in which the magnetically active upper layers of the disk coupled dynamically to the magnetically inert dead zone, resulting in a small but significant Reynolds stress, even though the Maxwell stress was zero. More recently, Turner & Sano (2008) suggested that the high resistivity characteristic of the dead zone is also a means for a magnetic field to diffuse into this region from the active layers. Moreover, these authors point out, an active MRI region is not an absolute prerequisite for accretion: direct magnetic torques on much larger scales would also serve, and since they do not dissipate energy in a turbulent cascade, they are much less costly to maintain.

Terquem (2008) constructed global models of protostellar disks based on the α prescription. The presence of a dead zone was surprisingly nondisruptive, provided that α was not too small. Steady solutions were found for values of α in the dead zone as small as 10^{-3} times the active zone value. In these models, the dead zone was thicker and more massive than its surroundings, but because of the relatively weak α scalings, by less than an order of magnitude. These results, taken as a whole, suggest that an embedded dead zone in a protostellar disk is not necessarily disruptive to the accretion process.

8.3. Dust

In all of the discussions, I have been very negligent by not mentioning the effects of small dust grains (e.g., Sano et al. 2000). Let us see crudely why this is so.

Consider a mass M of protostellar disk gas, of which $10^{-2}M$ is in the form of spherical dust grains of radius r_d. If the dust grains have a density of 3 gm cm^{-3}, there are a total of

$$N_d = (10^{-2}/4\pi)(M/r_d^3)$$

grains in a volume V. The grains present a geometric cross section of $\sigma_d = \pi r_d^2$ (we ignore here the enhancement due to the induced charge [Draine & Sutin 1987]), and the electrons have an average radial velocity of some $v_e = 1.6(kT/m_e)^{1/2}$. The total dust recombination rate per unit volume is

$$(N_d/V)n_e\sigma_d v_e, \tag{170}$$

where n_e is the number density of electrons. This should be compared with a typical dielectronic recombination rate of $\beta \equiv 8.7 \times 10^{-6}\ \text{cm}^3\ \text{s}^{-1}$(Glassgold, Lucas, & Omont 1986; Gammie 1996). If x is the ionization fraction, then the ratio of dielectronic recombination to dust recombination is

$$\frac{n_e\beta V}{N_d\sigma_d v_e} \sim \left(\frac{x}{10^{-13}}\right)\frac{r_d}{T}, \tag{171}$$

where r_d is in cm and T is in K. For small dust grains ($\sim 10^{-5}$cm), this ratio is $\ll 1$, and dust recombination is overwhelmingly important. But the relative importance of the grains diminishes as they grow in size, especially in the cooler portions of the disk. Sano et al. (2000) concluded that in their model (based on cosmic-ray ionization) a typical protostellar disk would be MRI stable inside about 20 AU if small grains were present, except for the innermost regions, which would be thermally ionized.

The detailed effects of dust grains have been considered by many authors; a very good review and list of references is given in the recent paper of Salmeron & Wardle (2008). These authors have considered the vertical structure of the magnetic coupling in a minimum-mass solar nebula model, including the effects of Hall electromotive forces and ambipolar diffusion. They employed a sophisticated chemical network (Nishi, Nakano, & Umebayashi 1991) and small dust grains ranging in radius from 0.1 to 3 microns. At the fiducial locations of 5 and 10 AU, Salmeron & Wardle found good magnetic coupling over an impressive range of magnetic-field strengths provided for the 3 micron grains, a considerable improvement from the naive estimate above. For the 0.1 micron grains, the magnetic coupling dropped sharply (as did MRI growth rates) and was restricted to higher elevations above the disk plane.

9. Summary

Protostellar disks are gaseous systems dynamically dominated by Keplerian rotation. These disks are accreting onto the central protostar, so there must be a source of enhanced angular momentum transport present. In the early stages of the disk's life, this enhanced transport may well be due to self-gravity, with density waves largely responsible for moving angular momentum outward. Once the disk becomes observable, its mass is below the minimum needed to sustain self-gravitational spiral wave transport. The only established mechanism able to sustain high levels of angular momentum transport is MHD turbulence produced by the magnetorotational instability, or MRI.

The dynamic effects of magnetic fields on protostellar disks depend crucially on the degree of ionization, and more specifically on the electron fraction, that is present. This means that there is a direct link between the gross dynamic behavior of a protostellar disk and its detailed chemical profile. In principle, the full multifluid nature of the disk gas—neutrals, ions, electrons, and dust grains—must be grappled with at some level to elicit and understand the disk structure. Fortunately, only a very small ionization level is needed to couple the charged and neutral components, leaving (in essence), a single magnetized fluid. An electron fraction of 10^{-13} is a typical fiducial number for the threshold of magnetic coupling at 1 AU in the T Tauri phase of the Solar Nebula: roughly one electron per cubic millimeter of disk gas. Unfortunately, however, on scales of 10s of AUs, near the dense midplane the ionization of protostellar disks may not rise even to this minimal level of ionization. At the boundary between good and poor magnetic coupling, the MHD processes are complex, not well understood, and the domain of ongoing inquiry.

In regions of the disk where the magnetic coupling is sufficient, the combination of a weak (subthermal) magnetic field and Keplerian rotation leads to the MRI. In ideal MHD, the magnetic field behaves as though it were frozen into the conducting gas, and the presence of differential rotation produces an azimuthal magnetic field from a radial magnetic field. But this is not all that happens. If one tries to simulate this simple process on a computer, the laminar flow breaks down into a turbulent mess, even if the field is very weak.

The reason for this is another classical property of magnetic fields, that the force exerted by the lines of force on the background gaseous fluid is the sum of a pressurelike term and a tensionlike term. It is the latter tensionlike term that is important for an understanding of the MRI. When two nearby fluid elements are moved apart, even if only because of a random perturbation, the magnetic tension force acts precisely like a spring coupling the two masses. The fluid element on the inside rotates faster than the element on the outside and tries to speed it up. The outer element, in acquiring angular momentum from the inner element, finds itself too well endowed and spirals outward toward a higher orbit, where its excess angular momentum can be accommodated. On the other hand, the inner element, having lost angular momentum, finds itself at a deficit and must drop to a lower angular momentum orbit. This separation stretches the field lines that couple the elements, the magnetic tension goes up, and the process runs away. This is the MRI. Notice that angular momentum transport is at the very core of the linear instability, rather than the result of some sort of nonlinear mixing process.

But the MRI does, in fact, lead to rapid turbulent mixing of the disk gas as outwardly moving and inwardly moving fluid elements encounter one another and dissipate their energy. If this happens locally, then the ingredients for a classical α model of accretion are present. To the extent that many disk features are insensitive to, or independent of, the precise rms level of the disk fluctuations, these models can be of some practical utility. The classical formula relating the disk emissivity to the radius R is the most important instance of this.

In real protostellar disks, the level of ionization present is such that there are significant departures from the behavior of an ideal (perfectly conducting) MHD gas. In principle, the gas can become completely decoupled from the magnetic field and go over to a hydrodynamic system. Less dramatically, the current-bearing electrons can acquire a velocity significantly different from the dominant neutrals, since the former need to maintain a current density and magnetic field even as charged species become rarer and rarer. When this happens, the field lines are no longer frozen into the motion of the bulk of the (neutral) disk gas; they are frozen into the electrons, and the distinction becomes important. In the so-called Hall regime, the ions and neutrals move together distinctly from the electrons, and the ion motion relative to the electron-following field lines induces an additional electromotive force in the gas, beyond the self-induction responsible for simple field-line freezing. Hall MHD is likely to be important in protostellar disks on scales of 1 AU to 10s of AUs.

Dust grains are an ever-present complication for MHD disk modeling. Typically, solids make up about 1% of the mass of interstellar gas. An interstellar population of small grains (radius $\sim 10^{-5}$ cm) with a total mass fraction of a 1% would present an enormous collecting area of would-be gas-phase electrons. Putting charges on the grains effectively removes them because of the low mobility of the grains. Determining the disposition of the dust grains is thus a necessity for constructing MHD disk models. As the disk evolves, so too do the dust grains. They grow in size as they agglomerate, and they tend to settle toward the midplane if they are not stirred by turbulence or some other dynamic process. Larger dust grains are much less efficient in removing gas-phase electrons than are smaller grains, where "larger" means growth in radius of an order of magnitude or more.

Where does all this leave us? Clearly, we are a long way from framing a picture of protostellar disk evolution at the level of, say, classical stellar evolution. But for all the gaping uncertainties, a useful zeroth-order MHD-based picture of a protostellar disk in its T Tauri phase can be cobbled together.

Inside about 0.3 AU, a protostellar disk will be thermally ionized by direct exposure to the central source and self-consistently by the dissipation of MHD turbulence. We may expect vigorous MRI-induced MHD turbulence in this zone.

On scales of AUs, thermal ionization is no longer adequate to maintain the requisite levels of ionization to ensure MHD coupling, at least not near the high-density midplane. Depending on the X-ray luminosity of the central star, the spectrum of dust grains, and the abundance of gas-phase potassium and sodium (electron donors), there could be good coupling at low-density, higher-elevation disk altitudes. The dead zone at lower altitudes need not be devoid of all transport; density waves from an adjacent active layer or large-scale magnetic torques would each contribute their own stresses. It is even possible that some degree of coupling could be maintained down to the disk midplane, although this depends strongly on modeling assumptions. It seems likely, however, that on a scale of 1 AU to 10s of AUs, the level of MHD turbulence will be far less than in the disk's innermost regions.

Finally, on scales larger than 10s of AUs, the falling density lengthens recombination times and suggests a return to good MHD coupling. The typical size of a disk is many 100s of AUs, so the dead zone is small when viewed globally and may not have much of an impact in the overall accretion process (Terquem 2008). Perhaps, however, it is not a coincidence that the MHD "quiet zone" coincides with the region of planet formation in the Solar Nebula and (more speculatively) in other protostellar disks.

Appendix A

Begin with eq. (16):

$$\frac{\boldsymbol{J}}{c} \times \boldsymbol{B} = \boldsymbol{p}_{In} + \boldsymbol{p}_{en}, \tag{172}$$

where the right side of this equation is

$$n_I n \mu_{In} \langle \sigma_{nI} w_{nI} \rangle (\boldsymbol{v}_I - \boldsymbol{v}) + n_e n m_e \langle \sigma_{ne} w_{ne} \rangle [(\boldsymbol{v}_e - \boldsymbol{v}_I) + (\boldsymbol{v}_I - \boldsymbol{v})]. \tag{173}$$

In what follows, we will often need an estimate of the ratio of the ion and electron collision rates. Following the discussion in § 3, we will assume that

$$\frac{\langle \sigma_{nI} w_{nI} \rangle}{\langle \sigma_{ne} w_{ne} \rangle} = \left(\frac{m_e}{\epsilon \mu_{In}} \right)^{1/2}, \tag{174}$$

where $\epsilon < 1$ is inserted because the electron-neutral cross section is geometric, while the ion-neutral collision cross section is larger than geometric.

In (173), the first $(\boldsymbol{v}_I - \boldsymbol{v})$ term dominates the last by a factor of order $(\mu_{In}/\epsilon m_e)^{1/2}$. Hence

$$\boldsymbol{p}_{In} = \frac{\boldsymbol{J}}{c} \times \boldsymbol{B} - \boldsymbol{p}_{en} = \frac{\boldsymbol{J}}{c} \times \boldsymbol{B} + \frac{n m_e}{e} \langle \sigma_{ne} w_{ne} \rangle \boldsymbol{J}. \tag{175}$$

With the help of eq. (10), this implies that

$$\boldsymbol{v}_I - \boldsymbol{v} = \frac{\boldsymbol{J} \times \boldsymbol{B}}{c \gamma \rho \rho_I} + \underbrace{\sqrt{\frac{\epsilon m_e}{\mu_{nI}}} Z(\boldsymbol{v}_I - \boldsymbol{v}_e)}_{A}. \tag{176}$$

The final term has been marked with an "A" for future reference.

Recall eq. (20):

$$\begin{aligned} \boldsymbol{E} + \frac{1}{c} \left[\boldsymbol{v} + (\boldsymbol{v}_e - \boldsymbol{v}_I) + (\boldsymbol{v}_I - \boldsymbol{v}) \right] \times \boldsymbol{B} \\ + \frac{m_e \nu_{en}}{e} \left[(\boldsymbol{v}_e - \boldsymbol{v}_I) + (\boldsymbol{v}_I - \boldsymbol{v}) \right] = 0. \end{aligned} \tag{177}$$

In substituting eq. (176) for $\boldsymbol{v}_I - \boldsymbol{v}$ in (177), we may always drop the A term, since it is small compared with $\boldsymbol{v}_e - \boldsymbol{v}_I$.

Proceeding with the above substitution leads to

$$E + \frac{v}{c} \times B - \frac{J \times B}{en_e c}\Big[1 - \underbrace{\frac{m_e \nu_{en} n_e}{\gamma \rho \rho_I}}_{B}\Big] + \frac{(J \times B) \times B}{c^2 \gamma \rho \rho_I} - \frac{J}{\sigma_{cond}} = 0, \tag{178}$$

where σ_{cond} is defined in eq. (24). The B term in this equation may now clearly be dropped: it is of order $Z(\epsilon m_e \mu_{nI})^{1/2}/m_n$. This leads to

$$E + \frac{v}{c} \times B - \frac{J \times B}{en_e c} + \frac{(J \times B) \times B}{c^2 \gamma \rho \rho_I} - \frac{J}{\sigma_{cond}} = 0, \tag{179}$$

which is precisely eq. (23) in the text.

Appendix B

To estimate the order of departures from charge neutrality or the displacement currents, we will assume that the ∇ operator is $\sim 1/L$, where L is a characteristic length scale of the flow, and $\partial/\partial t$ is $\sim v/L$, where v is a characteristic velocity (say, the largest of the neutral, ion, or electron velocities).

For the electric field, I take $E \sim vB/c$, since we are interested only in problems where the inductive terms are comparable to, or larger than, the resistive damping. Then

$$\nabla \cdot E \sim E/L \sim vB/Lc \sim 4\pi vJ/c^2 \sim 4\pi e n_e v^2/c^2. \tag{180}$$

Hence the divergence of E is of order v^2/c^2 times the electron charge density (at most, since we assumed $v_I - v_e \sim v$ in the above). It may thus be ignored.

For the displacement current, the demonstration is almost a matter of direct inspection. If we take the curl of eq. (29) and use eq. (27), then on the right side of (29) the first term is $\sim B/L^2$, while the second, displacement, term is $\sim v^2 B/c^2$. It may thus be ignored.

References

Balbus, S. A. 2001, ApJ, 562, 909.

Balbus, S. A. 2003, ARAA, 41, 555.

Balbus, S. A., & Hawley, J. F. 1991, ApJ, 376, 214.

Balbus, S. A., & Hawley, J. F. 1992, ApJ, 392, 662.

Balbus, S. A., & Hawley, J. F. 1998, Rev. Mod. Phys., 70, 1.

Balbus, S. A., & Hawley, J. F. 2000, in From Dust to Terrestrial Planets, ed. W. Benz, R. Kallenbach, G. Lugmaier, & F. Podosek, ISSI Space Sciences Series No. 9, Space Science Reviews, 92, 39.

Balbus, S. A., & Papaloizou, J. C. B. 1999, ApJ, 521, 650.

Balbus, S. A., & Terquem, C. 2001, ApJ, 552, 235.

Blaes, O. M., & Balbus, S. A. 1994, ApJ, 421, 163.

Draine, B. T., Roberge, W. G., & Dalgarno A. 1983, ApJ, 264, 485.
Draine, B. T., & Sutin, B. 1987, ApJ, 320, 803.
Fleming, T., & Stone, J. M. 2003, ApJ, 585, 908.
Fleming, T., Stone, J. M., & Hawley, J. F. 2000, ApJ, 530, 464.
Fromang, S., Terquem, C., & Balbus, S. A. 2002, MNRAS, 329, 18.
Fromang, S., Papaloizou, J., Lesur, G., & Heinemann, T. 2007, A&A, 476, 1123.
Gammie, C. F. 1996, ApJ, 457, 355.
Glassgold, A. E., Feigelson, E. D., & Montmerle, T. 2000, in Protostars and Planets IV, ed. V. Mannings, A. P. Boss, & S. Russel (U. Arizona Press: Tucson), 429.
Glassgold, A. E., Krstić, P. S., & Schultz, D. R. 2005, ApJ, 621, 808.
Glassgold, A. E., Lucas, R., & Omont, A. 1986, AA, 157, 35.
Hartmann, L. 1998, Accretion Processes in Star Formation (Cambridge University Press: Cambridge).
Ilgner, M., & Nelson, R. P. 2004a, AA, 445, 205.
Ilgner, M., & Nelson, R. P. 2004b, AA, 455, 731.
Jackson, J. D. 1975, Classical Electrodynamics (2nd ed.) (Wiley: New York).
Ji, H., Burin, M., Schartman, E., & Goodman, J. 2006, Nature 444, 343.
Kunz, M. W., & Balbus, S. A. 2004, MNRAS, 348, 355.
Landau, L. D., & Lifshitz, E. M. 1959, Fluid Mechanics (Oxford: Pergamon Press).
Lesur, G., & Longaretti, P.-Y. 2007, MNRAS, 378, 1471.
Lin, D. N. C., & Papaloizou, J. 1980, MNRAS, 191, 37.
Lin, D. N. C., & Papaloizou, J. C. B. 1996, ARAA, 34, 703.
Lubow, S. H., Papaloizou, J. C. B., & Pringle, J. E. 1994, MNRAS, 267, 235.
Moffatt, H. K., Kida, S., & Ohkitani, K. 1994, J. Fluid Mech., 259, 241.
Nishi, R., Nakano, T., & Umebayashi, T. 1991, ApJ, 368, 181.
Papaloizou, J. C. B., & Lin, D. N. C. 1995, ARAA, 33, 505.
Pneuman, G. W., & Mitchell, T. P. 1965, Icarus, 4, 494.
Pringle, J. E. 1981, ARAA, 19, 137.
Reipurth, B., Jewitt, D., & Keil, K. (eds.). 2007, Protostars and Planets V (U. Arizona Press: Tucson).
Richard, D. 2001, PhD thesis, Univ. Paris 7.
Sano, T., Miyama, S. M., Umebayashi, T., & Nakano, T. 2000, ApJ, 543, 486.
Sano, T., & Stone, J. M. 2002a, ApJ, 570, 314.
Sano, T., & Stone, J. M. 2002b, ApJ, 577, 534.
Salmeron, R., & Wardle, M. 2008, MNRAS 388, 1223.
Shakura, N. I., & Sunyaev, R. A. 1973, AA, 24, 337.
Stone, J. M., & Balbus, S. A. 1996, ApJ, 464, 364.
Stone, J. M., Gammie, C. F., Balbus, S. A., & Hawley, J. F. 2000, in Protostars and Planets IV, ed. V. Mannings, A. Boss, & S. Russell, (U. Arizona Press: Tucson), 589.
Terquem, C. 2008, ApJ, 689, 532.
Turner, N. J., & Sano, T. 2008, ApJ, 679L, 131.
Velikhov, E. P. 1959, Sov. Phys. JETP, 36, 995.
Wardle, M. 1999, MNRAS, 307, 849.
Weidenschilling, S. J. 1977, Astrophys. & Space Science, 51, 153.
Zel'dovich, Ya. B., Ruzmaikin, A. A., & Sokoloff, D. D. 1983, Magnetic Fields in Astrophysics (Gordon and Breach: New York).

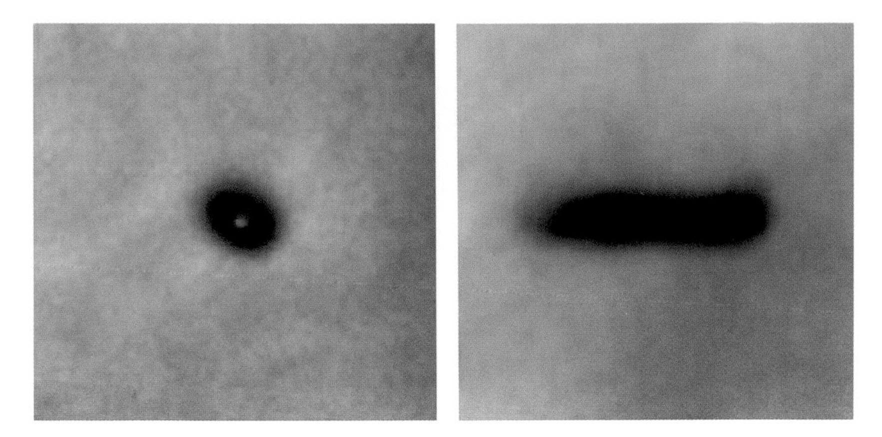

Plate 1. *HST* images of silhouette disks against the background of ionized emission from the Orion Nebula Cluster. Left: Orion 183-405; the disk radius is 264 AU. Right: Orion 114-426; the disk radius is 1,012 AU. Image credit NASA and McCaughrean and O'Dell (1996).

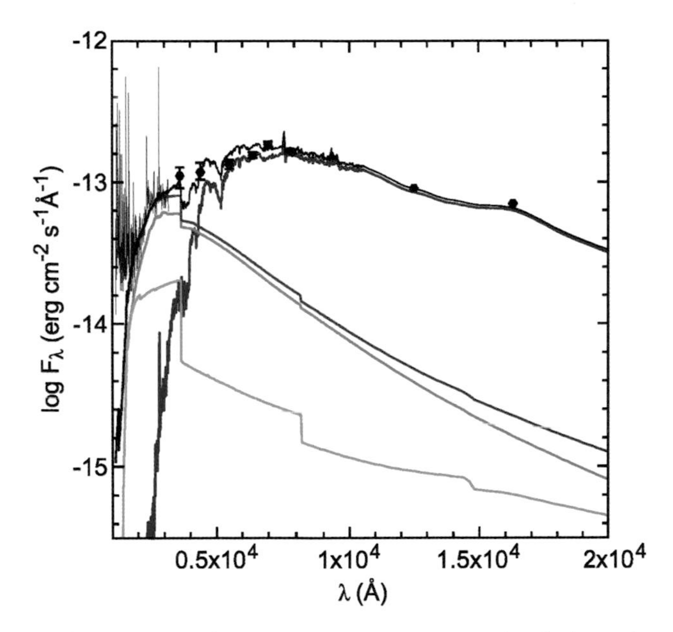

Plate 2. SED of BP Tau, including STIS UV spectra, compared with the predictions from the accretion-shock model. The different contributors to the emission are shown: preshock emission (cyan); photosphere heated by the shock (green); total emission from the accretion column, i.e., the sum of the emission from the preshock and the heated photosphere (magenta); photosphere (red); and photosphere plus accretion shock (black). From [103].

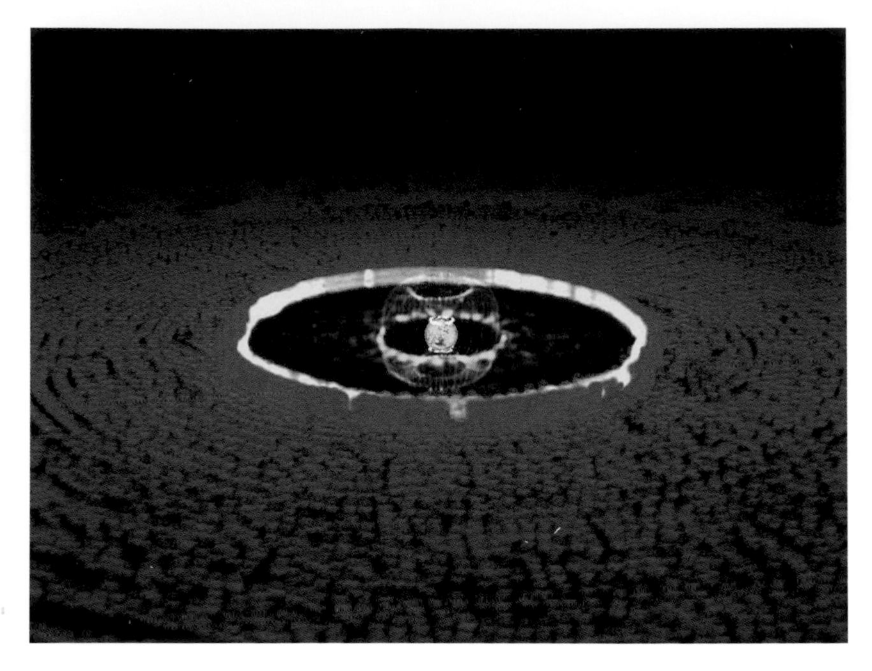

Plate 3. Artist is representation of the inner disk of a CTTS. The disk has a sharp transition at the dust- destruction radius, represented by a wall in the figure. Inside it, the dust-free disk is truncated at the magnetospheric radius, inside which matter falls onto the star along magnetic field lines. Artist: Luis Belerique.

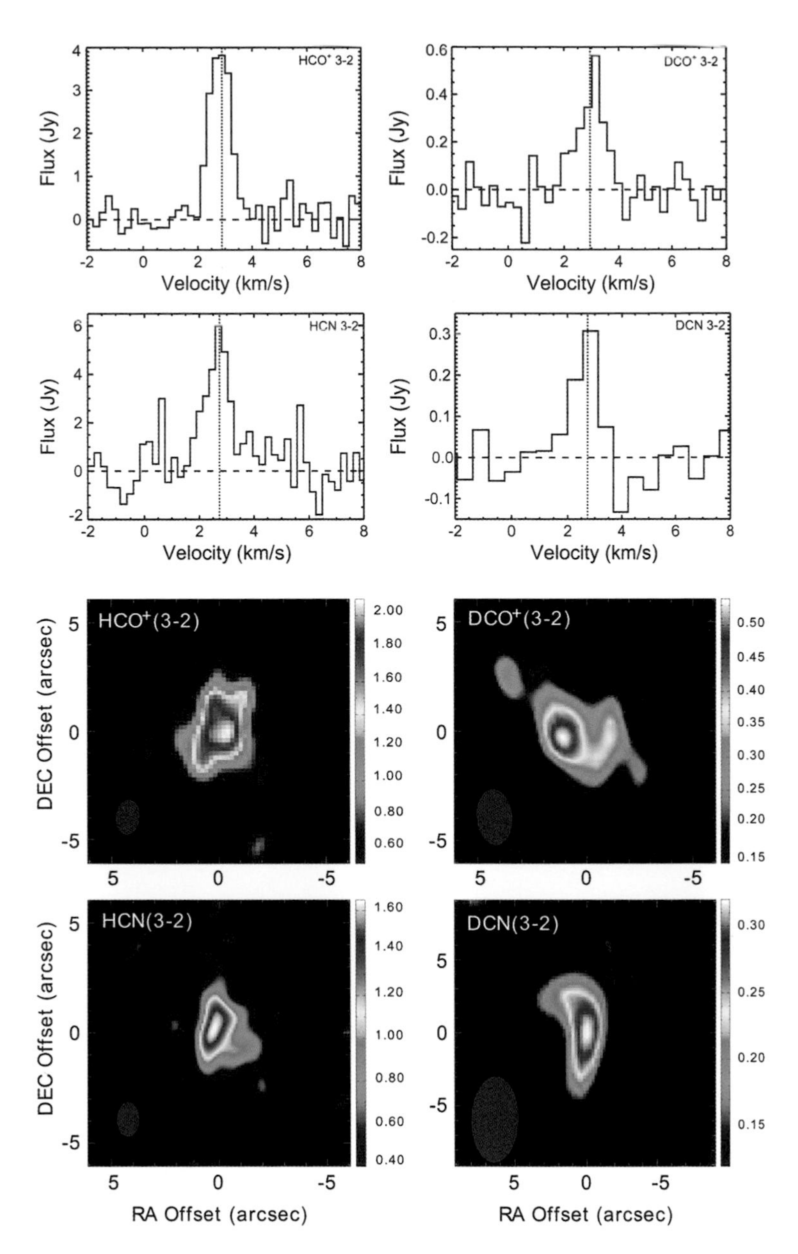

Plate 4. Sample spectra and integrated emission maps of the $J = 3-2$ transition of HCO^+, DCO^+, HCN, and DCN. Observations from the Submillimeter Array and figure published by Qi et al. (2008).

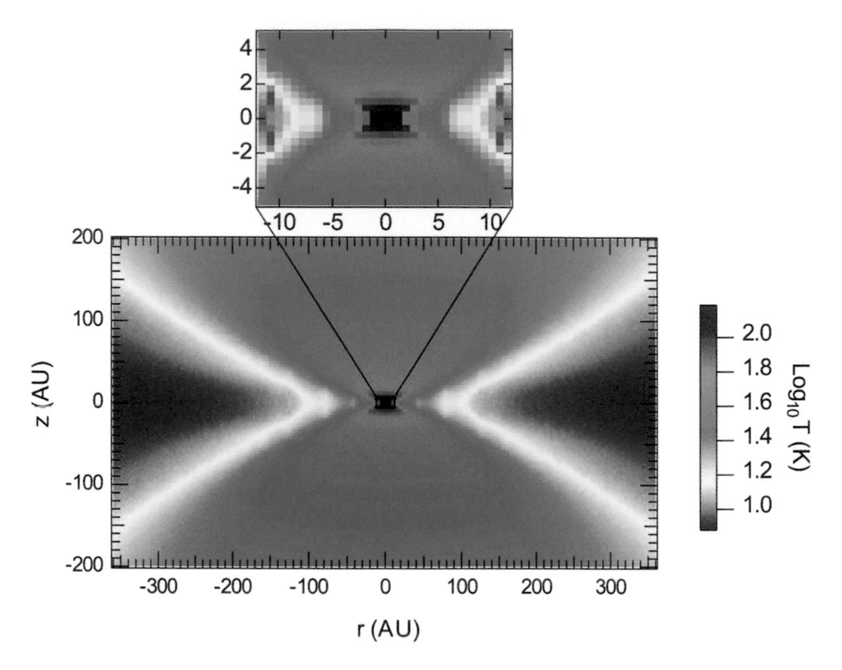

Plate 5. Predicted temperature structure for a typical T Tauri disk model by D'Alessio et al. (2001). Assumed parameters are 0.5 $M_{\odot}$, $\dot{M} = 3 \times 10^{-8}\ M_{\odot}/\mathrm{yr}$, and $\alpha = 0.01$.

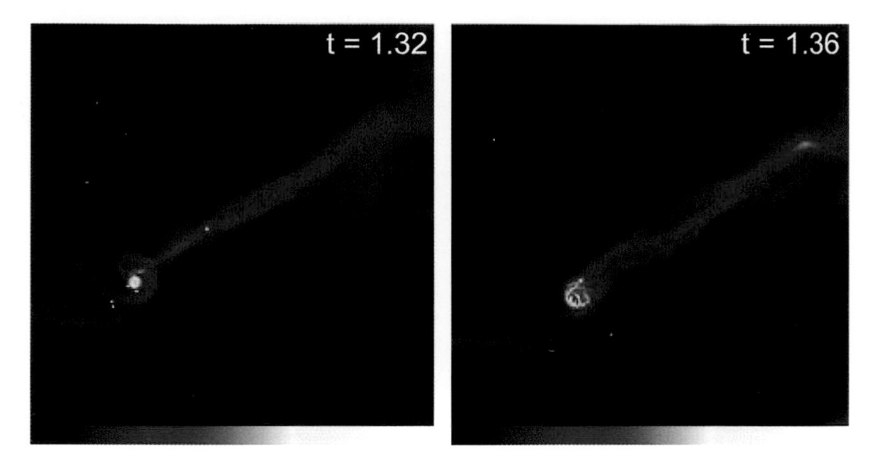

Plate 6. A density color scale of two snapshots from an isothermal smoothed-particle hydrodynamics (SPH) simulation for the collapse of an initially turbulent 50 $M_{\odot}$ cloud core. The frames zero in on a single large disk being fed by a stream. The frames are about 5,000 AU on a side, and the region illustrated contains a few $M_{\odot}$ of brown dwarfs, stars, and gas. The time stamp 5in the upper right is in units of the initial free-fall time of the original cloud core. Adapted from Bate and Bonnell (2005).

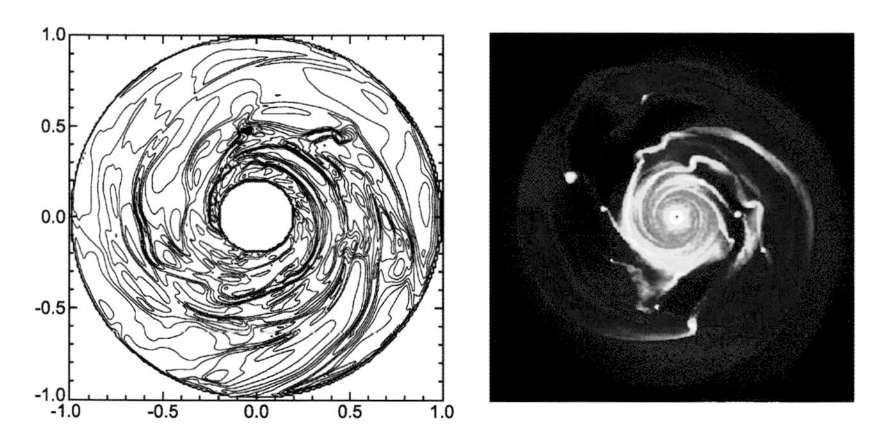

Plate 7. Two classic illustrations of the fragmentation of disks when they are evolved isothermally. The left panel, adapted from Boss (2000), shows density contours with a dense fragment near 12 o'clock at $(x, y) = (-0.05, 0.45)$. This disk has $M_d/M_s \approx 0.09$, an initial minimum $Q \approx 1.3$, and inner and outer radii of 4 and 20 AU. The right panel, a density color scale adapted from Mayer et al. (2002), is for a disk with $M_d/M_s \approx 0.1$, an initial minimum $Q \approx 1.4$, and initial inner and outer radii of 4 and 20 AU. In both cases, $M_s = 1\ M_{\odot}$, and the fragments have masses of 1 to 5 Jupiter masses.

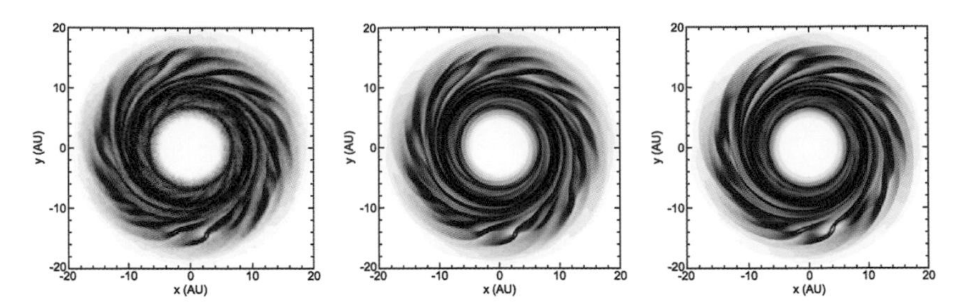

Plate 8. Equatorial plane density color scale after one disk orbit for high-resolution isothermal simulations based on the same highly unstable initial conditions with three different codes: the SPH code GASOLINE (left panel), the AMR Cartesian code FLASH (middle panel), and the cylindrical FG code of the Indiana University (IU) Hydro Group. All three codes produce similar dense structure. Adapted from results of the Wengen test (Gawryszczak et al., private communication) with images courtesy of A. C. Boley.

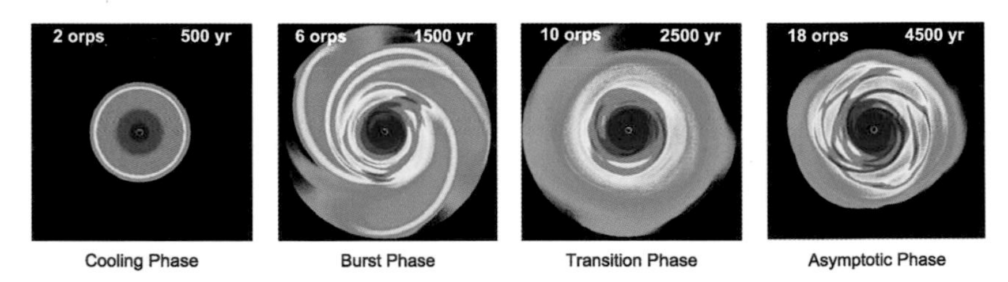

Plate 9. Logarithmic color scales of midplane density at four times that illustrate the four phases of a $t_{cool} = 2$ orp simulation. The parameters of the initial model are described in the text. A similar image from 23.5 orp looks essentially the same as the 18 orp image shown. The asymptotic phase is quasi-steady. Adapted from Mejía et al (2005).

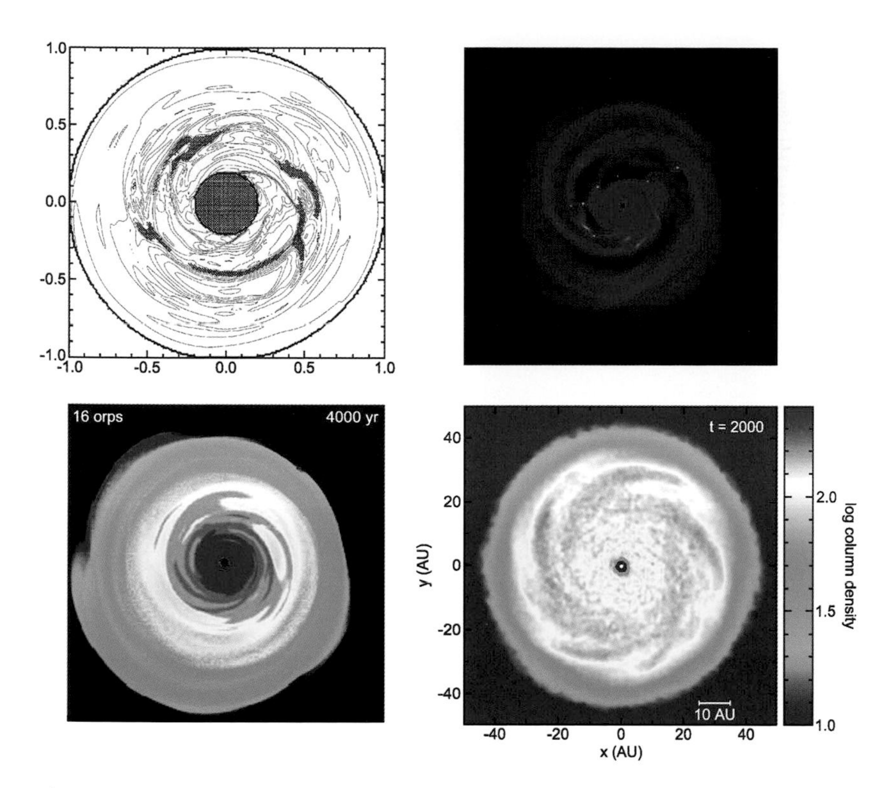

Plate 10. Equatorial plane strucutres for four simulations of roughly similar protoplanetary disks with realistic radiative cooling using different numerical schemes. The top left panel, adapted from Boss (2001), and the top right panel, adapted from Mayer et al. (2007b), develop very dense clumps. The bottom left panel, adapted from Boley et al. (2007a), and the bottom right panels adapted from Whitworth (2008), use the same initial disk as in Mejía et al (2005) and exhibit no fragmentation.

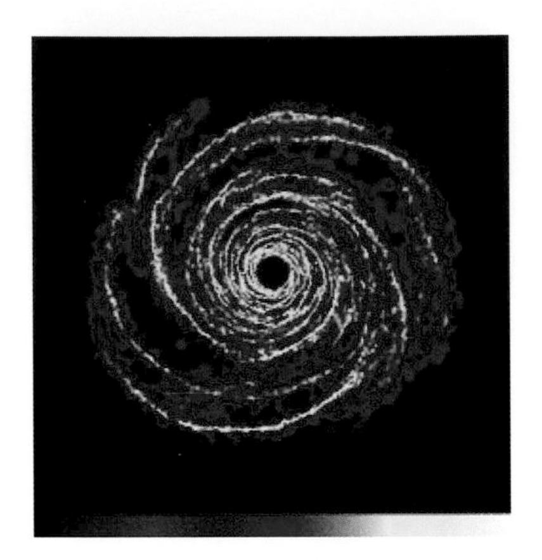

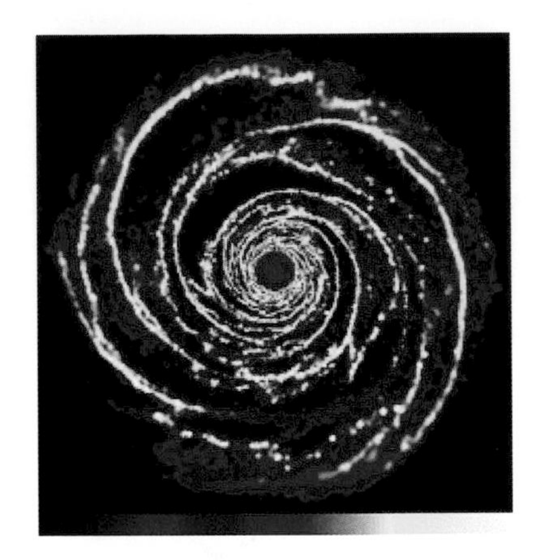

Plate 11. Color scales of particle surface density relative to gas density in GI-active disks. The left panel, adapted from Rice et al. (2004), illustrates the rapid concentration of particles of 1 meter radius into GI spirals. The right panel, adapted from Rice et al. (2006), where the self-gravity of the particles is included, shows that when particles of 50 cm radius are concentrated in GI spirals, they can become subject to GIs themselves, as shown by the dense knots of particles.

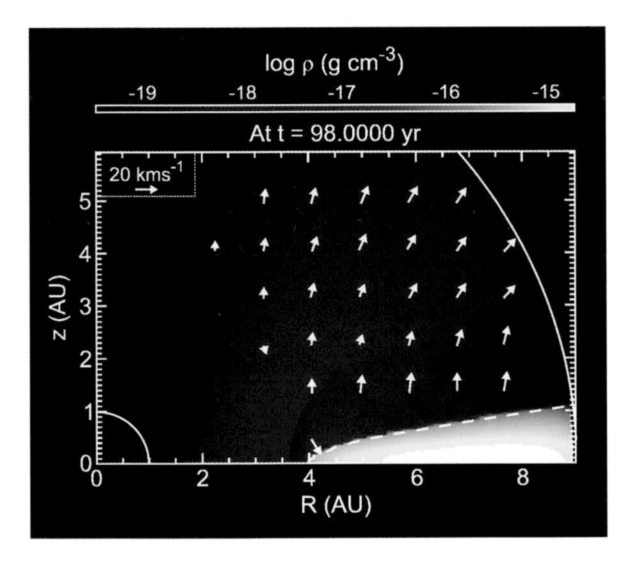

Plate 12. Snapshot of flow, with density color scale and flow vectors, based on simulations of the photoevaporation of a disk with a cleared inner hole ([7]).

7

ARIEH KÖNIGL AND RAQUEL SALMERON

THE EFFECTS OF LARGE-SCALE MAGNETIC FIELDS ON DISK FORMATION AND EVOLUTION

1. Introduction

In Chapter 6 it was shown that a sufficiently highly conducting Keplerian disk that is threaded by a weak magnetic field will be subject to the magnetorotational instability (MRI) and may evolve into a turbulent state in which the field is strongly amplified and has a small-scale, disordered configuration. This turbulence has been proposed as the origin of the effective viscosity that enables matter to accrete by transferring angular momentum radially out along the disk plane. In this chapter we focus on an alternative mode of angular momentum transport that can play an important role in protostellar disks, namely, vertical transport through the disk surfaces effected by the stresses of a large-scale, ordered magnetic field that threads the disk.

The possible existence of a comparatively strong, "open" magnetic field over much of the extent of at least some circumstellar disks around low- and intermediate-mass protostars is indicated by far-infrared (IR) and submillimeter polarization measurements, which have discovered an ordered, hourglass-shaped field morphology on subparsec scales in several molecular clouds (e.g., Schleuning 1998; Girart et al. 2006; Kirby 2009). The polarized radiation is attributed to thermal emission by spinning dust grains whose short axes are aligned along the magnetic field (e.g., Lazarian 2007). The detected hourglass morphology arises naturally in molecular cloud cores in which a large-scale magnetic field provides dynamic support against the core's self-gravity (see § 2.1). In this picture the field is interstellar in origin and is part of the Galactic magnetic-field distribution. As discussed in § 3, the inward gravitational force can become dominant, and the core then undergoes dynamic collapse to form a central protostar and a circumstellar disk. The magnetic field is dragged in by the infalling matter and could in principle lead to a large-scale "open" field configuration in the disk.

An ordered magnetic field that threads a disk can exert a magnetic torque that removes angular momentum from the interior gas. This angular momentum can be carried away along the field lines either by torsional Alfvén waves in a process known as magnetic braking (see § 2.2) or through a rotating outflow in what is known as a centrifugally driven wind (CDW; see § 4.3). These mechanisms could supplement or even entirely supplant the radial transport along the disk plane invoked in traditional disk models: by turbulent stresses, as in the MRI scenario mentioned above, or through gravitational torques in a self-gravitating disk, as described in chapter 5. In the case of radial transport, the angular momentum removed from the bulk of the matter is deposited into a small amount of gas at the outer edge of the disk. In the case of vertical transport, this angular momentum is deposited into a small fraction of the disk mass (the tenuous surface layers of the disk) that is removed as a CDW or else (when magnetic braking operates) into the ambient medium through which the torsional Alfvén waves propagate. The introduction of this new transport channel has profound implications for the structure and properties of disks in which it is a major contributor to the angular momentum budget and potentially also for the strong connection that has been found between accretion and outflow phenomena in young stellar objects. This is discussed in § 4, where we also consider how to determine which of the two possible angular momentum transport modes (radial or vertical) operates at any given location in a magnetically threaded disk and whether these two modes can coexist.

Of course, large-scale, ordered magnetic fields can also be produced in situ by a dynamo process; we consider this alternative possibility for the origin of the disk field in § 4.2. In the case of the Sun, high-resolution observations made at extreme ultraviolet and soft X-ray wavelengths and transformed into spectacular false-color images have revealed a complex web of organized structures that appear as loops and prominences near the stellar surface but simplify to a more uniform distribution farther out (e.g., Balogh et al. 1995). There is growing evidence that Sun-like stars are already magnetically active in the protostellar phase and, in fact, generate fields that are 1,000 times stronger than that of the present-day Sun. The dynamic interaction between such a field and a surrounding accretion disk through which mass is being fed to the nascent star could have important evolutionary and observational consequences. We consider these in § 5. We conclude with a summary and a discussion of future research directions in § 6.

2. Magnetohydrodynamics of Magnetically Threaded Disks

Before we get into the specifics of the various topics outlined in § 1, we present a general discussion of the dynamic properties of ordered magnetic fields in relation to protostellar disks and of the main methods that have been applied to their study.

2.1. *Magnetic Forces*

The magnetic force per unit volume on a magnetized fluid element is given by

$$\frac{\boldsymbol{J} \times \boldsymbol{B}}{c} = -\nabla\left(\frac{B^2}{8\pi}\right) + \frac{\boldsymbol{B} \cdot \nabla \boldsymbol{B}}{4\pi}\,, \tag{1}$$

where $\boldsymbol{J}$ is the current density, $\boldsymbol{B}$ is the magnetic-field vector, and c is the speed of light, and we substituted

$$\boldsymbol{J} = \frac{c}{4\pi} \nabla \times \boldsymbol{B} \tag{2}$$

(neglecting the displacement current in Ampère's law on the assumption that all speeds are $\ll c$) and used a vector identity to obtain the two terms on the right-hand side of eq. (1). As was already noted in chapter 6, the first term represents the magnetic pressure force and the second one the force due to magnetic tension. To build intuition, it is useful to consider the magnetic-field lines. Because of the solenoidal (i.e., absence of monopoles) condition on the magnetic field,

$$\nabla \cdot \boldsymbol{B} = 0\,, \tag{3}$$

the flux Ψ of "open" field lines through the disk (the integral $\int \boldsymbol{B} \cdot \mathrm{d}\boldsymbol{S}$, where $\mathrm{d}\boldsymbol{S}$ is a disk surface-area element) is conserved. This is a noteworthy result: it says that even if the disk is resistive in its interior, the flux of magnetic-field lines that thread it (also referred to as the *poloidal flux*) will not be destroyed (although poloidal flux could be added to, or removed from, the disk through its outer and inner boundaries). In the magnetohydrodynamics (MHD) picture, the field lines can be thought of as rubber bands that exert tension of magnitude $B^2/4\pi$ directed along the field and pressure of magnitude $\mathrm{B}^2/8\pi$ normal to their local direction (e.g., Parker 1979).

The nonmagnetic forces acting on a cloud core or a disk are the force of gravity (mostly self-gravity in the case of a core and central-mass gravity in the case of a nearly Keplerian disk), $-\varrho\nabla\Phi$ (where ϱ is the mass density and Φ is the gravitational potential), and the thermal pressure force $-\nabla P$. There may be an additional force, associated with the momentum flux of MHD waves, in a turbulent system. Such turbulence could give rise to an effective viscosity, but, as is generally the case in astrophysical systems, the effects of microphysical shear viscosity would remain negligible (e.g., Frank et al. 2002). Protostellar disks typically have sufficiently low temperatures that the dominant forces in the disk plane are gravity and magnetic tension, whereas in the vertical direction the thermal pressure force is always important, balancing the downward force of gravity or of the magnetic pressure gradient.

These notions can be illustrated by considering the equilibrium configurations of cloud cores (e.g., Mouschovias 1976). Within the interstellar gas that ultimately ends up in the core, the magnetic-field lines are initially nearly parallel to each other, and the field is well coupled to the matter (with the magnetic flux being, in the jargon of MHD, "frozen" into the matter). Since a magnetic field exerts no force parallel to itself, matter can slide relatively easily along the field to form a flattened mass distribution. The oblate structure predicted by this scenario is consistent with the observed shapes of starless cores in the Orion giant molecular cloud (e.g., Tassis 2007).[1] It is convenient to describe this configuration with the help of a cylindrical coordinate system $\{r, \phi, z\}$ centered on the forming star, with the z-axis aligned with the initial magnetic-field direction. A vertical hydrostatic equilibrium in which the (upward) thermal (and possibly turbulent) pressure force balances the (downward) gravitational force is established on a dynamic (free-fall) timescale and is by and large maintained throughout the subsequent evolution of the core, as verified by numerical simulations (e.g., Fiedler & Mouschovias 1993; Galli & Shu 1993), including cases of filamentary clouds that are initially elongated in the field direction (e.g., Nakamura et al. 1995; Tomisaka 1996). Self-gravity acting in the radial direction tends to pull the field lines inward. The hourglass shape revealed by the polarization measurements is produced because these field lines remain anchored in the cloud envelope. This results in magnetic tension that is associated mainly with the term $(B_z/4\pi)dB_r/dz$ in eq. (1) and is in complete analogy with the force exerted by a stretched rubber band. The field morphologies revealed by the polarization measurements are interpreted as arising from the approximate balance between this force and radial gravity. This interpretation is supported by H I and OH Zeeman measurements of the line-of-sight field amplitude and by estimates of the plane-of-sky field strength using the measured dispersion in the field orientation (the Chandrasekhar-Fermi method), both of which typically imply roughly virialized cores (with

1. Note, however, that many cores do not exhibit a clear oblate structure and that, in fact, many of the cores in Taurus are apparently prolate and may have formed from the fragmentation of the filamentary clouds in which they are embedded (e.g., Di Francesco et al. 2007).

ordered and turbulent magnetic fields contributing approximately equally to the overall support of the cloud against self-gravity; e.g., Ward-Thompson et al. 2007).

In applying these ideas to the disks that form from the gravitational collapse of cores, one should bear in mind the following two points. First, the flux threading the disk is sufficiently strongly concentrated that the bending of the field lines between the midplane (where $B_r = B_\phi = 0$ by reflection symmetry) and the disk surface can be large enough to make $B_{r,\mathrm{s}}$ (where the subscript "s" denotes the surface) comparable to B_z (which, in turn, changes little between the midplane and the surface if the disk is thin). Consequently, magnetic squeezing by the z-gradient of the magnetic pressure associated with the B_r (and possibly also B_ϕ) field components can become comparable to, or even exceed, the downward force of gravity. This is indeed a key property of the wind-driving disk models discussed in § 4.4. Second, the density scale height h in the disk typically satisfies $h \ll r$, implying a rapid decrease of the density with z even as the magnetic-field amplitude changes little for $z \lesssim r$. Therefore, magnetic forces generally dominate all other forces on scales $\lesssim r$ above the disk surface, and the field there assumes a so-called force-free field configuration ($\boldsymbol{J} \times \boldsymbol{B} \approx 0$). According to eq. (1), in this case the magnetic tension force has to balance the magnetic pressure force, which points outward (as B_z^2 increases toward the center). This implies that the field lines in this region assume a vaselike, "concave-in" morphology (i.e., bending toward the vertical axis), in contradistinction to the hourglass-like, "convex-out" shape that they have inside the disk. This behavior accounts for the initial collimation of disk-driven MHD winds (see § 4.3).

2.2. Magnetic Braking

Plucked rubber bands (or strings) carry waves whose phase velocity is the square root of the ratio of the tension to the mass density. Using again the analogy to magnetic-field lines, one immediately obtains the phase speed of Alfvén waves, $v_\mathrm{A} = B_z/\sqrt{4\pi\varrho}$ (taking the background field to point in the $\hat{\boldsymbol{z}}$ direction). When the transverse magnetic field of the wave points in the azimuthal direction, the corresponding (torsional) Alfvén wave carries angular momentum. A rotating molecular cloud core or disk will twist the magnetic "rubber bands" that thread it. The degree of twisting fixes the pitch $|B_{\phi,\mathrm{s}}/B_z|$ of the field lines and is determined, in turn, by the "load" on the other end of the band (i.e., by the inertia of the external matter to which the field lines are coupled). This twisting represents a transfer of angular momentum from the core or disk to the external medium (subscript "ext"). One can thus determine $B_{\phi,\mathrm{s}}$ by equating the torque per unit area (normal to $\hat{\boldsymbol{z}}$) exerted on each of the two surfaces of the flattened core or disk, $rB_zB_{\phi,s}/4\pi$, to minus the rate per unit area of angular momentum carried by the waves from each surface and deposited in the (initially nonrotating) ambient medium, $-\varrho_{ext}r_{ext}^2\Omega_B v_{Aext}$, where Ω_B is the angular velocity of the field line (which is conserved along the field under ideal-MHD conditions; see § 4.3). Poloidal flux conservation along the field makes it possible to relate the radius r and the field B_z in the core or disk to the corresponding quantities in the external medium, $B_z r dr = B_{z,ext} r_{ext} dr_{ext}$, where the ambient field (assumed to be uniform on large scales) can be expressed in terms of the poloidal flux Ψ by $B_{z,ext} = \Psi/\pi r_{ext}^2$. One then obtains (cf. Basu & Mouschovias 1994)

$$B_{\phi,s} = -\frac{\Psi}{\pi r^2}\frac{v_{B\phi}}{v_{Aext}}\,, \tag{4}$$

where $v_{B\phi} = r\Omega_B$. The general meaning of $v_{B\phi}$ is discussed in § 2.4; unless Ohm diffusivity dominates in the core/disk, it can be identified with the mid-plane angular velocity of the particles into which the field lines are frozen (see eq. [65]).

2.3. Centrifugal Wind Driving

As noted in § 2.1, the dynamics just above the disk surface is magnetically dominated, i.e., the magnetic energy density there is larger than the thermal, gravitational, and kinetic energy densities of the gas. The comparatively large electrical conductivity in this region implies that the poloidal (r, z) gas velocity is parallel to the poloidal magnetic field (see § 4.3), and the bulk particle motions can be approximated as those of beads along rotating, rigid, massless wires (Henriksen & Rayburn 1971). This mechanical analogy is useful for deriving the criterion for the centrifugal launching of disk winds (Blandford & Payne 1982). We neglect thermal effects in this derivation and, correspondingly, regard the disk as being infinitely thin. Since the field geometry varies only on a scale $\sim r$ in the force-free zone, very close to the disk surface the field lines can be regarded as being nearly straight. If we consider thus a straight wire that intersects a Keplerian disk at a distance r_0 from the center and makes an angle θ to the disk normal, the balance of gravitational and centrifugal forces along the wire implies that in equilibrium, the effective potential $\Phi_{eff}(\gamma)$ satisfies

$$\frac{\partial\Phi_{eff}}{\partial\gamma} = \frac{\gamma+\sin\theta}{(1+2\gamma\sin\theta+\gamma^2)^{3/2}} - \gamma\sin^2\theta - \sin\theta = 0\,, \tag{5}$$

where the dimensionless variable $\gamma \equiv s/r_0$ measures the distance s along the wire. The equilibrium is unstable when $\partial^2\Phi_{eff}/\partial\gamma^2 < 0$ (corresponding to a local maximum of Φ_{eff}), which at $\gamma = 0$ occurs for $\theta > 30°$. Hence centrifugal driving sets in if

$$\frac{B_{r,s}}{B_z} > \frac{1}{\sqrt{3}} . \quad (6)$$

This wind-launching criterion plays a key role in wind-driving disk models. It is worth noting that in contrast to typical stellar winds, in which the outflowing gas must "climb out" of basically the entire gravitational potential well at the stellar surface, in the case of outflows from a rotationally supported, infinitely thin disk the depth of the (gravitational plus centrifugal) potential well is lower by a factor of 2 on account of the rotation, and gas can in principle escape to infinity without any added thermal push if there is a sufficiently strongly inclined, rigid channel (the magnetic-field lines). Real disks, however, are not completely cold and therefore have a finite thickness. When this is taken into account, the effective potential attains a maximum at some height above the disk surface, and thermal pressure forces are required to lift gas from the disk up to that point (e.g., Ogilvie 1997). This and some other properties of CDWs are considered in § 4.3.

2.4. Nonideal MHD

Unlike the approach of classical electrodynamics, in which it is common to consider the magnetic field as being generated by current flows according to the Biot-Savart law, in MHD practice it is often more illuminating to focus on the magnetic field and to regard the current density as a subordinate quantity that is determined, through eq. (2), by how the field is shaped by its interaction with matter (e.g., Parker 2007). The current, in turn, helps determine the neutral-fluid-frame (denoted by a prime) electric field $\boldsymbol{E}'$ according to Ohm's law,

$$\boldsymbol{J} = \boldsymbol{\sigma} \cdot \boldsymbol{E}' = \sigma_{\rm O} \boldsymbol{E}'_{\parallel} + \sigma_{\rm H} \hat{\boldsymbol{B}} \times \boldsymbol{E}'_{\perp} + \sigma_{\rm P} \boldsymbol{E}'_{\perp} , \quad (7)$$

where the conductivity has been expressed as a tensor to take account of the inherent anisotropy that an ordered magnetic field induces in the motions of charged particles. Here the subscripts $\parallel$ and $\perp$ denote vector components that are, respectively, parallel and perpendicular to the unit vector $\hat{\boldsymbol{B}}$, whereas $\sigma_{\rm O}$, $\sigma_{\rm H}$, and $\sigma_{\rm P}$ are, respectively, the Ohm, Hall, and Pedersen conductivity terms.

Under ideal-MHD conditions, the conductivity is effectively infinite, and the comoving electric field vanishes. This is an adequate approximation for describing the dynamics of a disk wind or of the medium surrounding the core/disk. However, within the core or disk the degree of ionization is generally low, so finite conductivity effects must be taken into account in the dynamic modeling. When the conductivity is low, each charged particle species (denoted by a subscript "j") develops a drift velocity $\boldsymbol{v}_{\rm d,j} \equiv \boldsymbol{v}_{\rm j} - \boldsymbol{v}$ with respect to the neutral-fluid velocity (which we approximate by the average fluid velocity

$\boldsymbol{v}$, as appropriate for a weakly ionized medium). The drift velocities can be calculated from the equations of motion of these species, each of which is well approximated by a steady-state balance between the Lorentz force and the drag force $\boldsymbol{F}_{\mathrm{nj}}$ exerted by collisions with the neutrals,

$$Z_{\mathrm{j}}e\left(\boldsymbol{E}' + \frac{\boldsymbol{v}_{\mathrm{d,j}}}{c} \times \boldsymbol{B}\right) = -\boldsymbol{F}_{\mathrm{nj}} = m_{\mathrm{j}}\gamma_{\mathrm{j}}\varrho\boldsymbol{v}_{\mathrm{d,j}}, \tag{8}$$

where Z_{j} is the (signed) particle charge in units of the electronic charge e, $\gamma_{\mathrm{j}} \equiv <\sigma v>_{\mathrm{j}} / (m_{\mathrm{j}} + m)$, and $<\sigma v>_j$ is the rate coefficient for collisional momentum transfer between particles of mass m_{j} and neutrals (of mass m and mass density ϱ). This collisional interaction in turn exerts a force $\boldsymbol{F}_{\mathrm{jn}} = -\boldsymbol{F}_{\mathrm{nj}}$ on the neutrals. The degree of coupling between a given charged species and the magnetic field is measured by the Hall parameter, defined as the ratio of the particle's gyrofrequency to its collision frequency $\nu_{\mathrm{jn}} = \gamma_{\mathrm{j}}\varrho$ with the neutrals:

$$\beta_{\mathrm{j}} \equiv \frac{|Z_{\mathrm{j}}|eB}{m_{\mathrm{j}}c}\frac{1}{\gamma_{\mathrm{j}}\varrho} \,. \tag{9}$$

In this expression, $B \equiv |\mathbf{B}|\, sgn\{B_z\}$ is the signed magnetic-field amplitude, with the sign introduced to keep the dependence of the Hall conductivity on the magnetic-field polarity (see eq. [13]). When $|\beta_{\mathrm{j}}| \gg 1$, the coupling is good, and the collision term can be neglected in comparison with the magnetic term; in this case the Lorentz force is approximately 0, corresponding to the near vanishing of the electric field in the charged particle's frame. On the other hand, when $|\beta_{\mathrm{j}}| \ll 1$, the coupling is poor; in this case the magnetic term in eq. (8) can be neglected.

By summing eq. (8) over the particle species, one obtains

$$\sum_{\mathrm{j}} n_{\mathrm{j}}\boldsymbol{F}_{\mathrm{jn}} = \frac{\boldsymbol{J} \times \boldsymbol{B}}{c} \,. \tag{10}$$

This shows explicitly that in a weakly ionized medium, the Lorentz force (which acts only on the charged particles) is transmitted to the bulk of the matter only through a collisional drag, which involves a relative motion between the charged and neutral components. Therefore, if magnetic forces are important in such a medium, its structure is inherently not static. This is exemplified by the behavior of a magnetically supported molecular cloud core (e.g., Shu et al. 1987). The magnetic field that threads the core is anchored in the comparatively well-ionized cloud envelope, but in the core's interior the degree of ionization is low, and the magnetic tension force is transmitted to the predominantly neutral gas through ion–neutral drag. Since the ions (taken in what follows to constitute a single species denoted by the subscript "i") are well coupled to the magnetic field and thus remain nearly fixed in space, the associated ambipolar diffusion drift entails an inward motion of the neutral particles toward the center of mass. If the evolution lasts longer than the ambipolar diffusion time ($\sim (R/v_A)^2 \gamma_i \rho_i$, which can be inferred from eqs. [14] and [18]), the central concentration will become large enough to cause the core to become gravitationally unstable, and dynamic collapse will ensue.

If one expresses the current density in terms of the charged particles' drifts,

$$J = \sum_j e n_j Z_j \boldsymbol{v}_{d,j} \tag{11}$$

(where the charged particles have number densities n_j and satisfy charge neutrality, $\sum_j n_j Z_j = 0$), and uses eq. (8), one can solve for the conductivity tensor components in eq. (7):

$$\sigma_O = \frac{ec}{B} \sum_j n_j |Z_j| \beta_j \,, \tag{12}$$

$$\sigma_H = \frac{ec}{B} \sum_j \frac{n_j |Z_j|}{1 + \beta_j^2} \,, \tag{13}$$

and

$$\sigma_P = \frac{ec}{B} \sum_j \frac{n_j |Z_j| \beta_j}{1 + \beta_j^2} \tag{14}$$

(e.g., Cowling 1976; Wardle & Ng 1999). Note that σ_H (and correspondingly the Hall term in Ohm's law) depends on an odd power of the magnetic-field amplitude and can therefore have either a positive or a negative sign (reflecting the magnetic-field polarity). This leads to qualitative differences in the behavior of the disk solutions in the Hall regime for positive and negative values of B_z.

The conductivity tensor formalism is useful for constructing realistic disk models in which the relative magnitudes of the different conductivities can change as a function of height even at a single radial location as a result of the variation in the density and in the dominant ionization mechanism as one moves between the disk surface and the midplane (see § 4.4). It is nevertheless instructive to relate this formalism to the classical diffusivity regimes considered in chapter 6. This is best done by solving for the fluid-frame electric field,

$$c\boldsymbol{E}' = \eta_O \nabla \times \boldsymbol{B} + \eta_H (\nabla \times \boldsymbol{B}) \times \hat{\boldsymbol{B}} + \eta_A (\nabla \times \boldsymbol{B})_\perp \,, \tag{15}$$

where the Ohm, Hall, and ambipolar diffusivities are given, respectively, by

$$\eta_O = \frac{c^2}{4\pi \sigma_O} \,, \tag{16}$$

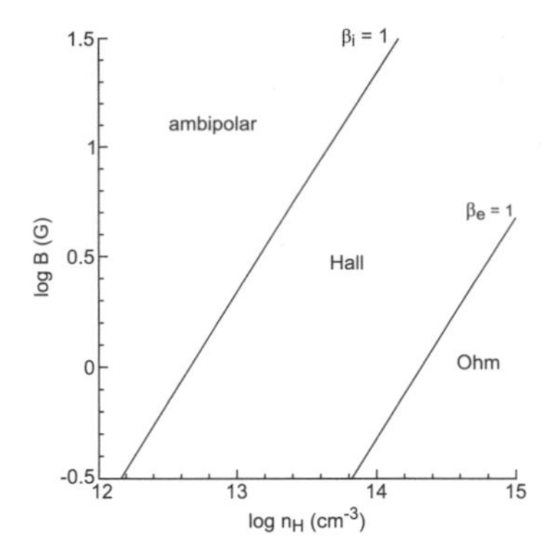

Fig. 7.1. Magnetic diffusivity regimes in the $\log n_{\rm H} - \log B$ plane for $T = 280\,{\rm K}$.

17
$$\eta_{\rm H} = \frac{c^2}{4\pi\sigma_\perp}\frac{\sigma_{\rm H}}{\sigma_\perp}\,,$$

and

18
$$\eta_{\rm A} = \frac{c^2}{4\pi\sigma_\perp}\frac{\sigma_{\rm P}}{\sigma_\perp} - \eta_{\rm O}\,,$$

with $\sigma_\perp \equiv (\sigma_{\rm H}^2 + \sigma_{\rm P}^2)^{1/2}$ (e.g., Mitchner & Kruger 1973; Nakano et al. 2002). The three distinct regimes are delineated by

19
$$\begin{aligned} &\sigma_{\rm O} \approx \sigma_{\rm P} \gg |\sigma_{\rm H}| \qquad &(\text{Ohm}),\\ &\sigma_{\rm O} \gg |\sigma_{\rm H}| \gg \sigma_{\rm P} \qquad &(\text{Hall}),\\ &\sigma_{\rm O} \gg \sigma_{\rm P} \gg |\sigma_{\rm H}| \qquad &(\text{ambipolar})\,. \end{aligned}$$

When ions and electrons (subscript "e") are the only charged species, one has $\eta_{\rm H} = \beta_{\rm e}\,\eta_{\rm O}$ and $\eta_{\rm A} = \beta_{\rm i}\,\beta_{\rm e}\,\eta_{\rm O}$, with the respective Hall parameters given numerically by $\beta_{\rm i} = q\beta_{\rm e} \approx 0.46\,(B/n_{13})$ (e.g., Draine et al. 1983, where $q \approx 1.3 \times 10^{-3}\,T_2^{1/2}$, $n_{13} = n_{\rm H}/(10^{13}\,{\rm cm}^{-3})$, B is measured in Gauss, $T_2 = T/(10^2\,{\rm K})$, and the mean ion mass is $m_{\rm i} = 30\,m_{\rm H}$). In this case the three classical regimes correspond, respectively, to the Hall parameter ranges $|\beta_{\rm i}| \ll |\beta_{\rm e}| \ll 1$ (Ohm), $|\beta_{\rm i}| \ll 1 \ll |\beta_{\rm e}|$ (Hall), and $1 \ll |\beta_{\rm i}| \ll |\beta_{\rm e}|$ (ambipolar).

Fig. 7.1 shows the regions of ambipolar, Hall, and Ohm dominance in the $\log n_{\rm H} - \log B$ plane for a weakly ionized gas.[2] Molecular cloud cores and the outer regions ($\gtrsim 10\,{\rm AU}$) of protostellar disks typically correspond to the ambipolar regime, as do the disk surface regions at smaller radii; disk midplanes on scales $\sim$ 1–10 AU are often dominated by Hall diffusivity, whereas

2. For a complementary figure in the $\log n_{\rm H} - \log T$ plane, see Fig. 6.2 of chapter 6.

Ohm diffusivity may characterize the midplanes of disks on smaller scales. In the innermost ($\lesssim 0.1$ AU) regions of the disk, where the temperature increases above $\sim 10^3$ K and the gas becomes collisionally ionized (e.g., Gammie 1996; Li 1996a), anomalous Ohm diffusivity (the enhanced drag between positive and negative charge carriers due to scattering off electromagnetic waves generated by current-driven plasma instabilities) might play a role. Note that the precise extent of the different regimes depends on the radial profile of the disk column density, since this profile determines the degree of penetration of the ionizing radiation or cosmic rays (see § 4.4). For a given mass-accretion rate, the column density depends on the nature of the angular momentum transport mechanism. In particular, transport by a large-scale, ordered magnetic field is generally more efficient than transport by a small-scale, disordered field,[3] resulting in higher inflow speeds and correspondingly lower column densities and higher degrees of ionization in CDW-mediated accretion than in MRI-based turbulent disks.

The behavior of the electric field is governed by Faraday's law,

$$\frac{\partial \boldsymbol{B}}{\partial t} = -c\nabla \times \boldsymbol{E} \,, \tag{20}$$

where $\boldsymbol{E} = \boldsymbol{E}' - \boldsymbol{v} \times \boldsymbol{B}/c$ is the lab-frame electric field. In view of eq. (15), when the resistive term in the expression for $\boldsymbol{E}'$ can be neglected and the only charge carriers are ions and electrons, one can express the ambipolar and Hall contributions to $c\boldsymbol{E}$ as $(\boldsymbol{v} - \boldsymbol{v}_\mathrm{i}) \times \boldsymbol{B}$ and $(\boldsymbol{v}_\mathrm{i} - \boldsymbol{v}_\mathrm{e}) \times \boldsymbol{B}$, respectively (see Königl 1989 and chapter 6), yielding $\partial \boldsymbol{B}/\partial t = \nabla \times (\boldsymbol{v}_\mathrm{e} \times \boldsymbol{B})$, which indicates that in this case the field lines are frozen into the electrons (the particle species with the highest mobility $|Z_\mathrm{j}|e/m_\mathrm{j}$). It is also seen that the ideal-MHD limit $\boldsymbol{E} = -\boldsymbol{v} \times \boldsymbol{B}/c$ is approached when the ion-neutral and ion-electron drift speeds are small in comparison with the bulk speed. It further follows that in the ambipolar regime, the ions and electrons drift together, relative to the neutrals (i.e., $|\boldsymbol{v}_\mathrm{i} - \boldsymbol{v}_\mathrm{e}| \ll |\boldsymbol{v}_\mathrm{i} - \boldsymbol{v}|$), so the field is effectively frozen also into the ions, whereas in the Hall regime the ions and neutrals essentially move together, and the electrons drift relative to them ($|\boldsymbol{v}_\mathrm{i} - \boldsymbol{v}_\mathrm{e}| \gg |\boldsymbol{v}_\mathrm{i} - \boldsymbol{v}|$). More generally, one can define an effective velocity $\boldsymbol{v}_B$ for the poloidal flux surfaces that applies also in the Ohm regime, when these surfaces are no longer frozen into any particle species (although, as noted in § 2.1, they continue to maintain their identity). If we focus on the midplane, where $B = B_z$, this is done through the relation $c\boldsymbol{E} = -\boldsymbol{v}_B \times \boldsymbol{B}$ (cf. Umebayashi & Nakano 1986). As we just observed,

3. This can be seen by representing the $r\phi$ turbulent stress component as αP (where the constant α is typically < 1; cf. Balbus & Papaloizou 1999), so the ratio of the torques exerted by the ordered field and by the turbulent stress is $(-B_z B_{\phi,s}/4\pi \langle P \rangle)(r/\alpha H)$ (where $\langle\rangle$ denotes vertical averaging over the disk half-thickness H), which is typically $\gg 1$.

in the absence of Ohm diffusivity (and still when only ion and electron charges are considered) this relation is satisfied if one substitutes $\boldsymbol{v}_{\rm e}$ for the velocity, which verifies that $\boldsymbol{v}_B = \boldsymbol{v}_{\rm e}$ in this case. However, in the Ohm regime the azimuthal field-line velocity

$$v_{B\phi,0} = -cE_{r,0}/B_0 \tag{21}$$

can differ from that of the most mobile charged-particle component, and similarly for the radial flux-surface velocity

$$v_{Br,0} = cE_{\phi,0}/B_0 \ , \tag{22}$$

where the subscript 0 denotes the midplane.

A departure from ideal MHD may lead to energy dissipation at a rate (per unit volume) $\boldsymbol{J} \cdot \boldsymbol{E}'$. As can be inferred from eqs. (7) and (15), both the Ohm and ambipolar terms in Ohm's law contribute to Joule heating, but not the Hall term. It may perhaps seem puzzling that energy is dissipated in the ambipolar regime even though the field lines remain frozen to the charged particles, but one can directly associate the heating in this case with the ion-neutral collisional drag: $\boldsymbol{J} \cdot \boldsymbol{E}' \approx n_{\rm i} \boldsymbol{v}_{\rm d,i} \cdot \boldsymbol{F}_{\rm in}$ (assuming $Z_{\rm i} \gg 1$; see eqs. [8] and [11]). Joule dissipation would be the main internal heating mechanism in disk regions where angular momentum transport is dominated by a large-scale, ordered field. Ambipolar heating, in particular, could also play an important role in the thermal structure of disk-driven winds (see § 4.3).

Just as we arrived at a combination of physical variables that measures the degree of coupling between a charged particle and the magnetic field (namely, the Hall parameter; eq. [9]), we will find it useful to identify an analogous measure for the neutrals. It turns out that an appropriate parameter combination of this type for Keplerian disks, which plays a similar role in MRI-based systems (where the neutrals couple to a small-scale, disorderd field) and in wind-driving disks (where the coupling is to a large-scale, ordered field), is the Elsasser number

$$\Lambda \equiv \frac{v_{\rm A}^2}{\Omega_{\rm K} \eta_\perp} \ , \tag{23}$$

where

$$\eta_\perp = \frac{c^2}{4\pi\sigma_\perp} \tag{24}$$

is the "perpendicular" magnetic diffusivity and $\Omega_{\rm K}$ is the Keplerian angular velocity. This parameter is $\gg$1 and $\ll$1 for strong and weak neutral–field coupling, respectively.[4] The Elsasser number is used in planetary dynamo theory to measure the ratio of Lorentz to Coriolis forces. Since both the MRI mechanism and the field-mediated vertical angular momentum transport involve magnetic coupling to Keplerian rotation, it is not surprising that this parameter also arises in the disk context. In the three main conductivity regimes it reduces to

$$\Lambda \Rightarrow \begin{cases} \frac{\gamma_i \rho_i}{\Omega_K} \equiv \Upsilon & \text{(ambipolar)}, \\ \beta_i \Upsilon & \text{(Hall)}, \\ \beta_e \beta_i \Upsilon & \text{(Ohm)}, \end{cases} \tag{25}$$

where we again assume, for simplicity, that the only charged particles are ions and electrons. The ambipolar limit has a clear physical meaning: it represents the ratio Υ of the Keplerian rotation time to the neutral–ion momentum-exchange time. This parameter has emerged as the natural measure of the field–matter coupling in the wind-driving disk models of Wardle & Königl (1993; see § 4.4), as well as in studies of the linear (e.g., Blaes & Balbus 1994) and nonlinear (e.g., Mac Low et al. 1995; Brandenburg et al. 1995; Hawley & Stone 1998) evolution of the MRI in such disks. Indeed, since the ions are well coupled to the field in this limit ($|\beta_i| \gg 1$), the neutrals will be well coupled to the field if their momentum exchange with the charged particles (which is dominated by their interaction with the comparatively massive ions) occurs on a timescale that is short in comparison with the dynamic time Ω_K^{-1} (corresponding to $\Upsilon > 1$). In the Hall regime, Λ is equal to this ratio of timescales multiplied by $|\beta_i|$. This product has figured prominently in the classification of wind-driving disk solutions (Königl et al. 2010), as well as in linear studies of disk MRI in this regime (e.g., Wardle 1999; Balbus & Terquem 2001). In this case, too, it has a clear physical meaning. In contradistinction to the ambipolar regime, the ions are not well coupled to the field in the Hall limit ($|\beta_i| \ll 1$). In order for the neutrals to be well coupled to the field, it is, therefore, not sufficient for them to be well coupled to the ions ($\Upsilon > 1$); rather, the product $|\beta_i|\Upsilon$ must be > 1 in this case. In the Ohm regime, even the electrons are not well coupled to the field ($|\beta_e| \ll 1$), so now the product $\beta_e \beta_i \Upsilon$ has to exceed 1 to ensure an adequate coupling of the neutrals to the field. The condition $\Lambda \gtrsim 1$ in fact characterizes both the linear (e.g., Jin 1996) and the nonlinear (e.g., Sano & Stone 2002) behavior of the MRI in this regime, and

4. The parameter Λ is distinct from the Lundquist number $S \equiv v_A L/\eta_O$ and from the magnetic Reynolds number $Re_M \equiv VL/\eta_O$ (where V and L are characteristic speed and length scale, respectively), which have been used in similar contexts in the literature. It was first introduced into the protostellar disk literature in this form (but using a different symbol) by Wardle (1999).

the Elsasser number is again a natural parameter for classifying viable wind-driving disk solutions in this case (with each subregime corresponding to a distinct lower bound on Λ; see Königl et al. 2010).

2.5. Similarity Solutions

Even the most simplified models of magnetically threaded protostellar disks lead to systems of nonlinear partial differential equations (PDEs). In some cases, such as the disk–star interaction considered in § 5, it has proved necessary to carry out full-fledged numerical simulations to get definitive answers about the expected behavior. However, for other aspects of the problem, such as disk formation (§ 3) and the wind-driving disk structure (§ 4), it has been possible to make analytic (or, rather, semianalytic) progress by looking for similarity solutions, an approach that converts the PDEs into ordinary differential equations (ODEs; e.g., Landau & Lifshitz 1987). This type of solution does not incorporate either inner or outer radial boundaries, so it can be justified only at radii that are sufficiently far removed from the disk's actual edges. However, the basic properties revealed by these solutions have invariably been confirmed by simulations, and the analytic approach has the advantage of making the qualitative behavior more transparent and of making it possible to better investigate the parameter dependence of the solutions and to incorporate more physics and dynamic range than is yet possible numerically. The semianalytic and numerical approaches have thus played complementary roles in shedding light on these questions.

In the case of the disk-formation problem, the separate dependences on the spatial scale r and the time t are subsumed into a single dependence on the dimensionless combination r/Ct by taking advantage of the fact that the isothermal sound speed C is nearly uniform (on account of efficient cooling by dust grains) in cloud cores and in the outer regions of circumstellar disks. The self-similarity of the derived solutions is expressed by the fact that they depend solely on the ratio r/t and not on the separate values of r and t. In the case of the stationary, axisymmetric wind-driving disk models, the separate dependences on the spatial coordinates z and r are subsumed into a single dependence on the dimensionless combination z/r (or, equivalently, on $z(r, r_0)/r_0$, where $z(r, r_0)$ describes the shape of a magnetic-field line that intersects the midplane at r_0). In these radially self-similar models, all physical variables scale as power laws of the spherical radius $R = (z^2 + r^2)^{1/2}$, and the differential equations depend only on the polar angle $\theta = \text{arc cot}\,(z/r)$. Further details are given in §§ 3 and 4, respectively.

3. Disk Formation and Early Evolution

Rotationally supported circumstellar disks evidently originate in the collapse of self-gravitating, rotating, molecular cloud cores. Molecular line observations (e.g., Goodman et al. 1993; Kane & Clemens 1997) have established that a majority of dense ($\gtrsim 10^4\,\mathrm{cm}^{-3}$) cloud cores show evidence of rotation, with angular velocities $\sim 3\times10^{-15}$–$10^{-13}\,\mathrm{s}^{-1}$ that tend to be uniform on scales of $\sim 0.1\,\mathrm{pc}$, and with specific angular momenta in the range $\sim 4\times10^{20}$–$3\times10^{22}\,\mathrm{cm}^2\,\mathrm{s}^{-1}$. The cores can transfer angular momentum to the ambient gas by magnetic braking (§ 2.2), and this mechanism also acts to align their angular momentum vectors with the local large-scale magnetic field (e.g., Machida et al. 2006). This alignment occurs on a dynamic timescale and hence can be achieved even in cores whose lifetimes are not much longer than that (as in certain models of the turbulent interstellar medium; e.g., Elmegreen 2000). The dynamic collapse might occur as a result of mass rearrangement in the core during the ambipolar diffusion time (e.g., Mouschovias et al. 2006; see § 2.4) or sooner if the core is close to the critical mass for collapse (the effective Jeans mass) from the start (e.g., Elmegreen 2007). In this section we consider the core-collapse problem in the context of angular momentum transport by a large-scale, ordered magnetic field; an alternative scenario involving gravitational torques is discussed in chapter 5.

Once dynamic collapse is initiated and a core goes into a near-free-fall state, the specific angular momentum is expected to be approximately conserved, resulting in a progressive increase in the centrifugal force that eventually halts the collapse and gives rise to a rotationally supported disk on scales $\sim 10^2\,\mathrm{AU}$. These expectations are consistent with the results of molecular-line interferometric observations of contracting cloud cores (e.g., Ohashi et al. 1997; Belloche et al. 2002). In this picture, the disk rotation axis should be aligned with the direction of the large-scale magnetic field that threads the cloud. Observations have not yet yielded a clear-cut answer to whether this is indeed the case in reality (e.g., Ménard & Duchêne 2004; Vink et al. 2005), and it is conceivable that the field in some cases is too weak to align the core's rotation axis or even control its contraction, or that additional processes (such as fragmentation or disk warping) play a role. In what follows, we nevertheless continue to pursue the implications of the basic magnetically supported cloud picture.

3.1. Modeling Framework

Since the gravitational collapse time is much shorter than the local ambipolar diffusion time in the core, the magnetic-field lines at first move in with the infalling matter. However, once the central mass begins to grow, ambipolar diffusion becomes important within the gravitational sphere of influence of the central mass (Ciolek & Königl 1998; Contopoulos et al. 1998). When the incoming matter enters this region, it decouples from the field and continues moving inward. The decoupling front, in turn, moves outward and steepens into an ambipolar diffusion shock. The existence of this C-type MHD shock was first predicted by Li & McKee (1996). The transition from a nearly freely falling, collapsing core to a quasi-stationary, rotationally supported disk involves a strong deceleration in a centrifugal shock. This shock typically occurs at a smaller radius than the ambipolar diffusion shock and is hydrodynamic, rather than hydromagnetic, in nature.

On the basis of the arguments presented in § 2.1, it should be a good approximation to assume that the gas rapidly establishes force equilibrium along the field and therefore to consider only motions in the radial direction.[5] One can obtain semianalytic solutions for this effectively one-dimensional (1D) time-dependent problem by postulating $r - t$ self-similarity, with a similarity variable

26 $$x \equiv \frac{r}{C t}$$

(see § 2.5). This modeling approach is motivated by the fact that core collapse is a multiscale problem, which is expected to assume a self-similar form away from the outer and inner boundaries and not too close to the onset time (e.g., Penston 1969; Larson 1969; Hunter 1977; Shu 1977). This has been verified by numerical and semianalytic treatments of restricted core-collapse problems—with/without rotation and with/without magnetic fields. Although the constancy of the isothermal sound speed C that underlies the ansatz (26) is not strictly maintained throughout the solution domain (in particular, C scales roughly as $r^{-1/2}$ in large portions of the disk that forms around the central star), this is of little consequence for the results since thermal stresses do not play a major role in the dynamics of the collapsing core. For a typical sound speed $C = 0.19\ \mathrm{km\,s^{-1}}$, $x = 1 \Leftrightarrow \{400, 4{,}000\}$ AU at $t = \{10^4,\ 10^5\}$ yr.

5. In reality, mass can also be added to the system from the polar directions. Numerical simulations of axisymmetric collapse in which new mass is added only through vertical infall have tended to produce disk-to-star mass ratios ~ 1, much higher than typically observed. A better agreement with observations may, however, be obtained when nonaxisymmetric density perturbations and resultant gravitational torques are included in the calculations (e.g., Vorobyov & Basu 2007).

Krasnopolsky & Königl (2002, hereafter KK02) constructed self-similar solutions of rotating magnetic molecular cloud cores that are subject to ambipolar diffusion. These solutions reveal many of the basic features of star and disk formation in the core-collapse scenario and are discussed in the remainder of this section. To incorporate ambipolar diffusion into the self-similarity formulation, it is necessary to assume that the ion density scales as the square root of the neutral density: $\varrho_{\rm i} = \mathcal{K}\varrho^{1/2}$. As discussed in KK02, this should be a good approximation for the core-collapse problem: it applies on both ends of a density range spanning ~ 8 orders of magnitude, which corresponds roughly to radial scales ~ 10–10^4 AU, with $\mathcal{K}$ varying by only ~ 1 order of magnitude across this interval.

To allow mass to accumulate at the center in a 1D rotational collapse, an angular momentum transport mechanism must be present. KK02 assumed that vertical transport through magnetic braking continues to operate also during the collapse phase of the core evolution. To incorporate this mechanism into the self-similar model, it is necessary to assume that $v_{\rm Aext}$, the Alfvén speed in the external medium, is a constant.[6] KK02 verified that in their derived solutions, magnetic braking indeed dominates the most likely alternative angular momentum transport mechanisms—MRI-induced turbulence and gravitational torques. However, they also found that angular momentum transport by a CDW arises naturally (and may dominate) in the Keplerian disk that forms in their fiducial solution (see § 3.4).

3.2. Basic Equations

The mass- and momentum-conservation relations in their differential form are given by

$$\frac{\partial \varrho}{\partial t} + \frac{1}{r}\frac{\partial}{\partial r}(r\varrho v_r) = -\frac{\partial}{\partial z}(\varrho v_z) \qquad \text{(mass)}, \tag{27}$$

$$\varrho\frac{\partial v_r}{\partial t} + \varrho v_r\frac{\partial v_r}{\partial r} = \varrho g_r - C^2\frac{\partial \varrho}{\partial r} + \varrho\frac{v_\phi^2}{r} + \frac{B_z}{4\pi}\frac{\partial B_r}{\partial z} - \frac{\partial}{\partial r}\left(\frac{B_z^2}{8\pi}\right)$$
$$-\frac{1}{8\pi r^2}\frac{\partial}{\partial r}(rB_\phi)^2 - \varrho v_z\frac{\partial v_r}{\partial z} \qquad \text{(radial momentum)}, \tag{28}$$

$$\frac{\varrho}{r}\frac{\partial}{\partial t}(rv_\phi) + \frac{\varrho v_r}{r}\frac{\partial}{\partial r}(rv_\phi) = \frac{B_z}{4\pi}\frac{\partial B_\phi}{\partial z} + \frac{B_r}{4\pi r}\frac{\partial}{\partial r}(rB_\phi)$$
$$-\varrho v_z\frac{\partial}{\partial z}(rv_\phi) \qquad \text{(angular momentum)}, \tag{29}$$

6. A nearly constant value $v_{\rm Aext} \approx 1\ {\rm km\ s^{-1}}$ is, in fact, indicated in molecular clouds in the density range $\sim 10^3$–$10^7\ {\rm cm^{-3}}$ (e.g., Crutcher 1999).

and

30

$$C^2\frac{\partial \varrho}{\partial z} = \varrho g_z - \frac{\partial}{\partial z}\left(\frac{B_\phi^2}{8\pi} + \frac{B_r^2}{8\pi}\right) + \frac{B_r}{4\pi}\frac{\partial B_z}{\partial r} \quad \text{(vertical hydrostatic equilibrium)},$$

where g_r and g_z are, respectively, the radial and vertical components of the gravitational acceleration. We have not yet dropped the terms that involve the vertical velocity component (except in the vertical momentum equation, which we assume takes its hydrostatic form).

We now integrate these equations over the core/disk thickness $2H$. We use the thin-disk approximation ($H(r) \ll r$) and assume that the density, radial velocity, azimuthal velocity, and radial gravity are constant with height, and that so also is B_z (except when it is explicitly differentiated with respect to z, in which case we substitute $\partial B_z/\partial z = -r^{-1}(\partial/\partial r)(rB_r)$ from $\nabla \cdot \boldsymbol{B} = 0$). We also assume that $B_r(r,z) = B_{r,\mathrm{s}}(r)\,[z/H(r)]$ (and similarly for B_ϕ) and, after deriving expressions that are valid to order $(H/r)^2$, further simplify by dropping all terms $\mathcal{O}(H/r)$ except in the combination $[B_{r\mathrm{s}} - H(\partial B_z/\partial r)]$ (see KK02 for details). We thus obtain the vertically integrated conservation equations:

31

$$\frac{\partial \Sigma}{\partial t} + \frac{1}{r}\frac{\partial}{\partial r}(r\Sigma v_r) = -\frac{1}{2\pi r}\frac{\partial \dot{M}_\mathrm{w}}{\partial r} \quad \text{(mass)},$$

32

$$\frac{\partial v_r}{\partial t} + v_r\frac{\partial v_r}{\partial r} = g_r - \frac{C^2}{\Sigma}\frac{\partial \Sigma}{\partial r} + \frac{B_z}{2\pi\Sigma}\left(B_{r,\mathrm{s}} - H\frac{\partial B_z}{\partial r}\right) + \frac{J^2}{r^3} \quad \text{(radial momentum)},$$

33

$$\frac{\partial J}{\partial t} + v_r\frac{\partial J}{\partial r} = \frac{rB_zB_{\phi\mathrm{s}}}{2\pi\Sigma} \quad \text{(angular momentum)},$$

and

34

$$\frac{\Sigma C^2}{2H} = \frac{\pi}{2}G\Sigma^2 + \frac{GM_*\varrho H^2}{2r^3} + \frac{1}{8\pi}\left(B_{\phi,\mathrm{s}}^2 + B_{r,\mathrm{s}}^2 - B_{r,\mathrm{s}}\,H\frac{\partial B_z}{\partial r}\right)$$

$$\text{(vertical hydrostatic equilibrium)},$$

where $\Sigma = 2\varrho H$ is the surface mass density, $J = rv_\phi$ is the specific angular momentum of the matter, G is the gravitational constant, and M_* is the mass of the central protostar. In the integrated equations we have implemented the 1D flow approximation by setting $v_z = 0$, but we retained the term on the right-hand side of eq. (31) to allow for mass loss through a disk wind (at a rate $\dot{M}_\mathrm{w}$). Such a wind could carry angular momentum, but we did not include vertical particle angular momentum transport in eq. (33) since most of a disk wind's angular momentum initially resides in the magnetic field (see § 4.3). In any case, we proceed to solve the equations by assuming at first that no wind is present and that the only mechanism of angular momentum transport is magnetic braking.

The dominant charge carriers in the precollapse core are ions and electrons. Adopting this composition, we approximate the ion equation of motion in the ambipolar diffusion limit by

$$\varrho \nu_{\rm ni} \boldsymbol{v}_{\rm d} = \frac{1}{4\pi} (\nabla \times \boldsymbol{B}) \times \boldsymbol{B} \,, \tag{35}$$

where $\nu_{\rm ni}$ is the neutral-ion collision frequency (see eqs. [8] and [10]). This relation yields the components of the drift velocity:

$$v_{{\rm d},\phi} = \frac{B_z B_{\phi,{\rm s}}}{2\pi \nu_{\rm ni} \Sigma} \tag{36}$$

and

$$v_{{\rm d},r} = \frac{B_z}{2\pi \nu_{\rm ni} \Sigma} \left(B_{r,{\rm s}} - H \frac{\partial B_z}{\partial r} \right) . \tag{37}$$

Magnetic braking is incorporated via eq. (4), in which we identify $v_{B\phi}$ with $v_{{\rm i},\phi}$ (see § 2.4). Expressing $v_{{\rm i},\phi}$ in terms of v_ϕ and $v_{{\rm d},\phi}$ (eq. [36]) and imposing a cap ($\delta \lesssim 1$) on $|B_{\phi,{\rm s}}/B_z|$ (to account for the possible development of a kink instability above the disk surface), one gets

$$B_{\phi,{\rm s}} = -\min \left[\frac{\Psi}{\pi r^2} \frac{v_\phi}{v_{\rm Aext}} \left(1 + \frac{\Psi B_z}{2\pi^2 r^2 \nu_{\rm ni} \Sigma v_{\rm Aext}} \right)^{-1} , \, \delta B_z \right] . \tag{38}$$

The flux-conservation relation $\partial \Psi / \partial t = -2\pi r v_{Br} B_z$ (obtained from eqs. [20] and [22]), which describes the advection of poloidal flux by the infalling matter, can be written in this limit as

$$\frac{\partial \Psi}{\partial t} = -2\pi r v_{{\rm i},r} B_z = -2\pi r \left(v_r + v_{{\rm d},r} \right) B_z \,, \tag{39}$$

with $v_{{\rm d},r}$ given by eq. (37). The surface value of B_r that appears in the latter equation can be determined in the limit of a potential field ($\nabla \times \mathbf{B} = 0$) outside an infinitely thin disk from an r integral of the midplane value of B_z:

$$B_{r,{\rm s}} = \int_0^\infty {\rm d}k \, J_1(kr) \int_0^\infty {\rm d}r' r' [B_z(r') - B_{\rm ref}] J_0(kr') \tag{40}$$

(e.g., Ciolek & Mouschovias 1993), where J_0 and J_1 are Bessel functions of order 0 and 1, respectively, and $B_{\rm ref}$ is the uniform ambient field at infinity (which is henceforth neglected). Although the current-free limit of the force-free medium outside the core/disk is not exact, because the presence of a B_ϕ component, which typically involves a distributed poloidal electric current, is required for angular momentum transport above and below the core/disk surfaces, the magnitude of $|B_{\phi,{\rm s}}/B_z|$ (as deduced from eq. [38]) is typically small at the start of the collapse (e.g., Basu & Mouschovias 1994). Furthermore, even after this ratio increases to ~ 1 as matter and field become concentrated near the center, the magnitude of the enclosed (poloidal) current ($I = (c/2)r|B_\phi|$) increases relatively slowly with decreasing r (scaling as $r^{-1/4}$ in the circumstellar disk; see eq. [55]), so overall we expect expression (40) to be a fair approximation. A similar integral (with $\Sigma(r)$ replacing B_z) can be written for g_r (Toomre 1963). These integrals can be approximated by their monopole terms

$$B_{r,s} = \frac{\Psi(r,t)}{2\pi r^2} \tag{41}$$

and

$$g_r = -\frac{GM(r,t)}{r^2} \tag{42}$$

(cf. Li & Shu 1997), where $M(r,t)$ is the mass enclosed within a radius r at time t. When $B_z(r)$ (or $\Sigma(r)$) scales as r^{-q}, one still obtains the monopole expression for $B_{r,s}$ (or g_r), but with a q-dependent coefficient $\mathcal{O}(1)$ (Ciolek & Königl 1998).

3.3. *Self-Similarity Formulation*

The various physical quantities of the problem can be expressed as dimensionless functions of the similarity variable x (eq. [26]) in the following fashion:

$$H(r,t) = Ct\,h(x)\,, \quad \Sigma(r,t) = (C/2\pi Gt)\,\sigma(x)\,, \tag{43}$$

$$v_r(r,t) = C\,u(x)\,, \quad v_\phi(r,t) = C\,w(x)\,, \tag{44}$$

$$g_r(r,t) = (C/t)\,g(x)\,, \quad J(r,t) = C^2 t\,j(x)\,, \tag{45}$$

$$M(r,t) = (C^3 t/G)\,m(x)\,, \quad \dot{M}_a(r,t) = (C^3/G)\,\dot{m}(x)\,, \tag{46}$$

and

$$\boldsymbol{B}(r,t) = (C/G^{1/2}t)\,\boldsymbol{b}(x)\,, \Psi(r,t) = (2\pi C^3 t/G^{1/2})\psi(x), \tag{47}$$

where $\dot{M}_a$ is the mass-accretion rate.

The vertical force-balance relation (34) yields a quadratic equation for the normalized disk half-thickness h,

$$\left(\frac{\sigma m_*}{x^3} - b_{r,s}\frac{db_z}{dx}\right)h^2 + \left(b_{r,s}^2 + b_{\phi,s}^2 + \sigma^2\right)h - 2\sigma = 0\,, \tag{48}$$

whose solution is

$$h = \frac{\hat{\sigma}x^3}{2\hat{m}_*}\left[-1 + \left(1 + \frac{8\hat{m}_*}{x^3\hat{\sigma}^2}\right)^{1/2}\right], \tag{49}$$

where $\hat{m}_* \equiv m_* - x^3 b_{r,s}(db_z/dx)/\sigma$ and $\hat{\sigma} \equiv \sigma + (b_{r,s}^2 + b_{\phi,s}^2)/\sigma$.

The initial ($t = 0$) conditions, which in the self-similar model also represent the outer asymptotic ($r \to \infty$) values, correspond to a collapsing core just before it forms a central-point mass. On the basis of previous analytic and numerical work, KK02 adopted

$$\sigma \to \frac{A}{x},\ b_z \to \frac{\sigma}{\mu_\infty},\ u \to u_\infty,\ w \to w_\infty \ \text{as}\ x \to \infty\,, \tag{50}$$

with $A = 3$, $\mu_\infty = 2.9$, and $u_\infty = -1$. From the constituent equations one can also derive the inner asymptotic behavior (corresponding to $r \to 0$ at a fixed t) in the Keplerian disk:

$$\dot{m} = m = m_*, \tag{51}$$

$$j = m_*^{1/2} x^{1/2}\,, \tag{52}$$

$$-u = (m_*/\sigma_1) x^{1/2}\,, \tag{53}$$

$$\sigma = \frac{(2\eta/3\delta)(2m_*)^{1/2}}{[1 + (2\tau/3\delta)^{-2}]^{1/2}} x^{-3/2}$$

$$\equiv \sigma_1 x^{-3/2}\,, \tag{54}$$

$$b_z = -b_{\phi,s}/\delta = [m_*{}^{3/4}/(2\delta)^{1/2}] x^{-5/4}\,, \tag{55}$$

$$b_{r,s} = \psi/x^2 = (4/3) b_z\,, \tag{56}$$

and

$$h = \{2/[1 + (2\tau/3\delta)^2] m_*\}^{1/2} x^{3/2}\,, \tag{57}$$

where $\tau \equiv (4\pi G\varrho)^{1/2}/\nu_{\rm ni}$.

The equations are solved as a boundary-value problem using the above asymptotic relations. A solution is determined by the values of the four model parameters: τ, $\delta = |B_{\phi s}/B_z|$, $w_\infty = v_\phi(t = 0)/C$, and $\alpha \equiv C/v_{\rm Aext}$. The scaling parameter m_* for the central mass and the mass-accretion rate is obtained as an eigenvalue of the problem.

3.4. Self-Similar Collapse Solutions

Fig. 7.2 shows a fiducial solution corresponding to the parameter combination $\tau = 1$, $\delta = 1$, $w(0) = 0.73$, and $\alpha = 0.8$ (which yields $m_* = 4.7$). In this case the initial rotation is not very fast and the braking is moderate, leading to the formation of a disk (with outer boundary at the centrifugal shock radius $x_{\rm c} = 1.3 \times 10^{-2}$) within the ambipolar diffusion (AD) region (delimited by the AD shock radius $x_{\rm a} = 0.41 \approx 30\, x_{\rm c}$). One can distinguish the following main flow regimes:

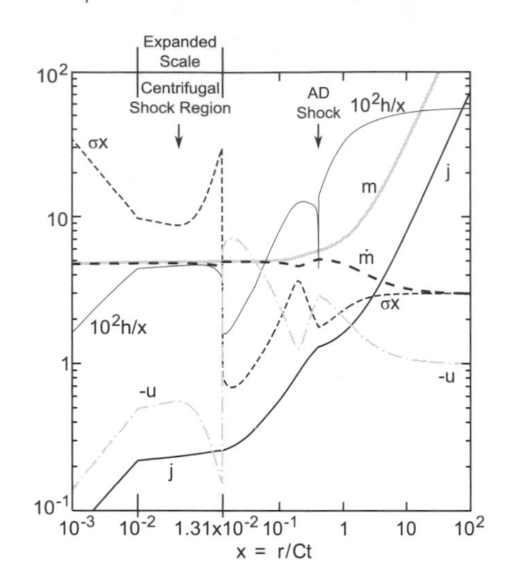

Fig. 7.2. Behavior of key normalized variables in the fiducial self-similar disk-formation solution as a function of the similarity variable x (see eqs. [43]–[46]).

- Outer region ($x > x_a$): ideal-MHD infall.
- AD shock: resolved as a continuous transition (but may in some cases contain a viscous subshock); KK02 estimated $x_a \approx \sqrt{2}\tau/\mu_\infty$.
- Ambipolar diffusion-dominated infall ($x_c < x < x_a$): near free fall controlled by the central star's gravity.
- Centrifugal shock: its location depends sensitively on the diffusivity parameter τ, which affects the amount of magnetic braking for $x < x_a$; KK02 estimated $x_c \approx (m_* w_\infty^2/A^2)\exp\{-(2^{3/2}m_*/\mu_\infty)^{1/2}\tau^{-3/2}\}$.
- Keplerian disk ($x<x_c$): asymptotic behavior (eqs. [51]–[57]) is approached after a transition zone representing a massive ring (of width $\sim 0.1\,x_c$ and mass $\sim$ 8% of the disk mass within x_c, which itself is $\lesssim$ 5% of m_*).

The inner asymptotic solution implies that at any given time the disk satisfies $\dot{M}_a(r) = \text{const}$, $\Sigma(r) \propto r^{-3/2}$, $B \propto r^{-5/4}$, and $B_{r,s}/B_z = 4/3$.[7] If the derived power-law dependence of B_z on r is used in eq. (40), one obtains $B_{r,s}/B_z = 1.428$, slightly less than the result found with the monopole approximation (41), but still representing a strong bending of the field lines (by $\sim$55° from the normal) that is significantly larger than the minimum of 30° required to launch a centrifugally driven wind from a "cold" Keplerian disk (eq. [6]). This implies that protostellar disks formed in this fashion are likely to drive disk

7. The surface-density scaling is the same as that inferred for the minimum-mass solar nebula (e.g., Weidenschilling 1977). Note, however, that this scaling is also predicted for a self-similar Keplerian disk with an α viscosity (Tsuribe 1999), as well as in certain models in which gravitational torques dominate the angular momentum transport (e.g., Lin & Pringle 1987; Voroboyov & Basu 2007).

outflows over much of their radial extents, at least during early times when they still accumulate mass from the collapsing core. Interestingly, the steady-state, radially self-similar CDW solution of Blandford & Payne (1982) also yields a radial magnetic-field scaling $\propto r^{-5/4}$. As discussed by KK02, this suggests that angular momentum transport by a CDW can be formally incorporated into the disk-formation solution, which would modify some of the details of the solution (such as the mass-accretion rate onto the star, which could be reduced by up to a factor of $\sim$ 3) although not its basic properties. Interpreted as a wind-driving disk model, the asymptotic solution corresponds to a weakly coupled disk configuration (using the nomenclature introduced in § 4.4).

Despite being based on a number of simplifying assumptions, the fiducial solution demonstrates that vertical angular momentum transport along interstellar magnetic-field lines can be sufficiently efficient to allow most of the mass of a collapsing molecular cloud core to end up (with effectively no angular momentum) at the center, with the central mass dominating the dynamics well beyond the outer edge of the disk even while the inflow is still in progress. The vertical transport is thus seen to resolve the so-called angular momentum problem in star formation (although the exact value of the protostar's angular momentum is determined by processes near the stellar surface that are not included in this model; see § 5). To the extent that self-similarity is a good approximation to the situation in such cores, it is conceivable that T Tauri (Class II) protostellar systems, whose disk masses are typically inferred to be $\lesssim$ 10% of the central mass, have had a similarly low disk-to-star mass ratio also during their earlier (Class 0 and Class I) evolutionary phases. This possibility remains to be tested by observations.

The fiducial solution also reveals that the AD shock, even though it is located well outside the region where the centrifugal force becomes important, helps enhance the efficiency of angular momentum transport through the magnetic-field amplification that it induces. The revitalization of ambipolar diffusion behind the AD shock in turn goes a long way toward resolving the magnetic-flux problem in star formation (the several-orders-of-magnitude discrepancy between the empirical upper limit on the magnetic flux of a protostar and the flux associated with the corresponding mass in the precollapse core), although it is conceivable that Ohm diffusivity in the innermost regions of the disk also plays an important role in this process (e.g., Shu et al. 2006; Tassis & Mouschovias 2007). The details of the flux detachment outside the Ohm regime can be modified if one takes account of the fact that the flux is strictly frozen into the electrons but not necessarily into the ions (i.e., if one also includes the Hall term in Ohm's law; see § 2.4). In particular, Tassis &

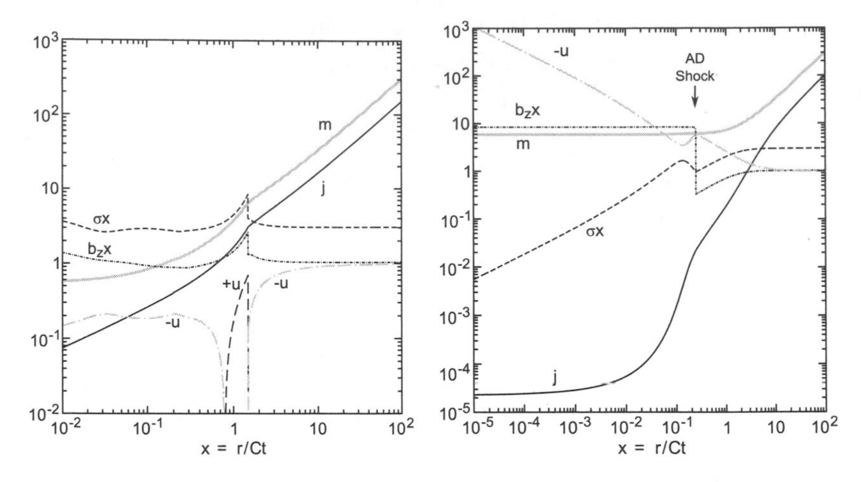

Fig. 7.3. Behavior of key normalized variables in representative fast-rotation (left) and strong-braking (right) solutions of the collapse of a rotating, magnetic, molecular cloud core as a function of the similarity variable *x* (see eqs. [43]–[47]).

Mouschovias (2005) found that a quasi-periodic series of outward-propagating shocks may develop in this case.

The shocks that form in the collapsing core, and in particular the centrifugal shock (which is the strongest), will process the incoming material and may have implications for the composition of protoplanetary disks (e.g., the annealing of silicate dust; see Harker & Desch 2002 and chapter 4). Note in this connection that the shock velocity $v_{\rm sh}$ relative to the intercepted matter has roughly the free-fall magnitude $\propto (M/r)^{1/2}$, which is constant for any given value (including, in particular, x_c) of the similarity variable. The postshock temperature (which scales as $v_{\rm sh}^{1/2}$) will thus not vary with time if the inflow remains self-similar; the postshock density, on the other hand, will decrease with time as the shock moves to larger radii.

By modifying the model parameters, one can study the range of possible behaviors in real systems. Fig. 7.3 shows two limiting cases, which bracket the fiducial solution (cf. Fig. 7.2).

The fast-rotation case differs from the fiducial solution primarily in having a large initial-rotation parameter ($w(0) = 1.5$). It has the following distinguishing features:

- The centrifugal shock is located within the self-gravity-dominated (and ideal-MHD) region; a back-flowing region is present just behind the shock.
- The central mass is comparatively small ($m_* = 0.5$), giving rise to a non-Keplerian outer disk region.

- The ideal-MHD/ambipolar-diffusion transition occurs behind the centrifugal shock and is gradual rather than sharp.

The strong-braking case ($\tau = 0.5, \delta = 10, w(0) = 1$, and $\alpha = 10$, yielding $m_* = 5.9$) is characterized by large values of the braking parameters δ and α. It is distinguished by having the following feature:

- There is no centrifugal shock (or circumstellar disk); essentially all the angular momentum is removed well before the inflowing gas reaches the center, and the $x \to 0$ behavior resembles that of the nonrotating collapse solution of Contopoulos et al. (1998).

This case may be relevant to the interpretation of slowly rotating young stars that show no evidence of a circumstellar disk (e.g., Stassun et al. 1999, 2001), which would seem puzzling if all protostars were born with significant rotation that could only be reduced as a result of their interaction with a disk (see § 5). In the strong-braking interpretation, both the slow rotation and the absence of a disk have the same cause. It is also worth noting in this connection that the physical conditions that underlie the fiducial solution presented above—namely, a highly flattened, magnetized core whose braking is controlled by the inertia of a comparatively low-density ambient gas—may not be generally applicable. In particular, it has been argued (see Mellon & Li 2009 and references therein) that magnetized molecular clouds may retain a significant amount of mass outside the collapsing core that would dominate the braking process and could place the core in the strong-braking regime where disk formation would be suppressed. Although there may be ways to alleviate or even circumvent this problem (e.g., Hennebelle & Ciardi 2009; Duffin & Pudritz 2009), there is obviously a need for additional observational and theoretical work to clarify the physical conditions that characterize core collapse in real clouds.

Although the semianalytic self-similarity model described in this section is useful for identifying important features of disk formation from the collapse of rotating cores, the intricacies of this process can be studied only numerically. Newly developed 3D, nonideal-MHD, nested-grid codes (e.g., Machida et al. 2007) are already starting to provide new insights into this problem.

4. Wind-Driving Protostellar Disks

In this section we discuss wind-driving disk models. We begin by highlighting the observational evidence for a close link between outflows and accretion disks in protostellar systems and the interpretation of this connection in terms of a large-scale magnetic field that mediates the accretion and outflow processes. We next outline the theory of centrifugally driven winds, describe the incorporation of CDWs into equilibrium disk models, and conclude by addressing the stability of such disk/wind configurations.

4.1. The Disk–Wind Connection

We give a brief survey of the observational findings on protostellar outflows and their connection to accretion disks. Besides the explicit references cited, the reader may consult some of the many review articles written on this topic over the years (e.g., Cabrit 2007; Calvet et al. 2000; Dutrey et al. 2007; Königl & Pudritz 2000; Millan-Gabet et al. 2007; Najita et al. 2007; Ray et al. 2007; Watson et al. 2007).

BIPOLAR OUTFLOWS AND JETS. Bipolar molecular outflows and narrow atomic (but sometimes also molecular) jets are ubiquitous phenomena in protostars, with about 1,000 collimated outflows of all sorts already known (e.g., Bally et al. 2007). The bipolar lobes are usually understood to represent ambient molecular material that has been swept up by the much faster, highly supersonic jets and winds that emanate from the central star/disk system (e.g., Arce et al. 2007). Jets associated with low-bolometric-luminosity ($L_{bol} < 10^3 L_\odot$) protostars have velocities in the range ~ 150–$400\ \mathrm{km\ s^{-1}}$, large (>20) Mach numbers, and inferred mass outflow rates $\sim 10^{-9}$–$10^{-7}\ M_\odot\ \mathrm{yr}^{-1}$. They are collimated on scales of a few 10s of AUs and exhibit opening angles as small as ~ 3–$5°$ on scales of $10^3 - 10^4$ AU. High-resolution observations of optically visible jets from classical T Tauri stars (CTTSs) reveal an onionlike morphology, with the regions closer to the axis having higher velocities and excitations and appearing to be more collimated. Detailed optical/near-IR spectral diagnostic techniques have been developed and applied to classical (Herbig-Haro) protostellar jets, making it possible to directly estimate the neutral densities in the forbidden-line emission regions. Recent results (Podio et al. 2006) yield an average mass outflow rate of $5 \times 10^{-8}\ M_\odot\ \mathrm{yr}^{-1}$, markedly higher than previous estimates. Outflows have also been detected by optical observations of intermediate-mass ($\sim$2–10 $M_\odot$) Herbig Ae/Be stars and other high-luminosity sources. The jet speeds and mass outflow rates in these protostars are larger by a factor $\sim$2–3 and $\sim$10–100, respectively, than in the low-L_{bol} objects.

THE LINK TO ACCRETION DISKS. Strong correlations are found between the presence of outflow signatures (P-Cygni line profiles, forbidden-line emission, thermal radio radiation, well-developed molecular lobes) and accretion diagnostics (ultraviolet [UV], IR, and millimeter emission excesses, inverse P-Cygni line profiles) in T Tauri stars (e.g., Hartigan et al. 1995). Such correlations evidently extend smoothly to protostars with masses of $\sim 10\ M_{\odot}$. A related finding is that the apparent decline in outflow activity with stellar age follows a trend similar to that exhibited by disk frequency and inferred mass-accretion rate. In addition, correlations of the type $\dot{M} \propto L_{\rm bol}^{q}$ (with $q \sim 0.6$–0.7) have been found in both low-$L_{\rm bol}$ and high-$L_{\rm bol}$ protostars for mass accretion rates as well as for mass outflow rates in ionized jets and in bipolar molecular lobes. Furthermore, CTTS-like accretion and outflow phenomena have now been detected also in very low-mass stars and brown dwarfs (e.g., Mohanty et al. 2005).

These findings strongly suggest that outflows are powered by accretion and that the same basic physical mechanism operates in both low- (down to nearly the planetary mass limit) and intermediate-mass protostars, and possibly also in some higher-mass objects (e.g., Shepherd 2005). The accretion proceeds through circumstellar disks, which can be directly probed by means of high-spectral- and spatial-resolution (in particular, interferometric) observations at submillimeter, millimeter, mid-IR, near-IR, and optical wavelengths. The disks appear to be rotationally supported (for $r \lesssim 100$ AU), and when the rotation law can be determined, it is usually consistent with being Keplerian ($v_{\phi} \propto r^{-1/2}$). Recent data indicate that $M_{\rm disk} \lesssim M_{*}$ at least up to $M_{*} \sim 20\ M_{\odot}$.

Strong evidence for a disk origin for the observed outflows is available for FU Orionis outbursts in rapidly accreting young protostars (e.g., Hartmann & Kenyon 1996). It is inferred that the emission during an outburst (of typical duration $\sim 10^{2}$ yr) originates in a rotating disk and that the outflow represents a wind that accelerates from the disk surface (with $\dot{M}_{\rm w}/\dot{M}_{\rm a} \sim 0.1$; $\dot{M}_{\rm a} \sim 10^{-4}\ M_{\odot}\ {\rm yr}^{-1}$). It has been suggested (Hartmann 1997) that most of the mass accumulation and ejection in low-mass protostars occurs during recurrent outbursts of this type. Interestingly, estimates in CTTSs (e.g., Kurosawa et al. 2006; Ray et al. 2007) indicate that $\dot{M}_{\rm w}/\dot{M}_{\rm a}$ has a similar value (i.e., ~ 0.1) also during the quiescent phases of these protostars.

4.2. Magnetic Driving of Protostellar Outflows

The most widely accepted explanation of protostellar outflows is that they tap the rotational kinetic energy of the disk (and/or central object) and are accelerated centrifugally from the disk (or stellar) surface by the stress of a large-scale, ordered magnetic field that threads the source. In this subsection we review the arguments that have led to this picture and list a few alternatives for the origin of the field.

OUTFLOW DRIVING FORCES. The momentum discharges inferred from observations of protostellar jets are compatible with the values deduced for the bipolar molecular outflows but are typically a factor $\sim 10^2$–10^3 higher than the radiation-pressure thrust L_{bol}/c produced by the stellar luminosity, ruling out radiative acceleration as a dominant driving mechanism in low- and intermediate-mass protostars (e.g., Richer et al. 2000). Radiative effects could, however, be important in driving photoevaporative disk outflows, particularly in high-L_{bol} systems (e.g., Hollenbach et al. 1994; see chapter 8). Disk heating by a luminous protostar (such as a Herbig Be star) could potentially also induce a line-driven wind from the inner disk (Drew et al. 1998). Another effect is radiation pressure on the dusty outer regions of a disk wind, which could act to decollimate the streamlines even in comparatively low-L_{bol} systems (see § 4.3 and cf. Königl & Kartje 1994).

Thermal pressure acceleration is commonly discounted as a dominant mechanism since the requisite high (effectively virial) temperatures are not generally observed at the base of the flow. However, it has been suggested that under suitable conditions, the thermal energy released in shocks at the boundary layer between the disk and the star could be efficiently converted into outflow kinetic energy (Torbett 1984; Soker & Regev 2003). Even if this is not a dominant mechanism, thermal pressure effects are nevertheless important in the mass loading of hydromagnetic winds (see § 4.3) and could potentially also play a significant role in the initial acceleration of disk outflows of this type (e.g., Pesenti et al. 2004).

A general result for protostellar jets can be obtained by combining

1. $\dot{M}_{jet} v_{jet} \sim 10^2 - 10^3\, L_{bol}/c$,
2. $L_{jet} \sim \dot{M}_{jet} v_{jet}^2$,
3. $v_{jet} \sim 10^{-3}\, c$, and
4. $L_{bol} \sim L_{acc}\ (\sim GM_* \dot{M}_a / R_*)$

(where R_* is the stellar radius), which implies that on average, the jet kinetic luminosity is related to the released accretion luminosity by $L_{jet} \sim 0.1$–$1\, L_{acc}$.[8] Such a high ejection efficiency is most naturally understood if the jets are driven hydromagnetically (see eqs. [68] and [70]). This mechanism also provides a natural explanation of the strong collimation exhibited by protostellar jets (see § 4.3).

8. Note that in protostars that are beyond the main accretion phase, L_{bol} can be expected to exceed L_{acc} (e.g., Tilling et al. 2008), which would increase the inferred value of L_{jet}/L_{acc} in these sources.

ORIGIN OF LARGE-SCALE DISK MAGNETIC FIELDS. Perhaps the most likely origin of a large-scale, open magnetic field that can launch a CDW from a protostellar disk is the interstellar field that supports the natal molecular cloud core. We have already considered the observational evidence for such a core-threading field (§ 1), its dynamic effect on the core (§ 2.1), and how it is advected inward by the inflowing matter during gravitational collapse (§ 3). However, to complete the argument, it is still necessary to demonstrate that once a rotationally supported disk is established, magnetic flux that is brought in to its outer edge remains sufficiently well coupled to the inflowing matter to be distributed along the disk plane. A potential problem arises, however, in turbulent disks in which only radial angular momentum transport is present. In particular, in a disk of characteristic radius R and half-thickness $H \ll R$ in which angular momentum transport is due to an effective turbulent viscosity $\nu_{\rm turb}$, inward dragging of magnetic-field lines is possible only if the effective turbulent magnetic diffusivity $\eta_{\rm turb}$ satisfies $\eta_{\rm turb} \lesssim (H/R)\, \nu_{\rm turb}$,[9] which may not occur naturally in such a system (Lubow et al. 1994a).[10] The introduction of vertical angular momentum transport along a large-scale magnetic field offers a straightforward way out of this potential dilemma by decoupling these two processes: magnetic diffusivity is determined by the ionization state and composition of the disk (or else by the local turbulence, if one is present) and no longer has the same underlying physical mechanism as the angular momentum transport. The disk-formation models presented in § 3 provide concrete examples of how open magnetic-field lines can be self-consistently incorporated into a disk in which they constitute the dominant angular momentum transport channel (see also Spruit & Uzdensky 2005).

An alternative possibility is that the outflow is driven by the stellar dynamo-generated field. This could happen either along field lines that have been effectively severed after they penetrated the disk, as in the X-wind scenario (e.g., Shu et al. 2000), or along field lines that are still attached to the star. However, even in the latter case the outflow is envisioned to result from an interaction between the stellar magnetic field and the disk (see § 5). It has also been

9. This follows from a comparison of the mass inflow speed, $\sim \nu_{\rm turb}/R$, as inferred from the angular momentum conservation equation, with the field diffusion speed relative to the gas, $\sim \eta_{\rm turb}/H$, as inferred from the flux-conservation equation.

10. Recent numerical simulations have indicated that the turbulent η is, in fact, of the order of the turbulent ν in disks where the turbulence derives from the magnetorotational instability (Guan & Gammie 2009; Lesur & Longaretti 2009). Note, however, that efficient inward advection of magnetic flux may nevertheless be possible in this case on account of the expected suppression of the instability near the disk surfaces (e.g., Rothstein & Lovelace 2008).

proposed that a large-scale, open field could be generated by a disk dynamo (e.g., Tout & Pringle 1996; von Rekowski et al. 2003; Blackman & Tan 2004; Pudritz et al. 2007; Uzdensky & Goodman 2008). In this connection it is worth noting that even if the disk dynamo generates small-scale, closed magnetic loops, it is conceivable, if we extrapolate from the situation on the Sun, that some of these loops could be dynamically extended to sufficiently long distances (in particular, beyond the respective Alfvén surfaces) for them to become effectively open (Blandford & Payne 1982).

Direct evidence for the presence of magnetic fields in either the outflow or the disk has been scant, but this may not be surprising in view of the considerable observational challenges. Among the few actual measurements are the strong circular polarization that was detected on scales of $\sim$20 AU in T Tau S (Ray et al. 1997), which was interpreted as a field of at least several gauss that was advected from the origin by the associated outflow,[11] and the meteoritic evidence that points to the presence in the protosolar nebula of a $\sim$1 G field at $r \sim 3$ AU (Levy & Sonett 1978). In view of the strong indications for disk launched outflows in FU Orionis systems, it is also worth mentioning the Zeeman-signature least-square deconvolution measurement in FU Ori (Donati et al. 2005), which was interpreted as a $\sim$1 kG field (with $|B_\phi| \sim B_z/2$) originating on scales of $\sim$0.05 AU in the associated disk (with the direction of B_ϕ being consistent with its origin in shearing of the poloidal field by the disk differential rotation). Further support for a magnetic disk–outflow connection has been inferred from measurements of apparent rotations in a number of protostellar jets, although, as noted in § 4.3, this evidence is still controversial.

4.3. Centrifugally Driven Winds

Our focus here is on the role of CDWs as an angular momentum transport mechanism. We survey their main characteristics and concentrate on those aspects that are relevant to the construction of combined disk/wind models, the topic of the next subsection.

WIND EQUATIONS. The qualitative basis for such outflows was considered in § 2.3. In a formal treatment, the winds are analyzed by using the equations of time-independent, ideal MHD.[12] These equations are as follows:

11. Circular polarization on much larger physical scales was detected along the HH 135–136 outflow, where it was similarly interpreted as evidence for a helical magnetic field (Chrysostomou et al. 2007).

12. The ideal-MHD assumption is justified by the fact that the charged particles' drift speeds rapidly become much lower than the bulk speed as the gas accelerates away from the disk surface; see § 2.4.

$$\nabla \cdot (\varrho \boldsymbol{v}) = 0 \qquad \text{(mass conservation [continuity equation])}, \tag{58}$$

$$\varrho \boldsymbol{v} \cdot \nabla \boldsymbol{v} = -\nabla P - \varrho \nabla \Phi + \frac{1}{4\pi} (\nabla \times \boldsymbol{B}) \times \boldsymbol{B} \quad \text{(momentum conservation [force equation])}, \tag{59}$$

where we take the thermal pressure to be given by the perfect-gas law

$$P = \frac{k_{\rm B} T \rho}{\mu m_{\rm H}} \tag{60}$$

(where T is the temperature, $k_{\rm B}$ is Boltzmann's constant, μ is the molecular weight, and $m_{\rm H}$ is the hydrogen atom mass),

$$\nabla \times (\boldsymbol{v} \times \boldsymbol{B}) = 0 \qquad \text{(induction equation [magnetic-field evolution])} \tag{61}$$

(cf. eq. [20]), and the solenoidal condition (eq. [3]). In general, one also needs to specify an entropy-conservation equation (balance of heating and cooling). For simplicity, however, we specialize to isothermal flows (spatially uniform isothermal sound speed $C = \sqrt{P/\varrho}$).

GENERIC PROPERTIES. If we concentrate on axisymmetric flows, it is convenient to decompose the magnetic field into poloidal and azimuthal components, $\boldsymbol{B} = \boldsymbol{B}_{\rm p} + \boldsymbol{B}_\phi$, with $\boldsymbol{B}_{\rm p} = (\nabla A \times \hat{\boldsymbol{\phi}})/r$, where the poloidal flux function $A(r, z)$ can be expressed in terms of the poloidal magnetic flux by $A = \Psi/2\pi$. The flux function is constant along $\boldsymbol{B}$ ($\boldsymbol{B} \cdot \nabla A = 0$) and thus can be used to label the field lines. The solution of the induction equation is given by

$$\boldsymbol{v} \times \boldsymbol{B} = \nabla \chi \ , \tag{62}$$

which shows that the electric field $\boldsymbol{E} = -(\boldsymbol{v} \times \boldsymbol{B})/c$ is derivable from an electrostatic potential (χ). Since $\partial \chi / \partial \phi = 0$ because of the axisymmetry, it follows that $\boldsymbol{v}_{\rm p} \parallel \boldsymbol{B}_{\rm p}$. Equivalently,

$$\varrho \boldsymbol{v}_{\rm p} = k \boldsymbol{B}_{\rm p} \ , \tag{63}$$

where k is a flux-surface constant ($\boldsymbol{B} \cdot \nabla k = 0$, using $\nabla \cdot (\varrho \boldsymbol{v}) = 0$ and $\nabla \cdot \boldsymbol{B} = 0$). Physically, k is the wind mass-load function,

$$k = \frac{\varrho v_{\rm p}}{B_{\rm p}} = \frac{{\rm d}\dot{M}_{\rm w}}{{\rm d}\Psi} \ , \tag{64}$$

whose value is determined by the physical conditions near the top of the disk (more precisely, at the sonic, or slow-magnetosonic, critical surface; see discussion below). By taking the vector dot product of $\boldsymbol{B}_{\rm p}$ with eq. (62)

and using eq. (63), one finds that the field-line angular velocity $\Omega_B = \Omega - (kB_\phi/\varrho r)$, where Ω is the matter angular velocity, is also a flux-surface constant. By writing this relation as

$$B_\phi = \frac{\varrho r}{k}(\Omega - \Omega_B), \tag{65}$$

we see that Ω_B can be identified with the angular velocity of the matter at the point where $B_\phi = 0$ (the disk midplane, assuming an odd reflection symmetry).[13]

Using the ϕ component of the momentum-conservation eq. (59) and applying eq. (63) and the field-line constancy of k, one finds that

$$l = rv_\phi - \frac{rB_\phi}{4\pi k} \tag{66}$$

is a flux-surface constant as well, representing the conserved total specific angular momentum (matter plus electromagnetic contributions) along a poloidal field line (or streamline). In a CDW the magnetic component of l dominates the matter component near the disk surface, $\frac{1}{\varrho v_p}\frac{r|B_p B_\phi|}{4\pi} \gg rv_\phi$, whereas at large distances this inequality is reversed. The transfer of angular momentum from the field to the matter is the essence of the centrifugal acceleration process. This transfer also embodies the capacity of such a wind to act as an efficient angular momentum transport mechanism. In fact, as was already noted in § 2.2, $rB_{z,s}B_{\phi,s}/4\pi$ represents the magnetic torque per unit area that is exerted on each surface of the disk. As was also already discussed in connection with magnetic braking, the value of $B_{\phi,s}$ is determined by the conditions outside the disk, essentially by the inertia of the matter that absorbs the transported angular momentum and exerts a back torque on the disk. In the case of a CDW, $B_{\phi s}$ is effectively fixed by the regularity condition at the Alfvén critical surface, which is the largest cylindrical radius from which information (carried by Alfvén waves) can propagate back to the disk. This condition yields the value of the Alfvén lever arm r_A, which satisfies

$$l = \Omega_B r_A^2 . \tag{67}$$

The rate of angular momentum transport by the wind is thus $\sim \dot{M}_w \Omega_0 r_A^2$ (where we replaced Ω_B by the midplane value of Ω), whereas the rate at which angular momentum is advected inward by the accretion disk is $\sim \dot{M}_a \Omega_0 r_0^2$. Hence wind transport can enable accretion at a rate

13. Eqs. (62)–(65) also hold in a nonideal fluid so long as the field remains frozen into a certain particle species (such as the electrons); in this case $\boldsymbol{v}$ and Ω are replaced by the velocity and angular velocity, respectively, of that species (e.g., Königl 1989).

68
$$\dot{M}_{\rm a} \simeq (r_{\rm A}/r_0)^2 \, \dot{M}_{\rm w} \, .$$

CDW solutions can have $r_{\rm A}/r_0 \sim 3$ for reasonable parameters, indicating that such outflows could transport the bulk of the disk angular momentum in protostellar systems if $\dot{M}_{\rm w} \simeq 0.1 \, \dot{M}_{\rm a}$, which is consistent with the observationally inferred mass outflow rates.

By taking the vector dot product of $\boldsymbol{B}_{\rm p}$ with eq. (59), one finds that the specific energy

69
$$\mathcal{E} = \frac{1}{2} v^2 - \frac{r \Omega_B B_\phi B_{\rm p}}{4\pi \varrho v_{\rm p}} + w + \Phi \, ,$$

where $w = C^2 \ln(\varrho/\varrho_{\rm A})$ (with $\varrho_{\rm A} = 4\pi k^2$ being the density at the Alfvén surface) is the specific enthalpy, is constant along flux surfaces. This is the generalized Bernoulli equation, in which the magnetic term arises from the poloidal component of the Poynting flux $c \, \boldsymbol{E} \times \boldsymbol{B}/4\pi$. By using eqs. (66) and (67), one can rewrite this term as $\Omega_0 \, (\Omega_0 r_{\rm A}^2 - \Omega r^2)$, and by approximating $\mathcal{E} \approx v_{\rm p,\infty}^2/2$ as $r \to \infty$ and assuming that $(r_{\rm A}/r_0)^2 \gg 1$, one can estimate the value of the asymptotic poloidal speed as

70
$$v_{\rm p,\infty} \simeq 2^{1/2} \Omega_0 r_{\rm A} \, ,$$

or $v_{\rm p,\infty}/v_{\rm K} \approx 2^{1/2} r_{\rm A}/r_0$. This shows that such outflows are capable of attaining speeds that exceed (by a factor of up to a few) the Keplerian speed $v_{\rm K} = r_0 \Omega_0$ at their base and thus could in principle account for the measured velocities of protostellar jets.

By combining $\mathcal{E}$, Ω_B, and l, one can form the field-line constant

71
$$\mathcal{H} \equiv \mathcal{E} - \Omega_B l = \frac{1}{2} v^2 - r^2 \Omega_B \Omega + \Phi \, ,$$

where we omitted w from the explicit expression. Evaluating at r_0 gives $\mathcal{H} = -\frac{3}{2} v_{\rm K}^2 = -\frac{3}{2} (G M_* \Omega_0)^{2/3}$, and when this is combined with the form of $\mathcal{H}$ at large distances (r_∞), one gets

72
$$r_\infty v_\phi(r_\infty) \Omega_0 - \frac{3}{2} (G M_*)^{2/3} \Omega_0^{2/3} - \frac{1}{2} \left[v_{\rm p}^2(r_\infty) + v_\phi^2(r_\infty) \right] \approx 0 \, .$$

If one could measure the poloidal and azimuthal speeds at a location r_∞ in a protostellar jet and estimate the central mass M_*, one would be able to use eq. (72) (regarded as a cubic in $\Omega_0^{1/3}$) to infer the jet-launching radius $r_0 = (G M_*/\Omega_0^2)^{1/3}$ (Anderson et al. 2003). One could then also use eqs. (65) and (67) to estimate $|B_{\phi\infty}/B_{\rm p\infty}|$ and $r_{\rm A}/r_0$, respectively, and verify that they are significantly greater than 1, as required for self-consistency. This has already been attempted in several instances, where it was claimed that the results are consistent with a disk-driven outflow that originates on scales $\gtrsim 1$ AU and carries a significant fraction of the disk angular momentum (e.g., Ray et al. 2007; Coffey et al. 2007; Chrysostomou et al. 2008). However, the interpretation of the measurements has been controversial (e.g., Soker 2005; Cerqueira et al. 2006), and there have been some conflicting observational results (e.g., Cabrit et al. 2006), so the issue is not yet fully settled.

CRITICAL POINTS OF THE OUTFLOW. The critical points occur in stationary flows at the locations where the fluid velocity equals the speed of a backward-propagating disturbance. Since the disturbances propagate along the characteristics of the time-dependent equations, the critical points can be regarded as relics of the initial conditions in a time-dependent flow (Blandford & Payne 1982). If one takes the magnetic-field configuration as given and considers the poloidal flow along $\boldsymbol{B}_\mathrm{p}$, one can derive the locations of the critical points and the values of the critical speeds (i.e., the values of v_p at the critical points) by regarding the Bernoulli integral as a function of the spatial coordinate along the field line and of the density ($\mathcal{H} = \mathcal{H}(s, \varrho)$, where s is the arc length of the streamline) and deriving the extrema of $\mathcal{H}$ by setting $\partial\mathcal{H}/\partial\varrho = 0$ (which yields the critical speeds) and $\partial\mathcal{H}/\partial s = 0$ (which, together with the other relation, yields the locations of the critical points). The critical points obtained from the generalized Bernoulli eq. (69) correspond to v_p becoming equal to either the slow- or the fast-magnetosonic wave speeds (e.g., Sakurai 1985). In the full wind problem, the shape of the field lines must be determined as part of the solution by solving also the transfield (or Grad-Shafranov) equation, which involves the force balance across the flux surfaces. This equation introduces a critical point corresponding to $v_\mathrm{p} = v_\mathrm{Ap}$ (Okamoto 1975). However, the critical points of the combined Bernoulli and transfield equations are in general different from those obtained when these two equations are solved separately. The modified slow, Alfvén, and fast points occur on surfaces that correspond to the so-called limiting characteristics (or separatrices; e.g., Tsinganos et al. 1996; Bogovalov 1997). The relevance of the modified critical points was recognized already in the original radially self-similar CDW model constructed by Blandford & Payne (1982). The modified critical surfaces of the exact, semianalytic wind solutions obtained in this model are defined by the locations where the component of the flow velocity that is perpendicular to the directions of axisymmetry (i.e., $\hat{\boldsymbol{\phi}}$) and self-similarity (i.e., the spherical radius vector $\hat{\boldsymbol{R}}$) equals the MHD wave speed in that direction (the $\hat{\boldsymbol{\theta}}$ direction, using spherical coordinates $\{R, \theta, \phi\}$). The significance of the modified fast surface, which in general is located beyond its classical counterpart, is that the poloidal acceleration of the wind continues all the way up to it: initially (roughly until the flow reaches the Alfvén surface) the acceleration is primarily centrifugal, but farther out the pressure gradient of the azimuthal field component comes to dominate.

Example: The Slow-Magnetosonic Critical Surface. The slow-magnetosonic critical surface is relevant to the determination of the mass flux in a disk-driven wind. There is some uncertainty about this case since the location of the first critical point is typically very close to the disk surface (with $|z|$ being $\ll r$), so a priori it is not obvious that ideal MHD is already a good approximation there. Under nonideal-MHD conditions, all magnetic terms in a disturbance are formally wiped out by magnetic diffusivity on its backward propagation from spatial infinity, and one is left with pure sound waves (e.g., Ferreira & Pelletier 1995). Here we discuss the situation in which the critical point is encountered when the charged particles' drift speeds are already small enough in comparison with the bulk speed to justify employing ideal MHD. In this case the poloidal flow is parallel to the poloidal magnetic field (see eq. [63]), and the relevant wave speed is the slow-magnetosonic (sms) one. We consider the standard (rather than the modified) sms point, and our explicit derivation can hopefully serve to correct inaccurate statements that have appeared in some of the previous discussions of this topic in the literature.

In the magnetically dominated region above the disk surface, the shape of the field lines changes on the scale of the spherical radius R. Anticipating that the height $z_{\rm sms}$ of the sms point is $\ll R$, we approximate the shape of the field line just above the point $\{r_{\rm s},\ z_{\rm s}\}$ on the disk surface by a straight line (cf. § 2.3):

$$r = r_{\rm s} + s \sin\theta_{\rm s}\,, \quad z = z_{\rm s} + s\cos\theta_{\rm s}\,, \tag{73}$$

where the angle $\theta_{\rm s}$ gives the field-line inclination at the disk surface ($\sin\theta_{\rm s} = B_{r,{\rm s}}/B_{\rm p,s}$; $\tan\theta_{\rm s} = B_{r,{\rm s}}/B_z$). The Bernoulli integral (in the form of eq. [71]) then becomes, after substituting $\Phi = -GM_*/(r^2+z^2)^{1/2}$,

$$\mathcal{H}(s,\varrho) = \frac{k^2 B_{\rm p}^2}{2\varrho^2} + \frac{\Omega_B^2}{2}(r_{\rm s} + s\sin\theta_{\rm s})^2 \left(\frac{\Omega}{\Omega_B}\right)\left(\frac{\Omega}{\Omega_B} - 2\right) - \frac{GM_*}{[(r_{\rm s}+s\sin\theta_{\rm s})^2 + (z_{\rm s}+s\cos\theta_{\rm s})^2]^{1/2}} + C^2 \ln\left(\frac{\varrho}{\varrho_{\rm A}}\right), \tag{74}$$

where $\Omega(s,\varrho)$ is given by combining

$$\Omega = \frac{1 - \frac{r_{\rm A}^2 \varrho_{\rm A}}{r^2 \varrho}}{1 - \frac{\varrho_{\rm A}}{\varrho}}\, \Omega_B \tag{75}$$

(obtained from eqs. [65]–[67]) and eq. (73). In the hydrostatic approximation to the disk structure (see § 4.4), $\Omega_B = \Omega_K(r_s)$. We expect this equality to hold approximately also for the exact solution, in which $v_z \neq 0$, so that, in particular, $\Omega_B \approx \Omega$ in the region of interest.

Setting $\partial\mathcal{H}/\partial\varrho = 0$ yields the speed of an sms wave propagating along $\hat{\boldsymbol{B}}_p$:

$$v_{sms} = \frac{B_{p,s}}{B_s} C , \tag{76}$$

where $B_s = (B_{p,s}^2 + B_{\phi,s}^2)^{1/2}$ and we approximated $(B_{p,sms}/B_{sms})^2$ by $(B_{p,s}/B_s)^2$ and assumed $\varrho/\varrho_A \gg 1$ and $(\Omega/\Omega_B - 1)^2 \approx (r_A/r)^4(\varrho_A/\varrho)^2 \ll 1$ in the region between the top of the disk and the sms surface. Setting also $\partial H/\partial r = 0$ and approximating $r_{sms} \approx r_s$ then give the height of this point:

$$\frac{z_{sms}}{z_s} = \frac{3\tan^2\theta_s}{3\tan^2\theta_s - 1} . \tag{77}$$

Eq. (77) yields a meaningful result only if $\tan\theta_s > 1/\sqrt{3}$, i.e., if the field line is inclined at an angle $> 30°$ to the z axis. This is the CDW-launching condition in a Keplerian disk (eq. [6]) that was derived in § 2.3 using the mechanical analogy to a bead on a rigid wire. The relationship between these two derivations becomes clear when one notes that in the limit $(\Omega/\Omega_B - 1)^2 \ll 1$, the second term on the right-hand side of eq. (74) becomes equal to the centrifugal potential $-\Omega^2 r^2/2$, so that the second and third terms (which together dominate the right-hand side of this equation) are just the effective potential Φ_{eff} used in eq. (5).[14] The correspondence of the sms point to the maximum of Φ_{eff} was already noted in § 2.3.

The density at the sms point can be related to the density at the disk surface by evaluating the energy integral (74) at both z_s and z_{sms}. This yields

$$\frac{\varrho_{sms}}{\varrho_s} = \exp\left\{-\frac{1}{2} - \frac{1}{2}\left(\frac{z_s\Omega_K(z_s)}{C}\right)^2 \frac{1}{3\tan^2\theta_s - 1}\right\} , \tag{78}$$

where we assume $(v_{p,s}/v_{sms})^2 \ll 1$. The mass flux injected into the wind from the two sides of the disk is then

$$\frac{1}{2\pi r}\frac{d\dot{M}_w}{dr} = 2\varrho_{sms}v_{sms}\cos\theta_s = 2\varrho_{sms}C\frac{B_z}{B_s} \tag{79}$$

(e.g., Lovelace et al. 1995).

When θ_s approaches (and decreases below) 30°, the field-line curvature needs to be taken into account in the analysis. The height of the sonic point

14. Furthermore, the condition $\partial\mathcal{H}/\partial r = 0$ is equivalent to $\boldsymbol{B}_p \cdot \nabla\Phi_{eff} = 0$ in this limit (Campbell 2002), so the latter relation can also be used to obtain the location of the sms point in this case (e.g., Ogilvie 1997).

rapidly increases to $\sim R$, with the potential difference growing to $\sim GM_*/R$: the launching problem becomes essentially that of a thermally driven spherical wind (e.g., Levinson 2006).

It has been argued (Spruit 1996) that the field-line inclination increases systematically with radius along the disk surface and that only in a narrow radial range are the conditions favorable for driving a wind that both satisfies the CDW-launching condition and is not overloaded (and hence conceivably highly unstable; cf. Cao & Spruit 1994). This is an intriguing suggestion, given the fact that so far there is no observational evidence for an extended wind-driving region in protostellar disks. However, in principle it may be possible to launch outflows over a large radial range. In particular, as illustrated by the similarity solution presented in § 3, if the magnetic flux is advected inward by the accretion flow, the field-line inclination could be favorable for launching and need not change strongly along the disk. Furthermore, evidently the mass loading of stable disk/wind configurations actually decreases as θ_s is increased (see § 4.5).

EXACT WIND SOLUTIONS. As was already noted in § 2.5, the radial self-similarity approach has been used to construct exact global solutions of CDWs. The basic character of this model is revealed by the prototypal Blandford & Payne (1982) solution. The underlying assumption that all quantities scale as a power law of the spherical radius implies that all the critical surfaces are conical. Furthermore, all the relevant speeds (including fluid, sound, and Alfvén) must scale like the characteristic speed of the problem (v_K), i.e., as $R^{-1/2}$. The scaling of the magnetic-field amplitude can be inferred from the vertically integrated thin-disk equations presented in § 3.2, in which we now set $\partial/\partial t \equiv 0$. In particular, if the mass outflow from the disk has only a negligible effect on the accretion rate, then $\dot{M}_a = 2\pi r|v_r|\Sigma = \text{const}$ by eq. (31), and the angular momentum conservation eq. (33) in a Keplerian disk can be written as

$$\frac{1}{2}\dot{M}_a v_K = r^2|B_z B_{\phi,s}| \,. \tag{80}$$

Eq. (80) implies the similarity scaling $B \propto r^{-5/4}$, from which we infer by dimensional arguments ($v_A \propto B/\sqrt{\varrho}$) that $\varrho \propto r^{-3/2}$ in this case. It then follows that $\dot{M}_w \propto \ln r$ (cf. eq. [79]), which is consistent with the underlying assumption that only a small fraction of the inflowing mass leaves the disk over each decade in radius. More generally, one can define a mass ejection index $\xi > 0$ by

$$\xi \equiv \frac{\mathrm{d}\ln \dot{M}_a}{\mathrm{d}\ln r} \tag{81}$$

(e.g., Ferreira & Pelletier 1995) and deduce

$$\frac{\mathrm{dln}\, B}{\mathrm{dln}\, r} = \frac{\xi}{2} - \frac{5}{4}, \quad \frac{\mathrm{dln}\, \varrho}{\mathrm{dln}\, r} = \xi - \frac{3}{2} \tag{82}$$

(e.g., Contopoulos & Lovelace 1994), with $\xi \to 0$ corresponding to the Blandford & Payne (1982) solution.[15] The poloidal electric current scales as $I \propto rB_\phi \propto r^{(2\xi-1)/4}$. For $\xi > 1/2$, the flow is in the current-carrying regime, with the poloidal current density being antiparallel to the magnetic field. In this case the current tends to zero as the symmetry axis is approached, so such solutions should provide a good representation of the conditions near the axis of a highly collimated flow. Conversely, solutions with $\xi < 1/2$ correspond to the return-current regime (in which the poloidal current density is parallel to the field) and are most suitable at larger cylindrical distances. The Blandford & Payne (1982) solution is sometimes critiqued for having a singular behavior on the axis.[16] One should bear in mind, however, that even though the detailed global current distribution (which includes both current-carrying and return-current regimes) cannot be represented by the simplified self-similar solution, this flaw is not fundamental. Besides, given that the disk has a finite inner radius, the issue of the behavior of the flow at $r = 0$ is, to a certain degree, merely academic. Whereas the value of the ejection index is arbitrary in pure wind models, it becomes an eigenvalue of the self-similar solution (fixed by the regularity condition at the sonic, or sms, critical surface) when one considers a combined disk/wind model (e.g., Li 1996a).

Although the magnetic-field lines must bend away from the symmetry axis within the disk in order to satisfy the wind-launching condition (6) at the surface, as soon as the magnetically dominated region at the base of the wind is reached, they start to bend back toward the axis because of the magnetic tension force (see § 2.1). Further collimation is achieved in current-carrying jets by the hoop stress of the toroidal magnetic field (the term $-J_z B_\phi/c = -(1/8\pi r^2)\partial(rB_\phi)^2/\partial r$ in the radial force equation), the analog of a z-pinch in laboratory plasma experiments. The asymptotic behavior in general

15. Interestingly, one can also generalize the Blandford & Payne (1982) solution to a class of semianalytic but non-self-similar solutions, defined by the constancy of a certain function of the magnetic flux that controls the shape of the Alfvén surface (Pelletier & Pudritz 1992). The Blandford & Payne solution then separates wind configurations in which the initial field-line inclination increases progressively with radius from outflows that emerge from a bounded region and in which the initial field inclination decreases with r and the field lines converge into a cylindrical sheath.

16. It is in fact a double singularity, since on the axis itself there is an oppositely directed line current that exactly compensates for the distributed return current.

depends on the current distribution: if $I \to 0$ as $r \to \infty$, then the field lines are space-filling paraboloids, whereas if this limit for the current is finite, then the flow is collimated to cylinders (e.g., Heyvaerts & Norman 1989). In practice, however, self-similar wind solutions typically do not reach the asymptotic regime but instead self-focus (with streamlines intersecting on the axis) and terminate at a finite height (e.g., Vlahakis et al. 2000).

The cold self-similar wind solutions are specified by any two of the following three parameters:

$$\lambda \equiv \frac{l}{v_{\mathrm{Ks}}\, r_{\mathrm{s}}} \qquad \text{(normalized specific angular momentum)},$$

$$\kappa \equiv 4\pi k \frac{v_{\mathrm{Ks}}}{B_{z,\mathrm{s}}} \qquad \text{(normalized mass/magnetic-flux ratio)}, \tag{83}$$

$$b_{r,\mathrm{s}} \equiv \frac{B_{r,\mathrm{s}}}{B_{z,\mathrm{s}}} \qquad \text{(poloidal field inclination at the disk surface)},$$

where $b_{r,\mathrm{s}}$ must satisfy the constraint (6). A viable solution is further characterized by $\lambda > 1$ (typically $\gg 1$) and $\kappa < 1$ (typically $\ll 1$). A two-parameter choice yields a solution if the corresponding flow crosses the modified Alfvén critical surface; if the two parameters are κ and λ, then, for any given value of κ, this can happen only if λ exceeds a minimum value that approximately satisfies $\kappa\lambda_{\mathrm{min}}(\lambda_{\mathrm{min}} - 3)^{1/2} = 1$.[17] In § 4.4 we discuss joining a global wind solution of this type to a radially localized wind-driving disk solution. In this case the sonic (or sms) critical-point constraint imposed on the latter solution yields the value of κ, and, in turn, the Alfvén critical-point constraint on the wind solution fixes one of the disk-model parameters (for example, by providing the value of $B_{\phi,\mathrm{s}}/B_{z,\mathrm{s}} = -\kappa(\lambda - 1)$ from given values of κ and $b_{r,\mathrm{s}}$).[18]

OBSERVATIONAL IMPLICATIONS. The exact wind solutions and estimates of the physical conditions around young stellar objects make it possible to identify a number of physical characteristics of disk-driven protostellar winds that could have potentially significant observational implications. Although this topic is not directly within the scope of this book, we nevertheless briefly describe some of these properties here inasmuch as it may not always be

17. This expression differs slightly from the one that appeared in Blandford & Payne (1982).

18. Although self-similar CDW solutions in which the wind also passes through the fast-magnetosonic critical surface have been obtained, the implications of this additional constraint for a global disk/wind model remain unclear. Ferreira & Casse (2004), for instance, suggested that in order to cross this surface, the outflow must experience significant heating after it leaves the disk and that the added constraint is related to this requirement.

feasible in practice to distinguish between the observational signatures of a disk and a wind. Furthermore, invoking a disk-driven wind may sometimes provide a straightforward explanation of a disk observation that may otherwise seem puzzling. Besides the references explicitly listed at the end of the chapter, the reader may also consult the review articles by Königl & Ruden (1993) and Königl & Pudritz (2000) on this subject.

Centrifugal driving is found to be an efficient acceleration mechanism that leads to a rapid increase in the poloidal speed, and a correspondingly strong decrease in the wind density, above the disk surface (see Fig. 7.4).[19] This leads to a strong stratification, which, in fact, is most pronounced in the $\hat{\boldsymbol{z}}$ direction even in the R-self-similar wind models (Safier 1993b). The strong momentum flux in the wind also implies that an outflow originating from beyond the dust-sublimation radius could readily uplift dust from the disk (Safier 1993a). This may lead to viewing angle-dependent obscuration and shielding of the central continuum, whose effect would be distinct from that of molecular cloud gas undergoing gravitational collapse.[20] The stratified dust distribution could intercept a large portion of the central continuum radiation and reprocess it to the infrared, which could be relevant to the interpretation of the IR spectra of low- and intermediate-mass star/disk systems (e.g., Königl 1996; Tambovtseva & Grinin 2008). Furthermore, scattering by the uplifted dust could contribute to the observed polarization pattern in these objects.[21]

Although, as we already noted, ideal MHD should be an excellent approximation for modeling the dynamics of the flow (except perhaps right above the disk surface, where the nonideal formalism employed within the disk might still be relevant for determining the first critical point of the wind), the estimated degree of ionization at the base of a protostellar disk wind is typically low enough that energy dissipation induced by ion–neutral drag (ambipolar diffusion) could play an important role in the wind thermodynamics. This is because the volumetric heating rate scales as $|(\nabla \times \boldsymbol{B}) \times \boldsymbol{B}|^2/\gamma_i \varrho_i \varrho \propto 1/\gamma_i \varrho_i$ (see § 2.4), where the distributions of $\boldsymbol{B}$ and ϱ are regarded as being fixed by the flow dynamics, whereas adiabatic cooling, the most important

19. The model wind presented in Fig. 7.4 exhibits a comparatively weak collimation. The degree of collimation depends on the model parameters and is particularly sensitive to the mass loading of the outflow (e.g., Pudritz et al. 2007). Protostellar jets are inferred to correspond to more highly collimated disk winds, in which the enhanced mass loading might be attributed to the presence of a disk corona (e.g., Dougados et al. 2004).

20. In an application to active galactic nuclei, this effect was invoked by Königl & Kartje (1994) to account for the obscuring/absorbing "molecular torus" identified in the centers of Seyfert galaxies.

21. For an application of this model to the polarization properties of Seyfert galaxies, see Kartje (1995).

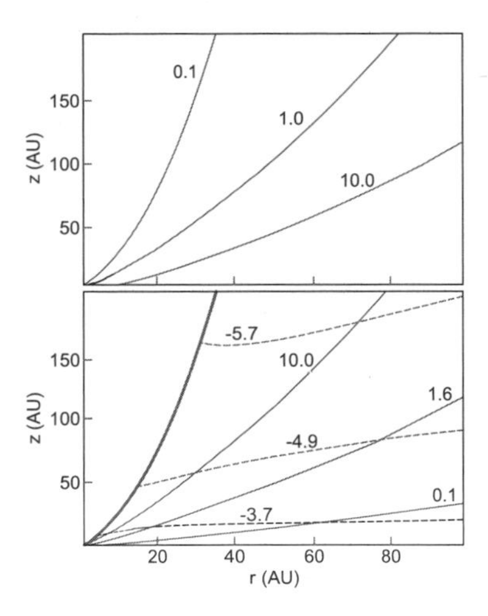

Fig. 7.4. Structure of a Blandford-Payne–type self-similar CDW from a protostellar disk (model C in Safier 1993a). Top: Meridional projections of the streamlines (labeled by the value in AUs of the radius where they intersect the disk). Bottom: Contours of $v_{p,\infty}/v_{K1}$ (solid lines), where $v_{p,\infty}$ is the asymptotic poloidal speed and v_{K1} is the Keplerian speed at 1 AU, and of $\log(\varrho/\varrho_1)$ (dashed lines), where ϱ_1 is the density at the base of the wind at 1 AU. The heavy line on the left indicates the streamline that originates at 0.1 AU.

temperature-lowering mechanism, remains relatively inefficient in a collimated disk outflow. Under these circumstances, the temperature might rise rapidly (Safier 1993a), although its precise terminal level depends on the relevant value of the collisional drag coefficient γ_i (Shang et al. 2002). This heating mechanism could contribute to the observed forbidden-line emission (Safier 1993b; Cabrit et al. 1999; Garcia et al. 2001a, 200lb) and thermal radio emission (Martin 1996) from CTTSs, although in both of these applications an additional source of heating might be needed, possibly associated with dissipation of weak shocks and turbulence (e.g., O'Brien et al. 2003; Shang et al. 2004).

On large scales, centrifugally driven outflows assume the structure of a collimated jet (most noticeable in the density contours) and a surrounding wide-angle wind (Shu et al. 1995; Li 1996b). This bears directly on the general morphology of protostellar sources, and in particular on the shapes of the wind/jet-driven outflow lobes that form in the surrounding molecular gas.

4.4. Equilibrium Disk Wind Models

In contrast with the wind zone, where the dynamics is well described by an effectively infinite conductivity, the accretion disk in protostellar systems must be modeled using nonideal MHD. The structure of the accretion flow thus depends critically on the properties of the conductivity tensor at each point, which are in turn determined by the spatial distribution of the degree of ionization (see eqs. [12]–[14]). We therefore start this subsection by reviewing the physical processes that affect disk ionization (see also chapter 3), which allow us to choose the relevant physical parameters for our models. In reality, the accretion flow itself has an influence on the ionization structure—through its effect on the disk column density (see § 2.4) or on the distribution of dust grains (see § 6), for example—which a truly self-consistent model must take into account. So far, only simpler models have been constructed, in which this influence is not fully accounted for and a variety of approximations are employed. We discuss these models next, describing the basic properties of the derived equilibrium solutions and distinguishing between *strongly coupled* and *weakly coupled* disk configurations.

DISK IONIZATION STRUCTURE. The dependence of the degree of field–matter coupling on the abundances of the ionized species in the disk (ions, electrons, and charged dust grains) is a direct outcome of the fact that this coupling is effected by the collisions of these particles with the much more abundant neutrals. The degree of ionization is given by the ratio of the sum of the positively (or, equivalently, negatively) charged particle densities to the neutral-particle density. It is calculated by balancing the ionization and recombination processes operating in the disk.

Interstellar (and possibly also protostellar) cosmic rays, as well as X-ray and far-UV radiation produced by a magnetically active protostar (e.g., Glassgold et al. 2000, 2005), are the main potential ionizing agents in protostellar environments. However, it is unclear how effective cosmic-ray ionization really is, since the low-energy particles most relevant for this purpose may be deflected by magnetized disk or stellar outflows or by magnetic mirroring near the disk surface, and they might also be scattered by magnetic turbulence within the disk. If present, they may dominate the midplane ionization even in the inner disk ($r \lesssim 1$ AU), where the surface density is larger than the X-ray attenuation length. But if cosmic rays are excluded from this region and ionized particles are not transported there by other means (e.g., Turner et al. 2007), the gas near the midplane might be ionized only at the low rate of decay of radioactive elements such as ^{40}K. On the other hand, free electrons are rapidly lost through recombination processes that occur both in the gas phase (where electrons recombine with molecular and metal ions via dissociative and radiative mechanisms, respectively) and on grain surfaces, and through sticking to dust particles (e.g., Oppenheimer & Dalgarno 1974; Nishi et al. 1991). The ionization equilibrium is, however, very sensitive to the abundance of metal atoms because they rapidly remove charges from molecular ions but then recombine much more slowly with electrons (e.g., Sano et al. 2000; Fromang et al. 2002).

Dust grains can also significantly affect the degree of field–matter coupling if they are mixed with the gas. They do so in two ways. First, they reduce the ionization fraction by absorbing charges from the gas and by providing additional recombination pathways for ions and electrons. Second, in high-density regions the grains themselves can become an important charged species (e.g., Nishi et al. 1991), leading to a reduction in the magnetic coupling because grains have much smaller Hall parameters than the much less massive electrons and ions. We note in this connection that if dust grains settle to the midplane, the ionization fraction may become sufficiently large to provide adequate magnetic coupling even in the inner ($\lesssim 1$ AU) disk regions (Wardle 2007). This conclusion could, however, be mitigated in the presence of turbulence, which might leave a residual population of small dust grains (carrying a significant fraction of the total grain charge) suspended in the disk (e.g., Nomura & Nakagawa 2006; Natta et al. 2007).

EXACT DISK SOLUTIONS. To gain physical insight into the distinguishing properties of wind-driving disks, we consider a simplified model originally formulated by Wardle & Königl (1993). In this model, the entire angular momentum of the accreted matter is assumed to be transported by a CDW. The disk is taken to be geometrically thin, vertically isothermal, stellar gravity dominated, and threaded by an open magnetic field (possessing an odd reflection symmetry about the midplane). Under the thin-disk approximation, the vertical magnetic-field component is taken to be uniform with height in the disk solution. The disk gas is assumed to be in the ambipolar diffusivity regime, with ions and electrons being the dominant charge carriers. In the original model, the ion density was taken to be constant with height (which could be a reasonable approximation in the outer regions of certain real systems). The main simplifications involve considering only a radially localized ($\Delta r \ll r$) disk region and (in accord with the thin-disk approximation) retaining the z derivatives but neglecting all r derivatives except those of v_ϕ (which scales as $r^{-1/2}$) and B_r (which appears in $\nabla \cdot \boldsymbol{B} = 0$). The neglect of the radial derivatives in the mass-conservation equation (eq. [58]) implies that the vertical mass flux ϱv_z is taken to be uniform with height in the disk solution.[22] This solution is extended through the first critical point (which is the thermal sonic point if the flow is still diffusive and the sms point if it is already in the ideal-MHD regime; see § 4.3) and then matched onto a global Blandford & Payne (1982) ideal-MHD wind solution. These simplifications do not compromise the physical essence of the results. In particular, the qualitative characteristics remain unchanged when the radially localized disk solution is generalized to a global configuration in which both the disk and the wind are described by a single self-similar model that includes both z and r derivatives (e.g., Li 1996a). Furthermore, solutions with similar properties are obtained when the disk is in the Hall or Ohm diffusivity regimes (e.g., Salmeron et al. 2011) and when the full conductivity tensor is used in conjunction with a realistic ionization profile (see Fig. 7.5).

In general, an equilibrium disk solution is specified by the following parameters:

1. The parameter $a_0 \equiv v_{A0}/C$, the midplane ratio of the Alfvén speed (based on the large-scale magnetic field) to the sound speed. This parameter measures the magnetic-field strength.[23]
2. The parameter C/v_K, which in a thin isothermal disk is equal to h_T/r, the ratio of the tidal (i.e., reflecting the vertical gravitational compression) density scale height to the disk radius. Although this parameter, which measures the geometric thinness of the disk, does not appear explicitly in the normalized structure equations, it nevertheless serves to constrain physically viable solutions (see eq. [84]).
3. The midplane ratios of the conductivity tensor components: $[\sigma_P/\sigma_\perp]_0$ (or $[\sigma_H/\sigma_\perp]_0$) and $[\sigma_\perp/\sigma_O]_0$. They characterize the conductivity regime of the gas (see § 2.4). When there are only two charged species (ions and electrons), one can equivalently specify the midplane Hall parameters β_{i0} and β_{e0}. In the inner disk regions the conductivity tensor components

22. This approximation has been critiqued (e.g., Ferreira 1997) for not allowing v_z to assume negative values within the disk, as it must do in cases (expected to be typical) in which the disk thickness decreases as the protostar is approached. This issue can be fully addressed only in the context of a global disk/wind model. However, it can be expected that any error introduced by this approximation will be minimized if the upward mass flux remains small enough for v_z to have only a weak effect on the behavior of the other variables within the disk. Under the assumption that $|v_r|$ is of the same order of magnitude as $|v_\phi - v_K|$, one can readily show that the condition for this to hold is that v_z/C remain $\ll 1$ everywhere within the disk. This can be checked a posteriori for each derived solution.

23. Note that a_0 is related to the midplane plasma beta parameter β_0 through $a_0 = (2/\gamma\beta_0)^{1/2}$, where γ denotes the adiabatic index of the fluid.

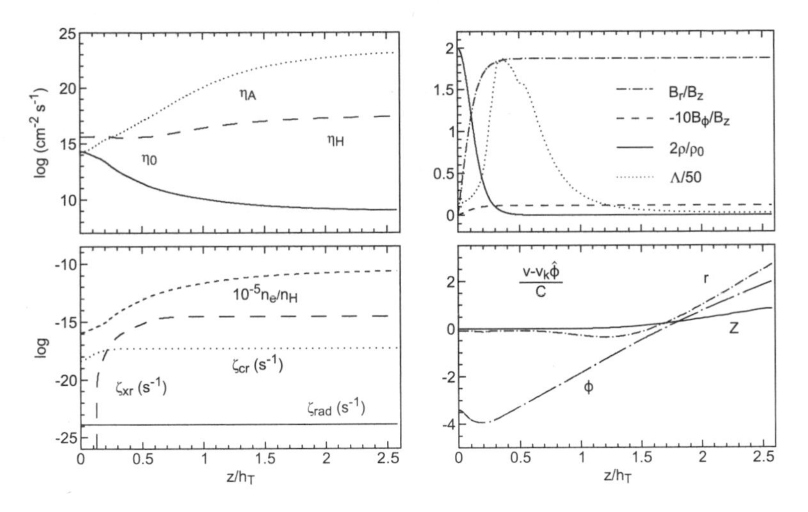

Fig. 7.5. Vertical structure of a radially localized wind-driving disk solution at 1 AU around a Sun-like protostar. The disk is assumed to have a column density of $\Sigma = 600\,\mathrm{g\,cm^{-2}}$, and its model parameters are $a_0 = 0.75$, $C/v_K = 0.1$, $\epsilon = 0.1$, and $\epsilon_B = 0$. The parameters of the matched radially self-similar CDW are $\kappa = 2.6 \times 10^{-6}$, $\lambda = 4.4 \times 10^3$, and $b_{r,s} = 1.6$. (Left) Top: Ambipolar, Hall, and Ohm diffusivities. Bottom: Ionization rates by cosmic rays (cr), X-rays (xr), and radioactivity (rad), and electron fraction n_e/n_H. (Right) Top: Radial and azimuthal magnetic-field components, mass density, and Elsasser number. Bottom: Velocity components. The mass-accretion rate for this model is $7 \times 10^{-6}\, M_\odot\,\mathrm{yr}^{-1}$, which is consistent with the inferred values for the early (Class 0/Class I) protostellar accretion phase.

typically vary with height, reflecting the ionization structure of the disk (see Fig. 7.5).

4. The midplane Elsasser number Λ_0 (see eqs. [23] and [25]), which measures the degree of coupling between the neutrals and the magnetic field.
5. The parameter $\epsilon \equiv -v_{r,0}/C$, the normalized inward radial speed at the midplane. Although the value of ϵ could in principle be negative (as, in fact, it is in certain viscous disk models; e.g., Takeuchi & Lin 2002), it is expected to remain > 0 when a large-scale magnetic field dominates the angular momentum transport.
6. The parameter $\epsilon_B \equiv -v_{Br,0}/C$, the normalized radial drift speed of the poloidal magnetic-field lines (see eq. [22]). This parameter vanishes in a strictly steady-state solution but is nonzero if, as expected, the magnetic-field lines drift radially on the long accretion timescale $r/|v_r|$. It is incorporated into the model through the z component

of eq. (20) (i.e., one sets $\partial B_z/\partial t \neq 0$ but keeps $\partial B_r/\partial t = 0$ and $\partial B_\phi/\partial t = 0$).[24]

We already remarked in § 4.3 on how the wind solution, through the Alfvén-surface regularity condition, can be used to constrain the disk solution when the two are matched. In particular, if the disk solution is used to fix the wind parameters κ (through the sonic/sms critical-point constraint) and $b_{r,s}$, the wind solution yields $B_{\phi,s}/B_z$, which in turn can be used to obtain the value of the disk parameter ϵ. Previous combined disk/wind models treated ϵ_B as a free parameter and typically set it equal to zero.[25] Under this approach, $b_{r,s}$ was fully determined by the conditions inside the disk. However, as we noted in § 3.2 (see also Ogilvie & Livio 2001), $b_{r,s}$ is in fact also determined by the conditions outside the disk and can be directly related to the distribution of B_z along the disk if the force-free field above the surface can be adequately approximated as being also current free (see eq. [40]). In a more general treatment of the disk/wind problem, this constraint can be used to fix the value of ϵ_B (see Teitler 2011).

Before we turn to the specific solution displayed in Fig. 7.5, it is instructive to list the dominant magnetic terms (under the thin-disk approximation) in the neutrals' force equation (eq. [59]) and review their main effects inside the disk. We have the following (see eqs. [28]–[30]):

- *Radial* component: $\frac{B_z}{4\pi}\frac{\mathrm{d}B_r}{\mathrm{d}z}$,
  representing the magnetic tension force that acts in opposition to central gravity
- *Azimuthal* component: $\frac{B_z}{4\pi}\frac{\mathrm{d}B_\phi}{\mathrm{d}z}$,
  representing the magnetic torque that transfers angular momentum from the matter to the field

24. It is worth noting, though, that the poloidal components of eq. (20) can be written as $\partial\Psi/\partial t = -2\pi r B_0 v_{Br,0}$, with $B_z = (1/2\pi r)\partial\Psi/\partial r$ and $B_r = -(1/2\pi r)\partial\Psi/\partial z$. Thus, if B_z changes because of the slow radial diffusion of the flux surfaces, so also will B_r. However, it can be argued (Königl et al. 2010) that B_r (and similarly also B_ϕ) can in principle change at any given radial location on the much shorter dynamic time r/v_ϕ through field-line shearing by the local velocity field. One can therefore assume that B_r and B_ϕ attain their equilibrium configurations on the short timescale and neglect, in comparison, the much slower variations that are expressed by the explicit time derivaties terms. The B_z component is distinct in that it can change only on the long radial drift time.

25. Wardle & Königl (1993) noted, however, that for physical consistency one has to require $\epsilon_B < \epsilon$. They also demonstrated that the solution variables (except for $B_{\phi,s}$ and $v_{r,0}$) are insensitive to the value of this parameter, as expected from the fact that the only modification to the equations introduced by varying ϵ_B involves changing the radial velocity of the reference frame in which the poloidal field lines are stationary.

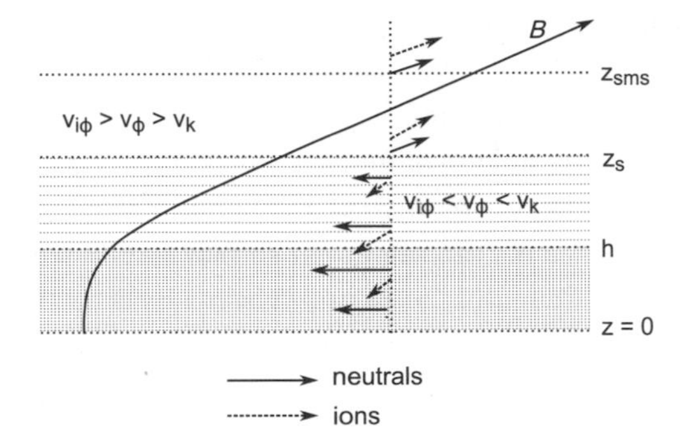

Fig. 7.6. Schematic diagram of the vertical structure of an ambipolar diffusion-dominated disk, showing a representative field line and the poloidal velocities of the neutral and the ionized gas components. Note that the poloidal velocity of the ions vanishes at the midplane ($z = 0$), consistent with the assumption that $\epsilon_B = 0$, and that it is small for both components at the top of the disk ($z = z_s$). The relation between the azimuthal speeds is also indicated.

- *Vertical* component: $-\frac{\mathrm{d}}{\mathrm{d}z}\frac{B_r^2+B_\phi^2}{8\pi}$,
  representing the magnetic squeezing of the disk (which acts in the same direction as the gravitational tidal force and in opposition to the thermal pressure-gradient force)

The solution shown in Fig. 7.5 describes the vertical structure of a disk that changes from being Hall dominated near the midplane to being ambipolar dominated near the surface. Although qualitative properties of the solution are not sensitive to the details of the diffusivity profile, it is heuristically useful to consider a distinct diffusivity regime. We therefore specialize in the following discussion to the "pure ambipolar" (with $\varrho_\mathrm{i}(z) = \mathrm{const}$) case treated in Wardle & Königl (1993). As illustrated in Fig. 7.6 (in which it is assumed that $\epsilon_\mathrm{B} = 0$), the disk can be vertically divided into three distinct zones:[26] a quasi-hydrostatic region near the midplane, where the bulk of the matter is concentrated and most of the field-line bending takes place, a transition zone where the inflow gradually diminishes with height, and an outflow region that corresponds to the base of the wind. The first two regions are characterized by a radial inflow and sub-Keplerian rotation, whereas the gas at the base of the wind flows out with $v_\phi > v_\mathrm{K}$.

- The *quasi-hydrostatic region* is matter dominated, with the ionized particles and magnetic field being carried around by the neutral material.

26. Our discussion pertains to disks in which $\Lambda > 1$ throughout their entire vertical extent, but it also applies to the $\Lambda > 1$ wind-driving surface layers of weakly coupled disks.

The ions are braked by a magnetic torque, which is transmitted to the neutral gas through the frictional drag; therefore, $v_{i\phi} < v_\phi$ in this region. The neutrals thus lose angular momentum to the field, and their back reaction leads to a buildup of the azimuthal field component away from the midplane. The loss of angular momentum enables the neutrals to drift toward the center, and in doing so they exert a radial drag on the field lines. This drag must be balanced by magnetic tension, so the field lines bend away from the rotation axis. This bending builds up the ratio B_r/B_z, as reqvired for consistency with the global field morphology. The ratio B_r/B_z needs to exceed $1/\sqrt{3}$ at the disk surface to launch a CDW. The magnetic tension force, transmitted through ion-neutral collisions, contributes to the radial support of the neutral gas and causes it to rotate at sub-Keplerian speeds.

- The growth of the radial and azimuthal field components on moving away from the midplane results in a magnetic pressure gradient that tends to compress the disk. The magnetic energy density comes to dominate the thermal and gravitational energy densities as the gas density decreases, marking the beginning of the *transition zone* (at $z \approx h$, where h is the density scale height). The field above this point is nearly force free and locally straight (see §§ 2.1 and 2.3).
- The field angular velocity Ω_B is a flux-surface constant (see eq. [65] and footnote 13; note, however, that this strictly holds only when ϵ_B is identically zero). The ion angular velocity $v_{i\phi}/r$ differs somewhat from Ω_B but still changes only slightly along the field. Since the field lines bend away from the symmetry axis, the cylindrical radius r, and hence $v_{i\phi}$, increase along any given field line, whereas v_ϕ decreases because of the near-Keplerian rotation law. Eventually a point is reached where $(v_{i\phi} - v_\phi)$ changes sign. At this point the magnetic stresses on the neutral gas are small, and its angular velocity is almost exactly Keplerian.[27] Above this point the field lines overtake the neutrals and transfer angular momentum back to the matter, and the ions start to push the neutrals out in both the radial and the vertical directions. This region can be regarded as the *base of the wind*, and one can accordingly identify the disk surface z_s with the location where v_ϕ becomes equal to v_K. The mass outflow rate is fixed by the density at the sonic/sms point (marked by z_{sms} in Fig. 7.6).

27. This result was used in the derivation of the slow-magnetosonic critical-surface properties in § 4.3.

The Hydrostatic Approximation. The structure equations of the radially localized disk model can be simplified by setting $v_z \approx 0$. Although this approximation is most appropriate for the quasi-hydrostatic region, one can nevertheless extend it to the disk surface to obtain useful algebraic constraints on viable disk solutions. For the pure ambipolar regime (and again with the assumption $\varrho_{\rm i} = {\rm const}$), they are given by (Wardle & Königl 1993; Königl 1997)

$$(2\Upsilon_0)^{-1/2} \lesssim a_0 \lesssim \sqrt{3} \lesssim \epsilon\Upsilon_0 \lesssim v_{\rm K}/2C \, , \tag{84}$$

where the parameter Υ_0 represents the midplane Elsassser number in the ambipolar limit (see eq. [25]). The four inequalities in eq. (84) have the following physical meaning (from left to right):

1. The disk remains sub-Keplerian everywhere below its surface.
2. The wind-launching condition ($b_{r,{\rm s}} > 1/\sqrt{3}$) is satisfied.
3. The top of the disk ($z_{\rm s}$) exceeds a density scale height ($h \approx (a_0/\epsilon\Upsilon_0)\, h_{\rm T}$), ensuring that the bulk of the disk material is nearly hydrostatic and that $\dot{M}_{\rm w}$ remains $\ll \dot{M}_{\rm a}$.[28]
4. The midplane Joule heating rate (see § 2.4) is less than the rate of gravitational potential energy release there ($\varrho_0 |v_{r,0}| v_{\rm k}^2/2r$ per unit volume), because the latter is the ultimate source of energy production in the disk.

Inequalities (1) and (2) together place a lower bound on the neutral–field coupling parameter Υ_0, for which a more detailed analysis yields the value of ~ 1 (see Königl et al. 2010). Inequalities (2) and (3), in turn, imply that $h/h_{\rm T} < 1$, i.e., that magnetic squeezing dominates tidal gravity. Analogous constraints on wind-driving disks can be obtained also in the Hall and Ohm diffusivity regimes. For example, in the Hall case the parameter space divides into four subregimes with distinct sets of constraints (Königl et al. 2010; Salmeron et al. 2011). In all these cases it is inferred that $\Upsilon_0 > 1$ and $h/h_{\rm T} < 1$, which indicates that these are generic properties of wind-driving disks. In practice one finds that successful full solutions typically have $a_0 \lesssim 1$ (the midplane magnetic pressure is smaller than the thermal pressure, but not greatly so) and $\epsilon \lesssim 1$ (the midplane inflow speed is not much smaller than the speed of sound).

The relations listed in eq. (84) have a few other interesting implications. For example, one can use inequality (3) to argue that wind-driving disks are stable to the fastest-growing linear mode of the MRI, because it implies that the vertical wavelength of this mode, $\sim v_{\rm A0}/\Omega_{\rm K}$ (see chapter 6), is larger than

28. These requirements also place upper limits on the ratio of the density at the sonic/sms point to the midplane density (see eq. [78] for the sms case).

magnetically reduced disk scale height $h = (a_0/\epsilon \Upsilon_0)C/\Omega_K$. But one can also argue that inequality (1) leads to a useful criterion for the onset of the MRI in the disk. Such a criterion is of interest since, as we already noted when the Elsasser number was introduced in § 2.4, the minimum-coupling condition for the development of MRI turbulence in a diffusive disk (represented by a lower bound on Λ) is evidently similar to that inferred for driving a disk wind. The question then arises whether, in a disk that is threaded by a large-scale, ordered field, both vertical (wind-related) and radial (MRI-related) angular momentum transport can occur at the same radial location; this can alternatively be phrased as a question about the maximum magnetic-field strength for the operation of the MRI. When inequality (1) is violated, the surface layers become super-Keplerian, implying outward-streaming motion that is unphysical in the context of a pure wind-driving disk. However, as elaborated in Salmeron et al. (2007), such motion could be associated with the two-channel MRI mode that underlies MRI-induced turbulence. If one regards the parameter combination $2\Upsilon a^2$ that figures in this inequality as a function of height in the disk rather than being evaluated at the midplane, one can infer from the fact that it scales as $B^2(\varrho_i/\varrho)$ that it generally increases with z. It is thus conceivable that inequality (1) (generalized in this manner) is violated near the midplane but is satisfied closer to the disk surface. Salmeron et al. (2007) developed disk models in which both radial turbulent transport and vertical transport associated with the mean field take place in the region where inequality (1) is violated, but only vertical transport (with ultimate deposition of the removed angular momentum in a disk wind) occurs at greater heights. They concluded, however, that significant radial overlap between these two transport mechanisms is unlikely to occur in real disks.

Weakly Coupled Disks. In the disk models considered so far, the minimum-coupling condition on the neutrals, $\Lambda \gtrsim 1$, was satisfied throughout the vertical extent of the disk. We refer to such disks as being *strongly coupled*. As we have just noted, however, the parameter values in a real disk could vary with height, reflecting the vertical stratification of the column density (which shields the ionizing radiation or cosmic rays) and the density. In particular, if the disk is in the ambipolar regime near the surface, in the Hall regime further down, and in the Ohm regime near the midplane, then Λ scales as ϱ_i, ϱ_i/ϱ, and ϱ_i/ϱ^2, respectively, on going from $z = z_s$ to $z = 0$ (see eq. [25]). The Elsasser number will thus increase with height on moving up from $z = 0$ as the gas becomes progressively more ionized and the density decreases. It will generally peak on reaching the ambipolar regime and will subsequently drop as ϱ_i (which typically scales as ϱ to a power between 0 and 0.5) decreases. The

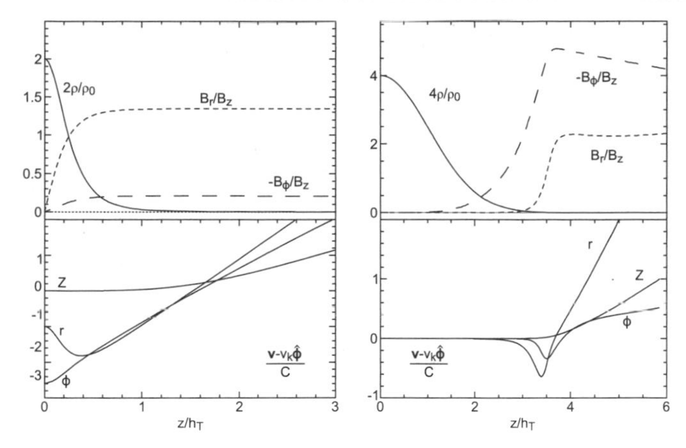

Fig. 7.7. Strongly coupled (left) versus weakly coupled (right) disk solutions.

gas will be weakly coupled in regions where Λ remains $\ll 1$. Weakly coupled disks have distinct properties from those of strongly coupled ones, as was first pointed out by Li (1996a).

As an example, consider a protostellar disk that is ionized solely by cosmic rays (with the ionization rate decreasing exponentially with depth into the disk with a characteristic attenuation column of $96\,\mathrm{g\,cm^{-2}}$) and in which the charge carriers are small, singly charged grains of equal mass (a reasonable approximation for the high-density inner regions of a real disk; e.g., Neufeld & Hollenbach 1994). Near the disk surface the gas is in the ambipolar regime, but if the disk half-column is $\gg 96\,\mathrm{g\,cm^{-2}}$, the degree of ionization near the midplane is very low, and the gas there is in the Ohm regime.[29] Fig. 7.7, taken from Wardle (1997), depicts two illustrative solutions obtained for different radii in this model. The solution on the left corresponds to a large-enough radius (and, correspondingly, a sufficiently low disk column) for the disk to be in the ambipolar regime throughout its vertical extent. By contrast, the solution on the right depicts a smaller radius where the column density is large and the disk is weakly ionized (and in the Ohm regime) near the midplane.

The main differences between these two solutions can be summarized as follows:

Strongly Coupled Disks:

- $v_{A0} \lesssim C$ (midplane magnetic pressure comparable to thermal pressure)
- $| < v_r > | \sim C$ (mean radial speed comparable to the speed of sound)

29. Because of the assumed equal mass of the positive and negative charge carriers, the Hall term in Ohm's law is identically zero (see eq. [13]).

- $B_{r,s} > |B_{\phi,s}|$ (with B_r increasing already at $z = 0$)

Weakly Coupled Disks:

- $v_{A0} \ll C$ (midplane magnetic field is highly subthermal)
- $| < V_r > | \ll C$ (mean inflow speed is highly subsonic)[30]
- $B_{r,s} < |B_{\phi,s}|$ (with B_r taking off only when Λ increases above ~ 1)[31]

There are two noteworthy features of weakly coupled disk solutions. First, even though the bulk of the disk volume is nearly inert, the disk possesses "active" surface layers where $\Lambda > 1$, from which a disk wind can be launched in the presence of a large-scale, ordered field (or in which MRI-induced turbulence can operate; see Gammie 1996). Second, in magnetically threaded disks angular momentum is transported vertically even in regions where Λ is still < 1 and $B_r \approx 0$. This is because a measurable azimuthal field component can already exist in these regions ($|B_\phi/B_r|$ can be $\gg 1$ when $\Lambda \ll 1$; see footnote 31), and we recall that the $z\phi$ stress exerted by the field is $\propto B_z \mathrm{d}B_\phi/\mathrm{d}z$. As a result, matter can continue to accrete even in the nominally inert disk regions, which would not be possible if only a small-scale, disordered field (i.e., MHD turbulence) were responsible for angular momentum transport (although in practice turbulent mixing of charges and field might enable accretion in these regions also in that case; e.g., Fleming & Stone 2003; Turner et al. 2007). This could have implications for the ongoing debate about the nature of dead zones in protostellar disks (e.g., Pudritz et al. 2007; see chapter 6).

4.5. Stability Considerations

It was noted in § 4.4 that wind-driving disks should be stable to the most rapidly growing linear mode of the MRI. On the other hand, by combining the wind-launching condition (6) with the condition for the onset of a radial interchange instability (e.g., Spruit et al. 1995), one can show (see Königl & Wardle 1996) that for disks to be unstable to radial interchange, the magnetic

30. These solutions can thus evade the short-lifetime criticism made against strongly coupled disk models, namely, that in the absence of a vigorous mass supply (which happens only during the early evolutionary phases of protostars), they would empty out on a relatively short timescale (e.g., Shu et al. 2008). Recall in this connection that the disk-formation models described in § 3.4 yielded asymptotic ($r \to 0$) solutions that correspond to weakly coupled disks. The short-lifetime conundrum could be avoided altogether even in a strongly coupled disk if angular momentum transport by a wind dominated only in the innermost disk regions, which is not inconsistent with existing observational data.

31. Using the hydrostatic approximation, one can derive a differential equation relating B_r and B_ϕ, which, assuming $\epsilon_B = 0$ and a vanishing Hall term, takes the form $\frac{\mathrm{d}B_r}{\mathrm{d}B_\phi} \approx -2\Lambda$ at the midplane.

term must be comparable to the gravitational term in the radial momentum equation. This is clearly not the case in the derived disk solutions, which are characterized by an inherently subthermal magnetic field (see, e.g., inequality (2) in eq. [84]). It is thus seen that when a large-scale magnetic field is responsible for the entire angular momentum removal from an accretion disk through a CDW, it automatically lies in a stability window in which it is strong enough not to be affected by the MRI but not so strong as to be subject to radial interchange.

There are, of course, other potential instabilities to which such disks might be susceptible (see Königl & Wardle 1996 for some examples), but here we focus on the question whether there might be an inherent aspect of the wind-related angular momentum transport that could render the disk/wind system unstable.[32] Using approximate equilibrium models, Lubow et al. (1994b) and Cao & Spruit (2002) suggested that an inherent instability of this sort may, in fact, exist (see also Campbell 2009). They attributed this instability to the sensitivity of the outflowing mass flux to changes in the field-line inclination at the disk surface (θ_s), according to the following feedback loop:

$$\begin{aligned}
|v_r| \text{ increases } &\Rightarrow \tan\theta_s \text{ increases} \\
&\Rightarrow \text{the wind is loaded by higher-density gas} \\
&\Rightarrow \dot{M}_w \text{ and the removed angular momentum } \dot{M}_w\, l \text{ increase} \\
&\Rightarrow |v_r| \text{ increases even more.}
\end{aligned}$$

The issue was reexamined by Königl (2004), who used the disk/wind model of Wardle & Königl (1993) and appealed to the fact that the stability properties generally change at a turning point of the equilibrium curve in the solution parameter space. Fig. 7.8 depicts such equilibrium curves, labeled by their vertically constant Elsasser-number values $\Lambda = \Upsilon$, in the Blandford & Payne (1982) $\kappa - \lambda$ wind parameter space. The lower branches of these curves end on the long-dashed curve, below which the outflows remain sub-Alfvénic, whereas the upper branches end on the short-dashed curve, to the right of which the surface layers of the disk are super-Keplerian. Any particular solution is determined by the value of the disk parameter $a_0 = (2/\gamma\beta_0)^{1/2}$, which increases along each $\Lambda = \text{const}$ curve from its minimum value on the super-Keplerian boundary (see inequality [1] in eq. [84]). It is seen that each curve

32. A distinct, but still relevant, question is whether the wind by itself may be unstable. Numerical simulations that treated the disk as providing fixed boundary conditions for the outflow indicated that the wind should not be disrupted, even in the presence of nonaxisymmetric perturbations (e.g., Anderson et al. 2006; Pudritz et al. 2007). However, these simulations did not account for the existence of a feedback between the wind and the disk.

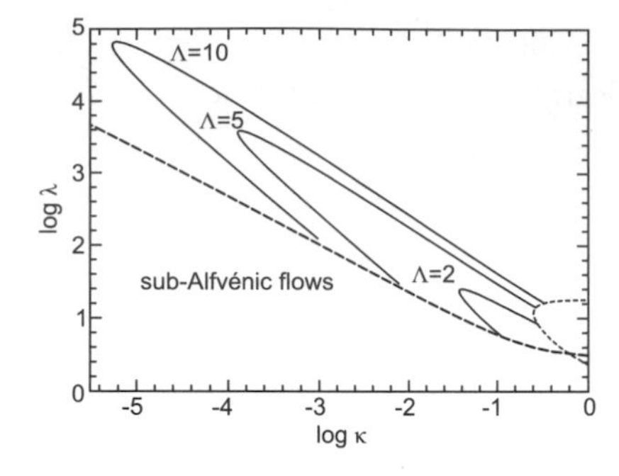

Fig. 7.8. Mapping of the wind-driving disk solutions (for several values of the neutral–field coupling parameter Λ) onto the self-similar wind solution space (defined by the values of the mass-loading parameter κ and angular momentum parameter λ). The field-strength parameter a_0 increases on moving counterclockwise along a given curve.

exhibits a turning point.[33] This behavior can be understood as follows. The vertical hydrostatic-equilibrium equation implies that $b_{r,s}^2 \approx 2/a_0^2$ (assuming $B_{r,s}^2 \gg B_{\phi,s}^2$), and since $|B_\phi|$ evolves with B_r (see footnote 31), it follows that $|B_{\phi,s}|/B_z$, and hence the magnitude of the angular momentum that the outflow must carry away, increase with decreasing a_0. Initially, as a_0 decreases from ~ 1, $b_{r,s} = B_{r,s}/B_z$ increases rapidly, and the corresponding increase in the cylindrical radius of the Alfvén surface (the effective lever arm for the back torque exerted by the outflow on the disk, which scales as $\lambda^{1/2}$) increases the value of λ and leads to a reduction in the ratio of the mass outflow to the mass inflow rates ($\dot{M}_w/\dot{M}_a \propto 1/(\lambda - 1)$). However, as a_0 continues to decrease, the rate of increase of $b_{r,s}$ declines while that of $|B_{\phi,s}|/B_z$ increases, and eventually the mass outflow rate must start to increase (with λ going down) to keep up with the angular momentum removal requirements. The transition between these two modes of enhanced angular momentum transport, predominantly by the lengthening of the lever arm (on the lower branch) versus mainly by a higher mass-loss rate (on the upper branch) occurs at the turning point of the solution curve. One can determine which of the two solution branches is stable and which is not by considering the competition between radial advection and diffusion on the magnetic-flux evolution (best done in the $a_0 - \lambda$ plane), in analogy with thermal stability arguments that analyze the relative magnitude of heating and cooling on each side of the thermal equilibrium curve. In this way one finds that the upper branch of the curves in Fig. 7.8 is unstable: this is the

33. This feature of the solution curves is also found in the corresponding plots in the $a_0 - \lambda$ plane, where now κ varies along each $\Lambda = \text{const}$ curve.

branch along which an increase in the angular momentum transport is accomplished through a higher mass outflow rate, precisely the behavior invoked in the heuristic instability argument reproduced above. Since, however, there is an alternative way of increasing the angular momentum transport (namely, through a lengthening of the effective lever arm), there is also a stable branch of the solution curve. Physically, the reason that a perturbation that reduces the field inclination to the disk does not necessarily trigger an instability is that an increase in $b_{r,s}$ also results in greater field-line tension, which tends to oppose the inward poloidal field bending. Whether a given solution branch is stable is determined by the extent to which this stabilizing effect can overcome the destabilizing influence of increased angular momentum removal brought about by the field-line bending.

It is possible to argue (see Königl 2004) that real protostellar systems likely correspond to the stable branches of the solution curves. Global numerical simulations could, however, shed more light on this question, as they have already started to do. For example, Casse & Keppens (2002) solved the nonideal-MHD equations (assuming axisymmetry and a polytropic gas) and demonstrated jet launching by resistive disks. These results are significant on two main counts: first, they corroborate the basic picture of a magnetically threaded diffusive disk centrifugally driving a wind that removes the bulk of its angular momentum; second, they indicate that these configurations are stable inasmuch as they reach a quasi-steady state. Subsequent simulations have extended this work while confirming its main conclusions (e.g., Kuwabara et al. 2005; Meliani et al. 2006; Zanni et al. 2007). Three-dimensional simulations are the natural next step.

5. Disk–Star Magnetic Coupling

We now turn to the best-established example of a large-scale field in protostars, namely, the stellar dynamo-generated field, and consider the dynamic interaction between this field and the surrounding accretion disk. This is a rich topic that has implications for the structure of the innermost disk region, the stellar rotation and the way mass reaches the stellar surface, protostellar jets, and the observational signatures of the region where most of the gravitational potential energy is liberated. Owing to the nontrivial topology of the stellar (and possibly also disk-generated) magnetic field and to the complexity of the field-mediated interaction between the star and the disk even for simple flux distributions, our understanding is far from complete, and new progress is driven largely by improved numerical simulations. We start this section by briefly surveying the relevant observations and then describe the basic physical processes that underlie this interaction and the results of some of the recent numerical work. Relevant additional details are provided in chapter 2.

5.1. Phenomenology

We discuss the observational manifestations of two phenomena in which a large-scale, ordered magnetic field has been implicated: the relationship between protostellar rotation and accretion disks, and magnetic disk truncation and resultant field-channeled accretion. The theoretical framework that relates these two phenomena is outlined in the next subsection.

Stellar Rotation and Accretion. It is now well established (on the basis of $\sim 1,700$ measured rotation periods) that about 50% of Sun-like protostars ($M_* \sim 0.4$–$1.2\, M_\odot$) undergo significant rotational braking during their pre-main-sequence (PMS) contraction to the zero-age main sequence (ZAMS), and that the objects whose specific angular momenta decrease with time are essentially the ones that are already slowly rotating at an early (~ 1 Myr) age (e.g., Herbst et al. 2007). A clue to the braking mechanism is provided by the finding of a clear correlation between having a comparatively long period ($\gtrsim$2 d) and exhibiting an accretion-disk signature (near-IR or mid-IR excess). This trend is particularly noticeable in higher-mass ($\gtrsim 0.25\, M_\odot$) protostars, but it is also found in very low-mass stars and brown dwarfs (in which, however, the braking efficiency is evidently lower). The connection with disks is supported by the fact that the inferred maximum stellar braking times ($\sim$ 5–10 Myr) are comparable to the maximum apparent lifetimes of gaseous accretion disks (which are $\sim$1 Myr for $\sim$50% of the stars and $\sim$ a maximum of 5 Myr for almost all stars). These results imply that the dominant braking mechanism in protostars is directly tied to active disk accretion. In the absence of such accretion (or after the disks disperse), PMS stars nearly conserve specific angular momentum and therefore spin up as they contract to the ZAMS. As we discuss in the next subsection, stellar magnetic fields likely play a key role in the braking process.

Disk Truncation and Field-Channeled Accretion. As was originally inferred in the case of accreting magnetized neutron stars and white dwarfs, a sufficiently strong protostellar magnetic field can be expected to truncate the accretion flow, with the truncation radius increasing with B_* and decreasing with $\dot{M}_a$ (see eqs. [85] and [86]). In this picture, the intercepted matter "climbs" onto the field lines and is magnetically channeled to some finite stellar latitude.

By the time the matter reaches the stellar surface, it is streaming along the field lines with a near-free-fall speed and is therefore stopped in an accretion shock. The basic elements of this scenario are supported by observations of CTTSs (but also of lower-mass brown dwarfs and of higher-mass Herbig Ae/Be stars, although so far these have been less well studied). The main findings have been the following (see Bouvier et al. 2007):

- The common occurrence of inverse P Cygni profiles, with redshifted absorption reaching several hundred km s^{-1}.
- Observed hydrogen and Na I line profiles that can be adequately modeled in this picture; the detection in the UV-near-IR spectral range of predicted statistical correlations between the line fluxes and the inferred mass-accretion rate.
- Observed spectral energy distributions of optical and UV excesses that are successfully reproduced by accretion-shock models.
- Periodic visual-flux variations due to "hot spots" (interpreted as non-axisymmetric accretion shocks on the stellar surface) that are observed only in actively accreting CTTSs (but not in the weak-line T Tauri stars [WTTSs], which lack the signatures of vigorous accretion). However, in contrast to "cool spots" (the analogs of sunspots), which last for $\sim 10^2$–10^3 rotations, the hot-spot periodicity persists only for a few rotation periods, indicating a highly nonsteady configuration. This inference is supported by the detection of high-amplitude irregular flux variations.
- Strong near-IR "veiling" variability that can be interpreted in the context of this picture as arising in the interaction region between the disk and an inclined stellar magnetosphere, with possible contributions also from the accretion shock and from the shock-irradiated zone in the inner disk.

Zeeman-broadening measurements in a growing number of CTTSs have yielded an intensity-averaged mean surface magnetic-field strength of $\sim$ 2.5 kG,[34] with the field inferred to reach $\sim$4–6 kG in some regions. Furthermore, circular polarization measurements in lines associated predominantly with the accretion shock have revealed rotational modulation and have demonstrated that the field is highly organized (with peak value $\sim$2.5 kG) in the shock region (covering $< 5\%$ of the stellar surface). These results are consistent with the model predictions. However, no net polarization has been found in

34. A similar value has now been inferred also in a protostar that is at an earlier (Class I) evolutionary phase (Johns-Krull et al. 2009).

photospheric absorption lines, implying a complicated surface-field topology (with a global dipole component $\lesssim 0.1$ kG). A likely physical picture is that there exists a spectrum of magnetic-flux loops of various lengths that extend from the stellar surface, with a fraction reaching to $\sim$ 5–20 R_* and intercepting the disk (see Feigelson et al. 2007). This picture is supported by analyses of X-ray flares from young stars (e.g., Favata et al. 2005), as well as by direct radio imaging of the large-scale magnetic structures (e.g., Loinard et al. 2005) and by spectropolarimetric reconstructions of the magnetospheric topology (e.g., Donati et al. 2007).

Numerical studies of magnetospheric accretion indicate that disk mass-loaded outflows (which are often predicted to be nonsteady) could be produced in the course of the star–disk interaction (see § 5.2). There is now observational evidence for accretion-induced winds in CTTSs emanating from both the inner disk and the star (e.g., Edwards et al. 2006). The stellar component has been inferred to move on radial trajectories and to undergo full acceleration (up to a few hundred km s^{-1}) in the stellar vicinity. Although this component may well be launched along stellar magnetic-field lines, the absence of evidence for it in WTTSs suggests that it, too, is triggered by the interaction with the disk.

5.2. Disk–Star Coupling Models

Basic Concepts. Spherical accretion with $\dot{M}_a = 4\pi R^2 \varrho(R)|V_R(R)|$ at the free-fall speed $v_{ff}(R) = (2GM_*/R)^{1/2}$ onto a star of mass M_* will be stopped by the magnetic stresses of the stellar magnetic field at a distance where the ram pressure of the flow (ϱv_{ff}^2) becomes comparable to the magnetic pressure of the stellar field ($B^2/8\pi$). For simplicity, we assume that the field at that location can be approximated by its dipolar component (corresponding to a dipole moment μ_*). For accretion in the equatorial plane, $B(r) = \mu_*/r^3 = B_* R_*^3/r^3$, and one obtains the nominal accretion Alfvén radius

$$r_A = \frac{\mu_*^{4/7}}{(2GM_*)^{1/7}\dot{M}_a^{2/7}} . \tag{85}$$

Although the radial ram pressure of the flow is relatively small for disk accretion, one can still define in this case a magnetospheric boundary radius r_m from the requirement that the total material and magnetic stresses (or energy densities) be comparable, i.e., $B^2/8\pi \approx \varrho v_\phi^2 + P \approx \varrho v_\phi^2$. By approximating $v_\phi \approx v_K$, one obtains an expression similar to the one given by eq. (85). This radius determines the region within which the stellar magnetic field controls the flow dynamics and provides a lower bound on the inner disk radius.

To estimate the radius where matter leaves the disk, note that the physical process of disk truncation requires that the torque exerted by the stellar field that penetrates the disk be comparable to the rate at which angular momentum is transported inward by the accretion flow, $-r^2 B_z B_{\phi,\mathrm{s}} \approx \dot{M}_{\mathrm{a}} \mathrm{d}(r^2 \Omega)/\mathrm{d}r$. Taking $|B_{\phi,\mathrm{s}}| \approx B_z$ (where we use the approximate maximum value of the azimuthal field component at the disk surface; see Uzdensky et al. 2002a and cf. eq. [38]) and setting $|\mathrm{d}\Omega/\mathrm{d}r| \sim \Omega_{\mathrm{K}}/r$ yield an estimate of the disk-truncation radius r_{d} that is of the order of r_{A},

86 $$r_{\mathrm{d}} = k_{\mathrm{A}}\, r_{\mathrm{A}} \,.$$

Semianalytic models (e.g., Ghosh & Lamb 1979) and numerical simulations (e.g., Long et al. 2005) yield $k_{\mathrm{A}} \approx 0.5$.

If the star is rotating with angular velocity Ω_*, one can define the corotation radius by setting $\Omega_{\mathrm{K}}(r) = \Omega_*$, which gives

87 $$r_{\mathrm{co}} = \left(\frac{GM_*}{\Omega_*^2}\right)^{1/3} \,.$$

The interaction of the disk with the stellar magnetic field naturally divides into two qualitatively different regimes depending on the ratio $r_{\mathrm{d}}/r_{\mathrm{co}}$:

$$r_{\mathrm{d}} \lesssim r_{\mathrm{co}} \qquad \text{(funnel-flow regime)},$$
$$r_{\mathrm{d}} \gtrsim r_{\mathrm{co}} \qquad \text{(propeller regime)}\,.$$

In the funnel-flow regime the disk angular velocity is higher than that of the star, and matter can climb onto the stellar magnetic-field lines that thread the disk and reach the stellar surface. By contrast, in the propeller regime the stellar angular velocity exceeds $\Omega_{\mathrm{K}}(r_{\mathrm{d}})$: when disk matter becomes attached to the stellar field lines, its angular momentum increases above the rotational equilibrium value for that radius, and it moves outward. Numerical simulations have found that if the disk effective viscosity and magnetic diffusivity are relatively high, most of the incoming matter is expelled from the system in this case, both in the form of a wide-angle CDW from the inner regions of the disk and as a strong, collimated, magnetically dominated outflow along the open stellar field lines near the axis (e.g., Romanova et al. 2005; Ustyugova et al. 2006).[35] This mechanism could be relevant to the initial spindown (on a time scale $< 10^6$ yr) of CTTSs.

35. A similar outflow pattern, comprising a conical disk wind and a higher-velocity, low-density jet component, was found also in simulations of the funnel-flow regime (e.g., Romanova et al. 2009).

Funnel-Flow Regime. Ghosh & Lamb (1979) proposed that a "disk-locked" state in which the torque exerted by the field lines that thread the disk (as well as by the material that reaches the star) could keep the star rotating in equilibrium (spinning neither up nor down). In this picture, the field lines that connect to the disk within the corotation radius (as well as the accreted material) tend to spin the star up, whereas the field lines that connect at $r > r_{co}$ have the opposite effect. This scenario could potentially explain the relatively low rotation rates observed in CTTSs (e.g., Königl 1991).[36]

This picture, however, was challenged on the grounds that the twisting of the magnetic-field lines that thread the disk at $r > r_{co}$ will tend to open them up, thereby reducing the spindown torque on the star by more than an order of magnitude compared with the original calculation (Matt & Pudritz 2004, 2005). This argument is based on the fact that stellar field lines that connect to the disk at $r \neq r_{co}$ are twisted by the differential rotation between their respective footpoints. Initially, the twisting leads to an increase in $|B_{\phi,s}|$, but then the built-up magnetic stress causes the field lines to elongate rapidly in a direction making an angle $\sim 60°$ to the rotation axis (e.g., Lynden-Bell & Boily 1994). During this phase $|B_{\phi,s}|$ decreases as the field-line twist travels out to the apex of the elongating field line, where the field is weakest, a process that can be understood in terms of torque balance along the field line (e.g., Parker 1979). As the twist angle approaches a certain critical value ($\sim$ 4 rad for a star linked to a Keplerian disk), the expansion accelerates, and the magnetic field formally reaches a singular state (a "finite time singularity"), which in practice means that it opens up. Although the twisting can be countered by magnetic diffusivity in the disk, this is possible only if the steady-state surface azimuthal field amplitude $(2\pi r \mathcal{S} \Delta\Omega / c^2) B_{z,s}$, where $\mathcal{S}$ is the vertically integrated electrical conductivity (treated as a scalar) and $\Delta\Omega$ is the differential rotation rate, does not exceed the maximum value of $|B_{\phi,s}|$ in the absence of diffusivity (e.g., Uzdensky et al. 2002a). The conductivity in the innermost regions of protostellar disks is probably too high for a steady state to be feasible except in the immediate vicinity of r_{co} (see also Zweibel et al. 2006), and in some proposed models (e.g., the X-wind; see Shu et al. 2000) it has, in fact, been postulated that the stellar field lines can effectively couple to the disk only in that narrow region.

Numerical simulations, however, have verified that when the star has a strong-enough field to disrupt the disk at a distance of a few stellar radii and channel accreting matter along field lines, it can maintain an equilibrium

36. Naively, one would expect a star accreting from a rotationally supported disk to rotate near breakup speed; in reality, CTTSs rotate, on average, at about 1/10th of this speed.

disk-locking state in which Ω_* is close to the disk angular velocity at the truncation radius (Long et al. 2005). In these simulations it was found that in equilibrium, $r_{co}/r_d \sim 1.2$–1.5, very close to the prediction of the Ghosh & Lamb (1979) model. Furthermore, as envisioned in the latter model, closed magnetic-field lines that link the star and the disk exert the dominant stresses in this interaction. However, in contrast to the original picture, this linkage does in fact occur primarily near r_{co}. The torque balance is achieved in part through field-line stretching (which mimics the connection to material at $r > r_{co}$ in the original model) and by magnetically driven outflows (which, however, are found to remain comparatively weak in the simulations). Field-line opening is not a major impediment to this process, in part because opened field lines tend to reconnect, especially if the departure from axisymmetry is not large.[37]

The question whether (or under what circumstances) the magnetic transfer of angular momentum to the disk is the most efficient way of attaining protostellar spin equilibrium, however, is still unresolved. For example, Matt & Pudritz (2008) suggested that stellar winds driven along open magnetic-field lines could dominate the braking torque on the star. These winds are inferred to be powered by the accretion process, but it remains to be determined how this could happen in practice and whether the proposed outflows are, in fact, related to the stellar winds already identified observationally. It is also important to bear in mind that the star–disk magnetic linkage mechanism may, in reality, be more complex than the simplified picture outlined above. Some of these expected complications are already being investigated with the help of 3D MHD codes, including the effects of misalignment between the rotation and magnetic axes, of an off-centered dipole, and of higher-order multipole moments (e.g., Romanova et al. 2003, 2004, 2008; Long et al. 2007, 2008; Kulkarni & Romanova 2009).[38] Furthermore, the field topology and the nature of the interaction could be modified if the disk itself contains a large-scale magnetic field (e.g., Hirose et al. 1997; Ferreira et al. 2000, 2006; Miller & Stone 2000; von Rekowski & Brandenburg 2004, 2006).

37. The idea that twisted field lines that open up could subsequently reconnect, leading to a repetitive cycle of inflation and reconnection and resulting in the star–disk linkage being steady only in a time-averaged sense, was first proposed by Aly & Kuijpers (1990) and received support from subsequent investigations (e.g., Uzdensky et al. 2002b; Romanova et al. 2002).

38. Given the importance of the question of how matter crosses magnetic-field lines in this problem, it would be helpful to examine these effects using an explicitly resistive numerical code. So far, however, only the axisymmetric version of this scenario has been investigated in this way (e.g., Bessolaz et al. 2008).

NONSTEADY ACCRETION. The magnetic interaction between stars and disks could be variable on a timescale as short as Ω_*^{-1}. One possibility, which could naturally give rise to observable hot spots, is for the system to lack axisymmetry—either because of a misalignment between the magnetic (e.g., dipole) and rotation axes (e.g., Romanova et al. 2004, 2008; Kulkarni & Romanova 2009) or because of the intrinsic structure of the magnetic field (e.g., von Rekowski & Brandenburg 2004, 2006). A longer variablity timescale is implied by the suggestion of Goodson & Winglee (1999) that the truncation radius would oscillate on the diffusion timescale of the field into the disk (resulting in field-line reconnection events and episodic polar ejections) if the diffusivity at the inner edge of the disk were relatively low (so $|(\nu_{B,r}/\nu_r) - 1| < 1$). A different type of variability is implied by the star/disk dynamo model of von Rekowski & Brandenburg (2006). In this model, the magnetic-field geometry changes at irregular time intervals (with magnetic polarity reversals occurring mostly on timescales of less than a day), and the star/disk system alternates between magnetically connected and disconnected states. The model predicts strong outflows, both from the inner disk and from the stellar surface, but typically only a small fraction of the disk-accretion flow reaches the stellar surface. In fact, there is an anticorrelation between the stellar magnetic-field strength and the accretion rate, and material that reaches the stellar surface comes in at a low velocity (von Rekowski & Piskunov 2006). It remains to be determined whether these two aspects of the model are consistent with the polarization data and the evidence for accretion shocks (see § 5.1).

6. Conclusion

The discussion in this chapter can be summarized as follows:

- There is strong observational evidence for a disk–wind connection in protostars. Large-scale, ordered magnetic fields have been implicated theoretically as the most likely driving mechanism of the observed winds and jets. The ubiquity of the outflows may be related to the fact that centrifugally driven winds (CDWs) are a potentially efficient means of transporting angular momentum from the disk.
- Ordered magnetic fields could arise in protostellar disks on account of (i) advection of interstellar field by the accretion flow, (ii) dynamo action in the disk, and (iii) interaction with the stellar magnetic field.
- Semianalytic MHD models have been able to account for the basic structure of diffusive disks that drive CDWs from their surfaces, as well as for the formation of such systems in the collapse of rotating

molecular cloud cores. Some of these models already incorporate a realistic disk-ionization and conductivity structure. These studies have established that vertical angular momentum transport by a CDW or through torsional Alfvén waves (magnetic braking) could in principle be the main angular momentum removal mechanism in protostellar disks and have determined the parameter regime where wind transport can be expected to dominate radial transport by MRI-induced turbulence. Further progress is being made by increasingly elaborate numerical simulations (involving nonideal-MHD codes) that have started to examine the global properties, time evolution, and dynamic stability of the magnetic disk/wind system.

- Robust observational evidence also exists for a magnetic interaction between CTTS disks and their respective protostars, including strong indications of a field-channeled flow onto the stellar surface. This interaction is likely to involve mass ejection and is thought to be responsible for the comparatively low rotation rates of CTTSs. Since the magnetic-field geometry in the interaction region is evidently quite complex and the interaction is likely time dependent, numerical simulations are an indispensable tool in the study of this problem.

Future advances in this area will probably arise from a combination of new observational findings, the refinement of current theoretical approaches, and the incorporation of additional physics into the models. On the observational side, the main challenge is still to demonstrate the existence of CDWs in protostars and to determine their spatial extent (spread out over most of the disk surface or occurring only near its inner edge, and, if the former, whether the launching region is nearly continuous or is confined to localized patches). Recent attempts to measure rotation in the outflows could potentially help answer this question, but, as noted in § 4.3, the results obtained so far are still inconclusive.

Regarding the further development of theoretical tools, the greatest impact would likely be produced by numerical simulations that study vertical angular momentum transport by either a CDW or magnetic braking with codes that include a realistic conductivity tensor and have full 3D and mesh-refinement capabilities. Such simulations should be able to clarify the relative roles of vertical and radial angular momentum transport and the possible interplay between them for relevant combinations of the disk-model parameters (see § 4.4). An interim step might be to solve for the evolution of a vertically integrated disk whose properties at any radial grid zone are determined from a vertical integration of a simplified version of the radially localized, steady-state disk model described in § 4.4. Time-dependent models of this type could examine the behavior of a magnetically threaded disk after its mass supply diminishes or stops altogether (corresponding to the protostellar system evolving into the optically revealed phase), which has previously been studied only in the context of α-viscosity models. As was noted above, a state-of-the-art, 3D, nonideal-MHD code is also crucial for investigating the stability of disk/wind systems and the various aspects of the star–disk field-mediated interaction. One could, however, also benefit from further development of the semianalytic models, which might include an extension of the ionization/conductivity scheme, a derivation of self-similar disk/wind solutions that allow for a radial drift of the poloidal magnetic field (see § 4.4), and a calculation of the predicted observational characteristics of wind-driving disks.

The mass fraction and size distribution of dust grains in the disk have a strong effect on its ionization and conductivity structure and on the degree of field–matter coupling (see § 4.4). Existing models incorporate the effect of dust in a somewhat ad hoc manner, by adopting an assumed distribution. In reality, the grain distribution is determined by the balance of several processes, including grain collisions due to relative velocities that develop as a result of Brownian motion, differential vertical-settling speeds, and turbulence, which can lead to either coagulation or fragmentation. Grains are also subject to a collisional drag force exerted by the gas and arising from the fact that the gas is subject to thermal and magnetic forces that do not affect the dust. This leads to vertical settling, as well as to radial migration, directed either inward or outward depending on whether the gas rotation is sub- or super-Keplerian, respectively. Furthermore, radial or vertical gas motions can affect sufficiently small grains through advection. Yet another effect is evaporation by the ambient radiation field, which could affect dust located at sufficiently high elevations and small radii. Some of these effects have already been incorporated into generic viscous-disk models (e.g., Dominik et al. 2007; Brauer et al. 2008), and one could similarly consider them in the context of a wind-driving disk model. In view of the fact that the latter model is characterized by a vertical outflow and by comparatively fast radial inflow speeds, one can expect to find new types of behavior in this case. In particular, grains located near the disk surfaces would either settle to the midplane if they are large enough or be uplifted from the disk if they are sufficiently small, whereas intermediate-size grains would first leave the disk and then reenter at a potentially much larger radius, from which they could be advected back inward. By including dust dynamics, one could examine whether the effect of dust on the gas motion (through its influence on the field–matter coupling) and the effect of gas on the grain motions (through gas–dust collisions) together place meaningful constraints on the resulting grain distribution. One could also investigate whether the predicted radial transport of intermediate-size grains from small to large radii and their possible thermal processing outside the disk could be relevant to the accumulating evidence for an outward transport of crystalline grains in the protosolar nebula and in other protostellar disks, and whether the implied dust distribution in the disk and the wind might have distinct observational signatures that could be tested by spectral and imaging techniques (see Millan-Gabet et al. 2007; § 4.3; and chapters 2 and 6).

Dust particles are thought to be the building blocks of planetesimals, and their distribution in the disk is thus a key ingredient of planet-formation models. In fact, the general properties of a protostellar disk are evidently relevant to planet formation in light of the growing evidence that the formation of giant planets, in particular, is strongly influenced by physical processes that occur when the disk is still predominantly gaseous. A disk threaded by a large-scale, ordered magnetic field could potentially have unique effects on planet growth and migration. One such effect is the generation (through magnetic resonances that are the analogs of Lindblad resonances) of a global torque that may reduce or even reverse the secular inward drift (the so-called Type I migration) predicted for low-mass planets, which has posed a conundrum for current theories of planet formation. As was demonstrated by Terquem (2003) and Fromang et al. (2005), a torque of this type could be produced if the disk had a comparatively strong (MRI-stable, but still subthermal) azimuthal field (with a nonzero vertical average of B_ϕ^2) that fell off sufficiently fast with radius ($\propto r^{-1} - r^{-2}$). A poloidal field component could in principle also contribute to this process (Muto et al. 2008). Given that a large-scale field with precisely these properties is expected in wind-driving protostellar disks (see §§ 3 and 4), this possibility clearly merits an explicit investigation in the context of the disk models considered in this chapter. The influence of the vertical channel of angular momentum transport and of the overall effect of an ordered, large-scale field on the disk structure in such systems (e.g., the reduction of the density scale height by magnetic squeezing) may also be worth examining in this connection.

References

J. J. Aly, J. Kuijpers: A&A **227**, 473 (1990).
J. M. Anderson, Z.-Y. Li, R. Krasnopolsky, R. D. Blandford: ApJ **590**, L107 (2003).

J. M. Anderson, Z.-Y. Li, R. Krasnopolsky, R. D. Blandford: ApJ **653**, L33 (2006).

H. G. Arce, D. Shepherd, F. Gueth, C.-F. Lee, R. Bachiller, A. Rosen, H. Beuther: Molecular Outflows in Low- and High-Mass Star-forming Regions. In *Protostars and Planets V*, ed. by B. Reipurth, D. Jewitt, and K. Keil (Univ. of Arizona Press, Tucson, 2007), p. 245.

S. A. Balbus, J. C. B. Papaloizou: ApJ **521**, 650 (1999).

S. A. Balbus, C. Terquem: ApJ **552**, 235 (2001).

J. Bally, B. Reipurth, C. J. Davis: Observations of Jets and Outflows from Young Stars. In *Protostars and Planets V*, ed. by B. Reipurth, D. Jewitt, and K. Keil (Univ. of Arizona Press, Tucson, 2007), p. 215.

A. Balogh, E. J. Smith, B. T. Tsurutani, D. J. Southwood, R. J. Forsyth, T. S. Horbury: Science **268**, 1007 (1995).

S. Basu, T. Ch. Mouschovias: ApJ **432**, 720 (1994).

A. Belloche, P. André, D. Despois, S. Blinder: A&A **393**, 927 (2002).

N. Bessolaz, C. Zanni, J. Ferreira, R. Keppens, J. Bouvier: A&A **478**, 155 (2008).

E. G. Blackman, J. C. Tan: Ap&SS **292**, 395 (2004).

O. M. Blaes, S. A. Balbus: ApJ **421**, 163 (1994).

R. D. Blandford, D. G. Payne: MNRAS **199**, 883 (1982).

S. V. Bogovalov: A&A **323**, 634 (1997).

J. Bouvier, S. H. P. Alencar, T. J. Harries, C. M. Johns-Krull, M. M. Romanova: Magnetospheric Accretion in Classical T Tauri Stars. In *Protostars and Planets V*, ed. by B. Reipurth, D. Jewitt, and K. Keil (Univ. of Arizona Press, Tucson, 2007), p. 479.

A. Brandenburg, Å. Nordlund, R. F. Stein, U. Torkelsson: ApJ **446**, 741 (1995).

F. Brauer, C. P. Dullemond, Th. Henning: A&A **480**, 859 (2008).

S. Cabrit: LNP **723**, 21 (2007).

S. Cabrit, J. Ferreira, A. C. Raga: A&A **343**, L61 (1999).

S. Cabrit, J. Pety, N. Pesenti, C. Dougados: A&A **452**, 897 (2006).

N. Calvet, L. Hartmann, S. E. Strom: Evolution of Disk Accretion. In *Protostars and Planets IV*, ed. by V. Mannings, A. P. Boss, S. S. Russell (Univ. of Arizona Press, Tucson, 2000), p. 377.

C. G. Campbell: MNRAS **336**, 999 (2002).

C. G. Campbell: MNRAS **392**, 271 (2009).

X. Cao, H. C. Spruit: A&A **287**, 80 (1994).

X. Cao, H. C. Spruit: A&A **385**, 289 (2002).

F. Casse, R. Keppens: ApJ **581**, 988 (2002).

A. H. Cerqueira, P. F. Velázquez, A. C. Raga, M. J. Vasconcelos, F. de Colle: A&A **448**, 231 (2006).

A. Chrysostomou, F. Bacciotti, B. Nisini, T. P. Ray, J. Eislöffel, C. J. Davis, M. Takami: A&A **482**, 575 (2008).

A. Chrysostomou, P. W. Lucas, J. H. Hough: Nature **450**, 71 (2007).

G. E. Ciolek, A. Königl: ApJ **504**, 257 (1998).

G. E. Ciolek, T. Ch. Mouschovias: ApJ **418**, 774 (1993).

D. Coffey, F. Bacciotti, T. P. Ray, J. Eislöffel, J. Woitas: ApJ **663**, 350 (2007).

I. Contopoulos, G. E. Ciolek, A. Königl: ApJ **504**, 247 (1998).

I. Contopoulos, R. V. E. Lovelace: ApJ **429**, 139 (1994).

T. G. Cowling: *Magnetohydrodynamics* (Adam Hilger, Bristol, 1976).

R. M. Crutcher: ApJ **520**, 706 (1999).

J. Di Francesco, N. J. Evans II, P. Caselli, P. C. Myers, Y. Shirley, Y. Aikawa, M. Tafalla: An Observational Perspective of Low-Mass Dense Cores I: Internal Physical and Chemical Properties. In *Protostars and Planets V*, ed. by B. Reipurth, D. Jewitt, and K. Keil (Univ. of Arizona Press, Tucson, 2007), p. 17.
C. Dominik, J. Blum, J. N. Cuzzi, G. Wurm: Growth of Dust as the Initial Step toward Planet Formation. In *Protostars and Planets V*, ed. by B. Reipurth, D. Jewitt, and K. Keil (Univ. of Arizona Press, Tucson, 2007), p. 783.
J.-F. Donati, F. Paletou, J. Bouvier, J. Ferreira: Nature **438**, 466 (2005).
J.-F. Donati et al.: MNRAS **380**, 1297 (2007).
C. Dougados, S. Cabrit, J. Ferreira, N. Pesenti, P. Garcia, D. O'Brien: Ap&SS **293**, 45 (2004).
B. T. Draine, W. G. Roberge, A. Dalgarno: ApJ **264**, 485 (1983).
J. E. Drew, D. Proga, J. M. Stone: MNRAS **296**, L6 (1998).
D. F. Duffin, P. R. Pudritz: ApJ **706**, L46 (2009).
J. A. Dutrey, S. Guilloteau, P. Ho: Interferometric Spectroimaging of Molecular Gas in Protoplanetary Disks. In *Protostars and Planets V*, ed. by B. Reipurth, D. Jewitt, and K. Keil (Univ. of Arizona Press, Tucson, 2007), p. 495.
S. Edwards, W. Fischer, L. Hillenbrand, J. Kwan: ApJ **646**, 319 (2006).
B. G. Elmegreen: ApJ **530**, 277 (2000).
B. G. Elmegreen: ApJ **668**, 1064 (2007).
F. Favata, E. Flaccomio, F. Reale, G. Micela, S. Sciortino, H. Shang, K. G. Stassun, E. D. Feigelson: ApJS **160**, 469 (2005).
E. Feigelson, L. Townsley, M. Güdel, K. Stassun: X-Ray Properties of Young Stars and Stellar Clusters. In *Protostars and Planets V*, ed. by B. Reipurth, D. Jewitt, and K. Keil (Univ. of Arizona Press, Tucson, 2007), p. 313.
J. Ferreira: A&A **319**, 340 (1997).
J. Ferreira, F. Casse: ApJ **601**, L139 (2004).
J. Ferreira, C. Dougados, S. Cabrit: A&A **453**, 785 (2006).
J. Ferreira, G. Pelletier: A&A **295**, 807 (1995).
J. Ferreira, G. Pelletier, S. Appl: MNRAS **312**, 387 (2000).
R. A. Fiedler, T. Ch. Mouschovias: ApJ **415**, 680 (1993).
T. Fleming, J. M. Stone: ApJ **585**, 908 (2003).
J. Frank, A. King, D. Raine: *Accretion Power in Astrophysics* (Cambridge Univ. Press, Cambridge, 2002).
S. Fromang, C. Terquem, S. A. Balbus: MNRAS **329**, 18 (2002).
S. Fromang, C. Terquem, R. P. Nelson: MNRAS **363**, 943 (2005).
D. Galli, F. H. Shu: ApJ **417**, 243 (1993).
C. F. Gammie: ApJ **457**, 355 (1996).
P. J. V. Garcia, J. Ferreira, S. Cabrit, L. Binette: A&A **377**, 589 (2001a).
P. J. V. Garcia, S. Cabrit, J. Ferreira, L. Binette: A&A **377**, 609 (2001b).
P. Ghosh, F. K. Lamb: ApJ **232**, 259 (1979).
J. M. Girart, R. Rao, D. P. Marrone: Science **313**, 812 (2006).
A. E. Glassgold, E. D. Feigelson, T. Montmerle: Effects of Energetic Radiation in Young Stellar Objects. In *Protostars and Planets IV*, ed. by V. Mannings, A. P. Boss, S. S. Russell (Univ. of Arizona Press, Tucson, 2000), p. 429.
A. E. Glassgold, E. D. Feigelson, T. Montmerle, S. Wolk: ASPC **341**, 165 (2005).
A. A. Goodman, P. J. Benson, G. A. Fuller, P. C. Myers: ApJ **406**, 528 (1993).
A. P. Goodson, R. M. Winglee: ApJ **524**, 159 (1999).

X. Guan, C. F. Gammie: ApJ **697**, 1901 (2009).

D. E. Harker, S. J. Desch: ApJ **565**, L109 (2002).

P. Hartigan, S. Edwards, L. Ghandour: ApJ **452**, 736 (1995).

L. Hartmann: The Observational Evidence for Accretion. In *Herbig-Haro Flows and the Birth of Stars*, ed. by B. Reipurth and C. Bertout (Kluwer, Dordrecht, 1997), p. 391.

L. Hartmann, S. J. Kenyon: ARA&A **34**, 207 (1996).

J. F. Hawley, J. M. Stone: ApJ **501**, 758 (1998).

P. Hennebelle, A. Ciardi: A&A **506**, L29 (2009).

R. N. Henriksen, D. R. Rayburn: MNRAS **152**, 323 (1971).

W. Herbst, J. Eislöffel, R. Mundt, A. Scholz: The Rotation of Young Low-Mass Stars and Brown Dwarfs. In *Protostars and Planets V*, ed. by B. Reipurth, D. Jewitt, and K. Keil (Univ. of Arizona Press, Tucson, 2007), p. 297.

J. Heyvaerts, C. Norman: ApJ **347**, 1055 (1989).

S. Hirose, Y. Uchida, K. Shibata, R. Matsumoto: PASJ **49**, 193 (1997).

D. Hollenbach, D. Johnstone, S. Lizano, F. Shu: ApJ **428**, 654 (1994).

C. Hunter: ApJ **218**, 834 (1977).

L. Jin: ApJ **457**, 798 (1996).

C. M. Johns-Krull, T. P. Greene, G. W. Doppmann, K. R. Covey: ApJ **700**, 1440 (2009).

B. D. Kane, D. P. Clemens: AJ **113**, 1799 (1997).

J. F. Kartje: ApJ **452**, 565 (1995).

L. Kirby: ApJ **694**, 1056 (2009).

A. Königl: ApJ **342**, 208 (1989).

A. Königl: ApJ **370**, L39 (1991).

A. Königl: LNP **465**, 282 (1996).

A. Königl: ASPC **121**, 551 (1997).

A. Königl: ApJ **617**, 1267 (2004).

A. Königl, J. F. Kartje: ApJ **434**, 446 (1994).

A. Königl, R. E. Pudritz: Disk Winds and the Accretion-Outflow Connection. In *Protostars and Planets IV*, ed. by V. Mannings, A. P. Boss, S. S. Russell (Univ. of Arizona Press, Tucson, 2000), p. 759.

A. Königl, S. P. Ruden: Origin of Outflows and Winds. In *Protostars and Planets III*, ed. by E. H. Levy, J. I. Lunine (Univ. of Arizona Press, Tucson, 1993), p. 641.

A. Königl, R. Salmeron, M. Wardle: MNRAS **401**, 479 (2010).

A. Königl, M. Wardle: MNRAS **279**, L61 (1996).

R. Krasnopolsky, A. Königl: ApJ **580**, 987 (2002).

A. K. Kulkarni, M. M. Romanova: MNRAS **398**, 701 (2009).

M. W. Kunz, S. A. Balbus: MNRAS **348**, 355 (2004).

R. Kurosawa, T. J. Harries, N. H. Symington: MNRAS **370**, 580 (2006).

T. Kuwabara, K. Shibata, T. Kudoh, R. Matsumoto: ApJ **621**, 921 (2005).

L. D. Landau, E. M. Lifshitz: *Fluid Mechanics*, 2nd ed. (Butterworth-Heinemann, Oxford, 1987).

R. B. Larson: MNRAS **145**, 271 (1969).

A. Lazarian: JQSRT **106**, 225 (2007).

G. Lesur, P.-Y. Longaretti: A&A **504**, 309 (2009).

A. Levinson: ApJ **648**, 510 (2006).

E. H. Levy, C. P. Sonett: Meteorite Magnetism and Early Solar System Magnetic Fields. In *Protostars and Planets*, ed. by T. Gehrels (Univ. of Arizona Press, Tucson, 1978), p. 516.

Z.-Y. Li: ApJ **465**, 855 (1996a).
Z.-Y. Li: ApJ **473**, 873 (1996b).
Z.-Y. Li, C. F. McKee: ApJ **464**, 373 (1996).
Z.-Y. Li, F. H. Shu: ApJ **475**, 237 (1997).
D. N. C. Lin, J. E. Pringle: MNRAS **225**, 607 (1987).
L. Loinard, A. J. Mioduszewski, L. F. Rodríguez, R. A. González, M. I. Rodríguez, R. M. Torres: ApJ **619**, 179 (2005).
M. Long, M. M. Romanova, R. V. E. Lovelace: ApJ **634**, 1214 (2005).
M. Long, M. M. Romanova, R. V. E. Lovelace: MNRAS **374**, 436 (2007).
M. Long, M. M. Romanova, R. V. E. Lovelace: MNRAS **386**, 1274 (2008).
R. V. E. Lovelace, M. M. Romanova, G. S. Bisnovatyi-Kogan: MNRAS **275**, 244 (1995).
S. H. Lubow, J. C. B. Papaloizou, J. E. Pringle: MNRAS **267**, 235 (1994a).
S. H. Lubow, J. C. B. Papaloizou, J. E. Pringle: MNRAS **268**, 1010 (1994b).
D. Lynden-Bell, C. Boily: MNRAS **267**, 146 (1994).
M. N. Machida, S. Inutsuka, T. Matsumoto: ApJ **670**, 1198 (2007).
M. N. Machida, T. Matsumoto, T. Hanawa, K. Tomisaka: ApJ **645**, 1227 (2006).
M.-M. Mac Low, M. L. Norman, A. Königl, M. Wardle: ApJ **442**, 726 (1995).
S. C. Martin: ApJ **473**, 1051 (1996).
S. Matt, R. E. Pudritz: ApJ **607**, L43 (2004).
S. Matt, R. E. Pudritz: MNRAS **356**, 167 (2005).
S. Matt, R. E. Pudritz: ApJ **681**, 391 (2008).
Z. Meliani, F. Casse, C. Sauty: A&A **460**, 1 (2006).
R. Mellon, Z.-Y. Li: ApJ, **698**, 922 (2009).
F. Ménard, G. Duchêne: A&A **425**, 973 (2004).
R. Millan-Gabet, F. Malbet, R. Akeson, C. Leinert, J. Monnier, R. Waters: The Circumstellar Environments of Young Stars at AU Scales. In *Protostars and Planets V*, ed. by B. Reipurth, D. Jewitt, and K. Keil (Univ. of Arizona Press, Tucson, 2007), p. 539.
K. A. Miller, J. M. Stone: ApJ **534**, 398 (2000).
M. Mitchner, C. H. Kruger Jr.: *Partially Ionized Gases* (Wiley, New York, 1973).
S. Mohanty, R. Jayawardhana, G. Basri: ApJ **626**, 498 (2005).
T. Ch. Mouschovias: ApJ **207**, 141 (1976).
T. Ch. Mouschovias, K. Tassis, M. W. Kunz: ApJ **646**, 1043 (2006).
T. Muto, M. N. Machida, S. Inutsuka: ApJ **679**, 813 (2008).
J. R. Najita, J. S. Carr, A. E. Glassgold, J. A. Valenti: Gaseous Inner Disks. In *Protostars and Planets V*, ed. by B. Reipurth, D. Jewitt, and K. Keil (Univ. of Arizona Press, Tucson, 2007), p. 507.
F. Nakamura, T. Hanawa, T. Nakano: ApJ **444**, 770 (1995).
T. Nakano, R. Nishi, T. Umebayashi: ApJ **573**, 199 (2002).
A. Natta, L. Testi, N. Calvet, Th. Henning, R. Waters, D. Wilner: Dust in Protoplanetary Disks: Properties and Evolution. In *Protostars and Planets V*, ed. by B. Reipurth, D. Jewitt, and K. Keil (Univ. of Arizona Press, Tucson, 2007), p. 767.
D. A. Neufeld, D. J. Hollenbach: ApJ **428**, 170 (1994).
R. Nishi, T. Nakano, T. Umebayashi: ApJ **368**, 181 (1991).
H. Nomura, Y. Nakagawa: ApJ **640**, 1099 (2006).
D. O'Brien, P. Garcia, J. Ferreira, S. Cabrit, L. Binette: Ap&SS **287**, 129 (2003).
G. I. Ogilvie: MNRAS **288**, 63 (1997).
G. I. Ogilvie, M. Livio: ApJ **553**, 158 (2001).

N. Ohashi, M. Hayashi, P. T. P. Ho, M. Momose: ApJ **475**, 211 (1997).
I. Okamoto: MNRAS **173**, 357 (1975).
M. Oppenheimer, A. Dalgarno: ApJ **192**, 29 (1974).
E. N. Parker: *Cosmical Magnetic Fields: Their Origin and Their Activity* (Clarendon Press, Oxford, 1979).
E. N. Parker: *Conversations on Electric and Magnetic Fields in the Cosmos* (Princeton Univ. Press, Princeton, 2007).
G. Pelletier, R. E. Pudritz: ApJ **394**, 117 (1992).
M. V. Penston: MNRAS **144**, 425 (1969).
N. Pesenti, C. Dougados, S. Cabrit, J. Ferreira, F. Casse, P. Garcia, D. O'Brien: A&A **416**, L9 (2004).
L. Podio, F. Bacciotti, B. Nisini, J. Eislöffel, F. Massi, T. Giannini, T. P. Ray: A&A **456**, 189 (2006).
R. E. Pudritz, R. Ouyed, Ch. Fendt, A. Brandenburg: Disk Winds, Jets, and Outflows: Theoretical and Computational Foundations. In *Protostars and Planets V*, ed. by B. Reipurth, D. Jewitt, and K. Keil (Univ. of Arizona Press, Tucson, 2007), p. 277.
T. Ray, C. Dougados, F. Bacciotti, J. Eislöffel, A. Chrysostomou: Toward Resolving the Outflow Engine: An Observational Perspective. In *Protostars and Planets V*, ed. by B. Reipurth, D. Jewitt, and K. Keil (Univ. of Arizona Press, Tucson, 2007), p. 231.
T. P. Ray, T. W. B. Muxlow, D. J. Axon, A. Brown, D. Corcoran, J. Dyson, R. Mundt: Nature **385**, 415 (1997).
J. Richer, D. Shepherd, S. Cabrit, R. Bachiller, E. Churchwell: Molecular Outflows from Young Stellar Objects. In *Protostars and Planets IV*, ed. by V. Mannings, A. P. Boss, S. S. Russell (Univ. of Arizona Press, Tucson, 2000), p. 867.
M. M. Romanova, A. K. Kulkarni, R. V. E. Lovelace: ApJ **673**, L171 (2008).
M. M. Romanova, G. V. Ustyugova, A. V. Koldoba, R. V. E. Lovelace: ApJ **578**, 420 (2002).
M. M. Romanova, G. V. Ustyugova, A. V. Koldoba, R. V. E. Lovelace: ApJ **610**, 920 (2004).
M. M. Romanova, G. V. Ustyugova, A. V. Koldoba, R. V. E. Lovelace: ApJ **635**, L165 (2005).
M. M. Romanova, G. V. Ustyugova, A. V. Koldoba, R. V. E. Lovelace: MNRAS **399**, 1802 (2009).
M. M. Romanova, G. V. Ustyugova, A. V. Koldoba, J. V. Wick, R. V. E. Lovelace: ApJ **595**, 1009 (2003).
D. M. Rothstein, R. V. E. Lovelace: ApJ **677**, 1221 (2008).
P. N. Safier: ApJ **408**, 115 (1993a).
P. N. Safier: ApJ **408**, 148 (1993b).
T. Sakurai: A&A **152**, 121 (1985).
R. Salmeron, A. Königl, M. Wardle: MNRAS **375**, 177 (2007).
R. Salmeron, A. Königl, M. Wardle: MNRAS (2011).
T. Sano, S. M. Miyama, T. Umebayashi, T. Nakano: ApJ **543**, 486 (2000).
T. Sano, J. M. Stone: ApJ **577**, 534 (2002).
D. A. Schleuning: ApJ **493**, 811 (1998).
H. Shang, A. E. Glassgold, F. H. Shu, S. Lizano: ApJ **564**, 853 (2002).
H. Shang, S. Lizano, A. E. Glassgold, F. H. Shu: ApJ **612**, L69 (2004).
D. Shepherd: Massive Molecular Outflows. In *Massive Star Birth: A Crossroads of Astrophysics*, ed. by R. Cesaroni et al. (Cambridge Univ. Press, Cambridge, 2005), p. 237.
F. H. Shu: ApJ **214**, 488 (1977).
F. H. Shu, F. C. Adams, S. Lizano: ARA&A **25**, 23 (1987).

F. H. Shu, D. Galli, S. Lizano, C. Mike: ApJ **647**, 382 (2006).
F. H. Shu, S. Lizano, D. Galli, M. J. Cai, S. Mohanty: ApJ **682**, L121 (2008).
F. H. Shu, J. R. Najita, E. C. Ostriker, H. Shang: ApJ **455**, L155 (1995).
F. H. Shu, J. R. Najita, H. Shang, Z.-Y. Li: X-Winds Theory and Observations. In *Protostars and Planets IV*, ed. by V. Mannings, A. P. Boss, S. S. Russell (Univ. of Arizona Press, Tucson, 2000), p. 789.
N. Soker: A&A **435**, 125 (2005).
N. Soker, O. Regev: ApJ **406**, 603 (2003).
H. C. Spruit: Magnetohydrodynamic Jets and Winds from Accretion Disks. In *Evolutionary Processes in Binary Stars*, NATO ASIC Proc. **477**, 249 (1996).
H. C. Spruit, R. Stehle, J. C. B. Papaloizou: ApJ **275**, 1223 (1995).
H. C. Spruit, D. A. Uzdensky: ApJ **629**, 960 (2005).
K. G. Stassun, R. D. Mathieu, T. Mazeh, F. J. Vrba: AJ **117**, 2941 (1999).
K. G. Stassun, R. D. Mathieu, F. J. Vrba, T. Mazeh, A. Henden: AJ **121**, 1003 (2001).
T. Takeuchi, D. N. C. Lin: ApJ **581**, 1344 (2002).
L. V. Tambovtseva, V. P. Grinin: AstL **34**, 231 (2008).
K. Tassis: MNRAS **379**, L50 (2007).
K. Tassis, Ch. Mouschovias: ApJ **618**, 783 (2005).
K. Tassis, Ch. Mouschovias: ApJ **660**, 388 (2007).
S. A. Teitler: ApJ (2011).
C. E. J. M. L. J. Terquem: MNRAS **341**, 1157 (2003).
I. Tilling, C. J. Clarke, J. E. Pringle, C. A. Tout: MNRAS **385**, 1530 (2008).
K. Tomisaka: PASJ **48**, 701 (1996).
A. Toomre: ApJ **138**, 385 (1963).
M. V. Torbett: ApJ **278**, 318 (1984).
C. A. Tout, J. E. Pringle: MNRAS **281**, 219 (1996).
K. Tsinganos, C. Sauty, G. Surlantzis, E. Trussoni, J. Contopoulos: MNRAS **283**, 811 (1996).
T. Tsuribe: ApJ **527**, 102 (1999).
N. J. Turner, T. Sano, N. Dziourkevitch: ApJ **659**, 729 (2007).
T. Umebayashi, T. Nakano: MNRAS **218**, 663 (1986).
G. V. Ustyugova, A. V. Koldoba, M. M. Romanova, R. V. E. Lovelace: ApJ **646**, 304 (2006).
D. Uzdensky, J. Goodman: ApJ **682**, 608 (2008).
D. Uzdensky, A. Königl, C. Litwin: ApJ **565**, 1191 (2002a).
D. Uzdensky, A. Königl, C. Litwin: ApJ **565**, 1205 (2002b).
J. S. Vink, J. E. Drew, T. J. Harries, R. D. Oudmaijer, Y. Unruh: MNRAS **359**, 1049 (2005).
N. Vlahakis, K. Tsinganos, C. Sauty, E. Trussoni: MNRAS **318**, 417 (2000).
B. von Rekowski, A. Brandenburg: A&A **420**, 17 (2004).
B. von Rekowski, A. Brandenburg: AN **327**, 53 (2006).
B. von Rekowski, A. Brandenburg, W. Dobler, A. Shukurov: A&A **398**, 825 (2003).
B. von Rekowski, N. Piskunov: AN **327**, 340 (2006).
E. I. Vorobyov, S. Basu: MNRAS **381**, 1009 (2007).
D. Ward-Thompson, P. André, R. Crutcher, D. Johnstone, T. Onishi, C. Wilson: An Observational Perspective of Low-Mass Dense Cores II: Evolution toward the Initial Mass Function. In *Protostars and Planets V*, ed. by B. Reipurth, D. Jewitt, and K. Keil (Univ. of Arizona Press, Tucson, 2007), p. 33
M. Wardle: ASPC **121**, 561 (1997).
M. Wardle: MNRAS **307**, 849 (1999).

M. Wardle: Ap&SS **311**, 35 (2007).

M. Wardle, A. Königl: ApJ **410**, 218 (1993).

M. Wardle, C. Ng: MNRAS **303**, 239 (1999).

A. M. Watson, K. R. Stapelfeldt, K. Wood, F. Ménard: Multiwavelength Imaging of Young Stellar Object Disks: Toward an Understanding of Disk Structure and Dust Evolution. In *Protostars and Planets V*, ed. by B. Reipurth, D. Jewitt, and K. Keil (Univ. of Arizona Press, Tucson, 2007), p. 523.

S. J. Weidenschilling: Ap&SS **51**, 153 (1977).

C. Zanni, A. Ferrari, R. Rosner, G. Bodo, S. Massaglia: A&A **469**, 811 (2007).

E. G. Zweibel, K. T. Hole, R. D. Mathieu: ApJ **649**, 879 (2006).

8

CATHIE CLARKE

THE DISPERSAL OF DISKS AROUND YOUNG STARS

1. Prologue

We have seen in chapter 2 that there is ample evidence that disks around young stars are accretion disks, and chapters 5–7 have shown that there are several promising mechanisms for driving the angular momentum transfer required for such accretion. A generic property of disks for which the angular momentum transport is a power-law function of radius is that the disk evolves in a self-similar fashion ([102], [66]) such that the ratio of disk mass to accretion rate is of the order of the system age. This then means that for most observed disks, a substantial fraction of the mass that they currently contain will end up on the star by the time the system reaches twice its current age.

This does not mean, however, that the total lifetime of the observable disk is necessarily set by the time that is required for virtually all its mass to drain onto the central star. Indeed, if one takes typical values for inferred disk masses and accretion rates (see chapter 2) and extrapolates forward using viscous-similarity solutions, one finds that the time that must elapse before a disk should become so depleted in mass that it becomes optically thin in the near infrared is in excess of 100 Myr. This contrasts strongly with the observed lifetimes of optically thick disks which are in the region of a few Myr ([64]). Evidently, then, some other process must intervene at some point to clear away observable disks on a much shorter timescale. Possible mechanisms for this final disk dispersal are discussed extensively in this chapter.

2. Introduction

Broadly speaking, young stars are surrounded by disks of gas and dust for periods of a few Myr (e.g., [64], [158]). The transition to diskless status, however, occurs on a much shorter timescale and appears to occur in a way that is remarkably synchronized across diagnostics that probe very different spatial regions of the disk (e.g., [87], [10]). The existence of these two timescales (for which I elaborate the observational evidence in § 6) provides one of the strongest constraints on disk evolution and narrows the field of contenders for possible disk-dispersal agents considerably.

In brief, the two mechanisms that are most promising in this regard involve either photoevaporation of the disk (by radiation from the central star or neighboring more luminous sources in clustered environments) or processes associated with the formation of planets (either the simple removal of dust opacity due to grain coagulation—a necessary precursor to eventual planet formation—or the tidal effect of a planet on the disk gas). In either of these scenarios, it should be stressed that these are effects that bring the disk's existence to an end; they are probably not the processes that remove most of the disk's mass over its few Myr lifetime, since this is almost certainly effected by accretion onto the star.[1] The reader should consult the chapters by Balbus and Durisen for a survey of mechanisms that can cause the necessary angular momentum redistribution for disk gas to accrete onto the star. My concern here is instead focused on the termination of the disk phase and the fascinating interplay between disk dispersal and the epoch of planet formation.

This survey of disk-dispersal mechanisms is inevitably biased toward areas where the theoretical framework is most fully developed. Despite considerable progress in modeling various stages of the planet-formation process in recent years, what is currently lacking is a framework in which the interplay between the secular evolution of a disk and the formation of planets can be understood (although semiempirical models ([79], [90]) provide a step in this direction). Although models can follow local disk evolution and trace various phases of the planet-formation process at a particular location, what is less clear, for example, is how the formation of one planet affects the formation of another (e.g., [14], [116], [117]). Consequently, it is difficult to address the question of how planet-formation can or cannot cause a rapid global clearing of the disk. In the case of disk photoevaporation, however, there has been considerable progress in recent years in coupling the results of radiative transfer modeling to simply parameterized models for disk secular evolution (see, for example, the reviews

1. Note, however, that the most recent estimates of Gorti & Hollenbach (private communication) suggest that photoevaporation via the fan ultraviolet (FUV) radiation from the central star may even be competitive with accretion. The observed census of extrasolar planets, however, suggests that the mass contained in planets is a small fraction of the likely initial disk mass and that planet formation, though providing a possible final clearing mechanism, is not the major sink for disk material.

in [77], [9]) , and it is thus possible to set out quite detailed predictions about the statistics of objects that are to be expected at various evolutionary stages.

My approach, therefore, is to concentrate first on setting out the theoretical background on the interaction between energetic radiation (X-rays, ultraviolet rays) and the disk and to describe the hydrodynamic flows that result when disks are irradiated. I then couple such models with secular-evolution models for the disk and assess how these models fare in reproducing a range of observational data, both with regard to statistics of objects in different evolutionary phases and to explaining individual objects (inner-hole sources) that are in an interesting state of partial clearing. At this point, I can reintroduce the possibility of clearing by planets, since although, as explained above, these models cannot make statistical predictions, it is nevertheless possible to assess whether individual objects can be explained by a suitably located planet-formation zone.

Before I embark on my survey of the physics of disk photoevaporation, it is worth emphasing that whether or not disks are dispersed by photoevaporation or planet formation (or, indeed, any other unsuspected agent), the issue of disk dispersal is a key one when one is considering the viability of the planet-formation process. Indeed, it may be argued that dispersal is more relevant to planet formation (or its curtailment) if it is not itself a by-product of planet formation, since dispersal then provides a temporal boundary condition on the planet-formation process. Notably, in the one context in which the agent of disk dispersal is unambiguous (i.e., the photoevaporation of disks by ultraviolet radiation from OB stars in the Orion Nebula Cluster), it is evident that the window of opportunity for planet formation is quite seriously curtailed, since the disk lifetimes in this case are extremely short. Ultimately, we would like a theory that, while satisfying all the available observational data derived from disks in different environments, can make predictions about the viability of planet formation as a function of central-object mass and environment. The reader will rapidly be able to judge that we are presently far from this situation. Nevertheless, this chapter attempts to assemble a number of necessary parts of the theoretical tool kit for studying this problem and to bring these models to bear, where possible, on observational data.

3. The Physics of Disk Photoevaporation

3.1. Extreme Ultraviolet Photoevaporation

THE INTERACTION OF EXTREME ULTRAVIOLET RADIATION WITH A GIVEN DENSITY FIELD. I here define extreme ultraviolet (EUV) radiation as comprising photons that are sufficiently energetic (i.e., with energy exceeding 13.6 eV) to be able to ionize atomic hydrogen from its ground state. Since the cross section for ionization falls with increasing photon energy above this threshold as E^{-3} (e.g., [115]), it follows that for a wide range of input spectra the dominant contribution to the ionization process derives from photons that are only modestly above the threshold energy. I defer a discussion of the magnitude, spectra, and origin of EUV photons in pre-main-sequence stars until § 5.1.

The ionization cross section for neutral hydrogen is 10^{-17} cm^2 per atom, which implies that the ionizing photons are all absorbed after passing through a neutral column density of $\sim 10^{17}$ cm^{-2} (obviously the total hydrogen column density between an ionizing source and the point where all the ionizing photons have been absorbed is much greater than this, since the bulk of material in this region is ionized). The physics of the interaction between EUV photons and atomic hydrogen is particularly simple, since the absorption of an ionizing photon leads to the ionization of a single hydrogen atom, and the remainder of the photon energy is deposited in the thermal energy of ions and electrons (and thus by collisions also into thermal energy of the neutrals). Typically, the temperature acquired in ionized regions is in the range 10,000 to 30,000 K. The production of electron-ion pairs is offset by recombinations, whose volumetric rate is simply the product of the electron and ion densities and a temperature-dependent recombination coefficient ($\alpha(T)$). Recombinations to the electronic ground state, however, are accompanied by the emission of a photon whose energy exceeds 13.6 eV and that is itself capable of causing fresh ionizations; such recombinations thus produce an isotropic diffuse field of ionizing photons. Thus it is only recombinations to excited electronic energy levels (case B recombination) that represent a net destruction of ionizing photons (since the lower-energy photons produced in the process are incapable of causing further ionizations). It is therefore useful to define a second recombination coefficient ($\alpha_B(T)$) that refers only to recombinations to excited electronic states. Therefore, in a state of ionization balance, the surface integral of the flux of ionizing radiation over any given surface is equal to the volumetric case B recombination rate integrated over the enclosed volume:

$$\int_S F_{ion} dS = \int_V \alpha_B(T) n_e n_{ion} dV. \tag{1}$$

Charge neutrality implies that $n_e = n_{ion}$, so this becomes

$$\int_S F_{ion} dS = \int_V \alpha_B(T) n_{ion}^2 dV, \tag{2}$$

where n_{ion} is essentially n_{H^+}, since hydrogen is the overwhelmingly dominant ion.

Note that F_{ion} in general comprises contributions from both the ionizing source (or sources) and the diffuse field produced by recombinations to the ground state. In practice, if the ionized region is limited by running out of ionizing photons (rather than running out of material, i.e., if it is ionization bounded rather than density bounded), then we can define an ionization front (IF) where F_{ion} is zero. If we define our region of integration such that it is bounded by the IF and by surfaces enclosing each of the ionizing sources (whose total ionizing luminosity is Φ_{ion} s^{-1}), then we can rewrite eq. (2) as

$$\Phi_{ion} = \int_V \alpha_B(T) n_{ion}^2 dV. \quad (3)$$

Eq. (3) can under some circumstances be used to determine the location of the IF without any necessity of computing the diffuse field. In order for this equation to be usable for this purpose, it is important that the system be ionization bounded in all directions from the source (i.e., no ionizing photons can escape the system). In addition, in order that this single equation can define the shape of a three-dimensional IF, it is necessary that the symmetries in the problem are such that the IF can be parameterized in terms of a single variable (e.g., the radius of a sphere or a penetration depth within a slab). It is also, of course, necessary that one know the density and temperature structure of the ionized region. Here, however, one can make a couple of approximations that ease this considerably. First, the recombination coefficient, α_B, is a sufficiently weak function of temperature, over the range of temperatures that are typically encountered, that one can approximate the region as being roughly isothermal, with a correspondingly uniform value of α_B ($= 2.6 \times 10^{-13}$ cm^3 s^{-1} for $T = 10^4$ K; [33]). Second, detailed computations of the ionization fraction in the vicinity of ionizing sources demonstrate that the gas is nearly 100% ionized within the ionized region and that the IF is a thin zone separating this from regions with nearly zero ionization fraction. Hence it suffices, when one is computing the right-hand side of eq. (3), to set n_{ion} equal to the total hydrogen density.

The best-known application of eq. (3) is to the simplest case of a single ionizing source located in a cloud with uniform hydrogen number density n. In this case (see [146]), we can define the Stromgren radius R_S (i.e., the radius of the IF) as satisfying

$$\Phi_{ion} = 4\pi \alpha_B n^2 R_S^3/3. \quad (4)$$

A closely related situation is that in which the density is a function of distance from the source (as occurs when ionization gives rise to hydrodynamic flows), in which case the location of the IF can be readily calculated from

$$\Phi_{ion} = 4\pi\alpha_B \int_0^{R_S} n(r)^2 r^2 dr. \tag{5}$$

Another commonly encountered situation is that of a single ionizing source and a density structure that is a function of polar angle, i.e., $n = n(\theta, \phi)$. In this case, obviously, the value of R_S is also a function of polar angle, being larger in regions of lower density. We may, in fact compute an approximate value for $R_S(\theta, \phi)$ by employing eq. (5) and using the radial profile of $n(\theta, \phi)$. This Stromgren-volume method, whereby the system is approximated over each small interval of solid angle as though it were a portion of a spherically symmetric distribution, is, however, only an approximation because it neglects the nonradial propagation of ionizing photons in the diffuse field. In the case where the density is angularly dependent, however, the diffuse field of ionizing photons (originating from recombinations to the ground state) will also cause a nonradial flux of ionizing photons (generally from dense to less dense regions), and it is therefore not strictly correct to treat each small interval of solid angle as a portion of an equivalent sphere. There is, however, a situation where this Stromgren-volume approach is approximately valid, namely, in the case where the density is so high that the mean free path for recombinations is small compared with the scale length for nonradial density variations. In this limit, called the on-the-spot approximation, one can then neglect the lateral transfer of photons in the diffuse field and use the Stromgren-volume method.

I have labored this point because it is central to our purpose here, which generally boils down to discovering which portions of an irradiated gas lie within the ionized region and hence which regions are going to be heated to $\sim 10^4$ K. I therefore conclude this brief survey of the physics of irradiation by EUV photons by applying these ideas to several situations that are relevant to irradiated disks.

Example: External Irradiation of a Disk, Radius r_d, by a Source at Distance d. In the case $r_d << d$, the incident radiation is close to being plane parallel, and hence, for every annulus in the disk, the problem is approximately one-dimensional (i.e., we consider only the net flux of ionizing radiation in the z direction, normal to the disk). We can then determine the height (H_{IF}) of the IF above the disk midplane, if the density structure of the ionized gas is known (see below), according to

$$\frac{e^{-\tau_d}\Phi_{ion}}{4\pi d^2} = \alpha_B \int_{H_{IF}}^{d} n_{ion}(z)^2 dz. \tag{6}$$

Here the first (exponential) term is an attenuation factor due to an assumed optical depth τ_d of dust between the source and the disk.

Example: Irradiation by a Central Star of a Disk with Inner Hole, Radius R_{in}. I consider the irradiation of the disk's inner rim (radius R_{in}) in the case that the optical depth due to attenuation by dust in the inner hole is τ_d. Given the density structure in the ionized rim of the disk, the penetration depth of the EUV radiation, Δ_{IF}, satisfies the recombination integral

$$\frac{e^{-\tau_d}\Phi_{ion}}{4\pi R_{in}^2} = \alpha_B \int_{R_{in}}^{R_{in}+\Delta_{IF}} n_{ion}^2 dR. \tag{7}$$

Note that we are using the Stromgren-volume method here (i.e., we are assuming that the location of the IF at the disk midplane is the same as it would be if the density distribution were spherically symmetric). This implicit neglect of the propagation of diffuse ionizing photons away from the disk midplane can be justified only in the case that the mean free path for recombination photons is much less than the disk scale height.

Example: Irradiation by a Central Star of a Disk without Inner Hole. This is actually a case where the Stromgren-volume approximation is not good for determining the location of the IF because it turns out that the mean free path for recombination photons is long. Explicitly, photons from the central star interact with a hot hydrostatic atmosphere above the inner disk, and the diffuse field of recombination photons from this atmosphere then irradiates the outer disk ([74]). Thus one would get entirely the wrong answer if one tried to locate the IF in the outer disk by merely considering the radial propagation of ionizing photons from the source. Thus this problem has to be solved as a problem in two-dimensional radiative transfer, which takes into account both the direct and the diffuse ionizing fields. I return to this problem again in § 5.3.

SELF-CONSISTENT DENSITY FIELDS IN THE PRESENCE OF EUV RADIATION. I showed above how eq. (3) can be adapted to determine the location of the IF if the density structure of the ionized region is specified. In general, the density in the ionized gas will self-adjust to a steady-state configuration—either one of hydrostatic equilibrium or a steady flow. Both of these are relevant to the photoevaporation problem since it often turns out that photons that will initiate a flow at large radius have to propagate through a hydrostatic photoionized region close to the star. In the case that we determine the density structure of flows, the fact that these expand at roughly the sound speed of ionized gas ($\sim$ 10 km s^{-1}) allows us to make an immediate rough estimate of the mass-loss rate, which is, after all, our main interest in the problem. I now consider a few simple cases.

Example: Ionization-Bounded Classical HII Region. In the absence of gravity or other external forces, hydrostatic equilibrium requires uniform pressure; therefore, the initial Stromgren radius—given by eq. (4) with n equal to its value in the undisturbed medium—is replaced by the corresponding expression where n takes a new (lower) value (n_{II}) such that the ionized gas is in pressure equilibrium with the surrounding medium, i.e., such that

$$n_I T_I = n_{II} T_{II}, \tag{8}$$

where the subscripts I and II refer to conditions in the neutral and ionized zones, respectively. In general, we are here interested only in equilibrium configurations and therefore need not directly concern ourselves with the large body of literature (e.g. [85], [143]) describing the expansion of regions of ionized hydrogen HII to their equilibrium state.

Example: Hydrostatic Atmosphere of an Irradiated Inner Disk. I now define an important radius, the gravitational radius for EUV photoevaporation, which is given by

$$r_{g_{II}} = \frac{GM}{c_{s_{II}}^2}, \tag{9}$$

where M is the stellar mass and $c_{s_{II}}$($\sim$ 10km s^{-1}) is the sound speed for photoionized gas. For a solar-mass star, $r_{g_{II}} \sim 10$ AU.

The radius $r_{g_{II}}$ is a measure of the region of the disk from which photoionized gas can escape the star's gravitational potential: when it is defined in this way, a free particle with velocity equal to the sound speed is unbound at radii $> 2r_{g_{II}}$. In practice, we shall see later that pressure-driven flows can originate from a radius that is several times less than $r_{g_{II}}$. Nevertheless, at a sufficiently small radius ($r << r_{g_{II}}$), ionized gas can settle into a hydrostatic atmosphere in which the density then varies with z (height above the midplane) according to

$$n_{ion}(z) = n_{II} e^{-(z^2 - H_{IF}^2)/2H^2}, \tag{10}$$

where $H = c_{s_{II}}/\Omega$ and Ω is the Keplerian angular frequency (e.g., [125]). This equation is valid for $z > H_{IF}$ (where H_{IF} is the height of the IF above the midplane) and we have normalized the expression such that n_{II} is the density of ionized material immediately above the ionization front.

If the disk outer radius, r_d, is less than $r_{g_{II}}$, then the recombination integral may be dominated by recombinations in the hydrostatic atmosphere, and hence one can estimate n_{II} by substituting (10) into eq. (6), yielding[2]

$$\frac{e^{-\tau_d}\Phi_{ion}}{4\pi d^2} \sim \alpha_B n_{II}^2 H. \tag{11}$$

Note that the structure of the ionized regions (i.e., both n_{II} and the scale height H) is independent of the properties of the underlying, un-ionized disk, since the ionized region self-adjusts to a hydrostatic structure in ionization balance with the incident flux. The underlying disk does, however, determine the location of the IF, which is situated so that the pressure in the underlying neutral disk matches the pressure in the ionized gas just above the IF, i.e., $n_I T_I = n_{II} T_{II}$ (eq. [8]).

Example: Irradiated-Disk Inner Hole. I consider the case where $R_{in} > r_{g_{II}}$, so that photoionized gas at the hole's inner edge can escape the star's gravitational field. We shall see later that this situation involves a progressive enlargement of the disk inner hole, but material that was ionized previously, when the disk radius was smaller, can expand normal to the disk plane and thus does not contribute to the recombination integral. Instead, this is dominated by an ionized rim of material at the disk's current inner edge. We can obtain a rough estimate of the density of ionized gas at the IF (n_{II}) by approximating the recombination integral ($\int n_{ion}^2 dR$) as $\sim n_{II}^2 \Delta_{IF}$, where Δ_{IF} is a measure of the radial scale length at the disk inner edge. It may readily be shown that a rotationally supported disk cannot be truncated on a radial scale length that is less than its pressure scale height ($H_I \sim c_{s_I}/\Omega$), since such a flow would violate the Rayleigh stability criterion ([98]). In fact, numerical calculations ([7]) show that $\Delta_{IF} \sim H_I$, so that one can estimate n_{II} by approximating the right-hand side of (7) by $n_{II}^2 H_I$ (note that H_I, being the pressure scale height in the neutral gas, is less than the value of H in the ionized gas e.g., eq. (10) by a factor equal to the square root of the temperature ratio in the two phases).

2. However, this is not a good approximation in the case that $r_d > r_{g_{II}}$, since the recombination integral to a point at radius $r < r_{g_{II}}$ in the disk is instead then dominated by material flowing off the outer disk, which is located at height $\sim r_d$ above the disk.

Putting all this together (in the case that we neglect dust attenuation in the hole), we have

$$n_{II}^2 H_I \propto \frac{\Phi_{ion}}{R_{in}^2}. \tag{12}$$

Thus $n_{II} \propto R_{in}^{-1.5}(H_I/R_{in})^{-0.5}$. Since gas flows off the rim at $c_{s_{II}}$, the flux of ions is just $n_{II}c_{s_{II}}$; given that the surface area of the rim is $2\pi R_{in}H \propto 2\pi R_{in}H_I$, we can then write

$$\dot{M} \propto R_{in}^{0.5}\left(\frac{H_I}{R_{in}}\right)^{0.5}. \tag{13}$$

I have expressed the scaling in this way since, although protostellar disks are likely to be modestly flared, H_I/R_{in} is believed to be fairly constant (and of order 0.1), so that I here emphasize the residual dependence on the size of the inner rim. It will be important in my discussion of the photoevaporation of systems with cleared inner holes that this rate increases as R_{in} increases (see § 6.1).

Example: Wind Flow from an Externally Irradiated Outer Disk. I now turn to the case of external irradiation of a disk at radii $> r_{g_{II}}$. Evidently, this is a flow situation, and since the flow is radial at large distances from the disk, it is possible to get a simple grasp of the problem by modeling the whole system as spherically symmetric (i.e., replacing the disk by a sphere of radius r_d).

In this dynamic outflow situation, material expands away from the IF in a spherical wind with velocity $v_{II} \sim c_{s_{II}}$. In fact, the flow remains roughly isothermal at all larger radii, since it turns out that photoheating can offset adiabatic cooling in the flow. It is also the case that where the flow velocity is transonic or supersonic, pressure gradients are relatively ineffective in accelerating the flow, and such flows are therefore of nearly constant velocity (this feature is demonstrated by the fact that the terminal velocity of an isothermal Parker wind is only a few times the sound speed; ([119]). Consequently, we can approximate the density structure of the flow by recalling that in a steady state the mass flow rate in the outflow is independent of radius: for a spherical wind with constant velocity, this implies that $n_{ion} \propto r^{-2}$. This means that in the right-hand side of eq. (6),

$$\int_{R_{IF}}^{\infty} n_{ion}^2 dr \sim n_{II}^2 R_{IF}. \tag{14}$$

Thus the magnitude of the ionizing flux irradiating the disk controls the structure of the ionized flow via the product $n_{II}^2 R_{IF}$. Further progress, however, requires another constraint on the individual values of n_{II} and R_{IF}.

One possibility is that the disk atmosphere is ionized virtually down to the disk surface. In this case, denoted an EUV-dominated flow by Johnstone et al. ([83]), $R_{IF} \sim r_d$, which thus immediately fixes n_{II} for a given incident ionizing flux. This is important because since $v_{II} \sim c_{s_{II}}$, the mass-loss rate in the wind is then simply

$$\dot{M}_{EUV} = 4\pi r_d^2 c_{s_{II}} m_H n_{II}, \tag{15}$$

where $m_{\rm H}$ is the mass of a hydrogen atom and n_{II} is determined from the equation of ionization balance (6 and 11). Note that since, for a given ionizing flux, $n_{II}^2 r_d$ is fixed, and $\dot{M}_{EUV} \propto n_{II} r_d^2$, $\dot{M}_{EUV} \propto r_d^{1.5}$.

A more complicated situation arises when the IF is not coincident with the disk surface. Here we need to mention the effect that will form the subject of the following section, i.e., the effect on the thermal properties of the disk of photons in the FUV spectral window (i.e., ultraviolet photons that are insufficiently energetic to ionize hydrogen). For our present purposes, we can simplify the effects of the FUV photons as producing a warm, neutral region, with sound speed c_{s_I}, number density n_I and flow speed v_I, that forms a sandwich between the disk and the IF. We can then treat the IF as an effective contact discontinuity between the FUV-heated and ionized flows. Applying mass and momentum balance across the contact discontinuity, we obtain

$$n_I v_I = n_{II} v_{II} \tag{16}$$

and

$$n_I (c_{s_I}^2 + v_I^2) = n_{II} (c_{s_{II}}^2 + v_{II}^2). \tag{17}$$

Since, as noted above, $v_{II} \sim c_{s_{II}}$ (i.e., the flow is transonic at the base of the ionized region), we obtain

$$v_I = (\frac{c_{s_I}^2}{2c_{s_{II}}}). \tag{18}$$

Since the FUV photons heat the gas to a lower temperature than that in the ionized region (i.e., $c_{s_I} < c_{s_{II}}$), it then follows that the flow from the neutral medium into the IF is subsonic with respect to the neutral medium.

These considerations always apply when one matches the ionized region onto the structure of the disk below it (note that in the hydrostatic case that we considered above, we just have that $v = 0$ in both regions, and so eq. [17] just boils down to the condition of thermal pressure balance at the IF). In the case that the warm neutral region is thin (i.e., of width $<< r_d$), the matching up of conditions does not much affect either the location of the IF or the density on the ionized side. Thus the mass flow rate (which is the critical quantity of interest) is determined completely by the EUV irradiation. As noted above, such flows are termed *EUV dominated* and are governed by eqs. (11) (with $R_{IF} = r_d$) and (15).

However, in cases where the FUV heating is relatively strong, the warm neutral region can become spatially thick, and so one does not know a priori what the location of the IF is. We shall see later that in this case the flow becomes supersonic before it reaches the IF, and hence there is no possibility of the presence of the IF being communicated to the underlying flow. In this case, therefore, we can argue on the grounds of causality that EUV photoionisation cannot initiate the flow, which instead results from FUV photoheating: although this FUV-induced wind ultimately passes through an IF at some point, it is the thermal interaction with the FUV that fixes the mass flow rate from the disk. However, we shall need to consider this thermal interaction in some detail (see § 3.2) before we can construct new expressions for the photoevaporation rate for FUV-dominated flows.

SUMMARY. I have presented a simplified scheme for the interaction of EUV radiation with cool disk material, whereby the ionized gas is more or less isothermal with temperature of $\sim 10^4$ K. Depending on the value of the sound speed of the ionized gas compared with the local escape velocity (i.e., as measured by the ratio of radius in the disk to ([$r_{g_{II}}$ eq. (9)]), the resulting structure will lie somewhere in the range between a state of hydrostatic equilibrium (for $r << r_{g_{II}}$) and that of roughly sonic flow from the IF. In the latter limit, knowledge of the density at the IF then allows one to immediately calculate the mass flow rate per unit area. The value of the density at the IF is determined by imposing the condition of ionization balance; this is a complex two-dimensional problem if the mean free path of ionizing photons produced by recombinations to the ground state is relatively long. There are, however, cases where the geometry of the problem means that the flux of ionizing photons into a region is dominated by those coming directly from the ionizing source. In such cases one can construct simple analytic estimates for the density at the IF in the ionized region, and hence, for the mass-loss rate per unit area. I have applied this simple analysis to the case of a disk irradiated by an external ionizing source and also to the case of direct irradiation by the central star of the inner rim of a disk with a cleared inner hole. I have emphasized that under some circumstances, we can compute the photoevaporation rate on the basis only of conditions in the ionized region (although conditions on the other side of the IF must self-adjust so as to satisfy suitable boundary conditions).

A final point is that the condition of ionization balance (eq. [3]) that we have exploited so much in this section is, strictly speaking, one that balances the flux of ionizing photons against recombinations in a fixed mass of gas. In flow situations, however, ionizing photons are also consumed in ionizing fresh neutral material that flows through the IF. The ratio of the rate of ionizing photons that are consumed per unit area in ionizing new material to the rate of recombinations per unit area is $n_{II} c_{s_{II}} / \alpha_B n_{II}^2 \lambda_{scale}$, where λ_{scale} is the scale length in the ionized region. If we assume that this ratio is small, we can equate the denominator with the flux of ionizing photons from the source and hence determine n_{II}, so that the ratio may then be written as $\sim c_{s_{II}} / \sqrt{\Phi_{ion} / 4\pi d^2 \alpha_B} \lambda_{scale}$. We can now substitute some typical values for ionizing fluxes encountered in rich clusters, as well as typical scale lengths in the ionized flow (which are of the order of the disk size) and can readily show that this ratio is small (i.e., < 0.1), thus justifying a posteriori the neglect of the flow's effect on the ionization balance.

3.2. Irradiation by FUV Photons

I define FUV photons as comprising those with energy in the range 6–13.6 eV, which thus, while not having sufficient energy to ionize hydrogen, can cause the photodissociation of molecular hydrogen and other molecules. Photodissociation of H_2 proceeds by the pumping of molecular hydrogen to excited electronic states, followed by fluorescence. In 85 to 90% of cases, this results in molecular hydrogen in excited vibrational states of the electronic ground state (see Fig. 8.1). In the remaining 10 to 15% of cases, however, the system is deposited into the vibrational continuum of the electronic ground state; i.e., to put it simply, a hydrogen molecule is converted into two hydrogen atoms.

Regions whose thermal properties and chemistry are dominated by the effect of irradiation by FUV photons are called photodissocation regions (henceforth PDRs). They are generally bounded by an IF on the side nearer the UV source and by a boundary with entirely molecular material on the other side. Thus (Fig. 8.2), depending on the ionization potentials and dissociation energies of various species, PDRs contain a succession of transitions from ionized to neutral to molecular states. Thus, for example, carbon has a lower ionization potential than hydrogen, and thus PDRs contain a layer of C^+; on the other hand, the lower dissociation energies of carbon and oxygen mean that they are still atomic in the region of molecular hydrogen. The bases of PDRs (where the visual extinction is in the range 3–8) are typified by conditions where

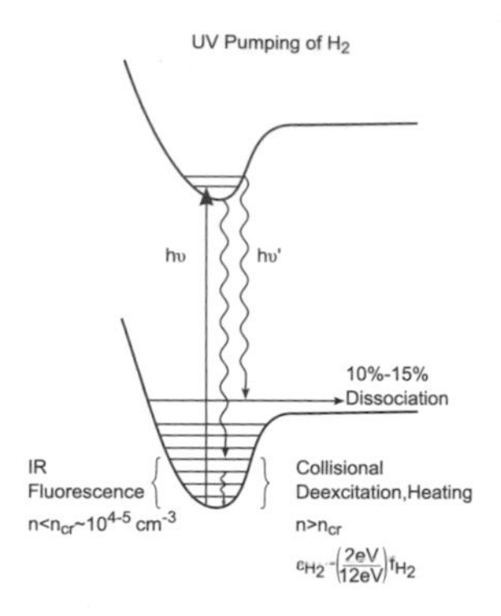

Fig. 8.1. Schematic diagram showing the sequence of excitation and deexcitation processes associated with the photodissociation of molecular hydrogen ([76]). The absorption of an FUV photon (frequency ν) can excite H_2 to the excited electronic state. As this radiatively deexcites (via the emission of a photon of frequency $\nu\prime$), the system is returned either to a vibrationally excited state of the electronic ground state or else to the vibrational continuum (resulting in dissociation). In the former case, the system is returned to the vibrational ground state either by IR fluorescence (at densities below the critical density) or else by collisional deexcitation (above the critical density). In the latter case, the surplus energy ($\sim$2 eV) goes into the thermal reservoir. Since each FUV photon has energy $\sim$12 eV, then if f_{H_2} is the fraction of FUV photons that pump H_2, then the fraction of the FUV luminosity transferred to the gas by H_2 pumping is $\epsilon_{H_2} \sim (2/12) f_{H_2}$.

the FUV is still sufficient to photodissociate species such as O_2, H_2O, and OH and where grain photoelectric heating is still an important thermal input.

The chemistry within PDRs is somewhat different from that in regions of the interstellar medium (ISM) that are shielded from FUV radiation and where reactions are often mediated by H_3^+ generated by interaction with cosmic rays. PDR chemistry is instead dominated by photodissociation and photoionization by FUV photons (see Fig. 8.1); moreover, the warm conditions produced by photoelectric grain heating can favor reactions that are otherwise unfavorable on account of activation barriers. I refer the reader to the chapter by Bergin for a more detailed discussion of PDR chemistry, as well as the excellent review by Hollenbach & Tielens ([76]). I will also have little to say about the rich array of spectral features and diagnostic information derivable from PDRs except in so far as they are of direct applicability to determining the mass flow rate in FUV-dominated disk winds. Instead, in this section I concentrate on setting out the physics that determines the thermal balance within PDRs and the physical extent of regions that are significantly heated by FUV photons.

The calculation of the extent of PDRs (i.e., the point at which hydrogen makes the transition from the mainly atomic to the mainly molecular state) involves the setting up of an equilibrium between the formation of H_2 per unit

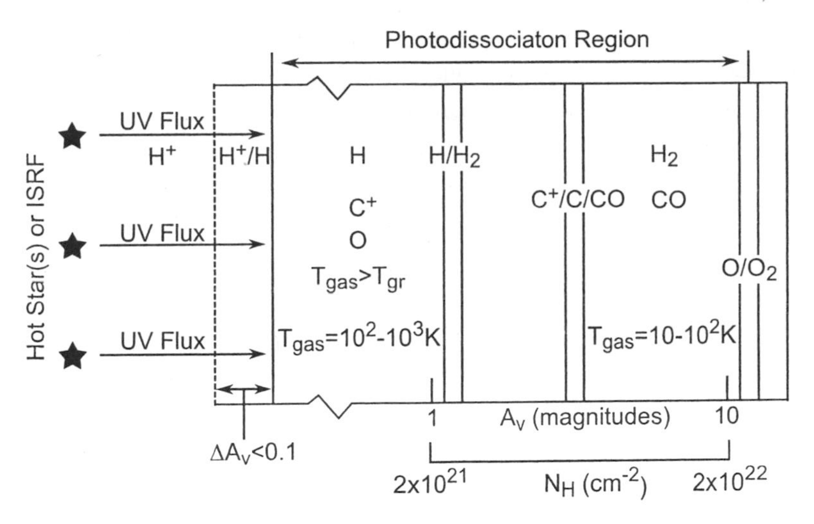

Fig. 8.2. Schematic representation of a PDR (here defined so that it extends to the point that oxygen is molecular), showing the sequence of ionic, atomic, and molecular species with increasing depth of ultraviolet penetration ([76]). The photons impinge the gas from hot stars or interstellar radiation field (ISRF) located to the left. Depth within the cloud is labelled in terms of column density (N_H) and magnitudes of extinction (A_ν).

volume on dust grains and the destruction of H_2 by photodissociation. The former can be equated ([75]) with the product of the atomic hydrogen number density (n_I), the number density of appropriate grains (n_{gr}), the grain cross section (σ_{gr}), and the thermal velocity of the atomic hydrogen (c_{s_I}).[3] We can thus write the rate of molecular hydrogen formation per unit volume as $\gamma_{H_2} n n_I$, where γ_{H_2} is a weak function of temperature ([84]). Here n is the total number density of hydrogen (in either molecular or atomic form) since it is assumed that the density of suitable grains is a fixed fraction of the total hydrogen number density. The rate of molecular hydrogen destruction by photodissociation can be written as $f_{shield}(N_{H_2})e^{-\tau_d} I_{diss}(0) n_{H_2}$ ([36]). Here $I_{diss}(0)$ represents the photodissociation rate per hydrogen molecule in the case that there is no intervening material between the source and the irradiated molecule, and n_{H_2} is the number density of molecular hydrogen. It is conventional to parameterize the strength of ultraviolet radiation fields in terms of G_0, the ratio of the FUV flux to its ambient value in the solar neighborhood, which is 1.6×10^{-3} erg cm^{-2} s^{-1}([63]). With this parameterization, the photodissociation rate per unshielded H_2 molecule can be written as $I_{diss}(0) = 4 \times 10^{-11} G_0$. The two

3. At low temperatures, the probability that a hydrogen atom that hits the grain both sticks and then migrates to find another hydrogen atom on the grain before evaporating is nearly unity.

preceding terms in the expression for the photodissociation rate represent the effects of dust attenuation and self-shielding by a column N_{H_2} of overlying molecular hydrogen, where the suppression due to self-shielding scales as $N_{H_2}^{-3/4}$ ([36]).

In order to locate the boundary of the molecular zone, we then equate these two expressions in the case that $n_{H_2} \sim n_I$. At this point, therefore, we can write

$$\gamma_{H_2} n = 4 \times 10^{-11} f_{shield}(N_{H_2}) e^{-\tau_d} G_0, \tag{19}$$

showing that the transition to the molecular state depends in general not only on local conditions but also on the overlying column density of molecular hydrogen. We can make progress here by considering two limits. First, there is the case where dust attenuation is important—so important, that is, that the transition to molecular hydrogen occurs while f_{shield} is still of order unity. The column density of dust that satisfies eq. (19) then depends only logarithmically on G_0/n, i.e., is fairly insensitive to G_0 and n. In fact, in this limit, it is found (for standard dust-to-gas ratios and grain size distributions) that the transition between molecular and atomic hydrogen occurs at a hydrogen column density of around 10^{21} cm^{-2}; henceforth I denote the critical gas column density corresponding to absorption of FUV radiation by dust as $10^{21}/\sigma_H$ cm^{-2}, where σ_H is the dust cross section per hydrogen atom, normalized to a standard value of 10^{-21} cm^2. Evidently, σ_H is less than unity either in situations of low dust-to-gas ratio (due, for example, to dust settling) or where there has been substantial grain growth. If, however, the ultraviolet field is relatively weak (or the density relatively high), then the column of molecular hydrogen lying in the surface $\sim 10^{21}/\sigma_H$ cm^{-2} of hydrogen nucleus column density may be high enough that f_{shield} at this point is much less than unity. In this case we can, in the limit, consider the transition to predominantly molecular hydrogen as being caused by H_2 self-shielding rather than dust attenuation, so that one can solve eq. (19) with the exponential (dust-attenuation) factor set equal to unity. In this case, the column of molecular hydrogen at the transition point scales as $(G_0/n)^{4/3}$. In practice, self-shielding sets the boundary of a PDR in the case that $G_0/n < 4 \times 10^{-2}$ cm^3.

In summary, then, for low G_0/n we are in a regime where the molecular boundary of the PDR is set by H_2 self-shielding and where the column of molecular hydrogen at the transition point scales as $(G_0/n)^{4/3}$. At higher G_0/n, however, dust attenuation takes over as the main determinant of the transition point, which occurs at a roughly constant total hydrogen column density of around 10^{21} cm^{-2}, independent of G_0 and n. Codes that study the structure of PDRs in various environments automatically take into account both regimes.

However, we shall find it particularly useful that in the case of FUV-dominated flows in the outer regions of protostellar disks, we are usually in the regime of high G_0/n, where, therefore, we can simply invoke the fact that the PDR consists of a zone of column density $\sim 10^{21}$ cm^{-2}.

It is now useful to consider the thermal structure of PDRs by examining the main heating and cooling processes. The following thumbnail sketch is based on the results of a number of one-dimensional PDR codes (see, e.g., [152], [35]). Although, of course, the mass in PDRs is dominated by the gas (by a factor $\sim$ 100 for typical solar ISM composition), it is the dust that processes most of the energy input and output. Grossly speaking, 95% of the incident FUV flux is absorbed by dust grains and the remainder by polycyclic aromatic hydrocarbons (PAHs), planar molecules comprising around 50 C atoms and associated aromatic CH bonds (see the chapter by Henning and Meeus for a discussion of the chemistry and spectral diagnostics of PAH molecules). This energy is then reradiated in the infrared in the form of continuum radiation from dust and via PAH bands, with the former dominating the radiative output by around a factor of 20. The energy radiated by the gas (which is dominated by fine structure lines of atomic species) is only around 0.1–1% of the energy processed by the dust. The gas is mainly heated collisionally, either by photoelectrons ejected by FUV photons impacting small dust grains or else by vibrationally excited H_2 produced via FUV pumping and flourescence.

Obviously, the relative dominance of these processes depends on depth within the PDR inasmuch as this controls the radiation field, the abundances of various molecular, atomic, and ionic species, and in reality (although not in many PDR models) the grain size distribution. In addition, the dust and gas thermal properties are also somewhat coupled by collisional energy transfer. All this means that the relative temperatures of the dust and gas depend on depth in the PDR. For example, Fig. 8.3 represents the thermal structure and abundances of various species computed in the case of a PDR with an ambient radiation field typical of Orion. In the surface layers (low A_V), the gas is hotter than the dust because of the relatively low efficiency of gas cooling (which is by ionic fine structure lines near the surface and via rotational lines of CO deeper in the PDR). At around an extinction of $A_V \sim 4$, however, the gas temperature falls as heating mechanisms cut out because of the attenuation of the FUV field. The dust, however, continues to be heated by midinfrared photons produced by overlying dust. Once the dust becomes slightly warmer than the gas, collisional energy transfer between the gas and dust then maintains close thermal coupling between the dust and gas at greater depths.

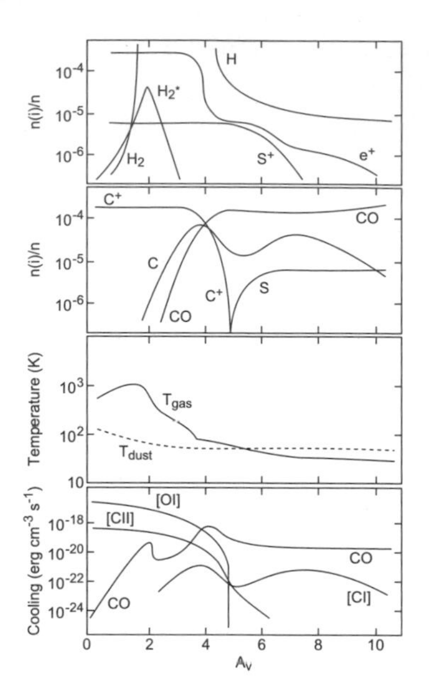

Fig. 8.3. The structure of the PDR in Orion ($n = 2.3 \times 10^5$ cm^{-3}, $G_0 = 10^5$) as computed by Tielens & Hollenbach ([153]). The upper panels denote the abundances of various species as a functon of A_V, the third panel the dust and gas temperatures, and the bottom panel the cooling in various gas lines.

From the point of view of the overall energy budget, the dust is the critical component of PDRs and, together with the PAHs, provides many of the useful observational diagnostics. Nevertheless, for our purposes here, where we want to compute the photoevaporation of irradiated disks, the key output of these calculations is the gas temperature as a function of depth. I show the results of a suite of PDR calculations performed by Adams et al. ([1]) for a variety of values of G_0 and n as a function of A_V in Fig. 8.4. These plots show that the gas temperature generally declines as one penetrates into the PDR (i.e., as A_V increases) and that the steep decline in temperature indeed coincides with A_V of slightly greater than unity (i.e., for standard σ_H, to a gas column density of order 10^{21} cm^{-2}), as expected in the case that dust attenuation controls the molecular boundary of the PDR. Over a fairly wide region of parameter space, the PDR is roughly isothermal at lower A_V, although this picture is complicated at the lowest A_V, where the temperature structure is not necessarily a monotonic function of n at fixed G_0. In the most self-consistent attempt to couple PDR thermal structure to the resulting flow structure to date, Adams et al. ([1]) combined the results of these calculations with the

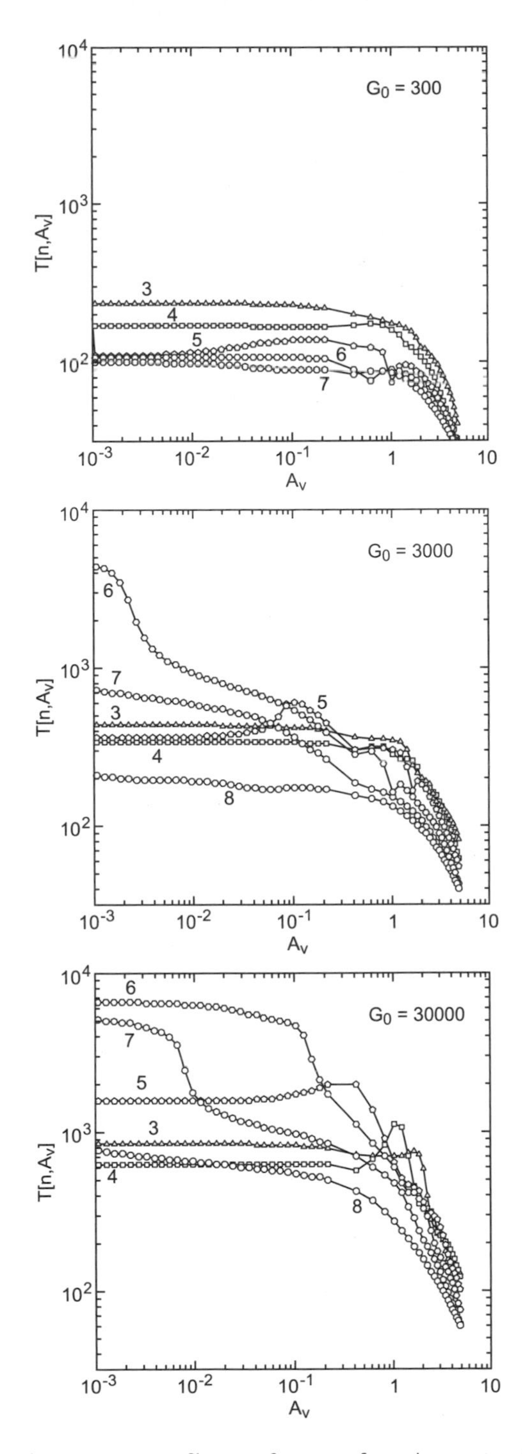

Fig. 8.4. Computed temperature profiles as a function of visual extinction, from Adams et al. ([1]). Each panel is labeled according to the strength of the FUV field (as measured by G_0) and each line is labeled by the logarithm of the gas number density.

requirements of steady-wind models. However, most analytic work to date on FUV-dominated flows ([83]) instead restricts itself to the simplest thermal structure, i.e., an isothermal, PDR (at ~1,000 K) extending to a column depth of $\sim 10^{21}$ cm^{-2}. I next consider how such a simple parameterization may be combined with the required boundary conditions at the IF so as to provide a simple analytic form for the structure (and photoevaporation mass-loss rate) of FUV-dominated flows.

COUPLED FLOWS WITH FUV AND EUV HEATING. For an incident spectrum consisting of both EUV and FUV photons, an irradiated disk will consist of a surface layer—heated and ionised by EUV radiation—overlying a warm layer heated and photodissociated by FUV radiation. Below the PDR, the disk is generally much cooler, being heated either by internal viscous dissipation or by irradiation by infrared photons produced by the reradiation of the central star's optical radiation field by dust in the surface layers of the disk. I have outlined above how, depending on the spatial thickness of the PDR produced by FUV heating, the mass-loss rate from an irradiated disk may be determined either by FUV heating or by EUV heating.

In order to place this on a more quantitative footing, it is instructive to consider the case of a simple, spherical, isothermal, pressure-driven wind in a point-mass potential, often termed a *Parker wind* ([119]). Fig. 8.5 illustrates the outflow velocity, normalized to the sound speed, plotted as a function of radius

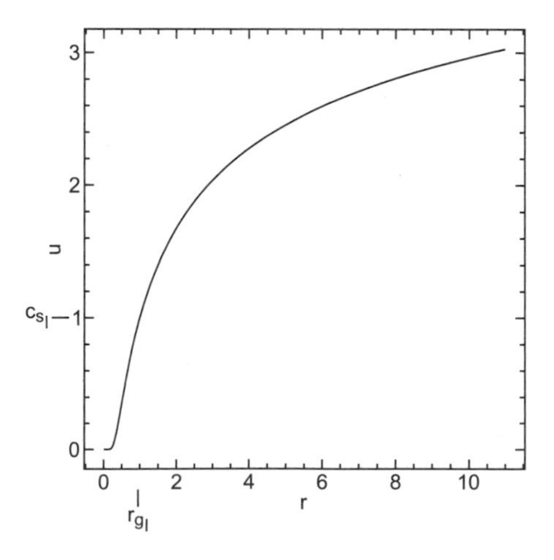

Fig. 8.5. Flow velocity (normalized to the sound speed) as a function of radius (normalized to the sonic radius; see text) for a steady spherical isothermal pressure-driven outflow (i.e., Parker wind; ([119]).

(normalized to the sonic radius, r_s, which is equal to $GM/2c_s^2$; note that r_s is therefore $0.5r_g$). This plot illustrates a feature mentioned above, namely that for $r > r_s$, the velocity rises rather gently with radius (yielding the physically plausible result that it is impossible to pass from flows that are subsonic to those that are highly supersonic by the action of thermal pressure forces alone). In the supersonic regime, therefore, the flow can be very roughly approximated as constant velocity and with density $\rho \propto r^{-2}$. In the subsonic regime, however, it is evident that the flow velocity increases steeply with radius.

I now consider the case in which we have two Parker winds, with sound speeds c_{s_I} and $c_{s_{II}}$ at small and large radii respectively. This situation corresponds to the case of the disk being irradiated by an external luminosity source. In this case the outer region is EUV heated and the inner region is FUV heated, while the IF coincides with the contact discontinuity between the two flows. I have already derived the conditions that must apply at the contact discontinuity between these flows (eqs. [16] and [17]) and note the requirement that the flow velocity into the IF on the neutral side is subsonic with respect to the neutral flow. Fig. 8.6 shows an example of the coupling between two Parker winds that could satisfy these conditions—the neutral wind is subsonic up to the IF, while the jump in velocity at the IF then renders the flow in the EUV region mildly supersonic. Evidently, for a flow of this sort to occur, it is necessary that the PDR be thin, i.e., that the IF lie within the sonic point for

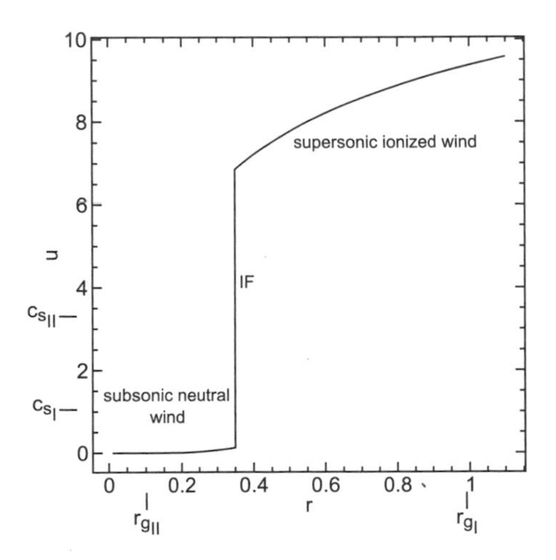

Fig. 8.6. Example of two coupled Parker winds (with $c_{s_{II}}^2 = 10 \times c_{s_I}^2$); the vertical axis is the flow speed normalized to the sound speed in the cooler gas, c_{s_I}. The neutral region is here thin, and the flow is subsonic in the neutral region. The velocity discontinuity at the IF satisfies eq. (18).

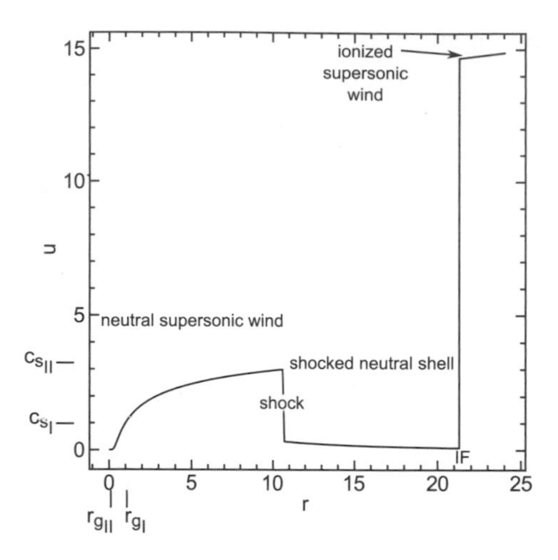

Fig. 8.7. Two coupled Parker winds (with $c_{s_{II}}^2 = 10 \times c_{s_I}^2$) where the neutral flow is supersonic and decelerated in an isothermal shock (at which point the Rankine-Hugoniot shock jump conditions require the product of the flow velocities on each side of the shock to equal $c_{s_I}^2$). Outside of the shock, the neutral gas flows subsonically through a shell with velocity $\propto r^{-2}$. At the IF, the velocities on each side of the shock satisfy eq. (18), with supersonic flow in the ionized region

the neutral flow. Such a flow is termed *EUV dominated* since, as mentioned above, the IF virtually coincides with the disk surface, and hence the mass flow rate is determined just by ionization balance in the EUV region.

However, if the IF lies beyond the sonic point in the neutral region, then we cannot simply match two Parker wind solutions, since then, by definition, the neutral flow would be supersonic at the IF. In this case, it is necessary that the neutral flow be brought to subsonic velocities by a shock that is intermediate in radius between the sonic point of the neutral flow and the IF (Fig. 8.7).

For a given incident radiation field, it is thus possible in principle to find a self-consistent solution, i.e., to determine the thermal structure of the steady-state winds in the two regimes and match them by a shock that satisfies the usual Rankine Hugoniot conditions. In general, this must be done numerically ([1]), but it turns out that there is a simple analytic case that is widely applicable to the irradiation of disks in environments like the Orion Nebula Cluster. In this approximation (see [83]), the FUV region is modeled as a shell of fixed column density N_{FUV} and sound speed (c_{s_I}). As noted above, in the case of relatively strong FUV fields, the column heated by FUV radiation is determined by a fixed optical depth in the dust rather than by molecular self-shielding.

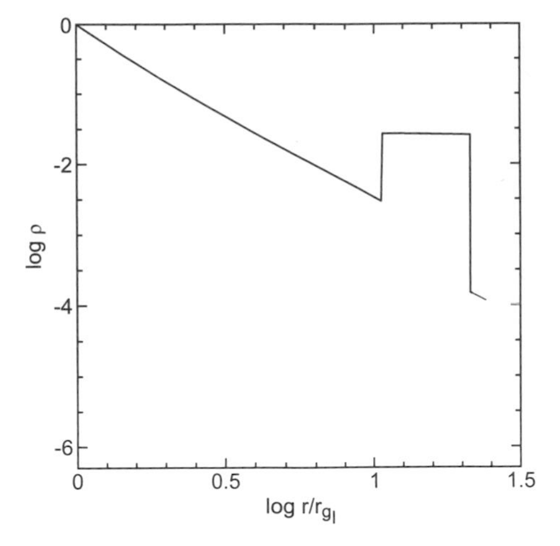

Fig. 8.8. Density field (arbitrary units) corresponding to the flow structure in Fig. 8.7. Note that the (subsonic) shocked region is of roughly constant density; for the case shown here (which is also usually the case in real photoevaporating disk systems, given the properties of the ambient FUV and EUV fields; [83]), most of the neutral column is associated with the base of the supersonic neutral flow (rather than the shocked shell).

Now, in the supersonic portion of the neutral wind, the flow velocity is roughly constant, and so, in a steady state, the density falloff in the wind is roughly $\propto r^{-2}$ (see Fig. 8.8). This means that the integrated column density is dominated by conditions at the base of the wind, i.e., at radius r_d. Thus, for a disk of given radius, the requirement of fixed N_{FUV} fixes the density at the base of the wind (n_d) such that $N_{FUV} \sim n_d r_d$, and therefore we find that $n_d \propto r_d^{-1}$. Since the flow velocity is $\sim c_{s_I}$, we can immediately write the mass flux in the wind,

$$\dot{M}_{FUV} = 4\pi r_d^2 c_{s_I} m_{\rm H} n_d, \tag{20}$$

where $m_{\rm H}$ is the mass per H atom. Substituting for n_d, one then obtains

$$\dot{M}_{FUV} \sim 4\pi m_{\rm H} N_{FUV} r_d \tag{21}$$

and thus establishes the important result that in this regime the mass-loss rate is a linear function of disk radius. (Contrast this with the corresponding result for EUV-dominated flows—eq. [15] and following discussion—where the mass-loss rate scales instead as $r_d^{1.5}$.)

In the FUV-dominated case, therefore, the EUV field is irrelevant to the magnitude of the mass loss, but nevertheless, the wind will at some point—following a shock, as argued above—become ionized. This is of observational importance since the proplyd phenomenon (see § 4.3) has been successfully interpreted as representing ionization fronts that are spatially offset from the disk surface because the EUV can penetrate only a finite distance into the FUV-driven wind. One can simply estimate the radius of the IF by invoking mass continuity (i.e., equating the product $4\pi r_{IF}^2 n_{II} m_H c_{s_{II}}$ with $\dot{M}_{FUV}$ as computed above) and also by requiring that the recombination integral match the ionizing flux (i.e., applying eqs. [3] and [6], which, for given ionizing flux, fix the product $n_{II}^2 r_{IF}$). Putting all this together, one obtains the result that $r_{IF} \propto (r_d^2 d^2/\Phi_{ion})^{1/3}$, or, inserting numerical values ([83]),

$$r_{IF} = 10^{15}\mathrm{cm}\left(\frac{r_{d_{14}}^2 d_{pc}^2}{\Phi_{ion_{49}}}\right)^{1/3}, \tag{22}$$

where $r_{d_{14}}$ and $\Phi_{ion_{49}}$ are the disk radius (normalized to 10^{14} cm) and the ionizing luminosity (normalized to 10^{49} s^{-1}). I shall return to eq. (22) in § 4.3 when I consider the match between the theory of FUV-driven winds and observations of proplyd structures where both the radius of the ionization front and that of the disk can be measured.

3.3. X-Ray Photoevaporation

I shall treat X-ray photoevaporation more briefly since at present it seems likely that the EUV and FUV are more important than X-rays in driving photoevaporative flows (see, however, the update to this conclusion in § 5.2). However, this conclusion is preliminary, pending the availability of calculations that self-consistently solve for the density and temperature structures of X-ray irradiated disks in two dimensions (see § 5.2) . I stress that the issue I consider here (the thermal interaction of X-rays with the upper layers of a disk, irradiated either by X-rays from the central star or by a diffuse external X-ray field) is quite distinct from the issue of how X-rays cause a low level of ionization in the cold midplane layers of the disk (see [59], [80], [52]). This latter issue may be of key importance for determining the viability of the magnetorotational instability (MRI) as an agent of angular momentum transfer at intermediate disk radii (see chapter by Balbus), and relies on the ability of hard X-rays to penetrate the large column densities required to reach the disk mid plane. Here, however, the main thermal input to the disk is from soft X-rays (i.e., in the few keV range) that are absorbed in the tenuous upper layers of the disk.

The dominant interaction between such soft X-rays and the disk is via photoionization of atoms and molecules and the subsequent thermalization of the energy contained in the kinetic energy of primary and secondary electrons ([92], [99], [104]). Unlike the case of irradiation by EUV photons (where the gas is roughly isothermal at 10^4 K), X-ray heating produces a range of temperatures, from coronal temperatures ($\sim 10^6$ K) in the uppermost, most tenuous layers down to a few hundred K in the dense conditions of X-ray PDRs. In optically thin gas, the character of an irradiated region is set by the ionization parameter (ξ), which is simply the ratio of the X-ray flux to the local number density of hydrogen nuclei:

$$\xi = \frac{L_X}{nd^2}. \tag{23}$$

In the case where there is significant absorption of the incident flux in overlying layers, it is found that a better measure of the effects of X-ray heating is via the modified parameter

$$\xi = \frac{L_X J_h}{nd^2} \tag{24}$$

where the attenuation factor J_h takes care of the effects of absorption integrated over the incident spectrum ([92], [59]).

It has been found that coronal gas is unimportant for the photoevaporation problem because of the very low das densities in such regions ($< 10^2$ cm^{-3}). As ξ is reduced, the gas enters a warm ($\sim 10^4$ K) atomic phase (ionization fraction $\sim$ 1–10%). At still lower values of ξ, one enters the regime of X-ray PDRs, with temperatures of around 1,000 K (see [104], [57], [54]). Here heating is still mainly caused by secondary electrons, although, where there is a significant fraction of hydrogen in molecular form, there are additional routes for transferring this energy into the thermal reservoir, involving molecular excitation and collisional deexcitation. In models of X-ray irradiated disks where heating by accretion is neglected, the warm atomic phase extends to columns of $\sim 10^{21}$ cm^{-2}; at greater depths within the disk, the gas cools to a few hundred degrees over a column of about 10^{22} cm^{-2} ([58]). At this point, the gas—which in overlying layers is mainly heated by X-rays—becomes thermally coupled to the dust.

Although these studies provide useful indications of the various physical regimes that are relevant to X-ray irradiated disks, it should be stressed that all studies to date are incomplete in some important respect. For example, the models of Glassgold et al. ([100], [101]) are one-dimensional and, moreover, do not iterate the density profile of the disk so as to achieve consistency with the computed temperature profiles. Alexander et al. ([5]) improved on this by iterating on the density profile so as to obtain hydrostatic equilibrium and also—in the case where the X-rays originate from the central star—implemented a pseudo-two dimensional approach. (Note that the problem is simplified in that, unlike the EUV case, where the diffuse field can be important, X-ray absorption rarely lead to reemission in the X-ray band, and thus one can simply calculate the X-ray attenuation and heating along various rays from the source through the disk). On the other hand, unlike Glassgold et al., Alexander et al. were unable to treat gas cooler than around 3,000 K and thus had to guess the extrapolation of their results into the cooler underlying regions of the disk. Most recently, Ercolano et al. ([43]) have performed the first genuinely three-dimensional calculations of X-ray irradiated disks via Monte Carlo radiative transfer. Here again, however, the code treats only regions that are warmer than 3,000 K and does not iterate on the disk density structure. Evidently, future studies are required that combine the various virtues of these approaches.

4. Extrinsic Disk Dispersal: The Role of Environmental Effects

Disk dispersal evidently does not require external effects. For example, there are many diskless stars in the archetypally sparse and quiescent star-forming environment of Taurus Aurigae ([12]), where none of the effects that I discuss in this section are likely to be relevant. On the other hand, processes that are associated with the radiation field or dynamic environment in which stars are formed can in principle shorten disk lifetimes relative to the outcome in an isolated system. Indeed, we shall see abundant evidence that this is the case in one well-studied local star-forming region (the core of the Orion Nebula Cluster). In the coming years, it will become possible to obtain good disk censuses for X-ray-selected low-mass stars in regions that are much more populous than Orion, and it is thus timely to consider how one expects the longevity of disks to vary with the size and mass scale of the parent cluster.

4.1. Star-Disk Interactions

If two stars of roughly equal mass pass one another on a mildly hyperbolic orbit with pericenter r_p, then, in the case that one of the stars is surrounded by a circumstellar disk, it is found that the majority of disk material beyond around $0.5 r_p$ becomes unbound ([31]).[4] The residual disk then recirculaties at around $0.4 r_p$. The disk-mass and viscous timescale is thus reduced by the encounter, so this could reduce its subsequent lifetime (see, e.g., [129]). At the stellar density of the core of the Orion Nebula Cluster (around 10^4 pc^{-3}), a disk of radius 100 AU would be expected to undergo an encounter in ~ 6

4. This figure is somewhat dependent on the mutual inclination of the disk and stars' orbital plane, with coplanar prograde passages being the most destructive, see also [121], [122], [106].

Myr, leading to the suggestion ([95]) that disk pruning by passing stars might be an important mechanism in Orion. Subsequently, Scally & Clarke ([137]) performed N-body simulations of the entirety of the cluster, keeping track of the closest encounter distances over each star's orbital history. Despite the complications of orbital mixing, the results implied that the distribution of closest encounter distances is remarkably close to what one would obtain by assuming that each star spent its life at its present orbital radius; evidently (Fig. 8.9) very few stars (as a fraction of the total cluster membership) are located in regions where one would expect disks of radius ~ 100 AU to interact. (Note that this conclusion does not necessarily extend to the most massive stars,

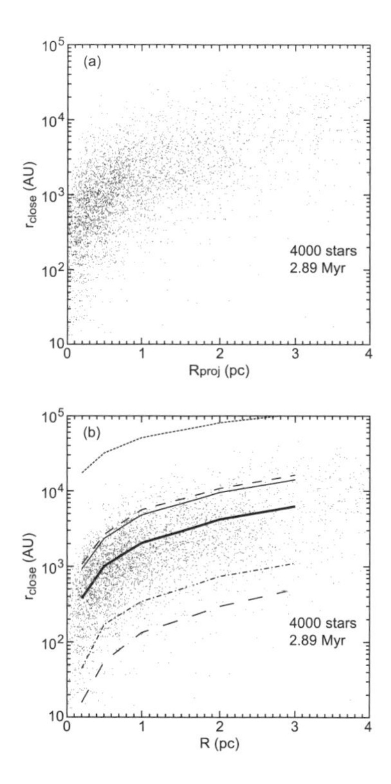

Fig. 8.9. Distribution of closest encounter distances for star-disk systems in the Orion Nebula Cluster, from the N-body simulations of Scally & Clarke ([137]). The upper panel is plotted as a function of projected position in the cluster, the lower as a function of distance from the clusters center. In the lower panel, the N-body results are compared with the results of a simple model in which it is assumed that stars spend all their lives at their instantaneous location (the thick line represents the median, and the thin- and long-dashed lines are 2σ and 3σ limits, respectively). The short-dashed line is the mean interstellar separation as a function of radius. Evidently the simple model is, in a statistical sense, a good representation of the N-body data.

which segregate to the dense regions of the cluster centre; [123], [106].) Overall, therefore, it would seem that although star-disk interactions indubitably must occur, they are unlikely to be major determinants of disk lifetime even in dense environments like the core of the Orion Nebula Cluster.

4.2. Diffuse X-Ray Field

Another negative result is obtained when one considers the effect of disk irradiation by the diffuse X-ray field associated with rich clusters such as M17 and the Rosette Nebula ([154]). This diffuse field is likely to be associated with the presence of early-type stars (earlier than O5) in these populous clusters, since a similar X-ray background is not present in the Orion Nebula, which does not contain stars of sufficiently high mass ([47]).

However, it would seem that the X-ray fluxes in these regions are simply too low to have a significant effect on the thermal structure of disks. The observed diffuse flux values (which are, incidentally, much lower than the X-ray flux from the central star that would irradiate the outer disk in the absence of attenuation by the inner disk), are more than three orders of magnitude lower than the flux at which X-ray irradiation is found to be a significant source of disk heating ([5]).

4.3. Ultraviolet Radiation from Massive Stars

In the case of ultraviolet radiation from massive stars, we have unambiguous observational evidence, in the Orion Nebula Cluster (henceforth ONC) that disks are destroyed by the combined EUV and FUV radiation produced by massive stars in the cluster (predominantly the cluster's most massive member, Θ^1C). The theoretical background describing the winds that are driven from disks in this case has been elaborated in § 3.2, and I shall call on a number of results from this section when discussing the modeling of observations.

PHOTOEVAPORATION IN ACTION: THE PROPLYDS IN THE ONC. The term *proplyd* was coined by O'Dell et al. ([112]) following their discovery of a number of bright, compact, but, thanks to the capabilities of the *Hubble Space Telescope* (*HST*) resolved sources of ionized emission within ~ 0.3 pc of $\Theta^1 C$ in the core of the ONC. The presence of ionized gas had also been detected in the radio continuum by Churchwell et al. ([28]) coincident with $H\alpha$ sources previously detected by [93]. Modeling of the radio emission indicated free-free emission from plasma at 10^4 K with electron densities of around $n_e \sim 10^6$ cm^{-3}. These sources were marginally resolved by Churchwell et al., implying typical dimensions of around 100 AU. This implies—in the absence of any confining mechanism—a mass outflow rate of $\sim n_e m_H r^2 c_{s_{II}} \sim 10^{-7} M_\odot \text{ yr}^{-1}$. A reservoir of material that can supply such outflow rates but not obscure the central star must have disk geometry. Subsequently, this interpretation (i.e., of an ionized outflow fed by a disk) was confirmed by direct imaging of silhouette disks (regions dark in nebula emission lines such as [OIII]) in the case of around 25 proplyds ([16]). Typical ratios of disk radius to proplyd radius (where the latter, r_{IF}, denotes the distance from the star to the IF) were found to be in the range 1/3–2/3, although this is probably an upper limit, given that many proplyds do not contain silhouette disks resolvable by the *HST* ([155]).

Although the bright ionized structures by which proplyds were first identified (see Fig. 8.10) demonstrate only the presence of emission below the Lyman break (i.e., EUV radiation), there are both theoretical and observational grounds for believing that FUV radiation plays an important role in proplyds. Observationally, Chen et al. ([26]) used Near Infrared Camera and Multi-Object Spectrograph (NICMOS) to demonstrate that emission in the 2.1μm S(1) line of molecular hydrogen is coincident with the disk, suggesting—since this line is likely to be FUV pumped—that FUV radiation is penetrating to the disk surface. This observation provided a possible explanation for the otherwise puzzling fact of the observed standoff between the IF and the disk

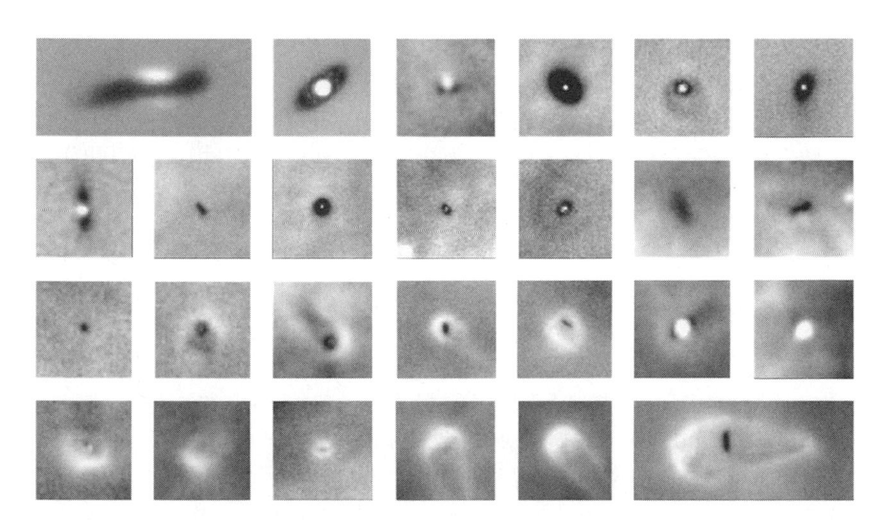

Fig. 8.10. A montage of silhouette disks (dark) and ionized structures (proplyds) imaged in the Orion Nebula Cluster. The morphology of the silhouettes relates simply to orientation on the sky. The ionized structures are generally cometary, being brightest on the side close to the dominant ionizing source and with a tail extending in the opposite direction. Note the lack of correlation between disk and IF orientation (McCaughrean [160]).

that presumably feeds it. Evidently this implies that some agent is lifting neutral material from the disk and transporting it out to the IF, but, equally evidently, this agent could not be the EUV radiation itself, since this is all absorbed at the IF. These considerations led Johnstone et al. ([83]) to elaborate a simple but elegant framework for understanding the proplyd phenomenon as an FUV-dominated flow (see § 3.2) where the location of the IF is set by the penetration of EUV photons into the neutral wind flow.

Observational Tests 1: Spatial Distribution of Proplyds. There are, according to this theory, several prerequisites for the proplyd phenomenon (here defined as being an offset ionisation front) to occur. First, the FUV field must be strong enough to sustain a significant column of warm material sandwiched between the disk and the IF. As described in § 3.2, the extent of PDR heating is set either by a total column of dust (in the case of strong FUV fields) or by molecular self-shielding (in the case of weak fields and/or dense gas). Thus a threshold value of the strength of the FUV field is required in order for the FUV to be able to heat the full column of $\sim 10^{21}$ cm^{-2} (as is assumed in the derivations of the mass-loss rate in the FUV regime, eq. [18]). Störzer & Hollenbach ([147]) investigated this issue and found that the FUV field should exceed $G_0 \sim 5 \times 10^4$ (for disks of size $\sim$ 100 AU). In the ONC, this condition is met in the inner $\sim$ 0.3 pc, in good agreement with the observed spatial distribution of proplyd sources in Orion ([16]). Second, the EUV field must not be so strong that the IF is coincident with the disk surface, since we here define proplyds as containing ionized structures that are well separated from the interior disk. For a disk of radius 100 AU and adopting the ionizing luminosity of $\Theta^1 C$ as $\sim 10^{49}$ s^{-1}, we deduce from eq. (22) that there is a very small region around Θ^1C (radius $\sim 5 \times 10^{17}$ cm) where the EUV field is so fierce that the flow is ionized right down to the disk surface. Similarly, as noted above, there should be another regime of EUV-dominated flow beyond $\sim$ 0.3 pc because of the FUV field being insufficiently strong in this region.

Observational Tests 2: Proplyd Sizes and Morphology. Eq. (22) sets out the expected dependence of the proplyd size (r_{IF}) on distance from the ultraviolet source, implying an expectation that $r_{IF} \propto d^{2/3}$. Johnstone et al. ([83]) showed that this is broadly compatible with the *HST* observations of Bally et al. ([15]), in that this shows an apparently mild increase of mean r_{IF} with increasing d. The more recent analysis of Vicente & Alves ([155]), however, which employs a larger sample of proplyds contained in archival *HST* data, shows no apparent correlation between the range of r_{IF} values and d. However, it is not clear that these data are necessarily in contradicition with the theory, given the considerable blurring of this correlation due both to projection effects and to the effect of a range of disk radii.[5]

A more definitive test of the theory is possible in those cases where a silhouette disk is resolved within the proplyd structure, since one can then compare the observed disk radius with that which the theory predicts on the basis of the observed values of r_{IF} and ultraviolet flux. The result of this exercise ([83]) yields generally fair agreement between observation and theory.

Emission-line imaging of proplyds shows that they are cometary-shaped objects whose tails generally (but not exclusively) point away from $\Theta^1 C$. In the theory outlined above, this is to be expected because it is assumed that the wind from the disk is quasi-spherical by the time it reaches the IF, and hence any asymmetry in the IF morphology derives from the different ionizing fluxes that are incident on the two sides of the flow. In fact, reasonable models of the observed morphologies of proplyds (e.g., [83], [67]) are obtained by assuming that whereas the hemisphere of the disk wind closest to $\Theta^1 C$ is directly irradiated by that star, the other hemisphere instead receives a diffuse EUV field from recombinations in the nebula. Some proplyds however do not display the expected symmetry with respect to the direction to $\Theta^1 C$. This in some cases suggests that other massive stars contribute significantly to the ionizing flux or else that some mechanism, possibly hydromagnetic in origin, causes the disk wind to retain a memory of the disk orientation.

Observational Tests 3: Disk Mass-Loss Rate. Early estimates of the mass-loss rate in proplyds were simply based on the sizes and electron densities inferred from radio observations ([28]), combined with the assumption of free expansion at the sound speed of ionized gas. Similar values are derived from the simplest models of FUV-driven winds (i.e., application of eq. [18] to systems with $r_d \sim 100$ AU). Although this agreement between theory and observation was in some ways welcome, these mass-loss rates are so high that they imply a problematically short lifetime for the parent reservoir. O'Dell ([111]) therefore suggested that the flows might be confined rather than freely expanding, a hypothesis that can be tested only with kinematic data. The follow-up study of Henney & O' Dell ([68]), involving the analysis of emission-line profiles in conjunction with emission-line imaging, effectively laid this possibility to rest and confirmed mass-loss rates of order $10^{-7} M_\odot$ yr^{-1}.

5. Note that $r_{IF} \propto r_d^{2/3}$ at a fixed distance from the source equation [22].

THE PROPLYD LIFETIME PROBLEM. One way in which the proplyd lifetime problem can be framed, which makes no reference to measured disk masses in the ONC, can be stated as follows: The age of the ONC is estimated to be about 2 Myr ([73]), and thus, if the measured/predicted mass-loss rates of a few $\times 10^{-7} M_\odot$ yr^{-1} are sustained at this level over the cluster lifetime, the total photoevaporated mass is around a solar mass. Since most stars in the ONC are less massive than this, and since long-lived dynamically stable disks are not possible if the disk mass exceeds the star's mass (see the chapter by Durisen), we have a contradiction. We therefore deduce that such rates were not sustained over the ~ 2 Myr lifetime of the cluster. Störzer & Hollenbach ([147]) suggested that the proplyd phenomenon may operate on a duty cycle, with proplyd mass loss switching on only when stars on rather radial orbits pass through the central FUV zone (roughly the inner 0.3 pc). The problem with this hypothesis is that the high incidence of proplyds within the core of the ONC ([16]) places extreme requirements on the anisotropy of stellar orbits, which are not in fact met by plausible dynamic models ([137]). On the other hand, another more obvious solution is to postulate that $\Theta^1 C$ switched on only rather recently (or perhaps more accurately that the cluster-forming gas became optically thin in the FUV quite recently). Since disk masses of up to ~ 0.1 or so of the stellar mass are likely to be long lived and dynamically stable, we need to reduce the above estimates of total photoevaporated mass by a factor of 10 and so would then require that the effective age of $\Theta^1 C$ be around 10% of the cluster age. Given that star formation in the ONC is likely to proceed over a few dynamic times (i.e., over a Myr or so), it is not particularly surprising that $\Theta^1 C$ should have switched on the order of 10^5 years ago, especially because one could make the argument that it was the formation of $\Theta^1 C$ (and the associated feedback effects) that terminated star formation in the ONC and revealed it as an optically visible cluster.

It becomes harder to argue this when, in addition, one uses information on disk masses in the ONC derived from submillimeter measurements. The majority of disks in the ONC have apparently low-submillimeter disk masses ($M_d < 0.02 M_\odot$; [107], [15], [39]),[6] which means that the timescale over which a disk can sustain its current mass-loss rate, $T = M_d/\dot{M}_{FUV}$, is two orders of magnitude less than the cluster age, if we adopt for $\dot{M}_{FUV}$ the values measured in the largest proplyds. This requires—unless we are witnessing the final brief phase of a long-lived population, which would imply an unreasonable degree of

6. Note, however, that these disk masses may be underestimated because of optical depth effects (see later) and uncertainties in the grain size distribution, dust-to-gas ratio, and submillimeter opacities: see the chapter by Henning and Meeus.

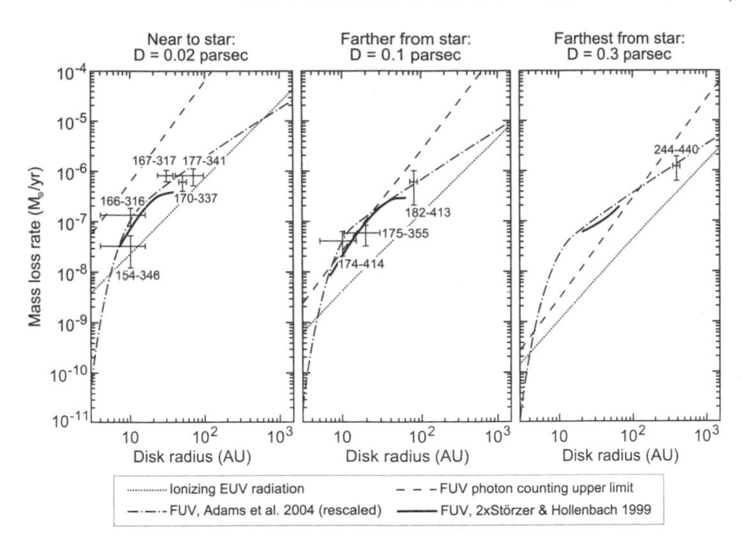

Fig. 8.11. Spectroscopic determinations of photoevaporation rates ([67], [68], [69]) as a function of projected distance to Θ^1C and inferred disk radius. Figure courtesy of W. Henney.

synchronization in disk evolution—that $\Theta^1 C$ would have to have been formed even more recently ($< 10^4$ years ago).

Clarke ([29]) suggested that this might be somewhat mitigated if one takes into account that whereas $\dot{M}_{FUV} \propto r_d$ for disks that are large enough that the sound speed of FUV-heated material exceeds the local escape velocity, the mass-loss rates are considerably lower in the case of small disks (smaller than a few 10s of AUs; [1]; see also observational data in Fig. 8.11). In principle, then, disks could lose mass copiously when large, but, when shorn by photoevaporation down to 10 AU or so, would lose mass at a much lower rate. According to this argument, the proplyd lifetime problem arises only when one combines $\dot{M}_{FUV}$ values from the largest disks and M_d values that are typical of the majority of (small) disks in the ONC.

This is an assertion that can in principle be examined by looking at submillimeter disk masses and spectroscopic mass-loss rates on a case-by-case basis. This exercise is complicated by optical depth effects and the limited sensitivity of millimeter/submillimeter measurements. On the former issue, it is found in the case of silhouette disks where the disk area is directly measurable that the submillimeter fluxes are compatible with such disks being optically thick ([29]). One thus obtains only a lower limit on the disk mass in these cases. In fact, recent estimates based on emission at longer wavelengths (i.e., 3 mm, which is less affected by optical depth effects) suggest that some (around 2%) of disks in the ONC are indeed quite massive ($> 0.1 M_\odot$; [39]). If these most massive disks are losing mass at the maximum rates determined for proplyds ($\sim 10^{-6} M_\odot \text{ yr}^{-1}$: see Fig. 8.11), then the resulting photoevaporation timescale is $\sim 10^5$ years (consistent with the notion that the age of Θ^1C is around 10^5 years or $\sim$ 10% of the cluster age). The smallest proplyds are losing mass at around $5 \times 10^{-8} M_\odot \text{ yr}^{-1}$ (Fig. 8.11). Here we do not know the disk masses on a case-by-case basis, since the majority of sources are not detected. Nevertheless, if measurements of sources that are individually undetected at 3 mm are simply stacked ([39]), one finds that the mean disk mass in the ONC is around $0.005 M_\odot$. This results combined with the above mass-loss rate, again yields an exhaustion timescale of 10^5 years.

We thus conclude that in all cases the measured mass-loss rates and disk masses are probably compatible with a scenario in which Θ^1C switched on $\sim 10^5$ years ago. By the same token, we would expect that after a further few $\times 10^5$ years, the disks in the ONC will have shrunk to the point where the mass-loss rate is very low and where they would not give rise to resolvable ionized structures. At this point, such truncated disks lose mass mainly by accretion, so that, given canonical viscosity models (with viscous α value ~ 0.01; [147]), they would be expected to drain away onto the central star after a further few $\times 10^5$ years. This means that in the vicinity of O stars, the window of opportunity for observing either the proplyd phenomenon or the subsequent phase of compact, optically thick discs should be relatively brief. Thus it would appear that we are observing the ONC at a special epoch, a conclusion supported by the very small numbers of proplyds that have been detected in other clusters ([144], [145], [18]).

4.4. Environmental Influences on Planet Formation

As discussed above, the observed and predicted rates of photoevaporative mass loss in the ONC are so high that it is likely that disks in this region will be photoevaporated down to small radii (less than $\sim$ 10 AU) over the next 10^5 years or so. This view is supported by the fact that the disks in the ONC appear to be significantly smaller than in quiescent regions like Taurus ([11], [134], [155]).[7]

7. Note that there is some disagreement about the interpretation of disk size data in the ONC. Rodmann ([134]) bases his analysis on the statistics of silhouette disks resolved by the *HST* and notes that only around 10% of sources are seen in silhouette at scales $>$ 90 AU. Vicente & Alves ([155]) instead assume that the paucity of large silhouette disks is a result of disks whose location or orientation is unfavorable to being seen in silhouette. They instead infer a disk size distribution on the (questionable) assumption that the ratio of disk radius to IF radius in proplyds with resolved disks is characteristic of the population as a whole, and hence they derive disk sizes by scaling proplyd sizes.

The outside-in photoevaporation of disks can be expected to have an impact on planet formation in several ways. One obvious implication is that it limits the window of opportunity for the creation of gas-giant planets, even though radial-velocity planets probably form in regions of the disk that are deep enough in the parent star's potential that photoevaporative mass-loss rates from such regions are rather low (see Fig. 8.11). Nevertheless, photoevaporation at larger radii provides an effective boundary condition on the disk's viscous evolution, cutting off the gas reservoir that would otherwise flow in and replenish the inner disk as it accreted onto the star. In essence, therefore, the effect of photoevaporation is to reduce the lifetime of gas in the inner disk. Although this would not affect the viability of planet formation by gravitational instability (which in any case occurs quickly while the disk is still strongly self-gravitating), it means that in core-accretion models, the solid core would have to be assembled very quickly, while the inner disk was still gas rich ([79]).

Throop & Bally ([151]), however, have argued that photoevaporation may even encourage terrestrial planet formation in the outer disk. They envisage that the concentration of dust toward the disk midplane through sedimentation ([159]) is further enhanced by the preferential removal of gas from the disk's upper layers by photoevaporation. Throop & Bally thus argue that photoevaporation facilitates the creation of a thin gas-poor central dust layer and thus, through the gravitational instability of this layer, the creation of planetesimals.

Before leaving this issue, we should ask, irrespective of whether photoevaporation is seen as a plus or a minus for planet formation, what fraction of stars are likely to be born in environments where external photoevaporation is significant. In the ONC, only around 10% of the stars are located in the (proplyd-producing) FUV region, but these stars dominate the integrated photoevaporative mass loss ([139]). Disks in the EUV region (beyond ~ 0.3 pc) can lose significant gas (i.e., more than a minimum-mass solar nebula over a few Myr) only if they are large (i.e., > 100 AU); since mass loss in the EUV regime is a steep function of disk radius ($\propto r_d^{1.5}$), the survival time of disks in the EUV region increases strongly as disks are pruned by photoevaporation. The main effect on planet-forming potential is thus restricted to stars in the inner (0.3 pc) core of the cluster.[8]

Adams and collaborators ([1], [46]), however, have emphasized that the majority of stars form in environments that are both sparser and less populous than the ONC. This has two effects, a reduction in ultraviolet fluxes and an

8. This conclusion is supported by the recent 24μm studies of NGC 2244, where it is found that the disk fraction is reduced only in regions within ~ 0.5 pc of an O star ([17]).

increase in the ratio of FUV to EUV emission, the latter reflecting the lower maximum stellar mass expected in smaller N clusters. Populations synthesis models ([46]) suggest that only about 25% of stars in the solar neighborhood should have suffered even a modest impact on their planet-forming potential ("modest" here being defined as photoevaporation down to $\sim$ 30 AU in 10 Myr). Another way to put this is that the fact that a star like the Sun bears planets can place only weak constraints on its birth environment, probably implying that it was formed in a cluster not exceeding several hundred members.

5. Intrinsic Disk Dispersal: Photoevaporation by Radiation from the Central Star

5.1. The Origin and Magnitude of Ultraviolet and X-Ray Emissions in T Tauri Stars

In order to assess the importance of disk dispersal by photoevaporation, it is clearly important to determine the X-ray and ultraviolet luminosities of low-mass pre-main-sequence stars. Less obviously, it is also vital to establish the origin of such emissions. In particular, we need to ascertain whether this radiation is related to the star itself (and its chromosphere/corona) or whether it derives from accretion. Radiation from the latter source will not give rise to "inner-hole sources" (see § 6.3) since a cleared inner region cannot be sustained if accretion is required in order to produce the photoevaporative radiation ([103], [135]).[9]

We can be fairly confident that X-rays are not predominantly associated with accretion because of the simple argument that, if anything, diskless stars (i.e., weak-line T Tauri stars) are more luminous in X-rays than are their disk-bearing counterparts ([124], [49]). It would seem likely that this slight deficit in the X-ray output of classical T Tauri stars is a consequence of X-ray absorption in the accretion columns ([60]) and that, regardless of a star's disk-bearing status, the X-rays originate in magnetic reconnection events in the stellar magnetosphere/corona. This emission can be fitted by a thermal bremsstrahlung spectrum peaking in the $1-2$ keV range ([48]); typical luminosities are $\sim 10^{30}$ erg s^{-1}, attaining values up to $10-100$ times this during flares ([157]), although the latter do not contribute significantly to the time-averaged luminosity, which is the important property for photoevaporation.

9. In principle, accretion-powered luminosity can evaporate the outer disk, since, depending on the efficiency with which accretion luminosity is transferred into kinetic energy of the wind flow, a given mass of material accreted onto the star can evaporate more than its own mass of loosely bound material at large radii.

In the case of ultraviolet radiation, it would seem likely that there are significant contributions both from accretion and from the stellar chromosphere ([32], [82]). However, photons below of the Lyman limit (i.e., EUV radiation) would be strongly absorbed in the accretion columns and would therefore be attenuated to levels even less than the tiny photospheric output of a low-mass star in the EUV ([4]). Therefore, the models described in § 5.3 rely on an additional (constant) source of EUV photons from an active chromosphere.

Of course, because of interstellar absorption by neutral hydrogen, such EUV emission is not directly detectable, and one therefore needs an indirect method to infer its presence. One method relies on analysis of the line spectrum in the observationally accessible region of the spectrum (i.e., the FUV) in order to derive the differential emission measure (DEM) that fits a number of lines in this spectral region ([21]). The DEM is a measure of the density structure of the emitting plasma integrated along the line of sight in various temperature ranges. This information can then be inserted into a spectral synthesis code in order to determine the EUV output that would be produced by such a plasma. This exercise ([5]) yields values in the range $\Phi_{ion} = 10^{42-43}\ s^{-1}$ for a small ensemble of classical T Tauri stars for which International Ultraviolet Explorer (IUE) and Space Telescope Imaging Spectrograph (STIS) spectra are available, with the greatest uncertainty deriving from the reddening correction applied to the FUV spectra. These values are similar to those estimated by Kamp & Sammar ([81]) using a simple argument (based on the observed relationship between age and activity indicators) to scale up the ultraviolet output of the Sun to that expected at the young age characterizing T Tauri stars.

Ideally, therefore, one would want to repeat this exercise for weak-line T Tauri stars in order to test whether this level of EUV output is sustained once disk accretion is shut off. Unfortunately, a more indirect approach is required, given the lack of weak-line T Tauri stars with FUV spectra of sufficient quality to perform a DEM analysis. Alexander et al., ([5]) noted from their spectral synthesis studies of those (classical) T Tauri stars with high-quality FUV spectra that the hardness of the ultraviolet spectrum (specifically the ratio of EUV to FUV) appears to correlate quite well with the ratio of the HeII 1640 to CIV 1550 lines (a reddening-independent indicator that is available from low-resolution spectra for a much larger sample of T Tauri stars). This ratio is at least as great in weak-line (diskless) T Tauri stars as in their classical counterparts and, if anything, appears to increase somewhat with system age. This supports the view that, as we require, the EUV emission is not produced by accretion.

Another possible diagnostic of the EUV radiation field is provided by midinfrared fine structure lines such as the [NeII] line at 12.8μm. Although there is some uncertainty whether this transition is instead excited by X-ray emission, the level of EUV output that would be required to excite this line is similar to the levels quoted above ([120]). However, this line cannot be used to assess correlations between the assumed EUV output and the evolutionary state of the disk, since the requirement that it be detectable above the disk's continuum introduces a strong bias toward systems with cleared inner holes (see § 6.3).

5.2. Dispersal by X-Rays from the Central Star

Since this article was written, there have been considerable theoretical developments to this subject which have reversed the previous conclusion that X-ray photoevaporation was unlikely to be significant. We are unable for reasons of space to give this topic the full discussion it now deserves and refer the reader to Ercolano et al. 2008, 2009, Owen et al. 2010, Ercolano and Owen 2010 ([43], [44], [113], [45]). Here we just summarize the main results. Note that we largely preserve the structure of § 3.1 in which we discuss observed clearing sources in terms of EUV photoevaporation. We however add footnotes to this discussion where the results of X-ray photoevaporation calculations are particularly relevant.

Prior to the X-ray hydrodynamical calculations of Owen et al. ([113]), a number of studies ([59], [5], [43], [44], [56] examined the structure of X-ray irradiated disks: of these, only [59] and [56] treated gas cooler than 3,000 K (the region of molecular dissociation by X-rays) whereas only [5], [44], and [56] iterated on the disk's density structure so as to produce irradiated structures in vertical hydrostatic equilibrium. Naturally such *static* structures cannot self-consistently be used to measure mass *flow* rates but one can make a rough estimate by locating the surface at which the gas is heated to the local escape temperature and then equating the mass loss rate per unit area with the product of the local density and sound speed. Gorti and Hollenbach ([56]) obtained very low resulting X-ray photoevaporation rates, a result that turned out to derive from their very hard input X-ray spectra (see discussion in [44]). The observationally motivated input spectra used by [5] and [44] however, yielded much higher mass loss rates per unit area: [5] did not extend their mass loss integrals to large radii, due to their distrust of their pseudo-two dimensional approach and thus missed the region of the disc (at 10s of AU), which largely contributed to the very high mass loss rates ($\sim 10^{-8} M_{\odot}\,\mathrm{yr}^{-1}$) inferred from the (three dimensional) calculations of [144].

Owen et al. ([113]) conducted hydrodynamic calculations in which the temperature of the gas (in regions where the column density to the central star $< 10^{22}\,\mathrm{cm}^{-1}$) is parameterized by the relationship between temperature and ionization parameter (eq. [23]) obtained from the hydrostatic solutions. This exercise confirmed that a vigorous wind (mass loss rate $\sim 10^{-8}\, M_{\odot}\,\mathrm{yr}^{-1}$) is driven from $\sim 3-60$ AU and that, moreover, such a mass loss rate is possible even in the case that there is a modest column of neutral material around the source that would preclude significant EUV photoevaporation. Note that the X-ray driven photoevaporation rate scales linearly with X-ray luminosity ([114]); the rates quoted above correspond to rather luminous X-rays sources so that—given the roughly two order of magnitude spread in the X-ray luminosities of T Tauri stars (Guedel et al. [61])—many stars will undergo clearing at a lower rate. Nevertheless, X-ray driven winds are likely to exceed those produced by the EUV in all but the weakest X-ray sources.

Most of the following discussion is based on EUV photoevaporation since this was the mechanism that was well understood at the original time of writing this article. Much, however, is qualitatively applicable to X-ray photoevaporation, provided that one bears in mind the following differences: i) the considerably higher over all rates for X-ray photoevaporation of EUV, which means that photoevaporation can start to modify disc evolution at much higher accretion rates than believed hitherto; ii) the rather lower temperatures for the X-ray heated wind cf the EUV wind ($\sim 6,000$ K cf $\sim 10,000$ K) (see footnote (11) which discusses the consequences for explaining neutral oxygen line diagnostics in T Tauri stars); and iii) the broader radial range, extending to 50–100 AU for X-ray photoevaporation. The latter is simply a result of the fact that X-rays (being more penetrative than EUV photons) are able to directly heat the outer disc; in the EUV case, by contrast, most of the irradiation of the disc at radii where it can drive a wind is provided by a diffuse field of photons produced by recombinations in the inner disk's atmosphere. This diffuse field is geometrically attenuated beyond several AUs and hence the EUV mass loss rate is sharply peaked within ~ 10 AU. The implications of this different irradiation geometry and mass loss profile are discussed in footnote (13).

5.3. Dispersal by EUV Radiation from the Central Star

The foundations of work on dispersal by EUV radiation were laid by the seminal study of Hollenbach et al. ([74]), who first investigated the process in the context of OB stars, where the ionizing fluxes (and hence the derived mass-loss rates) are much higher than in the case of T Tauri stars. At first, the rates obtained when the derived formulae were applied to T Tauri stars ($\sim 10^{-10} M_{\odot}$ yr^{-1}) were not considered to be of interest (but see § 6.1 for discussion of early ideas on the interplay between EUV photoevaporation and planet formation; [139]). This view was later revised in the light of Clarke et al. ([30]), who showed

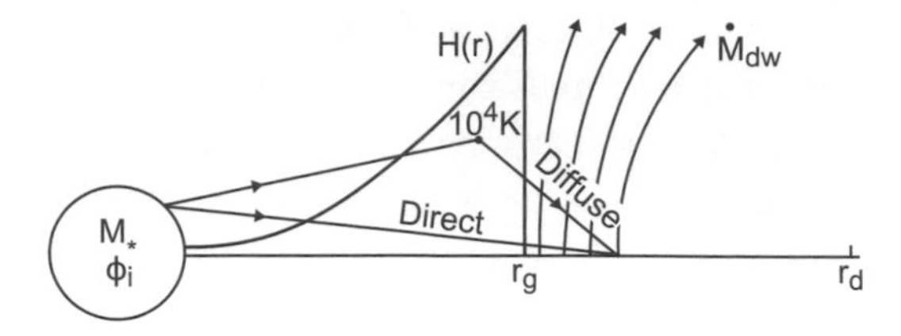

Fig. 8.12. Schematic of photoevaporation by EUV from the central star, showing both direct irradiation and irradiation by the diffuse field ([69]).

how even these low rates, when combined with viscous draining of the disk, could provide a promising mechanism for effecting the rapid turnoff of disk diagnostics at late times.

According to Hollenbach et al. ([74]), the structure of a disk exposed to EUV radiation from the central star can be divided into three regimes (see Fig. 8.12). In the inner two regimes, the ionized region forms a hydrostatic atmosphere with density structure governed by eq. (10). In the innermost regime, the ionizing flux irradiating the disk is dominated by that directly received from the source, so the density at the IF (n_{II} in eq. [10]) is determined by eq. (11), except that the plane-parallel flux of ionizing radiation from a distant source (Φ_{ion}/d^2 on the left-hand side of eq. [11]) is replaced by $\Phi_{ion} z_* / r^3$. This latter expression takes into account the angle of incidence of radiation emanating from a source at height z_* (which is of order of the stellar radius) above the disk plane. Consequently, $n_{II} \propto r^{-9/4}$ in this regime. However, the scale height H of this isothermal hydrostatic atmosphere scales as $r^{1.5}$, and thus, once $H \sim z_*$, the ionizing flux is no longer dominated by radiation coming directly from the star, but instead by recombinations from the overlying atmosphere. Hollenbach et al. solve numerically for the hydrostatic density structure in this regime, accounting for the diffuse field, and find that their results can be reasonably fitted by $n_{II} \propto \Phi_{ion}^{1/2} r^{-1.5}$. The third regime lies at $r > r_{g_{II}}$, where photoionized gas can now escape the disk in a wind. The lower density in the flow region compared with the hydrostatic region means that the diffuse field irradiating this outer region originates mainly from recombinations from the ionized atmosphere above $r_{g_{II}}$; this steeper reduction in the diffuse field with radius results in a steeper decline in n_{II} with radius also; numerically, $n_{II} \propto \Phi_{ion}^{1/2} r^{-5/2}$.

This latter quantity is important because the mass-loss rate per unit area from the region of the disk beyond $r_{g_{II}}$ is just $n_{II} c_{s_{II}} m_H$. Thus (on the assumption that mass loss occurs only beyond $r_{g_{II}}$, though see below)

25 $$\dot{M}_{EUV} = \int_{r_{g_{II}}}^{\infty} 2\pi r n_{II} c_{s_{II}} m_H dr.$$

We thus see that since $n_{II} \propto r^{-5/2}$, the mass loss is dominated by that occurring close to $r_{g_{II}}$. Putting this in numerical values, we obtain

26 $$\dot{M} = 4 \times 10^{-10} \left(\frac{\Phi_{ion}}{10^{41}\mathrm{s}^{-1}} \right)^{1/2} \left(\frac{M_*}{M_\odot} \right)^{1/2} M_\odot \mathrm{yr}^{-1}.$$

I note that the normalization for Φ_{ion} is 1 to 2 orders of magnitude less than the estimates for Φ_{ion} obtained by Alexander et al. ([6]) for classical T Tauri stars, but that this latter quantity is quite uncertain. However, the square-root dependence of the photoevaporation rate on Φ_{ion}, which typifies all recombination-limited situations, means that the resulting mass-loss rate is relatively insensitive to this value in any case.

The preceding description represents a semianalytic simplification of the problem that can be checked against radiation hydrodynamic calculations. For example, this description neglects dust. Dust attenuation, of course, reduces the incident flux (and hence the photoevaporation rate), but scattering by dust provides an extra mechanism for deflecting ionizing photons out to radii well beyond $r_{g_{II}}$. In fact, the radiative transfer calculations of Richling & Yorke ([130]) found that in their simulations of a disk irradiated by a star of mass $8M_\odot$, the latter effect is predominant, implying that dust boosts the above mass-loss estimates by a factor if 2 to 3 (with the extra mass loss occurring on scales of ~ 100 AU). However, as noted by [74], the optical depth in the dust simply scales as the square root of the ionizing luminosity; in the lower-luminosity sources we consider here, the ionized flow is expected to be optically thin in the dust. Another simplification is the neglect of angular momentum and nonvertical flow. Font et al. ([50]) addressed this issue with a two-dimensional hydrodynamic calculation that modeled only the ionized portion of the flow (as isothermal gas at 10^4 K),using the results of Hollenbach et al. to set the run of density at the base of the flow (which, in their calculations, which do not model the neutral disk, they set as the disk midplane). These calculations demonstrated that, as expected, pressure gradients deflect the flow from normal to the disk plane (at the flow base) to radial flow at large radii. They also showed that the disk in fact loses mass from radii well within $r_{g_{II}}$ (in to about 20% of $r_{g_{II}}$ or $\sim 1-2$ AU for a solar-mass star: see also [19], [97]). This may be readily understood inasmuch as the estimate that mass loss begins at $r_{g_{II}}$ is based on the expectations of a ballistic particle traveling at the sound speed and neglects the role of pressure gradients in providing modest acceleration to higher velocities. Nevertheless, these calculations confirm that the resulting mass loss is still reasonably well described by eq. (26) (i.e., just assuming sonic outflow at the flow base).

In addition, Font et al. were able to use the resulting two-dimensional velocity and density field to compute the expected line intensities and profiles of forbidden-line emission from ionized species (such as [SII]λ6731, [SII]λ6716, [NII]λ6583, and [OI]λ6300). As expected, a thermally driven ionized wind produces line widths (10's of km s^{-1}) in reasonable agreement with measurements of the low-velocity components of these lines. As far as line intensities are concerned, the model predicts, in the limit where collisional deexcitation of the lines can be ignored, that the line luminosity (resulting from collisional excitation) scales as $\int n^2 dV$. Given that $n \propto \Phi_{ion}^{1/2} r^{-3/2}$ close to r_{gII} (and more steeply beyond), this integral is dominated by conditions close to r_{gII} and so scales with $n(r_{gII})^2 r_{gI}^3$, which $\propto \Phi_{ion}$, irrespective of stellar mass (i.e., more massive stars have larger effective emitting volumes—larger gravitational radii—but lower densities at this radius, and these two effects exactly cancel at fixed Φ_{ion}). When one takes suppression of the lines by collisional deexcitation into account, this reduces the emission for lower-mass stars (higher $n(r_{gII})$) by an amount depending on the critical density of the line concerned. If we apply a plausible range of input values of Φ_{ion}, the model is in good agreement with the range of observed forbidden [SII] intensities, and their insensitivity to stellar masses[10] overpredicts the [NII] luminosities and is completely unable to reproduce the observed intensity of [OI]λ6300 emission. The reason for this is simply that the abundance of neutral oxygen in the ionized wind is low because of the similar ionization potentials of oxygen and hydrogen. We therefore conclude that some, but not all, of these lines can be generated in EUV-driven winds in T Tauri stars.[11]

Another line that may originate in a photoevaporative flow is the [NeII] 12.8μm line, which has already been mentioned as a possible means of quantifying the EUV radiation field experienced by the disk. The width of this line in the inner-hole source (§ 6.3) TW Hya is compatible either with a photoevaporating wind or with turbulent broadening in a warm disk atmosphere ([70]). Within the former interpretive framework, the mass-loss rate

10. Note that these arguments apply only for the range of stellar masses for which there is no strong systematic dependence of Φ_{ion} on stellar mass; for higher-mass stars (e.g., OB stars), the resulting line luminosities would be expected to be dominated instead by the value of Φ_{ion}.

11. Analogous work on line diagnostics in X-ray driven winds (Ercolano & Owen [45]) are in good agreement also with the neutral oxygen line intensities, this being simply a consequence of the rather cooler conditions in the X-ray heated gas, which means that oxygen is mainly neutral in the wind flow.

can be roughly estimated as the inferred mass of emitting gas divided by the timescale for a sonic ionized flow to traverse a scale $r_{g_{II}}$. The resulting photoevaporation rate ($\sim 10^{-10} M_{\odot}\ \mathrm{yr}^{-1}$) is very comparable with the theoretically expected value (eq. [26]).

5.4. Dispersal by FUV Radiation from the Central Star

I now come to the most difficult problem, the modeling of disk winds that are driven by FUV radiation from the central star. This problem is harder than the EUV case because of the much more complicated thermal structure of PDRs compared with ionized regions. It is also harder than the case of FUV irradiation from an external source because it is now a two-dimensional problem, where the inner disk can shield the outer disk from the FUV source. This sensitivity to shielding by the inner disk is considerably more important than in the EUV case, where we have seen above that the flow region is irradiated by diffuse emission from high latitudes and where the detailed disk shape is of minor importance.

Recently, Gorti & Hollenbach ([55], [56]) published an ambitious study to compute the thermal structure and photoevaporation of a disk subject to combined EUV, X-ray, and FUV emissions. The temperature at (r, z) in the disk is computed by calculating the fluxes of EUV, FUV, and X-ray fluxes incident from the central star, taking attenuation by dust and gas into account and thus using knowledge of the density and chemical structure of the disk inside r. Cooling is implemented by using the escape probability formalism in the z direction, through both the top and bottom of the disk. The density structure at r is then recomputed by assuming vertical hydrostatic equilibrium, and the thermal and chemical structures are then recalculated. This iteration is repeated until convergence and a self-consistent structure are obtained. Inevitably, the necessity of employing complex chemical and thermal networks, combined with the need to rezone in order to capture important processes, makes it very computationally expensive to achieve a converged temperature/density distribution. Fig. 8.13 shows results from this study for the temperature and density structure as a function of height at 31 AU. As expected, the temperature rises in response to FUV heating at altitudes where the visual extinction to the FUV source A_V is less than a few. We see that in the heated region, the small and large grains are, respectively, hotter and cooler than the gas and that the differences in temperature between small dust, large dust, and gas are important in computing the spectral features. From the point of view of computing the mass-loss rates, it is, of course, the gas temperature that is important, although I stress that this in turn depends on

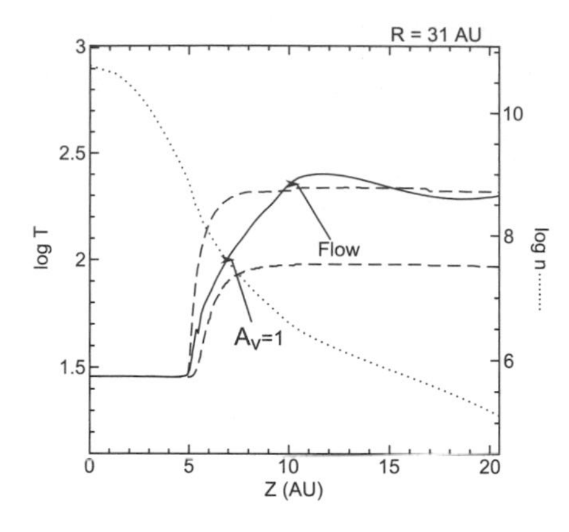

Fig. 8.13. Density (dotted, right-hand scale) and temperature (black, left-hand scale) of the disk as a function of height above the midplane at a radius of 31 AU, as computed by Gorti & Hollenbach (private communication). The solid line refers to the gas temperature, and the higher and lower dashed lines, respectively, to small and large dust grains. The origin of the mass outflow is indicated by "Flow," although, since the flow is subsonic to much larger radii, the gas density structure is approximately that of a hydrostatic atmosphere. Parameters for this solar mass model are X-ray and FUV luminosities of $6 \times 10^{-4} L_\odot$ and $0.28 L_\odot$, respectively.

the concentration of small grains in the upper atmosphere, since photoelectric heating by PAHs and small dust grains is an important thermal input to the gas. Although current models employ a fixed grain fraction and grain size distribution with height, future refinements could incorporate grain settling and thus determine these quantitities more self-consistently.

What is more problematic is the derivation of a mass-loss rate from these calculations. Gorti & Hollenbach ([49]), as a preliminary estimate, computed a nominal mass-loss rate from each cell by assuming that each such cell is the source of an isothermal Parker wind with this base density. At every radius, the cell with the highest nominal mass-loss rate is found, and the integrated mass-loss rate is then assumed to be equal to the nominal mass-loss rate from this cell. According to this estimate, in TW Hydra (which is a relatively luminous FUV source) the disk should be truncated down to a radius of ~ 120 AU in ~ 7 Myr, a result that is consistent with the apparent outer truncation of the gas disk (as evidenced by its CO emission) in this object.[12] This approximation, however, may overestimate the mass loss because the density of a hydrostatic structure falls off less steeply than if the gas were allowed to expand freely in a pressure-driven wind (although this difference is small as

12. The dust disk in TW Hydra is apparently much more extensive, which is explicable if the dust grains are relatively large (millimeter scale) and thus are left behind during the photoevaporation process ([3]).

long as the flow is highly subsonic). Other potential sources of error include the assumption that the outflowing gas remains isothermal at its base density, together with the neglect of advective terms in the solution of chemical equilibria. On the other hand, the complexity of the thermal structure means that it is much harder to find an analytic approximation to a self-consistent flow solution. Ultimately, therefore, it will be desirable to obtain FUV mass-loss rates via two-dimensional hydrodynamic calculations that are coupled, in some computationally feasible manner, to simplified modules that update the thermal structure.

For this reason, I focus entirely on EUV photoevaporation in the following section, since this is where we have the best quantitative estimates. I stress that this does not necessarily mean that FUV photoevaporation is negligible by comparison, and the reader should bear in mind that it might even turn out to be the prime agent of radiative disk dispersal.

6. EUV Photoevaporation and the Evolution of Protoplanetary Disks

6.1. The Secular Evolution of Disks Subject to EUV Photoevaporation

In T Tauri stars, the estimated EUV photoevaporation rates are low (a few $\times 10^{-10} M_{\odot}$ yr^{-1}; eq. [26]), implying that it would take $\sim$ 40 Myr to evaporate even a minimum-mass solar nebula. Since, in most clusters, the median disk lifetime is a few Myr ([64]), it is clear that the majority of gas in protostellar disks cannot end up being photoevaporated.

Shu et al. ([139]) nevertheless suggested that the process might have interesting consequences for planet formation. These authors noted that beyond Jupiter, the masses of the gaseous envelopes of the gas-giant planets decrease considerably. In the context of the core-accretion model for planet formation, this is often thought to reflect a lack of available gas in the outer Solar Nebula at late times. Shu et al. pointed out that Jupiter lies within $r_{g_{II}}$ and thus in a region of the disk that is not subject to direct photoevaporation. The outer gas giants, they argued, would, however, be accreting their envelopes from regions that were depleted by photoevaporation.

This attractive argument was turned on its head by Clarke et al. ([30]), who considered the operation of photoevaporation in a disk that was subject to secular evolution because of angular momentum redistribution (loosely speaking, viscous evolution; see the chapters by Balbus and Durisen). The net result of such evolution is always an inward accretion flow onto the star that, in the absence of disk replenishment, declines with time. The inferred accretion rates onto the star in some of the most actively accreting T Tauri stars are around $10^{-7} M_{\odot}$ yr^{-1}, but in some systems rates as low as $10^{-10} M_{\odot}$ yr^{-1} have been measured ([108]). It would therefore seem inevitable that whatever its prior accretion history, a disk must, en route to its ultimate disappearance, pass through the point where the accretion rate onto the star is around $10^{-10} M_\odot$ yr^{-1}. At this point, then, photoevaporation becomes competitive with flow onto the star.

It is important to recall that the dominant mass loss by EUV photoevaporation occurs within a factor of 2 or so of r_{gII} (i.e., in the range $5 - 10$ AU for low-mass stars). Thus at the stage that the accretion rate through the disk, $\dot{M}_{acc}$, becomes $\sim \dot{M}_{EUV}$ (eq. [26]), the disk inside this point becomes starved of resupply and therefore drains onto the star on its own viscous timescale ($t_\nu(r_{gII})$). Although we do not have a definitive model for the efficacy of viscous transport in various radial regimes (partly because of uncertainties surrounding the existence and extent of any dead zone; see the chapter by Balbus), the viscous timescale undoubtedly increases with radius in the disk. One way to quantify this without reference to the physical mechanism for angular momentum transport is to note that in a steady state, $\dot{M}_{acc}$ is independent of r, and so, since $\dot{M}_{acc} = M_{disk}(r)/t_\nu(r)$, we therefore deduce that the viscous timescale increases with radius in the same manner as the enclosed mass. Modeling of the submillimeter emission in pre-main-sequence disks suggests that the surface density scales as $\propto r^{-1}$ or $r^{-1.5}$ ([10], [11]), and hence one would deduce that $t_\nu(r) \propto M_{disk}(r) \propto r^{0.5}$ or $\propto r$.

The increase of viscous timescale with radius is important because r_{gII} is much less than the typical outer radii of T Tauri disks (which are constrained by their submillimeter emission to be at least 10s to 100s of AUs in extent; [11]). Thus, overall, the disk evolves on the viscous timescale of the outer disk, which is much longer than the viscous timescale at r_{gII}. The decoupling between inner and outer disks (which occurs at r_{gII} when $\dot{M}_{acc} \sim \dot{M}_{EUV}$) then introduces a second timescale into the problem: the disk age at this point is $\sim t_\nu(r_{out})$, but the clearing of the decoupled inner disk occurs on the much shorter timescale $t_\nu(r_{gII})$. Fig. 8.14 illustrates this two-timescale evolution in the models of [30], which combine EUV photoevaporation with a simple parameterization of the disk's viscous evolution. Note that before the point when $\dot{M}_{acc} \sim \dot{M}_{EUV}$, the evolution of the disk surface-density profile is almost unaffected by photoevaporation (i.e., almost all the mass flowing through the disk ends up on the central star). Once $\dot{M}_{acc} \sim \dot{M}_{EUV}$, however, the situation changes abruptly and the inner-disk drains, implying the rapid loss of inner-disk diagnostics (e.g., strong broad H_α associated with accretion onto the star or else near-infrared emission from the inner disk). Note that the absolute timescales depend on the viscosity parameterization, but that for any plausible prescription, the two timescales are quite different (here $t_\nu(r_{gII}) \sim 10^5$ years and $t_\nu(r_{out}) \sim 10^7$ years).

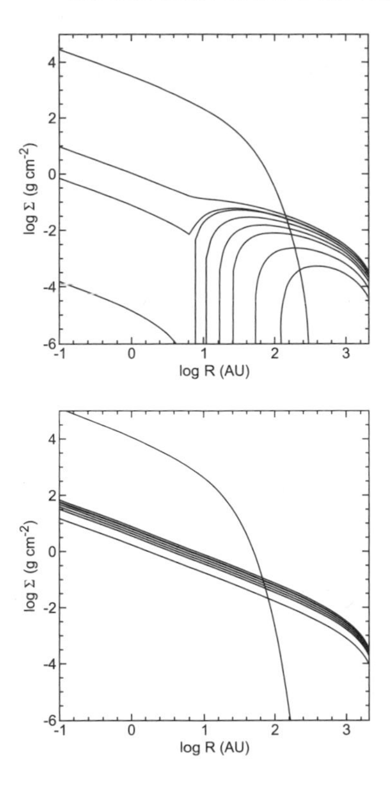

Fig. 8.14. Evolution of disk surface-density profile in the original ultraviolet switch model of Clarke et al. ([30]) (upper panel) compared with evolution of a control model without photoevaporation. The snapshots correspond to times $t = 0, 13, 14.1, 14.3, 16, 18, 20, 24$, and 28 Myr. Overall timescales depend on the parameterization of the disk viscosity, but the important points to note are (a) the rapid disappearance of the inner disk (on a timescale much less than the system age) and (b) the slow subsequent photoevaporation of the outer disk (i.e., on a timescale similar to the system age). In its original form, therefore, the model predicts a large population of disks with eroding inner holes.

The reason for laboring this point is that it has been recognized for many years (e.g., [141], [65]) that disk clearing indeed appears to be governed by two disparate timescales. This is illustrated by Fig. 8.15, which plots a sample of T Tauri stars in the infrared two-color plane from the data of [87]. The data clearly separate very neatly into two distinct populations: the disk-bearing classical T Tauri stars in the upper right and the weak-line T Tauri stars (lower left), whose infrared colors are just those expected for cool stellar photospheres (the few intermediate data points are now known to be binaries consisting of a mixture of classical and weak-line components). This clean separation would not be expected if the disks were to follow a power-law evolution with time, since then a number of disks would fall in the intermediate zone as

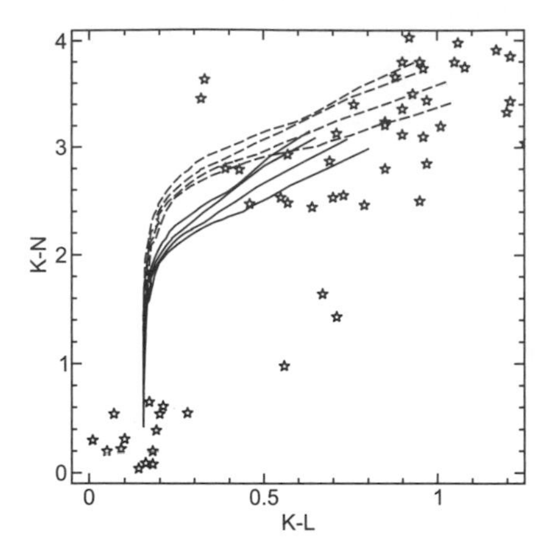

Fig. 8.15. Distribution of T Tauri stars in the K–L, K–N plane compared with theoretical tracks for magnetospheric clearing ([87], [13]). See text for details.

they became progressively optically thin. The tracks in Fig. 8.15 represent the expected evolution in the case of magnetically mediated accretion from a disk—although, in the models, magnetospheric clearing indeed causes a rapid decline in $K - L$ (which is a measure of emission from very small radii, ~ 0.1 AU), the clearing of the disk at the larger radii responsible for the $K - N$ excess is too slow and would result in the accumulation of systems in the upper left of this diagram that are not observed there ([13]).

Photoevaporation combined with EUV photoevaporation (dubbed "the ultraviolet switch" by [30]), however, readily reproduces the disposition of T Tauri stars in this plane, since the rapid draining of the inner disk causes systems to migrate rapidly from the classical to the weak-line T Tauri regime. Thus, quite contrary to the original supposition of Shu et al. ([139]), the inclusion of secular viscous evolution means that the rapid clearing actually occurs at small radii, in regions of the disk that are immune to direct photoevaporative loss.

Fig. 8.14, however, demonstrates that in the original models of Clarke et al. [30], the disk beyond $r_{g_{II}}$ clears much more slowly. This is because in order to be photoevaporated, material has to flow in on the viscous timescale of the outer disk before it reaches the small radii where direct photoevaporative mass loss is concentrated. The rest of the disk thus clears from the inside out, but slowly. In fact, the amount of time that the model systems spend with slowly eroding inner holes is comparable with their previous lifetime as classical T Tauri stars.

This behavior, however, is inconsistent with submillimeter measurements of weak-line T Tauri stars. In the above scenario, many systems that are classified as weak-line systems on account of their lack of diagnostics associated with inner disks and accretion should nevertheless exhibit submillimeter emission from their slowly eroding outer disks. In reality, such inner-hole sources are relatively rare; the most sensitive submillimeter measurements find evidence for outer dust disks in only $\sim$ 10% of weak-line stars ([38], [10]).

This shortcoming was remedied by the calculations of Alexander et al. ([7]), who revised the prescription for photoevaporative mass loss to include the effect of direct illumination of the inner edge of a disk with an optically thin inner hole. As argued in § 3.1, (see eq. [13]), one expects the mass-loss rate now to scale as $R_{in}^{0.5}(H/R)^{0.5}$, i.e., to increase with increasing R_{in}, an effect that can be attributed to the increase in the area of the exposed rim. Conversely, the mass-loss rate from the diffuse field decreases as the hole size increases.

These simple analytic arguments were verified by the two-dimensional hydrodynamic calculations of Alexander et al. ([7]) (Plate 12, which computed the thermal structure of the flow using a Stromgren-volume method; see § 3.1). These calculations therefore do not include any additional mass loss due to photoevaporation by the diffuse field. I nevertheless emphasize that once the inner disk has been cleared (by viscous draining when $\dot{M}_{acc} \sim \dot{M}_{EUV}$), the subsequent mass loss is clearly dominated by the effect of direct irradiation of the disk inner rim.

Fig. 8.16 illustrates the secular evolution of the disk when the photoevaporation prescription is revised to include direct irradiation of the disk inner rim. This differs from previous work ([30]) only after the clearing of the inner disk. Whereas previously the clearing of the outer disk was slow (since material had to be viscously conveyed to small radii before it was photoevaporated), the outer disk is now rapidly photoevaporated. I note that since the inner disk clears when $\dot{M}_{acc} \sim \dot{M}_{EUV}$, and that since now the photoevaporation rate increases as R_{in} grows, once the hole has grown to radii $>> r_{g_{II}}$, the wind mass-loss rate now far exceeds the accretion rate through the disk. In this limit, therefore, we have the situation where the photoevaporation front sweeps out through a disk with almost fixed surface-density distribution. One may readily show, in this limit, that if the disk surface-density profile scales as r^{-a} and if $H/R \propto r^b$, then (from eq. [13]), the timescale for enlarging the inner hole by a factor of 2 is $\propto R_{in}^{1.5-a-0.5b}$, and thus the number of sources per logarithmic bin of inner-hole size then also scales as $\propto R_{in}^{1.5-a-0.5b}$. Adopting plausible values of $a \sim 1-1.5$ and $b \sim 0.125-0.25$, we see that the resulting distribution is only weakly dependent on R_{in}; in other words, during the clearing process the system does not linger in a state with either very large or very small holes. Moreover, this also implies that the time to clear the outer disk is only modestly greater than the timescale for the initial viscous clearing of the inner disk.

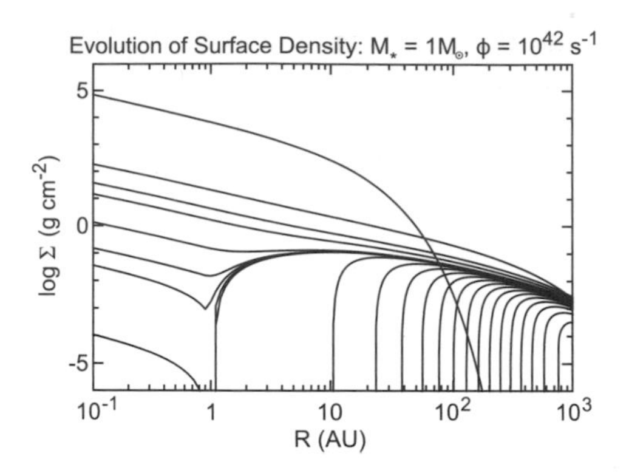

Fig. 8.16. Snapshots of surface-density evolution in the revised ultraviolet switch model ([8]) that also incorporates photoevaporation from direct illumination of the disk's inner rim at late times. Snapshots correspond to $t = 0, 2, 4, 5.9, 6.01, 6.02, 6.03, \ldots, 6.18$ Myr. As in Fig. 8.14, the absolute timescales depend on the viscous model, but the important distinction is that here the outer disk is eroded on a timescale similar to the clearing of the inner disk.

Finally, it should be stressed that all the foregoing discussion relates to the disk *gas*, which is the majority mass component and is directly subject to pressure-driven flows. The dominant opacity source (the dust) is also evaporated only if it is tightly coupled, by strong drag forces, to the gas. The study of Alexander & Armitage ([3]; see also [150]) showed that this is indeed the case for micron-sized grains. Grains on millimeter scales are left behind after photoevaporation.

6.2. Summary of Secular Evolution of EUV-Induced Photoevaporating Disks

When secular (viscous) evolution of a disk is combined with EUV-induced photoevaporative mass loss, the following three-stage evolution results:[13]

1. Normal viscous evolution with accretion through the disk at a rate ($\dot{M}_{acc}$) that comfortably exceeds $\dot{M}_{EUV}$.
2. Inner-disk clearing. Once $\dot{M}_{acc} \sim \dot{M}_{EUV}$, the inner disk drains onto the star on its own (short) viscous timescale, being starved of resupply from larger radius. The accretion rate onto the star thus falls from $\dot{M}_{EUV} \sim 10^{-10} \Phi_{ion_{41}}^{1/2} M_*^{1/2} M_\odot \text{ yr}^{-1}$ to essentially zero on this short timescale. The

13. X-ray photoevaporation follows a qualitatively similar three phase evolutionary pattern, with $\dot{M}_{EUV}$ replaced by the generally higher $\dot{M}_X \sim 10^{-8}(L_X/10^{30}\text{ergs}^{-1})M_\odot \text{ yr}^{-1}$. As in the EUV case, accretion onto the star only persists until the end of Phase II; since the X-ray wind is always heated by direct irradiation, there is no jump in $\dot{M}_X$ at the end of Phase II and thus the ratio of the durations of Phase III to Phase II is somewhat larger than in the EUV case. See Owen et al. ([113]).

size of the initial cleared inner region is a few $\times 0.1 GM_*/c_{s_{II}}^2$, i.e., of order a few AUs.

3. Hole growth. This phase lasts a few times the duration of phase 2. During this time, accretion onto the star remains zero, since a centrifugal barrier prevents photoevaporated gas from accreting onto the star. The outer disk is photoevaporated by progressive erosion of its inner rim.

I note, given that Φ_{ion} is probably an increasing function of stellar mass, that $\dot{M}_{EUV}$ is also expected to increase with stellar mass (at least in the T Tauri regime; there is some expectation that Φ_{ion} may decline somewhat around $3M_{\odot}$ as stars lose their convective envelopes and associated magnetic activity and a decline in X-ray luminosity is observed at this mass scale). In principle, therefore, one might expect to detect this effect as a lower locus in the observed distribution of accretion rate versus stellar mass, since the disk is predicted to empty rapidly (with a corresponding dearth of systems expected) for lower accretion rates. This is an unpromising avenue at present, however, since currently the lower locus of this distribution is set by observational detection thresholds ([110]).

6.3. Observations of Disk Clearing: Inner-Hole Sources

Here I summarize the observational situation regarding correlations between circumstellar diagnostics on various scales. It is worth bearing in mind here that the inner disk, being closer to the central star, is hotter and radiates predominantly in the near infrared, whereas cooler outer regions radiate at longer wavelengths.

Few weak-line T Tauri stars (i.e., those without either spectroscopic evidence for accretion onto the star or emission from an inner disk) are found to show disk diagnostics on any scales. The fraction of such stars with excess emission in *Spitzer* bands (in the range 3.6–70 μm) is around 5% to 15% ([118], [94], [71], [72]), and the fraction with submillimeter emission is a similarly low figure ([10], [38]).

These modest fractions mean that systems with weak/no inner disks, but with evidence for disk emission at a larger radius, are not the norm. A number of examples of such sources (variously dubbed *inner-hole* or, more presumptuously, *transition* systems) are now known, however. Naturally, these sources have attracted considerable interest because of the opportunity that they provide to study disk clearing observationally.

The four classic inner-hole sources (GM Auriga, TW Hydra, Co Ku Tau 4, and DM Tau) were all identified because of their unusual broadband spectral energy distributions (SEDs) (see Fig. 8.17), which showed a deficit of flux at

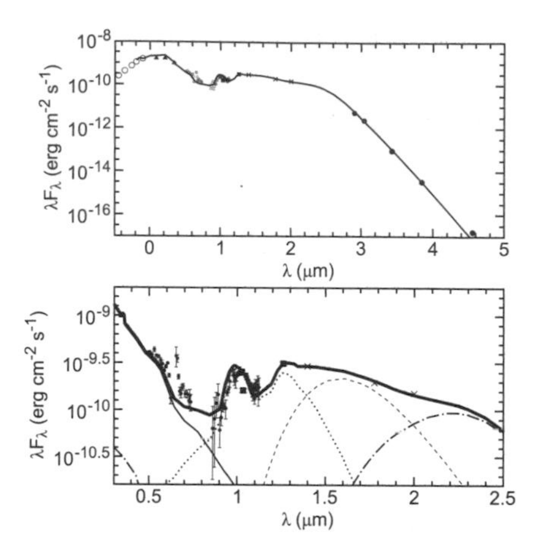

Fig. 8.17. Spectral energy distribution of the inner-hole source TW Hydra indicating the deficit in flux at near-infrared wavelengths ([24]). The model fit involves an optically thick outer disk, truncated at 4 AU, and an inner region containing optically thin dust. Note that the prominent 10 μm silicate feature in this object requires that the inner hole contain only small (submicron) grains.

near-infrared wavelengths compared with typical classical T Tauri stars but showed rising flux levels at longer infrared wavelengths ([20], [24], [34], [51]). Such systems were interpreted as cases where stars were surrounded by optically thick disks at large radii but were relatively devoid of material (or, more specifically, infrared opacity) at small radii. In each case it was possible to model the SED as a truncated optically thick disk combined with a varying amount of optically thin dust in the inner hole. It should also be noted that in order to obtain the rather steep rise in flux in the midinfrared that identified these sources, it was necessary to posit a rather abrupt change in disk properties at the disk-truncation radius.

Even these limited studies of a small number of sources suggested that these systems are quite heterogeneous. For example, whereas some showed evidence of accretion onto the star at a variety of rates—e.g., $\sim 10^{-8} M_\odot \text{ yr}^{-1}$ in GM Auriga ([62]) and $\sim 10^{-10} M_\odot \text{ yr}^{-1}$ in TW Hya ([24])—others did not, e.g. Co Ku Tau 4 ([34]). The properties of the dust within such inner holes are also varied: despite their similar gas-accretion rates, GM Aur and DM Tau have rather different near-infrared SEDs, suggesting that whereas the former does harbor a small quantity of small grains close to the star, the latter's inner hole is virtually clear of dust ([25]). As expected, the source that shows no evidence of gas accretion onto the star (Co Ku Tau 4) also shows no evidence of dust in the inner hole ([34]). One could therefore conclude from this small sample that whereas inner-hole sources are depleted in dust in their inner regions (as indeed must be the case in order to produce the shape of the SED that defines the class), they are not necessarily devoid of gas. Therefore, in some (but not necessarily all) inner-hole sources, the gas-to-dust ratio is high in the inner disk.

The observational situation regarding inner-hole sources is now evolving very rapidly. Therefore, now is not a good time to write an authoritative review of the subject. Several factors are driving this rapid accumulation of data, most notably the acquisition of midinfrared spectra by the *Spitzer Space Telescope* and, in the case of the closest sources, the opportunities afforded by the advent of near- and midinfrared interferometry to constrain the spatial location of various spectral components directly. High-resolution spectroscopy from the Infrared Spectrograph (IRS) on *Spitzer* also allows one to study issues such as the composition and crystallinity of the dust and the abundances of PAHs in these systems (note that the low midinfrared-continuum emission in these systems provides opportunities for such studies that are not possible in most classical T Tauri stars). Thus the last two or three years have seen an expansion both of the numbers of systems known and of the detailed characterization that is now possible in a handful of sources.

Turning first to the insights afforded by larger samples of inner-hole sources, I should first stress that this increase is a result of the larger number of sources for which high-quality midinfrared spectra are now available. The increase in sample size has not changed the basic result, already evident in early studies ([65], [87]) that such sources are relatively rare, representing around 10% of T Tauri stars. Thus it is still necessary to find a clearance mechanism that operates rapidly.

Larger samples allow one to study a range of circumstellar diagnostics in inner-hole sources and confirm the fact, already evident from the four classic sources listed above, that such sources exhibit a range of accretion rates onto the star ([140]). As a class, accretion rates are rather lower than in classical T Tauri stars in general ([109]). It is unclear, however, whether this should be interpreted causally or associatively. In other words, although the inner hole may in some cases be created by the cutoff of gas flow into the inner regions, it may also simply be the case that the process that creates the inner hole is one that operates at late times, when the accretion rate onto the star is in any case low. There is some evidence that the cleanest holes in terms of dust clearing (as measured by the 30:13 μm flux ratio) are those, such as Co Ku Tau 4, that lack accretion diagnostics ([22]).

As high-quality spectra become available at longer wavelengths, it is unsurprising that larger holes are being uncovered ([22]) (indeed, in some cases this has caused an upward revision of the hole size in the case of holes that were originally characterized with ground-based spectra supplemented by Infrared Astronomical Satellite (IRAS) data: estimates of the hole size in GM Auriga have increased from 6 to 24 AU in this way; [24], [25]). There is currently no consensus how the incidence of inner-hole systems depends on the mass of the central star, with claims for a higher incidence in both late-type ([94], [105]) and earlier-type systems ([22]). Interestingly enough, the holes in the most massive objects (i.e., those of F,G spectral type) are extremely large (15–50 AU). Given the limited statistics, however, it would be premature to draw conclusions about possible correlations between hole sizes and stellar mass.

I now turn to conclusions that can be drawn from detailed spectroscopy of individual objects. First, there seems to be considerable variety not only in the quantity of optically thin dust in the inner hole but also in its grain size distribution.[14] The latter is deduced through analysis of the shape of the silicate feature at 10 μm. Whereas in some sources this feature is sharply peaked, requiring a population of exclusively small (submicron) grains ([25], [34], [41]), in others the weakness of this feature requires grain growth of the optically thin dust to the $\sim$ 10 μm scale ([22], [53]). However, there is evidence that considerably larger grains are required in the outer disk of TW Hydra, where growth up to centimeter scales can be inferred from the millimeter continuum ([156]), and a deficiency of small grains in the outer disks of GM Aur and DM Tau has likewise been inferred ([25]). Clearly, therefore, models have to account for a radial segregation of the grain size distribution between the inner hole and the outer disk.

Ground-based near-infrared spectroscopy can also be used to study gas in inner-hole sources via the rovibrational lines of CO ([128], [136]). The kinematic modeling of such lines, in terms of Keplerian motion in a disk of known inclination, allows one to determine the location of the gas. Both TW Hydra and GM Auriga exhibit CO emission at small radii ($<$ 1 AU), a result that is unsurprising since these sources evidence accretion onto the star. Comparison with near-infrared interferometry (which images the optically thin dust in the inner hole; [40]) suggests that CO emission does not extend all the way into the dust-sublimation radius, a result that can be understood in terms of the CO molecules photodissociation in regions that are unshielded by dust from the ultraviolet radiation of the central star. In fact, the dust surface-density estimates of [40] suggest that the disk is marginally optically thick in the radial direction in the optical continuum. Although this does not necessarily prevent

14. Note that the fraction of crystalline grains in inner-hole sources is low ([34], [127]), with the sole exception of UX Tau A, where the presence of crystalline material in the disk rim at $>$ 50 AU poses a problem for theories of crystalline grain production ([41]).

energetic radiation from the central star from impacting the hole's rim (as is required in photoevaporation models), it does limit the effectiveness of agents such as photophoresis or radiation pressure on dust ([91], [27]) in clearing dust from the inner hole, because these require situations that are radially optically thin.

The inferred surface densities of CO can also be used, if they are radially extrapolated and combined with an assumed H_2-to-CO ratio, to estimate the total gas mass of the inner disk. In both GM Auriga and TW Hydra, the inferred gas masses are very low (of order 0.1 Earth mass); when they are combined with the measured accretion rates in these systems, this implies a very short accretion timescale and hence argues strongly that the inner disk continues to be fed by the outer disk. It also implies a rather high efficiency of angular momentum transport in the disk (i.e., viscous α parameter ([142]) of ~ 0.1). Perhaps most notably, comparison of the dust and gas surface densities (inferred, respectively, from interferometry and spectroscopy) suggests that the dust-to-gas ratio in these objects is very low. I note that the use of a conservatively low H_2-to-CO ratio means that, if anything, the total gas masses discussed here are probably underestimates. This therefore only strengthens the argument that the dust-to-gas ratio is very low in the inner holes of these sources.

Interferometry of the closest sources provides an additional check on the above deductions based on SED modeling. In TW Hydra, interferometric measurements in the millimeter and the near infrared ([78], [40]) confirm the results of SED modeling ([24]) and are consistent with an optically thin inner hole and a transition to an optically thick outer disk at 4 AU. However, it is unclear why midinfrared interferometry instead places the transition radius at less than 1 AU ([127]). In one source, IRS 48, interferometry revealed a hole in the dust continuum that was not deducible from the SED ([53]), a result that is probably ascribable to the unusually strong PAH emission in this object, which filled in what would otherwise have been the distinctive deficit in the midinfrared spectrum. This source raises the cautionary note, therefore, that not all inner-hole sources can necessarily be identified purely on the basis of their SEDs. It is thus worth bearing in mind that SEDs do not yield unique solutions for the radial distribution of dust; the complications introduced by issues such as grain composition and grain size distribution mean that it is strongly desirable that, wherever possible, such spectral information supplemented by spatially resolved observations.

6.4. Interpretation of Inner-Hole Sources

The fact that the dust-to-gas ratio is low in at least some inner-hole sources suggests that one possible explanation of the production of inner holes is simply via the loss of dust opacity. An obvious candidate for this is grain growth ([37]). Grain growth, however, is unlikely in those sources where modeling of the 10 μm silicate feature requires only small (submicron) grains in the inner hole. A more general objection is that whereas it is reasonable on this basis of expected timescales that grain growth should proceed from the inside out, it is not obvious why it should lead to the observed steep transition at the interface between the inner hole and the outer disk. Krauss et al. ([91]) have recently advanced the interesting hypothesis that grain growth may indeed lead to sharp dust features when one also includes a (grain-size-dependent) outward drift velocity due to photophoresis.[15]

Photoevaporation from EUV radiation combined with viscous evolution ([30], [7], [8]) appeared to be a promising explanation of inner-hole sources that could, in principle, be interpreted as a mixture of sources that are undergoing inner-hole clearing (stage 2 in § 6.2) and those undergoing progressive enlargement of the inner hole by evaporation from the outer disk's inner rim (stage 3). This may indeed account for some sources but cannot apply to all of them. First, in such models the accretion rate onto the star is always less than the photoevaporative wind mass-loss rate: accretion onto the star is a declining quantity during stage 2 and is essentially zero thereafter. If the canonical evaporation rates for T Tauri stars ($\dot{M}_{EUV} \sim$ a few times $10^{-10} M_{\odot}$ yr^{-1}) are taken at face value, then EUV photoevaporation can apply only to the inner-hole sources with very low accretion rates (such as Co Ku Tau 4 or TW Hydra), but not to systems with high accretion rates such as GM Auriga.[16] However, TW Hydra falls foul of another condition ([3]). In order to create an inner hole, the photoevaporation rate must also exceed the accretion rate through

15. Photophoresis is experienced by grains in gas disks that are radially optically thin and results from the temperature differential between the sides of the grain facing toward and away from the star. Gas molecules bombarding the grain and reevaporating with a speed given by the local grain temperature therefore impart an unbalanced (outward) impulse. The resulting acceleration is independent of grain size; since such grains are also subject to gas drag, their terminal velocity increases with grain size. The rate of dust clearing is set by the drift velocity of the smallest grains present in the disk and is therefore enhanced by grain growth. Note that this mechanism is viable only in disks that are optically thin in the radial direction, although recent near-infrared interferometric results ([40]) suggest that this is not the case in all sources.

16. Chiang & Murray-Clay ([27]) have argued that the large accretion rate in this system is compatible with the expectations of the accretion rate that would be driven by the MRI in an X-ray-irradiated inner rim (see the chapter by Balbus) given the large hole size in this object. This, however, does not explain what created the hole in the first place. The authors argue that the low dust content of the inner region of this disk is maintained by radiation pressure acting on dust. This, of course, requires that the disk be optically thin in the radial direction, which may not be the case according to recent near-infrared interferometric measurements ([40]); see a similar point in the previous footnote.

the outer disk. Naturally, this is not a quantity that is directly accessible observationally, but it can be estimated from the fact that in the asymptotic regime, accretion disks evolve in such a way that the accretion rate is roughly the current disk mass divided by the system age. Thus, for a star of age 10 Myr, for example, the disk mass once photoevaporation becomes significant should not exceed a few times $\sim 10^7 \times 10^{-10} M_\odot$, i.e., a few Jupiter masses. Thus the outer disk of TW Hydra is far too massive (nearly $0.1 M_\odot$) for photoevaporation, as currently understood, to have created the hole. Photoevaporation is therefore a viable explanation only for inner-hole sources for which both the accretion rate onto the star and the outer disk mass are low ([109], [9]). It is currently unclear what fraction of the holes discovered to date fall into this category.[17]

Another leading contender for the production of inner holes involves the presence of a satellite (planet or binary companion) at a small radius ([126]). This has a number of attractive features. First, tidal truncation by such a companion can produce the sharp edge to the outer disk that is required by SED modeling and interferometric observations. Second, depending on the mass of the companion, it can nevertheless allow an accretion flow into the inner disk ([101]); hydrodynamic and radiative transfer modeling of the effect of a putative planetary companion in GM Auriga provides a good fit to the observed SED in this source ([132]). Perhaps most encouragingly for this scenario, Rice et al. ([133]) have proposed a simple mechanism by which the tidally created disk hole can also cause the observed radial segregation of grain sizes. In this picture the smallest grains are so tightly coupled to the gas (i.e., with stopping time that is much less than the orbital time) that they simply follow the gaseous accretion streamers into the inner disk. Larger (millimeter-scale) grains are more loosely coupled (i.e., with stopping time of order of the orbital time) and can thus move radially with respect to the gas. It is well known that in the vicinity of a pressure maximum, the flow is respectively super- and sub-Keplerian on the inside/outside of the maximum; thus gas drag acting on grains in essentially Keplerian orbits causes them to spiral out (in) from regions respectively inside (outside) the pressure maximum. Consequently, such grains become concentrated in the pressure maximum ([148]), an effect

17. The much higher photoevaporation rates produced by X-rays mean that inner holes can have higher accretion rates during inner disc clearing (Phase II). However (as in the EUV case) discs in Phase III (clearing of outer disc through inner hole growth) cannot accrete on to the central star. Thus whereas small accreting holes may be compatible with X-ray photoevaporation, the population of large holes ($\sim$ 50 AU) with high accretion rates in the samples of Kim et al. and Espaillat et al. [88], [42]) are not compatible with photoevaporation models.

that has also been invoked to explain the concentration of larger grains in disk features such as spiral arms and vortices ([131], [89]). In the present context, it would simply mean that larger grains are left behind in the outer disk and would explain the observed paucity of large grains within inner holes.

Companions have very recently been discovered in the inner holes of both Co Ku Tau 4 ([81]) and TW Hydra ([138]). The former system turns out to be a nearly equal-mass binary with orbital parameters that are suitable for the production of a clean inner hole of the size inferred in this system. In TW Hydra, the companion is a 10 Jupiter-mass planet, which, however, is located at about one-tenth of the radius of the inner hole, i.e., far too close to have created the hole dynamically. (Although the planet could, of course, have migrated in across the hole due to tidal interaction with the outer disk, the inner hole should then have shrunk with the planet as it migrated). It is not obvious what the presence of this planet in TW Hydra is telling us about the creation of the inner hole, and it is possible that it is just an indication that TW Hydra is of an age ($\sim$ 10 Myr) when such planets should have formed if they were going to. The intriguing question now is whether there is a second planet, at larger radius, that could have dynamically created the inner hole in this object.

Despite the fact that several inner-hole sources may be nicely explained by the planet/binary hypothesis, a number of unanswered questions remain in this scenario. The creation of a single planet does not in itself provide a recipe for the rapid global dispersal of the disk, which, as I have stressed, is an observational prerequisite. Once a planet is formed, then, unless we can invoke some form of self-propagating planet formation ([13], [116], [117]), then there is no mechanism for clearing the rest of the disk. The timescale for draining the disk onto the central star by accretion at the observed rates is extremely long (about a Hubble time in the case of TW Hydra), but we know that by an age of only a few 10s of Myr, disk masses have reduced by orders of magnitude ([158]). It is not at all clear what the ultimate fate of a system like TW Hydra is.[18]

A related issue is that of how to explain some of the largest inner-hole sources (i.e., those at 50 AU or more). Certainly, any companion that could create such a hole would not be a planet formed by conventional core-accretion scenarios because the formation timescales would then be far too long ([99]). A binary companion or a planetary-mass companion formed early in the

18. Note that a putative planet would eventually be driven to small radii by continued interaction with the massive disk, but in the process the inner hole would close up again. Although this is possible—inner holes are temporary in some systems—it then means that the inner hole is not relevant to the ultimate question of what clears the disk.

disk's self-gravitating phase (see the chapter by Durisen) would be a possibility, but again the issue remains of what then disperses the residual outer disk.

Perhaps all this suggests—to end on a speculative note—that although companions undoubtedly clear the inner disks in some systems, one still needs the global, fast, inside-out dispersal that is offered by photoevaporation models. Current models based on EUV photoevaporation and viscous evolution may apply to some inner-hole sources but cannot be responsible for sources with high accretion rates and/or outer-disk masses. However, as I have stressed throughout this chapter, despite the great theoretical progress in the area of photoevaporation in recent years, there are a number of areas (most notably that of evaporation by the FUV and X-ray emissions from the central star) where we still lack a good theoretical model. Future attempts to develop these models need now to take the presence of inner holes into account, not just as a possible outcome of these models but also as an input condition. In other words, whatever the mechanism for producing inner-hole sources, the presence of such holes cannot but influence photoevaporation through the removal of material at small radii that would otherwise block the flux from the central star. The exploration of such ideas and the comparison of models with detailed characterization of the ever-growing data set of inner-hole sources are tasks for the future.

References

1. F. Adams, D. Hollenbach, G. Laughlin, U. Gorti, ApJ 611, 360 (2004).
2. R. Akeson et al., ApJ 635, 1173 (2005).
3. R. Alexander, P. Armitage, MNRAS 375, 500 (2007).
4. R. Alexander, C. Clarke, J. Pringle, MNRAS 348, 879 (2004).
5. R. Alexander, C. Clarke, J. Pringle, MNRAS 354, 71 (2004).
6. R. Alexander, C. Clarke, J. Pringle, MNRAS, 358, 283 (2005).
7. R. Alexander, C. Clarke, J. Pringle, MNRAS, 369, 216 (2006).
8. R. Alexander, C. Clarke, J. Pringle, MNRAS 369, 229 (2006).
9. R. Alexander, New Astronomy Reviews 52, 60 (2008).
10. S. Andrews , J. Williams, ApJ 631, 1134 (2005).
11. S. Andrews , J. Williams, ApJ 659, 705 (2007).
12. P. Armitage, C. Clarke, F. Palla, MNRAS 342, 1139 (2003).
13. P. Armitage, C. Clarke, C. Tout, MNRAS 304, 425 (1999).
14. P. Armitage, B. Hansen, Natur'e 402, 633 (1999).
15. J. Bally, L. Testi, A. Sargent, J. Carlstrom, AJ 116, 864 (1998).
16. J. Bally, C. R. O'Dell, M. McCaughrean, AJ 119, 2919 (2000).
17. Z. Balog, J. Muzerolle, G. Rieke, K. Su, E. Young, ApJ 660, 1532 (2007).
18. Z. Balog, G. Rieke, K. Su, J. Muzerolle, E. Young, ApJ 650, L83 (2006).
19. M. Begelman, C. McKee, G. Shields, ApJ 271, 70 (1983).

20. E. Bergin et al., ApJ 614, L33 (2004).
21. D. Brooks, V. Costa, M. Lago, A. Lanzafame, MNRAS 327, 177 (2001).
22. J. Brown et al., ApJ 664, L107 (2007).
23. J. Brown, G. Blake, C. Qi, C. Dullemond, D. Wilner, ApJ 675, L109 (2008).
24. N. Calvet, P. D'Alessio, L. Hartmann, D. Wilner, A. Walsh, M. Sitko, ApJ 568, 1008 (2002).
25. N. Calvet et al., ApJ 630, L185 (2005).
26. H. Chen et al., ApJ 492, L173 (1998).
27. E. Chiang, R. Murray-Clay, Nature Physics 3, 604 (2007).
28. E. Churchwell, M. Felli, D. Wood, M. Massi, ApJ 321, 516 (1987).
29. C. Clarke, MNRAS 376, 1350 (2007).
30. C. Clarke, A. Gendrin, M. Sotomayor, MNRAS 328, 485 (2001).
31. C. Clarke, J. Pringle, MNRAS 261, 190 (1993).
32. V. Costa, M. Lago, L. Norci, E. Meurs, A&A 354, 621 (2000).
33. A. Cox (ed.), *Allen's Astrophysical Quantities* (AIP Press, New York, 2000).
34. P. d'Alessio et al., ApJ 621, 461 (2005).
35. R. Diaz-Miller, J. Franco, S. Shore, ApJ 501, 192 (1998).
36. B. Draine, F. Bertoldi, ApJ 468, 269 (1996).
37. K. Dullemond, C. Dominik, A&A 434, 971 (2005).
38. G. Duvert, S. Guilloteau, F. Menard, M. Simon, A. Dutrey, A&A 355, 165 (2000).
39. J. Eisner, J. Carpenter, ApJ 641, 1162 (2006).
40. J. Eisner, E. Chiang, L. Hillenbrand, ApJ 637, L133 (2006).
41. C. Espaillat et al., ApJ 664, L111 (2007).
42. C. Espaillat et al., APJ 441, 457 (2010)
43. B. Ercolano, J. Drake, J. Raymond, C. Clarke, ApJ 688, 398 (2008)
44. B. Ercolano, C. Clarke, J. Drake, APJ 699, 1639 (2009)
45. B. Ercolano, J. Owen, MNRAS in press (2010)
46. M. Fatuzzo, F. Adams, ApJ 675, 1361 (2008).
47. E. Feigelson, J. Gaffney, G. Garmire, L. Hillenbrand, L. Townsley, ApJ 584, 911 (2003).
48. E. Feigelson, T. Montmerle, ARAA 37, 363 (1999).
49. E. Feigelson, L. Townseley, M. Güdel, K. Stassun, in *Protostars and Planets V* (Univ. Arizona Press, Tucson, ed. B. Reipurth, D. Jewitt, K. Keil, 2007), p. 313.
50. A. Font, I. McCarthy, D. Johnstone, D. Ballantyne, ApJ 607, 890 (2004).
51. W. Forrest et al., ApJS 154, 443 (2004).
52. S. Fromang, C. Terquem, S. Balbus, MNRAS 329, 18 (2002).
53. V. Geers et al., A&A 469, L35 (2007).
54. U. Gorti, D. Hollenbach, ApJ 613, 424 (2004).
55. U. Gorti, D. Hollenbach, ApJ 683, 287 (2008).
56. U. Gorti, D. Hollenbach, APJ 690, 1539 (2009)
57. A. Glassgold, J. Najita, in *Young Stars Near Earth: Progress and Prospects*, ASP Conf. Ser., Vol. 244 (Astron. Soc. Pac., San Francisco, ed. R. Jayawardhana, T. Greene, 2001), p. 251.
58. A. Glassgold, J. Najita, J. Igea, ApJ 615, 972 (2004).
59. A. Glassgold, J. Najita, J. Igea, ApJ 480, 344 (Erratum: ApJ 485, 920) (1997).
60. S. Gregory, K. Wood, M. Jardine, MNRAS 379, L35 (2007).
61. M. Guedel, et al., A&A 468, 529 (2007).
62. E. Gullbring, L. Hartmann, C. Briceño, N. Calvet, ApJ 492, 323 (1998).
63. H. Habing, Bull. Astron. Inst. Netherlands 19, 21 (1968).

64. K. Haisch, E. Lada, C. Lada, ApJ 553, L153 (2001).
65. P. Hartigan, L. Hartmann, S. Kenyon, S. Strom, M. Skrutskie, ApJ 354, L25 (1990).
66. L. Hartmann, N. Calvet, E. Gullbring, P. D'Alessio, ApJ 492, 323 (1998).
67. W. Henney, J. Arthur, AJ 116, 322 (1998).
68. W. Henney, C. R. O'Dell, AJ 118, 2350 (1999).
69. W. Henney, C. R. O'Dell, J. Meaburn, S. Garrington, J. Lopez, ApJ 566, 315 (2002).
70. G. Herczeg, J. Najita, L. Hillenbrand, I. Pascucci, ApJ 670, 509 (2007).
71. J. Hernandez et al., ApJ 662, 1067 (2007).
72. J. Hernandez et al., ApJ 671, 1784 (2007).
73. L. Hillenbrand, J. Carpenter, ApJ 540, 236 (2000).
74. D. Hollenbach, D. Johnstone, S. Lizano, F. Shu, ApJ 428, 654 (1994).
75. D. Hollenbach, E. Salpeter, ApJ 163, 155 (1971).
76. D. Hollenbach, A. Tielens, Rev. Mod. Phys. 71, 1 (1999).
77. D. Hollenbach, H. Yorke, D. Johnstone, in *Protostars and Planets IV* (Univ. Arizona Press, Tucson, ed. V. Mannings et al., 2000), p. 401.
78. A. Hughes, D. Wilner, N. Calvet, P. d'Alessio, M. Claussen, M. Hogerheijde, ApJ 664, 536 (2007)
79. S. Ida, D. Lin, ApJ 604, 388 (2004).
80. J. Igea, A. Glassgold, ApJ 518, 848 (1999).
81. M. Ireland, A. Kraus, ApJ submitted (2008).
82. C. Johns-Krull, J. Valenti, J. Linsky, ApJ 539, 815 (2000).
83. D. Johnstone, D. Hollenbach, J. Bally, ApJ 499, 758 (1998).
84. M. Jura, ApJ 197, 575 (1975).
85. F. Kahn, Bull. Astr. Inst. Neth. 12, 187 (1954).
86. I. Kamp, F. Sammar, A&A 427, 561 (2004).
87. S. Kenyon, L. Hartmann, ApJS 101, 117 (1995).
88. K. Kim et al., APJ 700, 1017 (2009).
89. H. Klahr, D. Lin, ApJ 632, 1113 (2005).
90. K. Kornet, P. Bodenheimer, M. Rozyczka, T. Stepinski, A&A 430, 1133 (2005).
91. O. Krauss, G. Wurm, O. Mousis, J.-M. Petit, J. Horner, Y. Alibert, A&A 462, 977 (2007).
92. J. Krolik, T. Kallman, ApJ 267, 610 (1983).
93. P. Lacques, J. Vidal, A&A 73, 97 (1979).
94. C. Lada et al., AJ 131, 1574 (2006).
95. R. Larson, in *Physical Processes in Fragmentation and Star Formation* (Kluwer Academic Publishers, Dordrecht, 1990), p. 389.
96. S. Lepp, R. McCray, ApJ 269, 560 (1983).
97. K. Liffman, Astr. Soc. Australia 20, 337 (2003).
98. D. Lin, J. Papaloizou, J. Faulkner, MNRAS 212, 105 (1985).
99. G. Lodato, E. Delgado-Donate, C. Clarke, MNRAS 364, L91 (2005).
100. S. Lubow, G. D'Angelo, ApJ 641, 526 (2006)
101. S. Lubow, M. Seibert, P. Artymowicz, ApJ 526, 1001 (1999).
102. D. Lynden-Bell, J. Pringle, MNRAS 168, 603 (1974).
103. I. Matsuyama, D. Johnstone, L. Hartmann, ApJ 582, 893 (2003).
104. P. Maloney, D. Hollenbach, A. Tielens, ApJ 466, 561 (1996).
105. C. McCabe, A. Ghez, L. Prato, G. Duchene, R. Fisher, C. Telesco, ApJ 636, 932 (2006).
106. N. Moeckel, J. Bally, ApJ 653, 437 (2006).
107. L. Mundy, L. Looney, E. Lada, ApJ 452, L137 (1995).
108. J. Muzerolle, K. Luhman, C. Briceno, L. Hartmann, N. Calvet, ApJ 625, 906 (2005).

109. J. Najita, S. Strom, J. Muzerolle, MNRQS 378, 369 (2007).
110. A. Natta, L. Testi, S. Randich, A&A 452, 245 (2006).
111. C. R. O' Dell, ApJ 115, 263 (1998).
112. C. R. O'Dell, Z. Wen, X. Hu, ApJ 410, 696 (1993).
113. J. Owen, B. Ercolano, C. Clarke, R. Alexander, MNRAS 401, 1415 (2010).
114. J. Owen, B. Ercolano, C. Clarke, in prep.
115. D. Osterbrock, *Astrophysics of Gaseous Nebulae* (Univ. Sci., Mill Valley, CA, 1989).
116. S. Paardekooper, G. Mellema, A&A 425, L9 (2004).
117. S. Paardekooper, G. Mellema, A&A 453, 1129 (2006).
118. D. Padgett et al., ApJ 645, 1283 (2006).
119. E. Parker, ApJ 139, 72 (1964).
120. I. Pascucci et al., ApJ 663, 383 (2007).
121. S. Pfalzner, ApJ 592, 986 (2003).
122. S. Pfalzner, P. Vogel, J. Scharwächter, C. Olczak, A&A 437, 967 (2005).
123. S. Pfalzner, C. Olczak, A. Eckart, A&A 454, 811 (2006).
124. T. Preibisch et al., ApJS 160, 401 (2005).
125. J. Pringle, ARAA 19, 137 (1981).
126. A. Quillen, E. Blackman, A. Frank, P. Varniere, ApJ 612, L137 (2004).
127. T. Ratzka, C. Leinert, T. Henning, J. Bouwman, C. Dullemond, W. Jaffe, A&A 471, 173 (2007).
128. T. Rettig, J. Haywood, T. Simon, S. Brittain, E. Gibb, ApJ 616, L163 (2004).
129. B. Reipurth, C. Clarke, AJ 122, 432 (2001).
130. S. Richling, H. Yorke, A&A 327, 317 (1997).
131. W. Rice, G. Lodato, J. Pringle, P. Armitage, I. Bonnell, MNRAS 355, 543 (2004).
132. W. Rice, K. Wood, P. Armitage, B. Whitney, J. Bjorkman, MNRAS 342, 79 (2003).
133. W. Rice, P. Armitage, K. Wood, G. Lodato, MNRAS 373, 1619 (2006).
134. J. Rodmann, Dust in Circumstellar Disks, diploma thesis, Univ. Potsdam, Germany (2002).
135. S. Ruden, ApJ 605, 880 (2004).
136. C. Salyk, G. Blake, A. Boogert, J. Brown, ApJ 655, L105 (2007).
137. A. Scally, C. Clarke, MNRAS 325, 449 (2001).
138. J. Setiawan, T. Henning, R. Launhardt, A. Müller, P. Weise, M. Kürster, Nature 451, 38 (2008).
139. F. Shu, D. Johnstone, D. Hollenbach, Icarus 106, 92 (1993).
140. A. Sicilia-Aguilar, L. Hartmann, G. Füresz, T. Henning, C. Dullemond, W. Brandner, AJ 132, 2135 (2006).
141. M. Skrutskie, D. Dutkevitch, S. Strom, S. Edwards, K. Strom, M. Shure, AJ 99, 1187 (1990).
142. N. Shakura, R. Sunyaev, A&A 24, 337 (1973).
143. L. Spitzer, *Physical Processes in the Interstellar Medium* (Wiley-Interscience, New York 1978).
144. K. Stapelfeldt, R. Sahai, M. Werner, J. Trauger, ASP Conf. Ser. 119 (1997), p. 131.
145. B. Stecklum et al., AJ 115, 767 (1998).
146. B. Strömgren, ApJ 89, 526 (1939).
147. H. Störzer, D. Hollenbach, ApJ 515, 669 (1999).
148. T. Takeuchi, D. Lin, ApJ 581, 1344 (2002).
149. T. Takeuchi, S. Miyama, D. Lin, ApJ 460, 832 (1996)
150. T. Takeuchi, C. Clarke, D. Lin, ApJ 627, 286 (2005).

151. H. Throop, J. Bally, ApJ 623, L149 (2005).
152. A. Tielens, D. Hollenbach, ApJ 291, 722 (1985).
153. A. Tielens, D. Hollenbach, ApJ 291, 747 (1985).
154. L. Townsley, E. Feigelson, T. Montmerle, P. Broos, Y.-H. Chu, G. Garmire, ApJ 593, 874 (2003).
155. S. Vicente, J. Alves, A&A 441, 195 (2005).
156. D. Wilner, P. D'Alessio, N. Calvet, M. Claussen, L. Hartmann, ApJ 626, L109 (2005).
157. S. Wolk, F. Harnden, E. Flaccomio, G. Micela, F. Favata, H. Shang, E. Feigelson, ApJS 160, 423 (2005).
158. M. Wyatt, W. Dent, J. Greaves, MNRAS 342, 876 (2003).
159. A. Youdin, F. Shu, ApJ 580, 494 (2002).
160. M. J. McCaughrean in Galaxies and Their Constituents at the Highest Angular Resolutions, Proc. IAU Symposium 205, ed. Q. T. Schilizzi, Manchester, UK, 236 (2001).

Acknowledgments

The authors acknowledge Luis Belerique for editing of the figures. The editor acknowledges partial support by grants POCI/CTE/AST/55691/2004 and Programa FACC of the Fundação Ciência e Tecnologia.

Nuria Calvet and Paola D'Alessio are indebted to Nathan Crockett, Catherine Espaillat, Jeffrey Fogel, Elise Furlan, Jesus Hernandez, Laura Ingleby, James Muzerolle, and Charlie Qi for providing us with scripts and figures adapted from their papers. Nuria Calvet acknowledges support from NASA Origins Grants NNG05GI26G and NNG06GJ32G and NNX08AH94G and JPL Grants 128887 and 1277575. Paola D'Alessio acknowledges grants from CONACyT, México.

Chapter 3 is based on work supported by the National Science Foundation under Grant No. 0707777 to E. Bergin. Its author is also grateful to the referee, Ewine van Dishoeck, and Joanna Brown for numerous helpful comments.

Thomas Henning and Gwendolyn Meeus thank their collaborators Roy van Boekel and Jeroen Bouwman (Heidelberg) and Cornelia Jäger and Harald Mutschke (Jena) for extensive discussions and for providing figures for this chapter. They thank the referee Ant Jones for carefully reading the chapter.

Richard Durisen would like to thank his former students A. C. Boley, K. Cai, J. N. Imamura, A. C. Mejía, S. Michael, M. K. Pickett, and T. Y. Steiman-Cameron for the pleasure of their collaborative work on GIs over the years. There are many other people to whom he owes a debt of gratitude, beginning with his earliest research mentors, M. Schwarzschild, J. L. Tassoul, J. P. Ostriker, and P. Bodenheimer. However, he refrains from trying to name everyone who has shared his on her insight and interest over the years for fear of a glaring omission. He is particularly grateful to L. Hartmann and T. Y. Steiman-Cameron for thoughtful reading of the manuscript. He was funded in part during the writing of this chapter by NASA grants NNG05GN11G and NNX08AK36G and by travel grants from the Max

Planck Institute of Astronomy in Heidelberg and the Max Planck Institute for Extraterrestrial Physics in Garching.

Steven A. Balbus would like to thank his collaborators over the past several years with whom and from whom he has learned much of what he knows on the subject of protostellar disks: O. Blaes, S. Fromang, C. Gammie, J. Hawley, J. Stone, and C. Terquem. He would also like to thank the referee for his/her important comments on the ionization equilibrium. This work was supported by a Chaire d'Excellence from the French Ministry of Higher Education and a grant from the Conseil Régional de l'Ile de France.

Arieh Königl and Raquel Salmeron have greatly benefited from their collaborations over the years with Glenn Ciolek, Ioannis Contopoulos, Ruben Krasnopolsky, Christof Litwin, Pedro Safier, Konstantinos Tassis, Seth Teitler, Dmitri Uzdensky, Nektarios Vlahakis, and, in particular, Mark Wardle on the topics discussed in their chapter. They are also grateful to Marina Romanova for helpful correspondence on disk–star magnetic coupling. Their work was supported in part by NSF Grant AST-0908184 (Königl) and Australian Research Council Grant DP0342844 (Salmeron).

Cathie Clarke is very grateful to David Hollenbach, whose painstaking and thorough reading of the manuscript resulted in many improvements, and also to Ulma Gorti for providing Fig. 8.13 in advance of publication. Thanks are also due to Richard Alexander and Barbara Ercolano for their useful input and to Will Henney for drawing her attention to Fig. 8.11.

Contributors

Steven A. Balbus
balbus@ens.fr
Laboratoire de Radio Astronomie
École Normale Supérieure
24 rue Lhomond
75231 Paris EDEX 05
France

Edwin A. Bergin
ebergin@umich.edu
Department of Astronomy
University of Michigan
830 Dennison Building, 500 Church Street
Ann Arbor, MI 48109
USA

Nuria Calvet
ncalvet@umich.edu
Department of Astronomy
University of Michigan
830 Dennison Building, 500 Church Street
Ann Arbor, MI 48109
USA

Cathie Clarke
cclarke@ast.cam.ac.uk
Institute of Astronomy
Madingley Rd
Cambridge, CB3 0HA
UK

Paola D'Alessio
p.dalessio@astrosmo.unam.mx
Centro de Radioastronomía y Astrofísica
Universidad Nacional Autónoma de México
58089 Morelia, Michoacán
México

Richard H. Durisen
durisen@astro.indiana.edu
Department of Astronomy
Indiana University
727 East Third Street
Bloomington, IN 47405
USA

Paulo J. V. Garcia
pgarcia@fe.up.pt
Universidade do Porto, Portugal
Institut de Planétologie et Astrophysique de Grenoble, France

Thomas Henning
henning@mpia.de,
Max Planck Institute for Astronomy
Königstuhl 17
D-69117 Heidelberg
Germany

Arieh Königl
arieh@oddjob.uchicago.edu
Department of Astronomy and Astrophysics
University of Chicago
5640 S. Ellis Ave.
Chicago, IL 60637
USA

Gwendolyn Meeus
gwendolyn.meeus@uam.es
Universidad Autónoma de Madrid
Departamento de Física Teórica C-XV
28049 Madrid
Spain

Antonella Natta
natta@arcetri.astro.it
Osservatorio di Arcetri
Largo Fermi 5
50124 Firenze
Italy

Raquel Salmeron
raquel@mso.anu.edu.au
Research School of Astronomy & Astrophysics and Research School of Earth Sciences
Australian National University
Canberra ACT 0200
Australia

Malcolm Walmsley
walmsley@arcetri.astro.it
Osservatorio di Arcetri
Largo Fermi 5
50124 Firenze
Italy